T5-BBL-799

A SURVEY OF FINITE MATHEMATICS

Marvin Marcus
The University of California at Santa Barbara

DOVER PUBLICATIONS, INC.
NEW YORK

Published in Canada by General Publishing Company, Ltd., 30 Lesmill Road, Don Mills, Toronto, Ontario.
Published in the United Kingdom by Constable and Company, Ltd., 3 The Lanchesters, 162–164 Fulham Palace Road, London W6 9ER.

This Dover edition, first published in 1993, is an unabridged, corrected republication of the work first published by the Houghton Mifflin Company, Boston, 1969.

Manufactured in the United States of America
Dover Publications, Inc., 31 East 2nd Street, Mineola, N.Y. 11501

Library of Congress Cataloging-in-Publication Data

Marcus, Marvin, 1927–
A survey of finite mathematics / Marvin Marcus.
p. cm.
Originally published: Boston : Houghton Mifflin, 1969. With corrections.
Includes index.
ISBN 0-486-67553-X (pbk.)
1. Mathematics. I. Title.
QA39.M346 1993
510—dc20 92-43550
CIP

Preface

The primary purpose of this book is for use as a text in courses usually entitled *Finite Mathematics* that have come into existence over the last few years in many colleges and universities. Such courses are normally one quarter or one semester in length and are intended for students from the social and biological sciences, business administration, and liberal arts (and even mathematics!). The first chapter of *A Survey* is specifically designed for this audience in both level and content. It is divided into seven sections covering basic mathematical concepts: logic; sets; functions; induction and combinatorics; partitions; probability; stochastic processes. This material can easily be covered in 30 lecture hours and constitutes the first third of the book.

The second two thirds of the book is devoted to the basic ideas from linear algebra and the theory of convex sets. The material from these disciplines constitute the fundamental mathematical tools used in the applications to linear programming, game theory, and Markov chains which appear in the third chapter.

The first section of Chapter 2 is a completely self-contained introduction to vectors and matrices motivated by simple examples from the social and biological sciences. The next four sections are somewhat unique for an elementary book. In these, the student is introduced to the elementary notions of combinatorial matrix theory: incidence matrices, systems of distinct representatives, and stochastic matrices. These topics have far-reaching applications in such diverse fields as communication networks, sociometric relations, operations research, and statistics. The last two sections of Chapter 2 are devoted to a development of the theory and applications of systems of linear equations.

The third and final chapter of the book, entitled *Convexity*, immediately starts with examples of simple linear programming problems. The basic geometry of convex sets, including the theory of maxima and minima for linear functions, appears in the first three sections of this chapter. The remainder of the chapter is devoted to game theory and Markov chains. The treatment of game theory uses techniques from earlier material on convex sets and matrix theory to solve matrix games. The section on Markov chains contains a complete treatment of the elementary aspects of this subject and includes numerous applications.

There are approximately 150 worked examples in the text. These cover a wide range of applications and form an integral part of the material. Most of these are routine, but a few require some thought. Each of the 19 sections of the book ends with a true-false quiz and a set of exercises. Altogether there are over 1200 exercises in the book. Many of the more difficult exercises are accompanied by "hints" for solutions that in some cases constitute complete analyses. Those exercises that are somewhat more difficult are marked with an asterisk. Exercises which are marked with a dagger contain results or definitions that are used elsewhere in the text. These exercises should at least be read and hopefully worked. The quizzes are intended to remind the reader of the essential points covered in the section. Experience at the author's institution indicates that the quizzes are highly effective teaching aids.

In general, then, this is a book on "applied mathematics". In its entirety it is suitable for a two-quarter or one-semester course, or a three quarter or two semester course if the pace is more leisurely. It is our belief that the material is appropriate and important for mathematics majors as well as students from other disciplines. The first chapter of *A Survey* has been used in manuscript form at the University of California, Santa Barbara for a freshman course for students from other departments. The material in Chapters 2 and 3 has been used for a course on discrete applications of matrix theory in a conference for college teachers sponsored on this campus by the National Science Foundation.

I would like to express my thanks to Miss Susan Katz, Miss Barbara Smith, and Mrs. Nancy Stuart for their invaluable assistance in the preparation of this manuscript. I am also very grateful to Mrs. Wanda Michalenko and Mrs. Delores Brannon for their extremely professional jobs in typing and assembling this manuscript. Professor B. N. Moyls of the University of British Columbia acted as a referee on this manuscript. His remarks and suggestions were of inestimable value.

Marvin Marcus

Contents

Chapter 3 Convexity

Numbering System

Each of the chapters is divided into sections. Thus the fifth section of Chapter 1 is Section 1.5. Definitions, theorems, and examples are numbered separately within each section. Thus Theorem 3.2 is the second theorem in the third section of the chapter in which it appears. Reference to a theorem (definition, example) is by its number alone when the reference is in the same chapter. If necessary, we give the section number or the chapter number in which the item appears.

Asterisks on exercises indicate that the problem is somewhat difficult. These exercises are usually accompanied by hints, or their solutions appear in the Answers. A limited number of exercises are marked with a dagger †. These exercises contain results that are used within the text and should be carefully read and worked out. This is particularly the case for Exercises 7–18 in Section 2.6 in which the properties of linear independence are developed. (Detailed solutions to this sequence of exercises are included in the Answers.)

fundamentals

chapter 1

1.1

Truth Tables and Applications

It is appropriate to begin a study of mathematics with a brief discussion of logic. In thinking about any organized body of information we should have some idea of the mechanical rules used in manipulating this information. This comment applies with equal validity to everyday situations. We are all familiar with rather obviously fallacious arguments. For example, any Communist advocates armed revolution; Mr. X advocates armed revolution; therefore Mr. X is a Communist. This argument is, of course, incorrect, for, we can all think of members of political groups who advocate armed revolution and who are, in fact, quite antagonistic to the Communist doctrine.

In general, the study of elementary logic has several purposes. Probably the most important of these purposes is to give precise meaning to such words and phrases as "and", "or", "not", "if . . . then", and "if and only if", which occur throughout any mathematical theory. Second, we want to make somewhat more precise the laws of inference and deduction that are constantly used in constructing a mathematical theory, i.e., we want to have at least a rudimentary idea of what constitutes a coherent mathematical argument. However, even an intimate knowledge of formal logic is no guarantee against error in a mathematical argument; such knowledge simply diminishes the chances of making certain "obvious" mistakes in reasoning. A third and more utilitarian justification for studying logic is its use in such practical applications as the analysis of switching networks.

The reader will recall that in elementary plane geometry one starts with certain primitive or "undefined" items such as "points", "lines", etc. There follows a set of "axioms" and "postulates" governing the relationships of these items, e.g., "two distinct points determine a line." One does not attempt to define "points", "lines", etc. in terms of simpler notions, nor to prove the axioms and postulates. These constitute the starting point of Euclidean geometry. In general, the basic ingredients of a mathematical theory are the following:

A. a set of undefined objects;

B. a certain set of statements or axioms relating these undefined objects;

C. a sequence of statements or theorems which concerns the undefined objects and which are obtained by the rules of logic.

In the development of a mathematical theory, we put together statements with *connectives* to obtain new statements. For example, if p and q represent statements, we may build up compound statements "p and q," "p or q," "not p," etc. We shall now introduce the connectives used in standard logical systems and develop the symbolism used to designate them. The first connective is the word "and"; the symbol used to denote this word is

$$\wedge. \tag{1}$$

The result of putting two statements together with the word "and" is referred to as a *conjunction.*

The second connective is the word "or," which is denoted by the symbol

$$\vee. \tag{2}$$

Joining two statements by the word "or" results in a statement referred to as a *disjunction.*

The word "not" is symbolized by

$$\sim \tag{3}$$

and inserting "not" at the beginning of a statement results in a statement called a *negation.*

The fourth basic symbol stands for "if . . . then" and the symbol is an arrow,

(4) $$\rightarrow.$$

An "if . . . then" statement is usually called an *implication.*

Thus, if p and q are statements, they may be connected symbolically by

$$p \wedge q,$$

read "p and q";

$$p \vee q,$$

read "p or q";

$$\sim p,$$

read "not p"; and finally

$$p \rightarrow q,$$

read "if p then q" or "p implies q."

We shall assume that to any meaningful statement it is possible to assign a *truth value*, namely true (T) or false (F). Observe that this is indeed an assumption. For, consider the statement, "The number of electrons in the universe exceeds 10^{1000}." Although this statement seems to make sense, it is not likely that we can decide whether it is true or false.

We can give meaning to the connectives described above by assigning truth values to each of the four statements $p \wedge q$, $p \vee q$, $\sim p$, and $p \rightarrow q$ as the truth values for p and q vary individually. We do this in a convenient tabular form known as a *truth table.* In each of the following tables, we think of p and q as each standing for entire sentences: p and q can each have one of two truth values, T or F. The last column of each truth table tells us the resulting truth value for the appropriate compound statement formed from p and q.

(5) $\wedge$:

p	q	$p \wedge q$
T	T	T
T	F	F
F	T	F
F	F	F

(6) $\vee$:

p	q	$p \vee q$
T	T	T
T	F	T
F	T	T
F	F	F

(7) $\sim$:

p	$\sim p$
T	F
F	T

(8) $\rightarrow$:

p	q	$p \rightarrow q$
T	T	T
T	F	F
F	T	T
F	F	T

Thus $p \wedge q$ is false unless both p and q are separately true; $p \vee q$ is true unless each of p and q is individually false. For example, if p is the statement "$2 + 2 = 5$" and q is the statement "dogs are animals," then the statement "$p \vee q$" is true. Table (7) for negation is self-explanatory and reasonable; for if p is true then $\sim p$ is false, and conversely, if p is false then $\sim p$ is true. From Table (8) we see that the statement $p \rightarrow q$ will be true unless p is true and q is false. In other words, we never want a true statement to imply a false one. The fact that $p \rightarrow q$ is true when p is false, regardless of the truth value of q, may require some additional explanation, for this is not the way implication is used in ordinary language. In conversation one usually has some causal connection in mind between p and q in an implication. Thus the statement, "If men are dogs, then women are cats" is meaningful, but is not one that would often be said. Of course, it is false that men are dogs and equally false that women are cats. Nevertheless we want every meaningful statement to have a definite truth value, either T or F, and Table (8) stipulates that "If men are dogs then women are cats" has truth value T. Another way of saying this is that Table (8) actually defines the connective $\rightarrow$. As another example, consider the implication $p \rightarrow q$ where p is the statement "n is a number greater than 17 and less than 3" and q is the statement "$n = 5$." Even though p is false we want the implication to be true, and this can be justified by observing that there is no number n (whether it is 5 or not) which is greater than 17 and less than 3. Although this may seem a little silly, it is important that establishing the truth of $p \rightarrow q$ not carry with it the burden of exhibiting a formal connection between p and q.

Using these elementary connectives we can formulate compound statements that are quite complicated.

Example 1.1 Construct a truth table for the statement $p \rightarrow (p \vee \sim p)$. In other words, we want to assign a truth value to the preceding statement for each of the two possible truth values for p.

p	$\sim p$	$p \vee \sim p$	$p \rightarrow (p \vee \sim p)$
T	F	T	T
F	T	T	T

We filled in the second column by using Table (7), the third column using Table (6) and the fourth column using Table (8). Thus, when the truth value of p is T then the truth value of $p \vee \sim p$ is T (Table (6), row 2), and hence the truth value of the compound statement $p \rightarrow (p \vee \sim p)$ is T by the first row of (8). Similarly, when the truth value of p is F then the truth value of $p \vee \sim p$ is T from the third row of Table (6), and the truth value of $p \rightarrow (p \vee \sim p)$ is T by the third row of Table (8).

A compound statement is said to be *valid* or a *tautology* if its truth value is T regardless of the truth values of its component statements. Thus from Example 1.1 we see that the implication $p \rightarrow (p \vee \sim p)$ is a tautology.

Another useful connective can be defined as follows: the compound statement

$$(p \rightarrow q) \wedge (q \rightarrow p) \tag{9}$$

will be abbreviated to

$$p \leftrightarrow q. \tag{10}$$

The formula (10) is read "p if and only if q." The statements p and q in (10) are sometimes said to be *equivalent*.

Example 1.2 Construct a truth table for the statement $p \leftrightarrow q$.

(11)

p	q	$p \rightarrow q$	$q \rightarrow p$	$p \leftrightarrow q$
T	T	T	T	T
T	F	F	T	F
F	T	T	F	F
F	F	T	T	T

The third and fourth columns of Table (11) are read from Table (8), e.g., in column 4, row 3 of (11), q has truth value T, p has truth value F, and from the second row of (8), we see that $q \to p$ has truth value F. The fifth column of (11) is obtained by joining the third and fourth columns with the connective $\wedge$ and using Table (5), e.g., when $p \to q$ has truth value T and $q \to p$ has truth value F, then $p \leftrightarrow q$ has truth value F, as one sees from the second row of (5).

Example 1.3 Show that the following compound statement is valid:

$$f\colon\ [(\sim p \vee q) \wedge (p \vee \sim q)] \leftrightarrow (p \leftrightarrow q).$$

Consider the table

p	q	$\sim p \vee q$	$p \vee \sim q$	$3 \wedge 4$	$p \leftrightarrow q$	f
T	T	T	T	T	T	T
T	F	F	T	F	F	T
F	T	T	F	F	F	T
F	F	T	T	T	T	T

where in column 5 we have written $3 \wedge 4$ to denote the conjunction of the statements in columns 3 and 4.

Example 1.4 Show that the following compound statement is valid:

$$f\colon\ (p \wedge (p \to q)) \to q.$$

p	q	$p \to q$	$p \wedge (p \to q)$	f
T	T	T	T	T
T	F	F	F	T
F	T	T	F	T
F	F	T	F	T

The fundamental assumption that we shall make about valid compound statements or tautologies is that *they represent correct arguments in any mathematical system*. In other words, it will be assumed that tautologies represent arguments which are acceptable in establishing the theorems in a mathematical theory. To illustrate this, consider the following kind of reasoning: "If p is the case, and whenever p is the case it follows that q is the case, then q must hold." Put more succinctly: "If p, and p implies q, then q." If we state this in logical symbolism, we obtain the compound statement

$$f\colon\ (p \wedge (p \rightarrow q)) \rightarrow q.$$

But we saw in Example 1.4 that f is a tautology, i.e., that f is "true," or has truth value T, whatever the truth values of p and q may be. The fact that $(p \wedge (p \rightarrow q)) \rightarrow q$ is a tautology is usually referred to as the *law of detachment* or, in somewhat more rarefied terms, "modus ponens." What we have actually done is set up the truth tables for implication and conjunction in such a way that they yield the law of detachment as a tautology.

As another example, consider the following statement: "If p always implies q and q fails to be the case, then p cannot hold." This is a very familiar and acceptable form of argument used not only in mathematics but in everyday life. It is known as an *indirect proof* or, in Latin, "reductio ad absurdum." For example, suppose it is the case that whenever it is raining I invariably carry my umbrella, and suppose I am not carrying my umbrella. Knowing these two facts, you can conclude that it is not raining. The symbolic statement of the method of indirect proof takes the following form:

$$f\colon\ ((p \rightarrow q) \wedge \sim q) \rightarrow \sim p.$$

Consider the truth table for f.

p	q	$p \rightarrow q$	$\sim q$	$\sim p$	$(p \rightarrow q) \wedge \sim q$	f
T	T	T	F	F	F	T
T	F	F	T	F	F	T
F	T	T	F	T	F	T
F	F	T	T	T	T	T

Thus the statement f is a tautology and, by our fundamental agreement, represents a correct argument.

A somewhat more subtle argument mentioned earlier is widely used: "If x is a Communist, then x advocates armed revolution. Moreover, x advocates armed revolution. It follows that x is a Communist." This argument is of course rubbish, since x could equally well be a Minute Man. The argument has the following form: "If p implies q, and q, then p." Denote this statement by f and consider the truth table:

$$f\colon\ ((p \rightarrow q) \wedge q) \rightarrow p.$$

p	q	$p \rightarrow q$	$(p \rightarrow q) \wedge q$	f
T	T	T	T	T
T	F	F	F	T
F	T	T	T	F
F	F	T	F	T

The fifth column of the truth table has an F in the third row; hence f cannot be a tautology. If we examine the third row we can see our error in reasoning. Thus, if p is false and q is true, then the implication $p \rightarrow q$ is true and the conjunction $(p \rightarrow q) \wedge q$ is also true. But then $((p \rightarrow q) \wedge q) \rightarrow p$ is false by Table (8), row 2.

A consequence of the law of detachment is the following general statement about tautologies. Suppose f and g are compound statements involving $p_1, p_2, \ldots, p_n$. If f is a tautology and $f \rightarrow g$ is a tautology, then g is a tautology. For if g were ever to have the truth value F then in order for the implication $f \rightarrow g$ to have truth value T (which it always does, since we are assuming it to be a tautology), f would of necessity have truth value F. But f is assumed to be a tautology and hence has truth value T for all the assignments of truth values to $p_1, p_2, \ldots, p_n$. It follows, then, that g can never have truth value F and must therefore be a tautology.

The implication $p \rightarrow q$ is sometimes called a *conditional*. Together with $p \rightarrow q$, we often consider the *contrapositive*, $\sim q \rightarrow \sim p$, and the *converse*, $q \rightarrow p$. (See Exercise 14.) If $p \rightarrow q$ has truth value T, we sometimes say that p is a *sufficient* condition for q or that q is a *necessary* condition in order for p to hold, or that p holds *only if* q holds. Thus, if $p \leftrightarrow q$ has truth value T, we say that p is a *necessary and sufficient* condition for q, or that p holds *if and only if* q holds. For example, the fact that $\sim(\sim p) \leftrightarrow p$ is a tautology can be expressed in this language by saying that $\sim(\sim p)$ is a necessary and sufficient condition for p. This particular tautology has the somewhat pompous name of the *law of double negation*, which we verify by means of a truth table.

p	$\sim p$	$\sim(\sim p)$	$p \leftrightarrow \sim(\sim p)$
T	F	T	T
F	T	F	T

In ordinary language, we often consider conjunctions or disjunctions of more than two statements. However, our logical symbolism only allows two statements at a time to be put together by these connectives. Nevertheless, we want to assign meaning to such statements as

(12) $$p_1 \wedge p_2 \wedge p_3.$$

There are two ways in which we can make (12) into a conjunction of two statements:

(13) $$p_1 \wedge (p_2 \wedge p_3)$$

and

(14) $$(p_1 \wedge p_2) \wedge p_3.$$

It is important that we know (13) and (14) have the same truth values for all truth values assigned to the individual statements p_1, p_2, and p_3. To see that (13) and (14) have this property, we can argue as follows: $p_2 \wedge p_3$ has truth value F unless both p_2 and p_3 have truth value T. The only way in which the conjunction $p_1 \wedge (p_2 \wedge p_3)$ can have truth value T is for p_1 and $(p_2 \wedge p_3)$ to both have truth value T. That is, (13) can have truth value T only when each of p_1, p_2, and p_3 has truth value T. Similarly one sees that (14) can have truth value T only when each of p_1, p_2, and p_3 has truth value T.

Suppose now that we have four statements p_1, p_2, p_3, and p_4, which are connected in pairs by conjunction, e.g.,

(15) $$p_1 \wedge (p_2 \wedge (p_3 \wedge p_4)).$$

According to our previous argument, $p_2 \wedge (p_3 \wedge p_4)$ has truth value F unless all three of p_2, p_3, and p_4 have truth value T. Thus one sees that (15) has truth value F unless p_i has truth value T, $i = 1, \ldots, 4$. In general, if we take any k propositions $p_1, p_2, \ldots, p_k$ and form a compound proposition using only the connective $\wedge$, it is clear that the compound will have truth value T only when every p_i has truth value T, $i = 1, \ldots, k$. Otherwise the compound proposition will have truth value F. This is the case for *any* association of these statements using the connective $\wedge$.

Theorem 1.1 *If f and g are two compound statements obtained by using only the connective $\wedge$ on all of the k propositions $p_1, \ldots, p_k$, then $f \leftrightarrow g$ is a tautology.*

Proof By the above discussion, f and g both have truth value T if all of $p_1, \ldots, p_k$ have truth value T and otherwise both have truth value F. Thus if one were to construct a truth table for $f \leftrightarrow g$, the columns for f and g would be identical and $f \leftrightarrow g$ would always have truth value T. ▮

By a similar argument, one can show that the same kind of theorem is available for the connective $\vee$.

Theorem 1.2 *If f and g are two compound statements obtained by using only the connective $\vee$ on all of the k propositions $p_1, \ldots, p_k$, then $f \leftrightarrow g$ is a tautology.*

Proof Both compound statements f and g will have truth value T unless every one of $p_1, \ldots, p_k$ has truth value F. Thus f and g have the same truth values for all choices of truth values for $p_1, \ldots, p_k$ and it follows that $f \leftrightarrow g$ is a tautology. ▮

Definition 1.1 ***Conjunction and disjunction of several propositions.*** Let $p_1, \ldots, p_k$ be propositions. Then

$$p_1 \wedge p_2 \wedge p_3 \wedge \cdots \wedge p_k \tag{16}$$

denotes any one of the equivalent compound propositions obtained from $p_1, \ldots, p_k$ using the connective $\wedge$. Similarly,

$$p_1 \vee p_2 \vee p_3 \vee \cdots \vee p_k \tag{17}$$

denotes any one of the equivalent compound propositions obtained from $p_1, \ldots, p_k$ using the connective $\vee$.

In discussing these multiple disjunctions and conjunctions it is important to realize that it is only the truth value of a compound statement that matters. For example, if f and g are two compound statements depending on the individual statements $p_1, \ldots, p_k$ and $f \leftrightarrow g$ is a tautology, then f may be substituted for g in any proposition involving g.

Next let p_1 and p_2 be statements and suppose we consider the compound statement

$$c\colon\ p_1 \wedge \sim p_2.$$

The proposition c has the truth value T when p_1 has truth value T and p_2 has truth value F. For the remaining three pairs of truth values for p_1 and p_2, c has truth value F. In other words, the truth table for

$p_1 \wedge \sim p_2$ is

p_1	p_2	$p_1 \wedge \sim p_2$
T	T	F
T	F	T
F	T	F
F	F	F

Next consider the statement

$$c_{13}\colon \sim p_1 \wedge p_2 \wedge \sim p_3.$$

(We have denoted the statement by c_{13} for reasons to be discussed in Theorem 1.3.) We know that the statement c_{13} has truth value T only when $\sim p_1$, p_2, and $\sim p_3$ individually have truth value T. But $\sim p_1$ and $\sim p_3$ have truth value T only when p_1 and p_3 have truth value F. Thus we see that $\sim p_1 \wedge p_2 \wedge \sim p_3$ will have truth value T when p_2 has truth value T and p_1 and p_3 have truth value F. For all other truth values assigned to p_1, p_2, and p_3, the statement c_{13} will have truth value F.

This example is typical of a general situation.

Theorem 1.3 *Let $p_1, \ldots, p_k$ be statements and let $i, j, \ldots, m$ be some selection of integers between* 1 *and k. Let $c_{ij\ldots m}$ be the compound statement*

$$p_1 \wedge p_2 \wedge \cdots \wedge \sim p_i \wedge \cdots \wedge \sim p_j \wedge \cdots \wedge \sim p_m \wedge \cdots \wedge p_k. \tag{18}$$

Then $c_{ij\ldots m}$ has truth value T *only when $p_i, p_j, \ldots, p_m$ all have truth value* F *and all the remaining statements have truth value* T.

Proof We first illustrate examples of (18) in the case of four statements. Thus, suppose $i = 2, j = 4$. Then c_{ij} is

$$c_{24}\colon p_1 \wedge \sim p_2 \wedge p_3 \wedge \sim p_4.$$

Again, if $i = 1, j = 2, m = 4$, then (18) becomes

$$c_{124}\colon \sim p_1 \wedge \sim p_2 \wedge p_3 \wedge \sim p_4.$$

To proceed to the proof of the theorem we first recall that the statement (18) has truth value T only when each of the individual statements appearing in the conjunction has truth value T. This requires that all of $\sim p_i, \sim p_j, \ldots, \sim p_m$, as well as the remaining statements, have truth value T. Thus $p_i, p_j, \ldots, p_m$ must have truth value F and the remaining statements must have truth value T. ▌

We let c_0 denote the statement

$$p_1 \wedge p_2 \wedge \cdots \wedge p_k$$

in which none of the statements is preceded by a negation sign.

Example 1.5 Construct a compound statement using p_1, p_2, and p_3 which has truth value T only when p_1 has truth value T and p_2 and p_3 have truth value F.

In this case, $i = 2, j = 3$, and the required statement is

$$c_{23}\colon p_1 \wedge \sim p_2 \wedge \sim p_3.$$

Illustrating Theorem 1.3, we have the following truth table for c_{23}.

p_1	p_2	p_3	$\sim p_2$	$\sim p_3$	c_{23}
T	T	T	F	F	F
T	T	F	F	T	F
T	F	T	T	F	F
T	F	F	T	T	T
F	T	T	F	F	F
F	T	F	F	T	F
F	F	T	T	F	F
F	F	F	T	T	F

Observe that the only row in which c_{23} has truth value T is the row corresponding to the choice of truth values T, F, F for p_1, p_2, and p_3, respectively.

Example 1.6 Find a compound proposition depending on p_1, p_2, p_3, and p_4 which has truth value T when p_1, p_2, p_3, and p_4 have truth values T, F, F, T and F, F, F, T, respectively, and which has truth value F for all other choices of truth values for p_1, p_2, p_3, and p_4. First consider the two statements c_{23} and c_{123}. According to Theorem 1.3, c_{23} has truth value T when p_1, p_2, p_3, and p_4 have truth values T, F, F, T, respectively, and otherwise has truth value F. By the same theorem c_{123} has truth value T only when p_1, p_2, p_3, and p_4

have truth values F, F, F, T, respectively. Now consider the compound statement

(19) $$c_{23} \vee c_{123}.$$

We know from the truth table for the connective $\vee$ (see Table (6)) that $c_{23} \vee c_{123}$ has truth value T when either c_{23} or c_{123} (or both) has truth value T. But as we just argued, this happens for precisely the truth values T, F, F, T and F, F, F, T for p_1, p_2, p_3, and p_4, respectively.

Example 1.7 Find a compound proposition depending on p_1, p_2, p_3, and p_4 which has truth value T when p_1, p_2, p_3, and p_4 have truth values F, F, T, T; F, T, F, T; and F, F, F, F, respectively, and otherwise has truth value F. To solve this problem, consider the statement

(20) $$f\colon c_{12} \vee c_{13} \vee c_{1234}.$$

Now, c_{12} is $\sim p_1 \wedge \sim p_2 \wedge p_3 \wedge p_4$ and, according to Theorem 1.3, c_{12} has truth value T for the choice of truth values F, F, T, T, and otherwise has truth value F. Similarly, c_{13} is $\sim p_1 \wedge p_2 \wedge \sim p_3 \wedge p_4$, which has truth value T for the choice of truth values F, T, F, T and otherwise has truth value F. Finally, c_{1234} is $\sim p_1 \wedge \sim p_2 \wedge \sim p_3 \wedge \sim p_4$ which has truth value T for the choice of truth values F, F, F, F and truth value F otherwise. The truth value of f is T if any one of c_{12}, c_{13}, c_{1234} has truth value T, and otherwise the truth value of f is F. Thus f has truth value T for precisely the choices of truth values F, F, T, T; F, T, F, T; and F, F, F, F, for p_1, p_2, p_3, and p_4, respectively.

The preceding examples suggest that it is possible to construct a proposition depending on statements $p_1, p_2, \ldots, p_k$ having arbitrarily preassigned truth values.

Theorem 1.4 *For given propositions $p_1, \ldots, p_k$, it is always possible to construct a compound proposition f depending on $p_1, \ldots, p_k$ which has preassigned truth values for stipulated choices of truth values for $p_1, \ldots, p_k$. In other words, given an arbitrary truth table of the form*

(21)

p_1	p_2	$\cdots$	p_k	f
T	T	$\cdots$	T	*
$\vdots$	$\vdots$		$\vdots$	$\vdots$
F	F	$\cdots$	F	*

in which the stars indicate an arbitrary selection of T *or* F, *there always exists a compound statement* f *having* (21) *as its truth table.*

Proof Examine the last column of the table (21) to find precisely those rows in which the truth value T appears. If there are none, simply take f to be the proposition $p_1 \wedge \sim p_1$ which trivially satisfies the conclusion of the theorem. (Why?) Now, each row of the table (21) corresponds to a choice of truth values for $p_1, \ldots, p_k$. Thus suppose that in one of these rows corresponding to a truth value T for f the statements $p_i, p_j, \ldots, p_m$ have truth value F, and the remaining statements have truth value T. Then the proposition $c_{ij\ldots m}$ of formula (18) in Theorem 1.3 will have truth value T for just this selection of truth values for $p_1, \ldots, p_k$, and otherwise will have truth value F. If we let f be the disjunction of every one of the statements $c_{ij\ldots m}$ corresponding to the rows with preassigned truth value T, then f will have truth value T whenever any one of the $c_{ij\ldots m}$ has truth value T and will otherwise have truth value F. This completes the proof. ▮

Example 1.8 Find a compound statement f depending on p_1, p_2, and p_3 which has the following truth table.

p_1	p_2	p_3	f
T	T	T	T
T	T	F	F
T	F	T	F
T	F	F	T
F	T	T	F
F	T	F	F
F	F	T	F
F	F	F	T

The answer according to Theorem 1.4 is the compound statement

$$f\colon c_0 \vee c_{23} \vee c_{123},$$

where c_0 is $p_1 \wedge p_2 \wedge p_3$, c_{23} is $p_1 \wedge \sim p_2 \wedge \sim p_3$, and c_{123} is $\sim p_1 \wedge \sim p_2 \wedge \sim p_3$.

Rather surprisingly, Theorem 1.4 can be effectively used in the design of switching networks. A "switch" is simply a device which allows precisely two possibilities: either the current flows when the switch is on, or the current does not flow when the switch is off. One can also think of a switch as a valve controlling the flow of water through a pipe. With any switch s, we can conceive of the "negation" of s which has precisely the opposite effect of s, i.e., the negation of s, which we denote informally by $\sim s$, is off when s is on, and on when s is off. One can imagine two valves in a pipe which have this property.

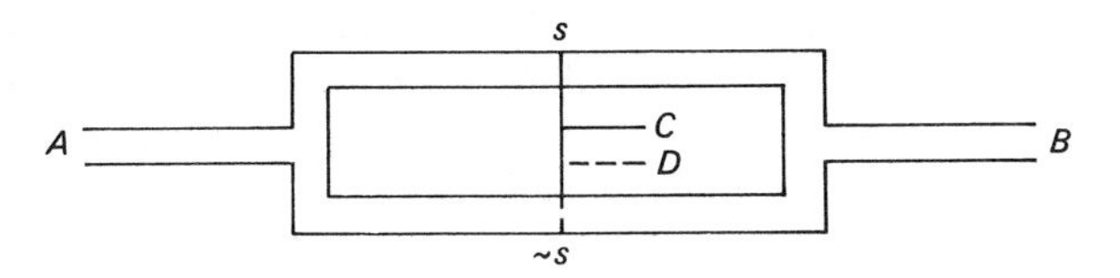

When the handle of these valves is in position C, s is closed and $\sim s$ is open, as shown by the solid line. When the handle is moved downward to position D, s opens and $\sim s$ closes, as shown by the dotted lines. Since the author's knowledge of plumbing is somewhat limited, we are prepared to accept the possibility that there exists a more efficient system.

The general problem in the design of switching or valve networks can be expressed as follows. We wish to control the flow between specified points A and B by a collection of k switches. These switches may be connected in series or in parallel (to be explained shortly). With various on and off configurations of the switches, we wish the flow either to take place or not to take place. Thus, for example, suppose we wish to design a "circuit" which controls the light in a stairway using a switch at the first floor and a switch at the second. To analyze this problem, let p_1 be the statement, "Switch 1 is in the on position," and p_2 the statement, "Switch 2 is in the on position." Let f be the statement, "The light is on." Then we would like the following to happen:

(22)

p_1	p_2	f
T	T	T
T	F	F
F	T	F
F	F	T

That is, we want the light to be on when both s_1 and s_2 are on (row 1 of Table (22)). If we change s_2 from on to off and leave s_1

at on, we want the light to go off (row 2). If we change s_1 from on to off and leave s_2 at on, we similarly want the light to go off (row 3). If the light is off (row 2 or row 3) we want the light to go back on when we change either the first or second switch. Such a change in row 2 or row 3 (changing the truth values of either p_1 or p_2) will produce either row 1 or row 4 and the light will be on. According to Theorem 1.4, we can construct a compound statement f which has (22) as its truth table:

$$f\colon\ (p_1 \wedge p_2) \vee (\sim p_1 \wedge \sim p_2). \tag{23}$$

In order that (23) have physical meaning in terms of the arrangement of switches, we must interpret the conjunction and disjunction of the statements p_1 and p_2 in terms of switch configurations. Consider the following diagram.

(24)

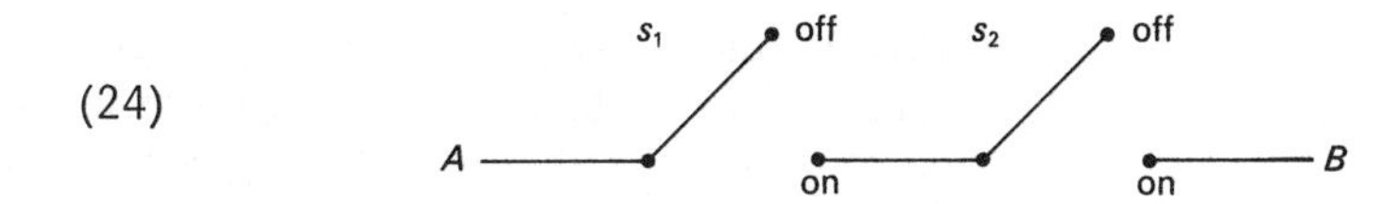

If switches s_1 and s_2 are both in the on position, it means that the statement $p_1 \wedge p_2$ is true; and if either s_1 or s_2 is off it means the statement $p_1 \wedge p_2$ is false. Clearly, the current flow will take place from A to B only when both s_1 and s_2 are on, i.e., the statement $p_1 \wedge p_2$ is true. Stated another way, the current does or does not flow from A to B precisely as the truth value of $p_1 \wedge p_2$ is T or F, respectively. If the switches are arranged as in diagram (24), we shall say they are in *series*.

Now consider the next diagram.

(25)

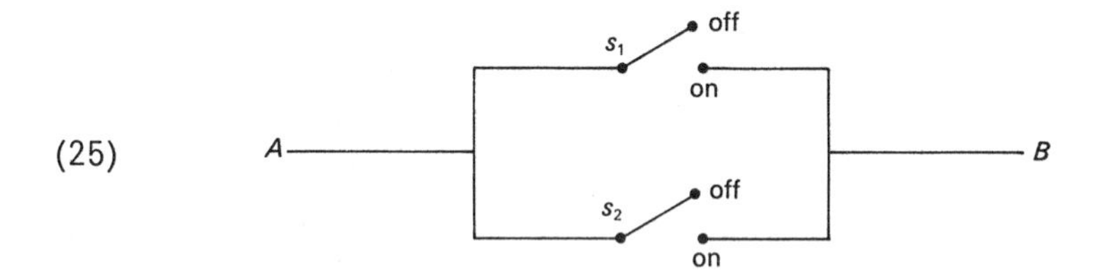

It is clear from the diagram (25) that the flow from A to B will take place if either s_1 or s_2 is on. In other words, the flow from A to B does or does not take place precisely according as the disjunction $p_1 \vee p_2$ has truth value T or F, respectively. Two switches that are arranged as in diagram (25) will be said to be in *parallel*. Let us construct a switching arrangement which will correspond to the compound statement (23).

(26)

The circled numbers 1 and 2 refer to switches 1 and 2, and the negated circles refer to the negations of the switches 1 and 2. Switches ① and ~① are coupled and switches ② and ~② are coupled, e.g., when ① is open, ~① is closed and when ~① is open, ① is closed. The diagram (26) corresponds to the fourth row of Table (22) and it is clear that the flow takes place from A to B. Now consider the diagram which corresponds to the third row of (22), when p_1 has truth value F and p_2 has truth value T. This configuration of switches corresponds to the following diagram.

(27)

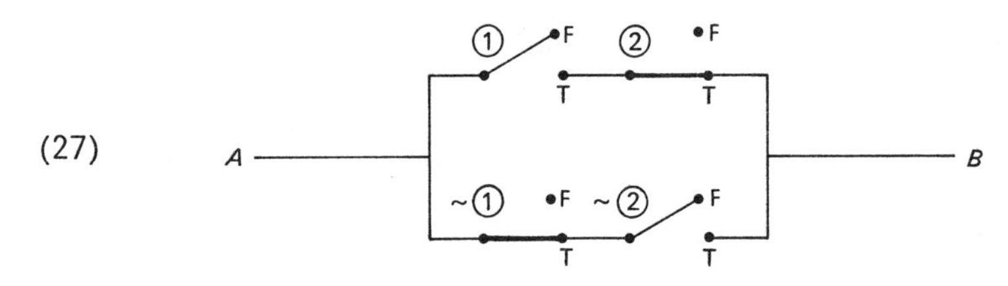

It is clear from the diagram (27) that no flow takes place from A to B, i.e., f has truth value F. Thus, we can control the light in the stairway by making the following arrangement of switches: s_1 and s_2 are connected in series, $\sim s_1$ and $\sim s_2$ are connected in series, and these two parts are connected in parallel. A practical wiring scheme is indicated in the following diagram.

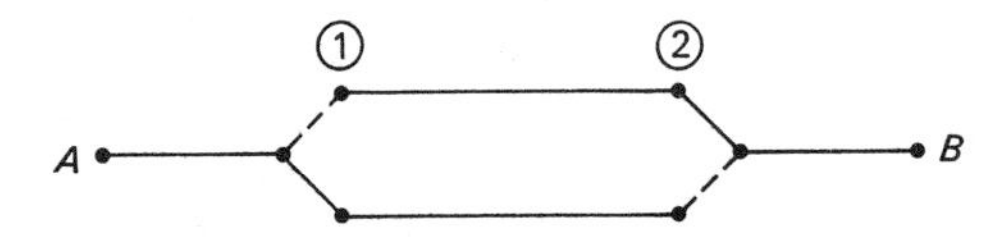

Quiz

Answer true or false.

1 The two statements p and $\sim p$ always have different truth values.

2 The disjunction of the two statements p and q always has truth value T.

3 The disjunction of the statements p and $\sim p$ always has truth value T.

4 The implication, "If $1 = 0$, then $2 = 1$," has truth value T.

5 The implication, "If $1 = 0$, then $2 = 2$," has truth value T.

6 A tautology depending on statements $p_1, p_2, \ldots, p_k$ is a compound statement which has truth value T only when each of $p_1, \ldots, p_k$ has truth value T.

7 The law of detachment describes the fact that the statement $(p \wedge (p \rightarrow q)) \rightarrow q$ is a tautology.

8 An argument which establishes a statement by "indirect proof" constitutes an application of the tautology $((p \rightarrow q) \wedge \sim q) \rightarrow \sim p$.

9 A truth table must always have at least four rows.

10 A truth table for a compound statement f formed from p_1, p_2, and p_3 has eight rows.

Exercises

1 Decide which of the following arguments are valid.

(a) If it is raining, I always carry my umbrella. It is not raining, and hence I do not carry my umbrella.

(b) Whenever the sun shines I play tennis or golf. I am playing squash, therefore the sun is not shining.

(c) A blue-eyed girl is always pretty or pleasant. Mary is blue-eyed and pleasant and it follows that she is not pretty.

(d) It is in the national interest that draft deferments be granted to scientists. The present draft law places scientists in first priority in terms of eligibility for the draft. It follows that the present draft law is erroneous.

(e) If the sun is not shining I cannot play tennis. Therefore, the sun is shining.

(f) I will purchase a new suit on Saturday if and only if a sale is in progress. A sale is not in progress on Saturday. It follows that I will purchase a new suit on Saturday.

(g) A necessary and sufficient condition that the whole number n be even is that it be divisible by 2. The whole number n is not divisible by 3. It follows that n is not even.

(h) If a whole number n is divisible by both 2 and 3 then n is divisible by 6. A necessary and sufficient condition that n be divisible by 6 is that it leaves a remainder of 0 upon division by 6. The whole number n is divisible by 3 but not by 2. It follows that n does not leave a remainder of 0 upon division by 6.

2 Verify that the following statements are tautologies by using truth tables.

(a) $(p \wedge q) \leftrightarrow (q \wedge p)$.

(b) $(p \vee (q \wedge r)) \leftrightarrow ((p \vee q) \wedge (p \vee r))$.

(c) $(p \wedge p) \leftrightarrow p$.

†(d) $((p \vee q) \vee r) \leftrightarrow (p \vee (q \vee r))$.

(e) $p \vee \sim p$.

†(f) $(\sim(p \wedge q)) \leftrightarrow (\sim p \vee \sim q)$.

†(g) $(\sim(p \vee q)) \leftrightarrow (\sim p \wedge \sim q)$.

(h) $(p \rightarrow q) \leftrightarrow (\sim q \rightarrow \sim p)$. (Thus a statement and its contrapositive are equivalent.)

3 Reformulate the following argument in symbolic form and decide on its validity. If Mary is pretty, then Mary is popular. Either Mary is popular or she is a good student. Hence, if Mary is a good student, then she is pretty.

4 Construct a compound statement depending on the statements p, q, and r which has truth value T when p, q, and r have truth values T, T, and F, respectively, and which has truth value F otherwise.

5 Find compound statements f corresponding to the following truth tables.

(a)

p_1	p_2	f
T	T	T
T	F	T
F	T	T
F	F	T

(b)

p_1	p_2	f
T	T	F
T	F	F
F	T	F
F	F	F

(c)

p_1	p_2	f
T	T	T
T	F	F
F	T	T
F	F	F

(d)

p_1	p_2	p_3	f
T	T	T	F
T	T	F	T
T	F	T	T
T	F	F	F
F	T	T	T
F	T	F	F
F	F	T	T
F	F	F	F

(e)

p_1	p_2	p_3	p_4	f
T	T	T	T	F
T	T	T	F	F
T	T	F	T	F
T	T	F	F	F
T	F	T	T	F
T	F	T	F	F
T	F	F	T	F
T	F	F	F	F
F	T	T	T	F
F	T	T	F	T
F	T	F	T	F
F	T	F	F	T
F	F	T	T	F
F	F	T	F	T
F	F	F	T	F
F	F	F	F	T

6 Let s_1, s_2, and s_3 be switches controlling the flow from A to B. Draw diagrams of networks which have the following properties.

(a) The flow takes place from A to B only when s_1, s_2, and s_3 are in the on position.

(b) The flow takes place from A to B only when s_1 is on and s_2 and s_3 are off, or when s_3 and s_2 are on and s_1 is off.

(c) The flow never takes place from A to B whatever the configuration of switches.

(d) The flow takes place from A to B for all configurations of switches.

7 Three valves, s_1, s_2, and s_3, are to be used to control the flow of water from points A to B in an irrigation system. Design a network using these valves (and their negations) which allows water to flow between A and B only under the following circumstances: when either s_1 and s_2 are in the on position and s_3 is in the off position; or when s_1 and s_3 are in the on position and s_2 is in the off position.

8 Let s_1, s_2, and s_3 be on-off switches in an electrical network. Construct a compound statement which corresponds to each of the following switching arrangements and draw a diagram showing how the switches are connected.

(a) The current flows if and only if all three switches are in the on position.

(b) The current flows if and only if s_1 and s_2 are in the on position and s_3 is in the off position.

(c) The current flows if and only if s_1 and s_2 are in the on position whatever the position of s_3 may be.

9 Let f be a compound statement constructed from $p_1, \ldots, p_k$. The *truth set* of f is just the set of those choices of truth values for $p_1, \ldots, p_k$ for which f has truth value T. For example, the truth set of f: $p_1 \vee p_2$ is just (T, T), (T, F), (F, T). Find the truth sets of each of the following compound statements:

(a) $p_1 \wedge p_2$,

(b) $p_1 \vee (p_2 \wedge p_3)$,

(c) $p_1 \rightarrow (p_2 \rightarrow p_3)$,

(d) $(p_1 \vee p_2) \rightarrow p_2$,

(e) $(p_1 \wedge p_2) \leftrightarrow \sim p_3$,

(f) $(p_1 \wedge p_2) \rightarrow \sim p_1$,

(g) $(p_1 \vee \sim p_2) \leftrightarrow \sim(\sim p_1 \wedge p_2)$.

10 A switching network to control the launching of an I.C.B.M. is to be designed so it can be managed by three second lieutenants. In conformity with democratic tradition, the Department of Defense makes the following assumption. If any two of the three second lieutentants decide to close their control switches, the missile will be launched. On the other hand, if any two of the three decide not to close their control switches, the missile will not be launched. Find an appropriate switching network.

†11 Find a statement equivalent to the statement $p \rightarrow q$ which only involves the connective $\vee$ and the negation $\sim$.

†12 Find a statement equivalent to the statement $p \leftrightarrow q$ which only involves the connective $\vee$ and the negation $\sim$.

13 Find a statement equivalent to the statement $p \wedge q$ which only involves the connective $\rightarrow$ and the negation $\sim$.

14 Let p denote the statement, "I study regularly"; q the statement, "I get good grades in school"; and r the statement, "I lie on the beach." Write down the following statements in symbolic form using p, q, and r.

(a) The converse of "If I study regularly, then I get good grades in school."

(b) The contrapositive of "If I get good grades in school, then I do not lie on the beach."

(c) The negation of "I lie on the beach and I do not study regularly."

(d) The statement, "I study regularly or I lie on the beach only if I get good grades in school."

1.2 Sets

In the beginning there were sets. By this we mean that the notion of a collection of items is a primitive or undefined concept. A set of objects is, after all, an abstraction in the sense that a number of individual items may have a physical reality, but the collection or totality of these items is an intellectual construct. The important property that a set must have is the following: it must be possible to decide whether a given item is in the set. For example, if N denotes the set of all girls with first name Nancy, then surely Lana is not in this set. Thus the notion of a set is indeed coextensive with the notion of its defining property. Consider the set of all blue-eyed blondes with long hair on the campus of the University of California at Santa Barbara. Until quite recently, most rational people would concur that Joe Smith is not in this set. On the other hand, unless the defining property is adequately described, it is no longer possible to make such a statement. We must be certain, then, that the sets we study are well-defined. Another way of saying this is that given a mathematical object x, and given a set S, it is always possible to decide whether or not x is in S.

Example 2.1 Among whole numbers, the set of even whole numbers is well-defined. For, given a whole number it is possible, at least in principle, to divide it by 2 and determine whether the division is exact.

Example 2.2 The set of all real numbers between 0 and 1 which have the digit 1 in the third place of their decimal expansions is a well-defined set, for it is always possible to find the decimal expansion of any particular real number r to three decimal places and then simply examine the third digit.

Example 2.3 Consider the "set" of all whole numbers that Pythagoras thought of in the first 20 years of his life. Although the preceding sounds as if a set is being defined, there is, in fact, no way for us to know even in principle whether a given integer qualifies for membership in this set.

There is a certain amount of notation which is very useful in the elementary theory of sets. If S is a set and x is an item which belongs to the set S, we write

$$x \in S. \tag{1}$$

Formula (1) is read, "x is an element of S" or "x is in S" or "x is a member of S" or "x belongs to S" or "x is a point in S." If it is not the case that x is a member of S, we write

$$x \notin S.$$

There are two somewhat different ways to describe a set. The first is simply to list the elements. For example, the set which consists of 0, 1, and the square root of 2 can be written

$$\{0, 1, \sqrt{2}\}. \tag{2}$$

The curly brackets in (2) suggest that 0, 1, and $\sqrt{2}$ are contained in the set. A second and generally more useful notation for a set is in terms of its "defining" property. Thus, suppose that we are discussing certain items x, and for each x there is a proposition depending on x and denoted by $p(x)$, which is verifiably either true or false. Then we write

$$\{x \mid p(x) \text{ is true}\} \tag{3}$$

to designate the set of all x for which the proposition $p(x)$ is true. For example, the set of even integers can be written

$$\{x \mid x = 2y,\ y \text{ is an integer}\}. \tag{4}$$

In (4), we have abbreviated slightly: (4) is read, "The set of all x such that x is twice y, where y is an integer." To put the preceding sentence in the form indicated in (3), we would have to say, somewhat ponderously, "The set of all x such that it is true that $x = 2y$ for some integer y."

In studying sets, we very often find it necessary to use the phrases: "For every x (in a set S) $p(x)$ is true," or "There exists an x (in S) for which $p(x)$ is true." Such phrases are called quantifications. The first is a *universal quantification* for $p(x)$, $x \in S$, and the second is an *existential quantification* for $p(x)$, $x \in S$. The standard notations for these items are respectively

(5) $$\bigwedge_{x \in S} (p(x) \text{ is true})$$

and

(6) $$\exists_{x \in S} (p(x) \text{ is true}).$$

Formula (5) is read, "For every x in S, $p(x)$ is true" or "For every x in S, $p(x)$ holds." Formula (6) is read, "There exists an x in S such that $p(x)$ is true," or "For at least one x in S, $p(x)$ holds." The symbols "$\bigwedge$" and "$\exists$" are called *quantifiers.*

There is an important relationship between (5) and (6). To say that (5) fails is the statement, "It is not true that for every $x \in S$, $p(x)$ holds." We mean by this: "For some $x \in S$, $p(x)$ does not hold." In symbols,

(7) $$\sim\Big(\bigwedge_{x \in S} p(x)\Big) \leftrightarrow \exists_{x \in S} \sim p(x).$$

In formula (7) we further abbreviate our notation so that instead of writing "$p(x)$ holds" or "$p(x)$ is true," we simply write $p(x)$.

Example 2.4 Find the negation of the statement, "Every redhead has a nasty temper and is easily sunburned."

Let S be the set of all redheads. Let $p_1(x)$ be the statement "x has a nasty temper" and $p_2(x)$ the statement "x is easily sunburned." Then the given statement has the following form:

(8) $$\bigwedge_{x \in S} (p_1(x) \wedge p_2(x)),$$

and hence from formula (7), the negation of (8) is equivalent to the statement

(9) $$\exists_{x \in S} \sim(p_1(x) \wedge p_2(x)).$$

We know (see Exercise 2(f), Section 1.1) that

$$(\sim(p_1(x) \wedge p_2(x))) \leftrightarrow (\sim p_1(x) \vee \sim p_2(x)).$$

Thus (9) is equivalent to

$$\exists_{x \in S} (\sim p_1(x) \vee \sim p_2(x)).$$

In words, the negation of (8) is, "There exists a redhead who does not have a nasty temper or who is not easily sunburned."

Example 2.5 Find the negation of the following statement: "Every x which is a member of the set S is a member of the set T."

This statement can be formulated in terms of the quantifier notation as follows. Let us assume that S and T are both contained in a larger set U called the *universe* which contains all of the elements in the present discussion. Then the statement becomes

(10)
$$\bigwedge_{x \in U} (x \in S \rightarrow x \in T).$$

According to (7), the negation of (10) is equivalent to

(11)
$$\exists_{x \in U} \sim(x \in S \rightarrow x \in T).$$

Now, in general, $p \rightarrow q$ is equivalent to the statement $\sim p \vee q$ (see Exercise 11, Section 1.1) and hence $\sim(p \rightarrow q)$ is equivalent to the statement $p \wedge \sim q$ (see Exercise 2(g), Section 1.1). Thus (11) is equivalent to the statement

(12)
$$\exists_{x \in U} (x \in S \wedge x \notin T).$$

In words, (12) is the statement, "There exists an x in U such that x is in S but not in T." If the statement (10) is true, then S is said to be a *subset* of T, and the notation for this is

(13)
$$S \subset T.$$

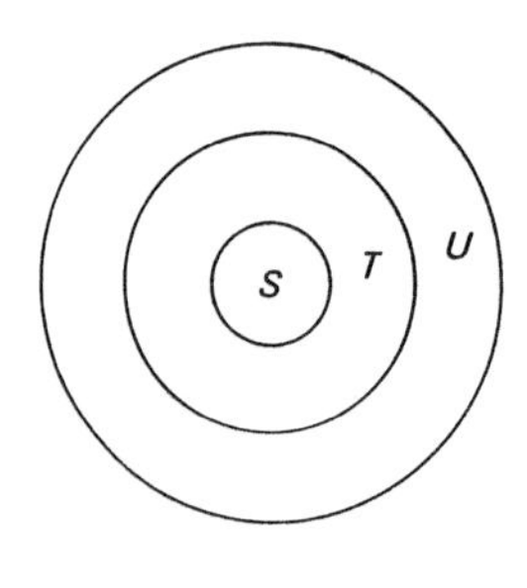

We can now introduce a convenient and informative graphical representation for sets called Venn diagrams (named after John Venn, an English logician). Thus let U be the universe and let S and T be subsets of U. Then the Venn diagram to the left indicates that S is a subset of T, i.e., that (13) holds. The smallest disc represents S; it is part of the disc representing T and both S and T are inside the "universal" disc representing U. It is not always necessary to draw the set U. Although Venn diagrams are helpful illustrations of relationsihps between sets, they are only illustrations, never proofs. Realize that the sets S and T are actually not the circles pictured in the diagram: the diagrams are only schematic and not literal. Moreover, the sets need not be pictured by circles.

Two sets S and T are said to be *equal*,

$$S = T,$$

if every element of S is an element of T and every element of T is an element of S. Once again, if S and T are made up of elements from a universe U, then $S = T$ is equivalent to the statement

$$\bigwedge_{x \in U} (x \in S \leftrightarrow x \in T),$$

i.e., for every $x \in U$ it is true that if $x \in S$ then $x \in T$ and conversely.

There are three operations on sets that correspond to the negation $\sim$ and the connectives $\wedge$ and $\vee$. Thus, let U be the universe, and let $\mathfrak{A}$ be a collection or family of subsets of U. Then

$$\bigcap_{S \in \mathfrak{A}} S \tag{14}$$

denotes the the totality of elements x in U which belong to every one of the sets in the family $\mathfrak{A}$. The set (14) is called the *intersection* of the sets in $\mathfrak{A}$. In symbols,

$$\bigcap_{S \in \mathfrak{A}} S = \left\{x \,\middle|\, \bigwedge_{S \in \mathfrak{A}} x \in S\right\}. \tag{15}$$

The equality (15) reads, "The intersection of the sets in the family $\mathfrak{A}$ is equal to the set of all elements which are members of every one of the sets S of $\mathfrak{A}$." If we have a family $\mathfrak{A}$ which consists of only a finite number of sets, $S_1, \ldots, S_p$, then the intersection of the sets in $\mathfrak{A}$ is written

$$S_1 \cap S_2 \cap \cdots \cap S_p. \tag{16}$$

To say that x is a member of the set (16) means that the statement

$$(x \in S_1) \wedge (x \in S_2) \wedge \cdots \wedge (x \in S_p)$$

is true. This shows the relation between the connective $\wedge$ and set intersection.

If $p = 2$ then the Venn diagram for $S_1 \cap S_2$ is

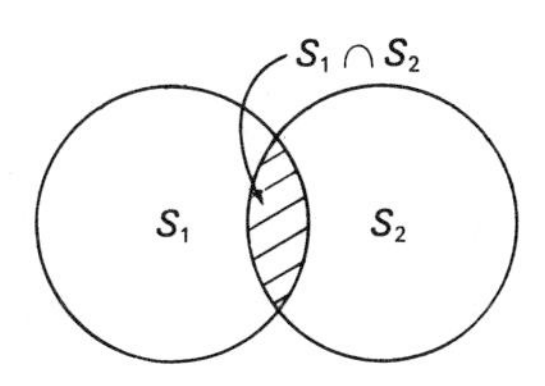

The shaded region indicates the set of points that belong to both S_1 and S_2.

If $p = 3$ then the Venn diagram for $S_1 \cap S_2 \cap S_3$ is

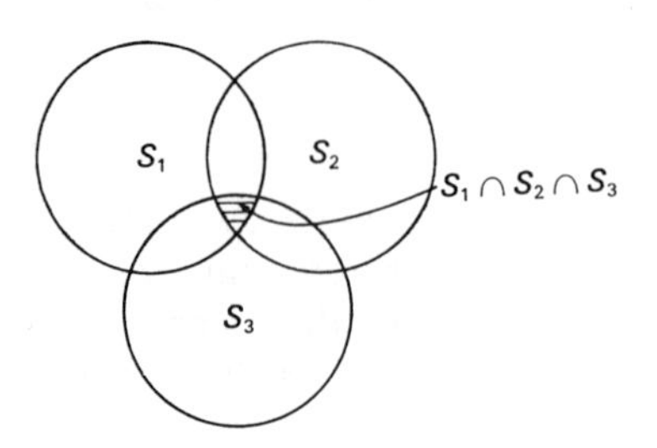

The shaded region indicates the set of points that S_1, S_2, and S_3 have in common.

Example 2.6 **(a)** Let the universe U be the set of all girls on the campus of the University of California, Santa Barbara. Let S_1 be the set of all blondes in U; let S_2 be the set of all blue-eyed girls in U. Then $S_1 \cap S_2$ is the set of all blue-eyed blonde girls on the campus of the University of California, Santa Barbara.
(b) Let $S_1 = \{1, 2, 4, 6, 8\}$ and $S_2 = \{2, 3, 4, 5, 6\}$. Then $S_1 \cap S_2 = \{2, 4, 6\}$. For, 2, 4, and 6 are precisely those elements which are common to S_1 and S_2.

Example 2.7 Let U be the set of all real numbers x, $0 \leq x \leq 1$. Let $\mathfrak{A}$ be the family of subsets S_n of U, where

$$S_n = \left\{x \,\middle|\, 0 \leq x < 1 - \frac{1}{n+1}\right\}, \qquad n = 1, 2, 3, \ldots .$$

(Recall that $a \leq b$ is read, "a is less than or equal to b" and $a < b$ is read, "a is strictly less than b.") Then the intersection of the sets in $\mathfrak{A}$ can be written

$$\bigcap_{n=1}^{\infty} S_n, \tag{17}$$

where the "∞" (read "infinity") symbol in (17) simply means that the intersection of *all* of the sets S_n, $n = 1, 2, \ldots$, is to be computed. We can make a pictorial representation of the sets S_n.

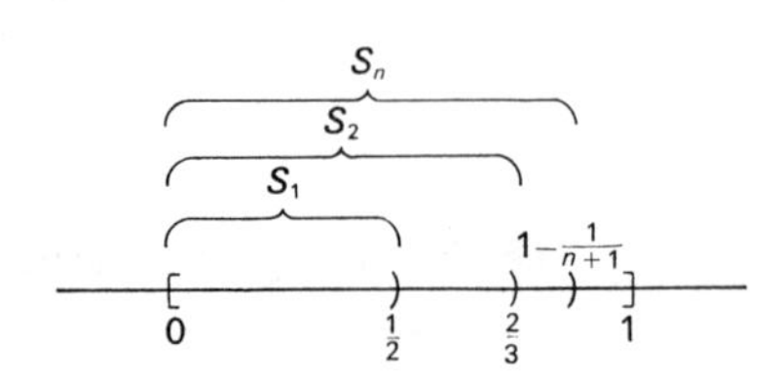

Since $S_1 \subset S_2,\ S_2 \subset S_3,\ \ldots,\ S_k \subset S_{k+1}, \ldots$, it follows that

$$S_1 \subset \bigcap_{n=1}^{\infty} S_n.$$

On the other hand, an element of $\bigcap_{n=1}^{\infty} S_n$ must by definition be in every one of the S_n, $n = 1, 2, \ldots$, and hence in particular must be in S_1. It follows that

$$S_1 = \bigcap_{n=1}^{\infty} S_n.$$

The operation on sets that corresponds to the connective $\vee$ can be defined succinctly as follows. Let U be the universe and let $\mathfrak{A}$ be a family or collection of subsets of U. Then the *union* of the sets in $\mathfrak{A}$ is the totality of elements x which belong to at least one of the sets S in $\mathfrak{A}$. The notation for the union is

$$\bigcup_{S \in \mathfrak{A}} S. \tag{18}$$

Using the "setmaker" notation, we have

$$\bigcup_{S \in \mathfrak{A}} S = \left\{x \,\middle|\, \underset{S \in \mathfrak{A}}{\exists}\ x \in S\right\}. \tag{19}$$

In words, (19) reads, "The union of the sets in the family $\mathfrak{A}$ is equal to the totality of elements x such that there exists at least one $S \in \mathfrak{A}$ for which $x \in S$." If the family $\mathfrak{A}$ consists of only a finite number of sets, $S_1, \ldots, S_p$, then the union is denoted by

$$S_1 \cup S_2 \cup \cdots \cup S_p. \tag{20}$$

If x is an element of the union (20) then the statement

$$(x \in S_1) \vee (x \in S_2) \vee \cdots \vee (x \in S_p)$$

is true, i.e., at least one of the statements $x \in S_1, \ldots, x \in S_p$ is true.

If $p = 2$ a Venn diagram for $S_1 \cup S_2$ is

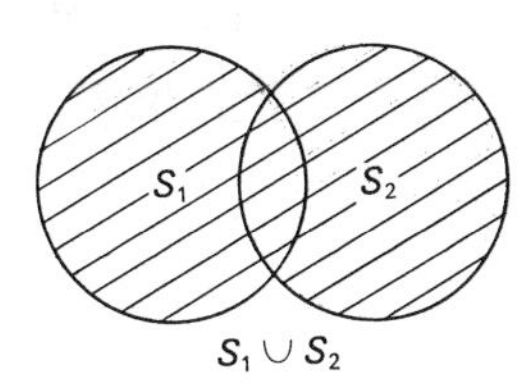

$S_1 \cup S_2$

The shaded region, $S_1 \cup S_2$, indicates the totality of points which belong to either one (or both) of S_1 and S_2.

If $p = 3$ then a Venn diagram for $S_1 \cup S_2 \cup S_3$ is

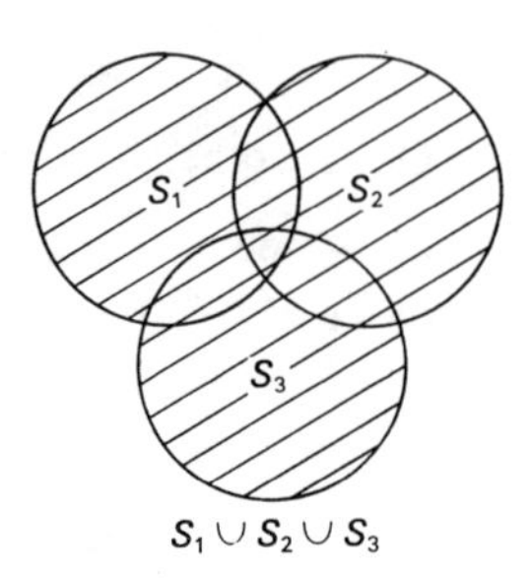

$S_1 \cup S_2 \cup S_3$

The shaded region depicting $S_1 \cup S_2 \cup S_3$ is the totality of points which belong to any one of S_1, S_2, and S_3.

Example 2.8 **(a)** Let $S_1 = \{1, 2, 3, 5, 7\}$, $S_2 = \{3, 4, 5, 6, 7\}$, and $S_3 = \{4, 6, 8, 9\}$. Then $S_1 \cup S_2 \cup S_3 = \{1, 2, 3, 4, 5, 6, 7, 8, 9\}$, because each of the numbers $1, \ldots, 9$ is in at least one of the S_i, $i = 1, 2, 3$.

(b) Consider the family $\mathfrak{A}$ in Example 2.7. Then we can write

$$\bigcup_{n=1}^{\infty} S_n$$

to denote $\bigcup_{S \in \mathfrak{A}} S$. It is easy to show that

$$\bigcup_{n=1}^{\infty} S_n = \{x \mid 0 \leq x < 1\}. \tag{21}$$

For, every set S_n is a subset of the right-hand side of (21) and it follows that

$$\bigcup_{n=1}^{\infty} S_n \subset \{x \mid 0 \leq x < 1\}.$$

On the other hand, if we take any number x satisfying $0 \leq x < 1$, we can find an integer k such that $0 \leq x \leq 1 - \dfrac{1}{k+1}$. For, it is clear that the points $1 - \dfrac{1}{k+1}$ get arbitrarily close to 1 as k increases. But it follows from this inequality that $x \in S_k$ and hence $x \in \bigcup_{n=1}^{\infty} S_n$. We have established that

$$\{x \mid 0 \leq x < 1\} \subset \bigcup_{n=1}^{\infty} S_n,$$

which, taken with the preceding inclusion, establishes (21).

Example 2.9 Let S_1, S_2, and S_3 be subsets of U. Show that

(22) $$S_1 \cup (S_2 \cup S_3) = (S_1 \cup S_2) \cup S_3.$$

The following sequence of equivalences formally establishes (22):

(23) $$\begin{aligned} x \in S_1 \cup (S_2 \cup S_3) &\leftrightarrow x \in S_1 \vee (x \in S_2 \cup S_3) \\ &\leftrightarrow x \in S_1 \vee (x \in S_2 \vee x \in S_3) \\ &\leftrightarrow (x \in S_1 \vee x \in S_2) \vee x \in S_3 \\ &\leftrightarrow (x \in S_1 \cup S_2) \vee x \in S_3 \\ &\leftrightarrow x \in (S_1 \cup S_2) \cup S_3. \end{aligned}$$

The calculation (23) is based on two facts. First, to say that an element x is in the union of sets S_1 and S_2 is to say that x is in at least one of the sets S_1 and S_2, i.e.,

$$x \in S_1 \cup S_2 \leftrightarrow (x \in S_1 \vee x \in S_2).$$

Second, the following compound statement is a tautology:

(24) $$((p_1 \vee p_2) \vee p_3) \leftrightarrow (p_1 \vee (p_2 \vee p_3))$$

(see Exercise 2(d), Section 1.1). In the present problem, p_i is the statement "$x \in S_i$", $i = 1, 2, 3$, and the third equivalence in (23) is an instance of the tautology (24). We can also convince ourselves immediately that (22) is correct by using Venn diagrams. The left side of (22) is the union of S_1 with the union of S_2 and S_3.

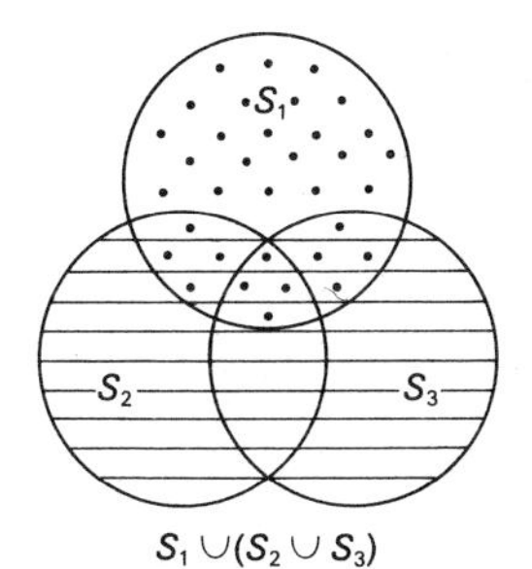

$S_1 \cup (S_2 \cup S_3)$

Thus $S_1 \cup (S_2 \cup S_3)$ is the shaded region taken together with the dotted region. The right side of (22) is the union of $S_1 \cup S_2$ with S_3.

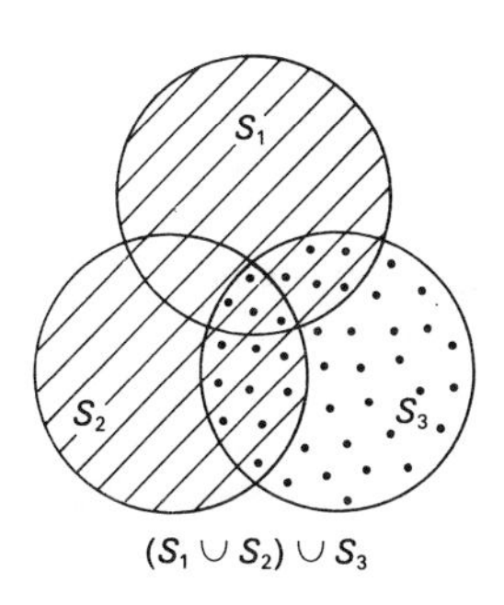

$(S_1 \cup S_2) \cup S_3$

Then $(S_1 \cup S_2) \cup S_3$ is the shaded region taken together with the dotted region. Clearly the two regions in the diagrams to the left are the same.

The set operation which corresponds to the logical operation of negation is called *complementation*. Let S be a subset of the universe U. Then the set of all x in U which are *not* in S, i.e., $\sim(x \in S)$, is denoted by S'. The set S' is called the *complement of S in U*. Thus, using the set-maker notation, we have

$$S' = \{x \mid x \in U \wedge x \notin S\}.$$

Observe that S' depends on the universe U. For example, if $S = \{2, 4, 6, 8, 10\}$ and $U = \{1, 2, \ldots, 10\}$, then $S' = \{1, 3, 5, 7, 9\}$.

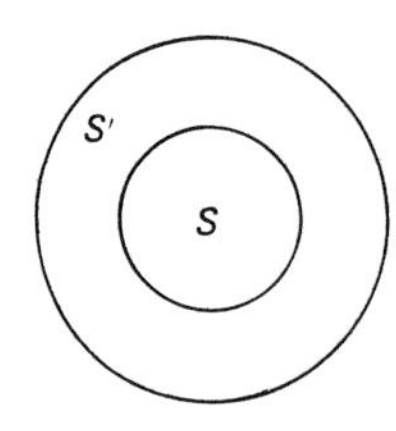

However, if $U = \{2, 3, \ldots, 10\}$, then $S' = \{3, 5, 7, 9\}$. The Venn diagram for complementation is particularly simple.

Observe that $(S')' = S$.

Suppose S_1 and S_2 are sets. We draw the Venn diagram for the complement of the union $S_1 \cup S_2$. This is represented by the shaded region outside the union of S_1 and S_2.

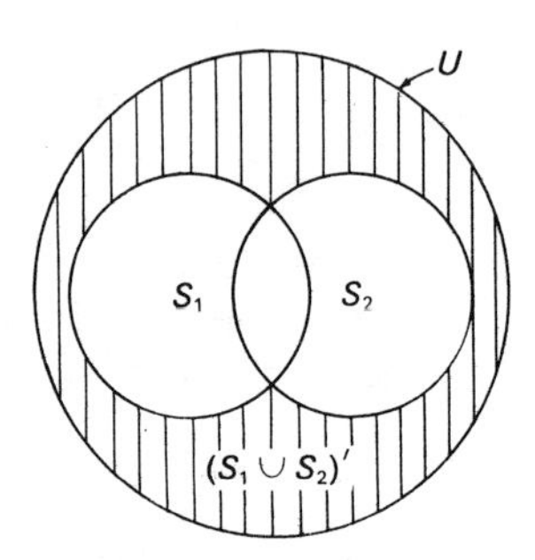

On the other hand, consider the Venn diagram for the intersection of the complements of S_1 and S_2.

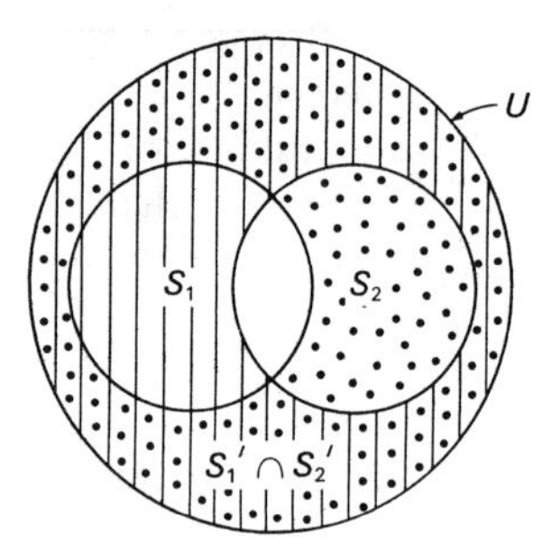

The dotted region depicts S_1' and the lined region S_2'. The region which is *both* lined and dotted is $S_1' \cap S_2'$. Clearly this region coincides with $(S_1 \cup S_2)'$ in the preceding diagram so that

$$(S_1 \cup S_2)' = S_1' \cap S_2'.$$

If we replace S_1 by S_1' and S_2 by S_2' in this equation, we have

$$(S_1' \cup S_2')' = (S_1')' \cap (S_2')'.$$

But $(S_1')' = S_1$ and $(S_2')' = S_2$. Hence

$$(S_1' \cup S_2')' = S_1 \cap S_2.$$

If we take the complement of both sides we have

$$S_1' \cup S_2' = (S_1 \cap S_2)'.$$

These considerations are in fact special cases of an interesting and important theorem relating complementation, union, and intersection.

Theorem 2.1 ***De Morgan's Laws*** *Let $\mathfrak{A}$ be a family of subsets of a universe U. Then*

$$\left(\bigcap_{S\in\mathfrak{A}} S\right)' = \bigcup_{S\in\mathfrak{A}} S' \tag{25}$$

and

$$\left(\bigcup_{S\in\mathfrak{A}} S\right)' = \bigcap_{S\in\mathfrak{A}} S'. \tag{26}$$

Proof We begin with (25). Let $x \in \left(\bigcap_{S\in\mathfrak{A}} S\right)'$. Then $\sim\left(x \in \bigcap_{S\in\mathfrak{A}} S\right)$, which in turn is equivalent to the statement

$$\sim \bigwedge_{S\in\mathfrak{A}} x \in S. \tag{27}$$

In other words, to say that $x \in \left(\bigcap_{S\in\mathfrak{A}} S\right)'$ means that it is not the case that x is an element of every set S in the family $\mathfrak{A}$. According to (7), (27) is equivalent to the statement

$$\exists_{S\in\mathfrak{A}} x \notin S,$$

that is,

$$\exists_{S\in\mathfrak{A}} x \in S', \tag{28}$$

or, in words, there exists at least one set $S \in \mathfrak{A}$ such that x is in the complement of S. According to (19), (28) is equivalent to the statement

$$x \in \bigcup_{S\in\mathfrak{A}} S',$$

i.e., x is in the complement of some set $S \in \mathfrak{A}$ if and only if x is in the union of all these complements. Since all of the statements in the preceding argument are equivalent, it follows that

$$x \in \left(\bigcap_{S\in\mathfrak{A}} S\right)' \leftrightarrow x \in \bigcup_{S\in\mathfrak{A}} S',$$

and hence (25) is established.

The argument used to prove (26) is based on a statement closely related to (7), namely, negating an existential quantifier is equivalent to applying a universal quantifier to the negation. More precisely, suppose T is a set and for each $y \in T$, $p(y)$ is a proposition. Then to assert the negation of

$$\exists_{y \in T} p(y)$$

is to say that it is false that there exists a y in T such that $p(y)$ holds. But this means that for every $y \in T$, $p(y)$ must fail. In other words,

$$\sim \exists_{y \in T} p(y) \leftrightarrow \bigwedge_{y \in T} \sim p(y). \tag{29}$$

The following string of equivalences establishes (26):

$$\begin{aligned} x \in \Big(\bigcup_{S \in \mathfrak{A}} S\Big)' &\leftrightarrow \sim \exists_{S \in \mathfrak{A}} x \in S \\ &\leftrightarrow \bigwedge_{S \in \mathfrak{A}} x \notin S \\ &\leftrightarrow \bigwedge_{S \in \mathfrak{A}} x \in S' \\ &\leftrightarrow x \in \bigcap_{S \in \mathfrak{A}} S'. \ \blacksquare \end{aligned}$$

If the family $\mathfrak{A}$ in Theorem 2.1 consists of a finite number of sets, $S_1, \ldots, S_n$, then (25) and (26) take the following forms:

$$(S_1 \cap \cdots \cap S_n)' = S_1' \cup \cdots \cup S_n';$$
$$(S_1 \cup \cdots \cup S_n)' = S_1' \cap \cdots \cap S_n'.$$

There are many other properties of union, intersection, and complementation. Some of these will be developed in Exercises 10, 11, and 12.

Example 2.10 Consider the family of sets $\mathfrak{A}$ of Example 2.7, but here take U to be the entire real line. Find the union of the complements of the sets $S_n \in \mathfrak{A}$, $n = 1, 2, \ldots$. It is required to find

$$\bigcup_{n=1}^{\infty} S_n'. \tag{30}$$

According to (25), (30) is the same as the set

$$\Big(\bigcap_{n=1}^{\infty} S_n\Big)'.$$

But we saw in Example 2.7 that

$$\bigcap_{n=1}^{\infty} S_n = S_1$$

and hence it follows that

$$\bigcup_{n=1}^{\infty} S_n' = S_1'.$$

Now, S_1 is the set of real numbers x satisfying

$$0 \le x < \tfrac{1}{2}.$$

The complement of this interval on the real line (recall that the universe U is the real line) is clearly

$$\{x \mid x < 0 \vee x \ge \tfrac{1}{2}\}. \tag{31}$$

The set (31) is the union of two infinite intervals:

$$\{x \mid x < 0\} \cup \{x \mid x \ge \tfrac{1}{2}\}.$$

Let U be the universe and let S be a subset of U. Consider the intersection of the sets S and S'. Then

$$\begin{aligned} x \in S \cap S' &\leftrightarrow x \in S \wedge x \in S' \\ &\leftrightarrow x \in S \wedge \sim(x \in S). \end{aligned}$$

Now, this latter statement is of the form

$$p \wedge \sim p,$$

which always has truth value F. (Why?) Thus it is never the case that an element $x \in U$ can be in $S \cap S'$. Nevertheless, we want to be able to form the intersection of any two sets. We are thereby led to the introduction of a fiction known as the "empty set." The notation for the *empty set*, also called the *null set*, is

$$\emptyset, \tag{32}$$

and it is formally defined as the subset of U with no members.

Example 2.11 Show that

$$\emptyset \subset S$$

for any subset S of U. It is required to show that for any $x \in U$ the implication

$$x \in \emptyset \rightarrow x \in S$$

has truth value T. The statement $x \in \emptyset$ always has truth value F and hence the implication

$$x \in \emptyset \rightarrow x \in S$$

always has truth value T. (Why?)

Example 2.12 Let X and Y be subsets of the universe U, and let $A = X \cap Y$, $B = X' \cap Y$, $C = X \cap Y'$. Then

$$X \cup Y = A \cup B \cup C$$

and

$$\begin{aligned} A \cap B &= A \cap C \\ &= B \cap C \\ &= \emptyset. \end{aligned}$$

This assertion is easy to illustrate using a Venn diagram.

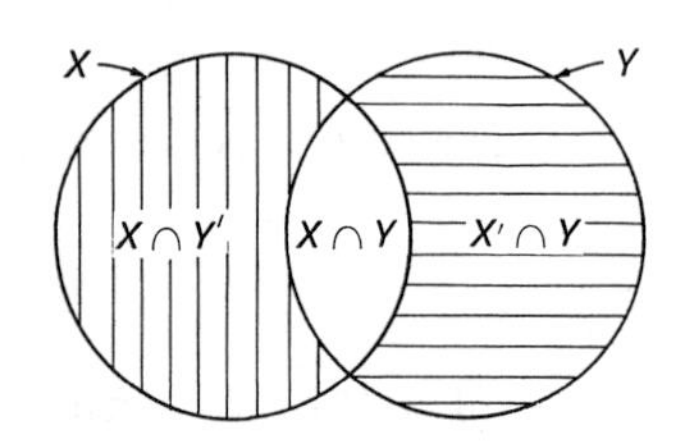

The vertically lined region is $C = X \cap Y'$, the clear region is $A = X \cap Y$, and the horizontally lined region is $B = X' \cap Y$. Obviously $A \cap B = A \cap C = B \cap C = \emptyset$ and $X \cup Y = A \cup B \cup C$. To prove the result formally, observe first that $A \subset X \cup Y$, $B \subset X \cup Y$, and $C \subset X \cup Y$ and hence $A \cup B \cup C \subset X \cup Y$. On the other hand, let p_1 be the statement $x \in X$ and p_2 the statement $x \in Y$. Then by definition

$$x \in (X \cup Y) \leftrightarrow p_1 \vee p_2. \tag{33}$$

We assert that the following compound statement is a tautology:

$$f\colon\ p_1 \vee p_2 \leftrightarrow ((p_1 \wedge p_2) \vee ({\sim}p_1 \wedge p_2) \vee (p_1 \wedge {\sim}p_2)).$$

For, consider the truth table corresponding to f.

p_1	p_2	${\sim}p_1$	${\sim}p_2$	$p_1 \vee p_2$	⑥ $p_1 \wedge p_2$	⑦ ${\sim}p_1 \wedge p_2$	⑧ $p_1 \wedge {\sim}p_2$	⑥ ∨ ⑦ ∨ ⑧	f
T	T	F	F	T	T	F	F	T	T
T	F	F	T	T	F	F	T	T	T
F	T	T	F	T	F	T	F	T	T
F	F	T	T	F	F	F	F	F	T

Thus from (33) we see that

$$\begin{aligned} x \in (X \cup Y) &\leftrightarrow (x \in X \wedge x \in Y) \vee (x \notin X \wedge x \in Y) \vee (x \in X \wedge x \notin Y) \\ &\leftrightarrow x \in ((X \cap Y) \cup (X' \cap Y) \cup (X \cap Y')). \end{aligned}$$

This establishes that $X \cup Y = A \cup B \cup C$. Now,

$$\begin{aligned} A \cap B &= (X \cap Y) \cap (X' \cap Y) \\ &= X \cap Y \cap X' \cap Y \\ &= X \cap X' \cap Y \\ &= \emptyset \cap Y \\ &= \emptyset, \end{aligned}$$

and similarly, $A \cap C = B \cap C = \emptyset$. What we have proved is that the union of the two sets X and Y can be written as the union of the three pairwise *disjoint* sets A, B, and C. To say that two sets A and B are disjoint simply means that $A \cap B$ is the empty set, i.e.,

$$A \cap B = \emptyset.$$

Example 2.12 suggests that it is possible to express a union of sets as a union of a number of disjoint sets.

Theorem 2.2 *Let S_1, S_2, and S_3 be subsets of the universe U. Let*

$$\begin{aligned} A_0 &= S_1 \cap S_2 \cap S_3, \\ A_1 &= S_1' \cap S_2 \cap S_3, \\ A_2 &= S_1 \cap S_2' \cap S_3, \\ A_3 &= S_1 \cap S_2 \cap S_3', \\ A_{12} &= S_1' \cap S_2' \cap S_3, \\ A_{13} &= S_1' \cap S_2 \cap S_3', \\ A_{23} &= S_1 \cap S_2' \cap S_3'. \end{aligned} \tag{34}$$

Then $S_1 \cup S_2 \cup S_3$ is the union of the sets in (34) *and moreover the sets in* (34) *are pairwise disjoint.*

Proof Let x be an arbitrary element in U and let p_i be the statement "$x \in S_i$", $i = 1, 2, 3$. To say that x belongs to $S_1 \cup S_2 \cup S_3$ is equivalent to the assertion that

$$p_1 \vee p_2 \vee p_3 \tag{35}$$

has truth value T. Recall the notation of Theorem 1.3, Section 1.1, and let c_0 be the proposition $p_1 \wedge p_2 \wedge p_3$. Similarly, let c_1 be $\sim p_1 \wedge p_2 \wedge p_3$; let c_2 be $p_1 \wedge \sim p_2 \wedge p_3$; ...; let c_{13} be $\sim p_1 \wedge p_2 \wedge \sim p_3$; and let c_{23} be $p_1 \wedge \sim p_2 \wedge \sim p_3$. Observe that $x \in A_0$ is equivalent to c_0 having truth value T. Similarly, $x \in A_1$ if and only if c_1 has truth value T and in general the statement that x is in a particular set in (34) is equivalent to the fact that the "c" proposition with a corresponding subscript has truth value T.

Now let f be the disjunction of all the statements $c_0, c_1, c_2, \ldots, c_{23}$. Then f can have truth value T if and only if at least one of the "c" statements has truth value T, i.e., if and only if x is an element of at least one of the sets in (34). Thus f has truth value T if and only if x is in the union of the sets in (34). On the other hand we claim that f has precisely the same truth table as (35). For, if (35) has truth value T then at least one of p_1, p_2, p_3 must have truth value T. For example, if p_1 and p_3 have truth value T and p_2 has truth value F then (35) has truth value T and so does $c_2 = p_1 \wedge \sim p_2 \wedge p_3$. But then f has truth value T. If (35) has truth value F then each of p_1, p_2 and p_3 must have truth value F and hence every one of the statements $c_0, c_1, \ldots, c_{23}$ must have truth value F. Then f has truth value F. We have proved that x is an element of the union of the sets (34) if and only if x is in $S_1 \cup S_2 \cup S_3$. Now any two of the sets (34) always involve at least one S_i and its complement S_i'. Hence the sets (34) are pairwise disjoint. ▌

Example 2.13 It is asserted by a toothpaste manufacturer that in a sample of 100 people using brands A, B, and C, the following statistics were obtained: 15 people used brands A, B, and C; 11 people used only brands B and C; 5 people used only brands A and C; 23 people used only brands A and B; 18 people used just brand C; no one used just brand B; and 2 people used just brand A. Question: Is this survey valid?

To answer this question, let S_1 be the set of people who use brand A, S_2 the set of people who use brand B, and S_3 the set of people who use brand C. It is clear that $S_1 \cap S_2 \cap S_3$ is the set of people who use all three brands, $S_1' \cap S_2 \cap S_3$ is the set of people who use only brands B and C, etc. Now, according to Theorem 2.2,

$$\begin{aligned} S_1 \cup S_2 \cup S_3 = {} & (S_1 \cap S_2 \cap S_3) \cup (S_1' \cap S_2 \cap S_3) \cup (S_1 \cap S_2' \cap S_3) \\ & \cup (S_1 \cap S_2 \cap S_3') \cup (S_1' \cap S_2' \cap S_3) \\ & \cup (S_1' \cap S_2 \cap S_3') \cup (S_1 \cap S_2' \cap S_3'). \end{aligned} \tag{36}$$

The sets on the right-hand side of (36) are pairwise disjoint, and thus if we add up the number of people in each one of these sets we should get the number of people in $S_1 \cup S_2 \cup S_3$, i.e., 100. This is because there is no overlap between any two of the sets and hence no one is ever counted twice. Adding the numbers, we have

$$15 + 11 + 5 + 23 + 18 + 0 + 2 = 74 \neq 100.$$

Therefore, the survey is invalid.

In order to extend the counting techniques we used in the preceding example, we make the following notational convention. Let U be the universe and let S be a subset of U which has only a finite number of elements in it. Then

$$\nu(S) \tag{37}$$

denotes the *number of elements* in S. It is obvious that if S_1 and S_2 are disjoint then $\nu(S_1 \cup S_2) = \nu(S_1) + \nu(S_2)$.

As an immediate consequence of Theorem 2.2 we have the following important result for counting the number of elements in the union of sets containing finitely many elements.

Theorem 2.3 *Let S_1, S_2, and S_3 be subsets of the universe U and assume each S_j has only finitely many elements in it. Then*

$$\nu(S_1 \cup S_2 \cup S_3)$$

is the sum of the seven numbers $\nu(A_0)$, $\nu(A_1)$, $\nu(A_2)$, $\nu(A_3), \ldots,$ $\nu(A_{23})$, where the sets $A_0, A_1, \ldots, A_{23}$ are precisely those given in (34).

Proof The result follows immediately from Theorem 2.2 which states that the sets (34) are pairwise disjoint and that their union is $S_1 \cup S_2 \cup S_3$. ▮

If we set $S_3 = \emptyset$ then $S_1 \cup S_2 \cup S_3$ becomes $S_1 \cup S_2$ and Theorem 2.3 specializes to

$$\nu(S_1 \cup S_2) = \nu(S_1 \cap S_2) + \nu(S_1' \cap S_2) + \nu(S_1 \cap S_2'). \tag{38}$$

For, $A_0 = A_1 = A_2 = A_{12} = \emptyset$ and $A_3 = S_1 \cap S_2$, $A_{13} = S_1' \cap S_2$, $A_{23} = S_1 \cap S_2'$.

Example 2.14 In a sample of 100 smokers of brands X and Y, it is found that 25 people smoke only brand X and 10 people smoke only brand Y. How many people smoke both brand X and brand Y? To answer this we use (38) to write

$$\nu(S_1 \cup S_2) = \nu(S_1 \cap S_2) + \nu(S_1' \cap S_2) + \nu(S_1 \cap S_2'),$$

where S_1 is the set of smokers who smoke brand X and S_2 is the set of smokers who smoke brand Y. We are given that

$$\nu(S_1' \cap S_2) = 10, \quad \nu(S_1 \cap S_2') = 25 \quad \text{and} \quad \nu(S_1 \cup S_2) = 100.$$

Hence

$$\begin{aligned}\nu(S_1 \cap S_2) &= (100 - 10) - 25 \\ &= 65.\end{aligned}$$

We can indicate the number of elements in each of these three pairwise disjoint sets in a simple Venn diagram.

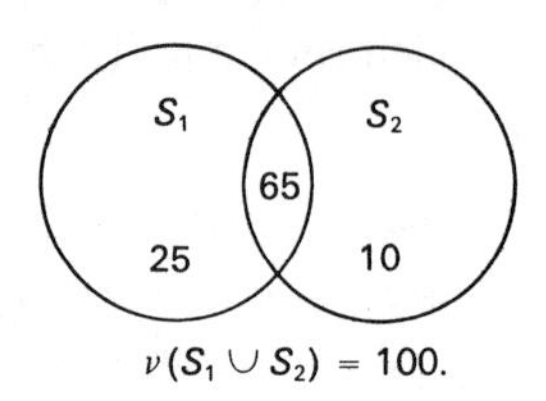

$\nu(S_1 \cup S_2) = 100.$

Two important formulas for counting the number of objects in a finite union of sets can be immediately derived from Theorem 2.3. Thus let S_1, S_2, and S_3 be subsets of the universe U, each consisting of a finite number of elements of U. Then we establish that

$$\nu(S_1 \cup S_2) = \nu(S_1) + \nu(S_2) - \nu(S_1 \cap S_2), \tag{39}$$

and

$$\nu(S_1 \cup S_2 \cup S_3) = \nu(S_1) + \nu(S_2) + \nu(S_3) - [\nu(S_1 \cap S_2) + \nu(S_1 \cap S_3) + \nu(S_2 \cap S_3)] + \nu(S_1 \cap S_2 \cap S_3). \tag{40}$$

To prove (39) observe that (see Exercise 16)

$$S_1 = (S_1 \cap S_2') \cup (S_1 \cap S_2), \tag{41}$$

and

$$S_2 = (S_1' \cap S_2) \cup (S_1 \cap S_2), \tag{42}$$

where

$$(S_1 \cap S_2') \cap (S_1 \cap S_2) = S_1 \cap S_2' \cap S_2 = \emptyset,$$

and similarly

$$(S_1' \cap S_2) \cap (S_1 \cap S_2) = S_1' \cap S_1 \cap S_2 = \emptyset.$$

Hence from (41) and (42), we have

$$\nu(S_1) = \nu(S_1 \cap S_2') + \nu(S_1 \cap S_2)$$

and

$$\nu(S_2) = \nu(S_1' \cap S_2) + \nu(S_1 \cap S_2).$$

Adding these last two equations, we have

$$\nu(S_1) + \nu(S_2) = \nu(S_1 \cap S_2') + \nu(S_1' \cap S_2) + 2\nu(S_1 \cap S_2). \tag{43}$$

However, we know from Theorem 2.3 (see (38)) that

$$\nu(S_1 \cap S_2') + \nu(S_1' \cap S_2) = \nu(S_1 \cup S_2) - \nu(S_1 \cap S_2).$$

Using this in (43), we see that

$$\nu(S_1 \cup S_2) = \nu(S_1) + \nu(S_2) - \nu(S_1 \cap S_2),$$

which is precisely (39). We can prove (40) by using (39) as follows:

$$\begin{aligned}
\nu(S_1 \cup S_2 \cup S_3) &= \nu(S_1 \cup (S_2 \cup S_3)) \\
&= \nu(S_1) + \nu(S_2 \cup S_3) - \nu(S_1 \cap (S_2 \cup S_3)) \\
&= \nu(S_1) + [\nu(S_2) + \nu(S_3) - \nu(S_2 \cap S_3)] - \nu((S_1 \cap S_2) \cup (S_1 \cap S_3)) \\
&= \nu(S_1) + [\nu(S_2) + \nu(S_3) - \nu(S_2 \cap S_3)] \\
&\quad - [\nu(S_1 \cap S_2) + \nu(S_1 \cap S_3) - \nu(S_1 \cap S_2 \cap S_1 \cap S_3)] \\
&= \nu(S_1) + \nu(S_2) + \nu(S_3) - [\nu(S_1 \cap S_2) + \nu(S_1 \cap S_3) \\
&\quad + \nu(S_2 \cap S_3)] + \nu(S_1 \cap S_2 \cap S_3).
\end{aligned}$$

In going from the second equality to the third above, we have used the result of Exercise 10(e).

Example 2.15 In a sample of 100 patients at a state mental hospital, it was found that 74 patients exhibited symptoms of schizophrenia, 17 exhibited symptoms of paranoia, and 25 patients were manic-depressives. Of these 100 patients, precisely 4 patients exhibited symptoms of all three mental diseases. Moreover, every patient had at least one of the three illnesses. How many patients exhibited symptoms of precisely two of the three mental diseases?

To solve this problem, let S_1, S_2, and S_3 be the sets of patients exhibiting symptoms of schizophrenia, paranoia, and manic-depression, respectively. The conditions of the problem state that

$$\begin{aligned}
\nu(S_1 \cup S_2 \cup S_3) &= 100; \\
\nu(S_1 \cap S_2 \cap S_3) &= 4; \\
\nu(S_1) &= 74; \\
\nu(S_2) &= 17; \\
\nu(S_3) &= 25.
\end{aligned}$$

We can draw a Venn diagram to construct a relatively simple solution to this problem. To begin, take U to be the union of the three sets: $U = S_1 \cup S_2 \cup S_3$.

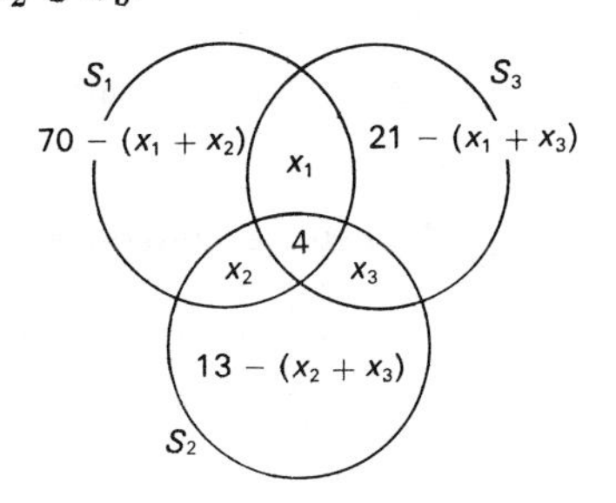

Now let $x_1 = \nu(S_1 \cap S_2' \cap S_3)$, $x_2 = \nu(S_1 \cap S_2 \cap S_3')$ and $x_3 = \nu(S_1' \cap S_2 \cap S_3)$. Since $\nu(S_1) = 74$ we can see from the diagram that $\nu(S_1 \cap S_2' \cap S_3') = 74 - (4 + x_1 + x_2) = 70 - (x_1 + x_2)$. The other numbers are similarly computed. Now adding up the total number of patients in each of the pairwise disjoint parts of the diagram we obtain:

$$\begin{aligned} 100 &= (70 - (x_1 + x_2)) + x_1 + 4 + x_2 + (21 - (x_1 + x_3)) + x_3 + (13 - (x_2 + x_3)) \\ &= 108 - (x_1 + x_2 + x_3) \end{aligned}$$

or

$$x_1 + x_2 + x_3 = 8. \tag{44}$$

But $S_1 \cap S_2 \cap S_3'$ is the set of patients that exhibit symptoms of the first two diseases but not the third. Similar remarks apply to $S_1' \cap S_2 \cap S_3$ and $S_1 \cap S_2' \cap S_3$. Hence $x_1 + x_2 + x_3$ is the number of patients that exhibit the symptoms of precisely two of the diseases. According to (44) this number is 8.

Quiz

Answer true or false.

1 The set of all real numbers between 0 and 1, each of which has a 5 in the fourth place in its decimal expansion, is a well-defined set.

2 Let S be a set and let $p_1(x)$ and $p_2(x)$ be propositions defined for each $x \in S$. Then

$$\sim \bigwedge_{x\in S} (p_1(x) \leftrightarrow p_2(x)) \leftrightarrow \exists_{x \in S} [(\sim p_1(x) \wedge p_2(x)) \vee (p_1(x) \wedge \sim p_2(x))].$$

3 If S_1 and S_2 are subsets of the universe U, then $(S_1 \cup S_2)' = S_1' \cap S_2'$.

4 If S_1 and S_2 are subsets of the universe U, then $(S_1 \cap S_2)' = S_1' \cap S_2'$.

5 $\emptyset \subset S$ for any set S.

6 $\{\emptyset\} \subset S$ for any set S, i.e., the set which consists of the empty set is a subset of any set S.

7 Let $a \in S$. Then $\nu(\{a, \{a\}\}) = 1$.

8 Let a and b be two elements in a set S. Then $\{a, \{a, b\}\} \cap \{a, b\} = \{a\}$.

9 Let a and b be two elements in a set S. Then $\nu(\{a, \{a, b\}\} \cup \{a, b\}) = 3$.

10 If S_1 and S_2 are finite sets, then $\nu(S_1 \cap S_2') = \nu(S_1) - \nu(S_2)$.

Exercises

1 Let the universe be $U = \{1, 2, 3, 4, 5, 6, 7, 8, 9, 10\}$ and let $S_1 = \{2, 4, 6, 8, 10\}$, $S_2 = \{3, 6, 9\}$, $S_3 = \{4, 5, 6, 7, 8\}$, and $S_4 = \{1, 3, 5, 7, 9, 10\}$. List the elements in each of the following sets:

(a) $S_1' \cap S_2'$; (b) $(S_1 \cup S_2)'$;
(c) $(S_1 \cap S_2)' \cap S_3'$; (d) $S_1 \cup S_2 \cup S_3$;
(e) $S_1' \cap S_2' \cap S_3' \cap S_4'$; (f) $S_4' \cap S_1 \cap S_2'$.

2 Let the universe U be the set of all even whole numbers between 2 and 100, including both 2 and 100. Let

$$S_k = \{x \mid x \in U \wedge x \text{ is divisible by } k\}.$$

List the elements in each of the following sets:

(a) $S_2 \cap S_7$;
(b) $S_3 \cap S_5 \cap S_{30}'$;
(c) $S_2' \cap S_{13}$;
(d) $S_2 \cap S_4 \cap S_8 \cap S_{16} \cap S_{32}$;
(e) $S_{100} \cap S_1$.

3 In each of the following Venn diagrams, shade the indicated regions in separate pictures:

(a)

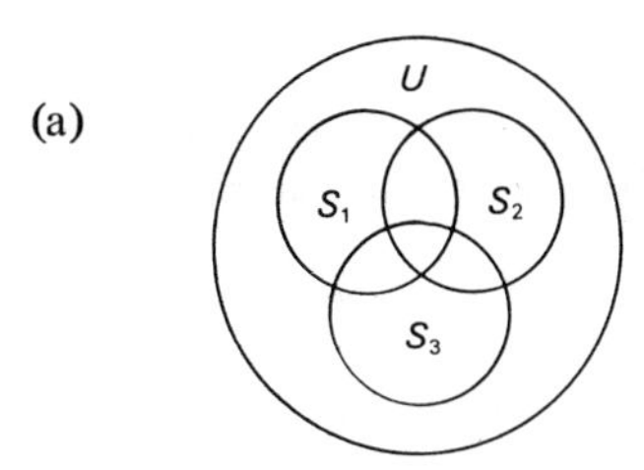

(i) S_1';
(ii) $S_1 \cap S_3$;
(iii) $S_1 \cap S_2 \cap S_3'$;
(iv) $S_2 \cap S_3 \cap U$;
(v) $(S_1 \cup \emptyset) \cup (S_1' \cap S_3)$;
(vi) $(S_1 \cup S_2) \cap S_3$;

(b)

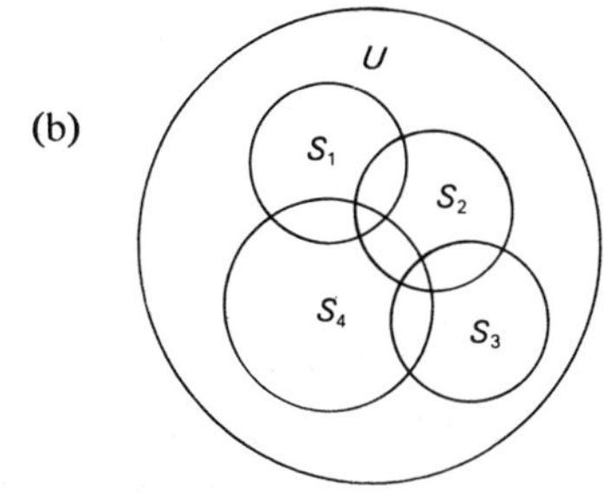

(i) $S_1 \cap S_4$;
(ii) $S_2 \cap S_3' \cap S_4$;
(iii) $(S_1 \cup S_2) \cap S_4'$;
(iv) $(S_2 \cup S_3) \cup (S_1 \cap S_4')$;
(v) $(S_1 \cup S_2 \cup S_3 \cup S_4)'$;
(vi) $((S_1' \cap S_2)' \cap S_3)'$.

4 In a typical sample of 100 selective service inductees it was found that

(a) 15 had flat feet;
(b) 10 had defective eyes;
(c) 20 chewed their fingernails;
(d) 6 had flat feet and defective eyes;
(e) 8 had flat feet and chewed their fingernails;
(f) 7 had defective eyes and chewed their fingernails;
(g) 4 had flat feet, defective eyes, and chewed their fingernails.

How many inductees had flat feet but not defective eyes or chewed fingernails? How many had none of the three afflictions?

5 Let W be the set of all women. For $x \in W$ let $p_1(x)$ be the statement, "x is a blonde;" $p_2(x)$ the statement, "x has blue eyes;" $p_3(x)$ the statement, "x is pretty;" $p_4(x)$ the statement, "x is overweight." Translate each of the following into an English sentence:

(a) $\underset{x \in W}{\exists} (p_2(x) \wedge p_3(x))$;

(b) $\underset{x \in W}{\bigwedge} (p_4(x) \rightarrow \sim p_3(x))$;

(c) $\sim \bigwedge_{x \in W} [(\sim p_2(x)) \vee (\sim p_3(x))]$;

(d) $\bigwedge_{x \in W} (p_3(x) \rightarrow \sim p_4(x))$;

(e) $\exists_{x \in W} (p_1(x) \wedge p_2(x) \wedge p_3(x) \wedge \sim p_4(x))$.

6 Let A denote the set of all animals; M the set of all men; W the set of all women; R the set of all rats; P the set of all people; D the set of all dogs. Express each of the following statements concerning sets as an English sentence:

(a) $P = M \cup W$; (b) $P \subset A$; (c) $R \cap M \neq \emptyset$;
(d) $R \cup D \subset A$; (e) $M \cap W = \emptyset$; (f) $A \cap D = D$.

7 The Department of Mathematics at U.C.S.B. has 220 undergraduate majors. The distribution of these students into the basic algebra, analysis, and geometry courses is as follows: 100 students take algebra; 130 take analysis; 140 take geometry; 60 take algebra and geometry; 60 take algebra and analysis; 80 take analysis and geometry; and 50 take algebra, analysis, and geometry. Draw an appropriate Venn diagram and determine:

(a) how many students take analysis but not geometry;
(b) how many students take algebra and geometry but not analysis;
(c) how many students take geometry but not algebra;
(d) how many students take analysis only;
(e) how many students take geometry only if they take analysis.

8 If X is a set, let sb(X) denote the family of all subsets of X.

(a) If $X = \{1, 2\}$, write down all the members of sb(X).
(b) How many elements are there in sb(sb(X))?

9 The Registrar has observed that 57% of the students have a 2 p.m. class; 52% have an 8 a.m. class; 42% have a 9 a.m. class; 35% have a 9 a.m. and a 2 p.m. class; 12% have an 8 a.m. and a 9 a.m. class; 9% have an 8 a.m. and a 2 p.m. class; 5% have an 8 a.m., a 9 a.m., and a 2 p.m. class.

(a) What percent of students have only a 9 a.m. class among these 3 hours?
(b) What percent of students have no classes during these 3 hours?

10 Let X, Y, and Z be subsets of the universe U. Prove the following statements by verifying set membership (see Example 2.9):

(a) $(X \cap Y) \subset X$;
(b) $X \subset (X \cup Y)$;
(c) $X \cap Y = Y \cap X$;
(d) $X \cup Y = Y \cup X$;
†(e) $X \cap (Y \cup Z) = (X \cap Y) \cup (X \cap Z)$;
(f) $X \cup (Y \cap Z) = (X \cup Y) \cap (X \cup Z)$;
(g) $X \cap X = X$;
(h) $(X \cap Y) \cap (X \cap Y \cap Z)' = X \cap Y \cap Z'$;
(i) $X \cap U = X$;
(j) $X \cup U = U$;
(k) $X \cap \emptyset = \emptyset$;

(l) $X \cup \emptyset = X$;
(m) $(X')' = X$;
(n) $((X')')' = X'$;
(o) $\emptyset' = U$;
(p) $U' = \emptyset$.

11 Let A and B be subsets of the universe U. Prove that the following conditions are all equivalent to the inclusion $A \subset B$:
(a) $A \cup B = B$; (b) $B \cap A = A$;
(c) $B' \subset A'$; (d) $A \cap B' = \emptyset$;
(e) $B \cup A' = U$.

12 Draw Venn diagrams which illustrate the following statements concerning the subsets X, Y, and Z of the universe U:
(a) $X \cap Y$;
(b) $X \cup Y$;
(c) $X \cap X' = \emptyset$;
(d) $X \cup X' = U$;
(e) $(X \cup Y)' = X' \cap Y'$;
(f) $(X \cap Y)' = X' \cup Y'$;
(g) $X \cup (Y \cap Z) = (X \cup Y) \cap (X \cup Z)$;
(h) $X \cap (Y \cup Z) = (X \cap Y) \cup (X \cap Z)$.

13 Let U be the set of all whole numbers. Define subsets X, Y, Z of U as follows:

$$X = \{n \mid n = 8t,\ t \in U\};$$
$$Y = \{n \mid n = 6t,\ t \in U\};$$
$$Z = \{n \mid n = 48t,\ t \in U\}.$$

Is it true that $Z = X \cap Y$? Explain your answer.

14 In a sample of 100 students, the following information was obtained in a survey:
(i) 12 take English, math, and physics;
(ii) 22 take just English and math;
(iii) 3 take just English and physics;
(iv) 7 take just math and physics;
(v) all 100 take at least one of English, math, and physics.

How many students take precisely one of the three subjects?

15 Let S_1, S_2, S_3, and S_4 be sets consisting of finitely many elements. Derive a formula analogous to (40) for $\nu(S_1 \cup S_2 \cup S_3 \cup S_4)$. (Hint: Apply (40) with S_3 replaced by $S_3 \cup S_4$.)

16 Establish formulas (41) and (42).

17 Let S_n denote the set of all positive fractions between 0 and 1 which have denominator n when expressed in lowest terms. What is

$$\bigcup_{n=1}^{\infty} S_n?$$

Are the sets S_n pairwise disjoint?

1.3

Functions

Let X and Y be two subsets of the universe U. We are interested in considering sets of pairs of elements in which the first member of the pair comes from X and the second from Y. A pair of elements is called an *ordered* pair if it possesses the important property that the first and second members are distinguishable by the order in which they appear.

Definition 3.1 ***Ordered pair*** If x and y are elements of U, then the notation for the *ordered pair* with first element x and second element y is

$$(x, y). \tag{1}$$

If (x_1, y_1) and (x_2, y_2) are ordered pairs then $(x_1, y_1) = (x_2, y_2)$ means that

$$x_1 = x_2, \ y_1 = y_2. \tag{2}$$

The notion of a set of ordered pairs is fundamental throughout mathematics.

Definition 3.2 ***Cartesian product, relation*** The *Cartesian product* of the sets X and Y is the totality of ordered pairs (x, y), where $x \in X$ and $y \in Y$. That is, the Cartesian product, denoted by $X \times Y$, is the set

$$X \times Y = \{z \mid z = (x, y), \ x \in X, \ y \in Y\}.$$

Any subset R of $X \times Y$ is called a *relation* on X to Y. The *domain* of a relation R is the totality of first elements of pairs in R. The *range* or *codomain* of R is the totality of second elements of pairs in R. The domain of R is denoted by

$$\text{dmn } R$$

and the range of R is denoted by

$$\text{rng } R.$$

Example 3.1 Find the range of the relation R on X to Y for which dmn $R = X = \{1, 2, 3, 4, 5, -5\}$ and $R = \{(x, y) \mid x \in X, \ y = x^2\}$. The pairs in R are (1, 1), (2, 4), (3, 9), (4, 16), (5, 25), (−5, 25). Hence

$$\text{rng } R = \{1, 4, 9, 16, 25\}.$$

Observe that we do not count 25 twice simply because it appears in two different pairs in the relation.

Example 3.2 Let X denote a set of five cities, $x_1, \ldots, x_5$. We define a relation R on X to X in terms of the following map by saying that $(x_i, x_j) \in R$ if and only if x_i and x_j are directly connected by a road (indicated by a straight line in the diagram (3)).

(3)

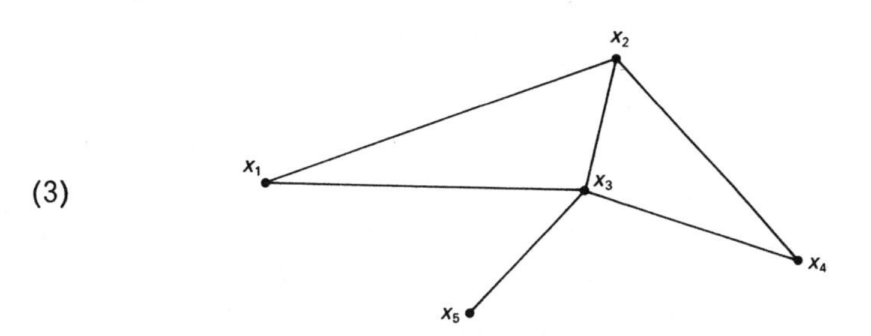

Find the elements of the relation R. The set R is given by

$$R = \{(x_1, x_2), (x_2, x_1), (x_1, x_3), (x_3, x_1), (x_2, x_3), (x_3, x_2), (x_2, x_4), (x_4, x_2), (x_3, x_4), (x_4, x_3), (x_3, x_5), (x_5, x_3)\}.$$

Observe that when x_i and x_j are connected by a road, then x_j and x_i are connected by the same road and hence the relation R is "symmetric", i.e., $(x_i, x_j) \in R$ implies that $(x_j, x_i) \in R$. Of course not all relations are symmetric in this sense. For, suppose we modify the map (3) by making the streets one-way.

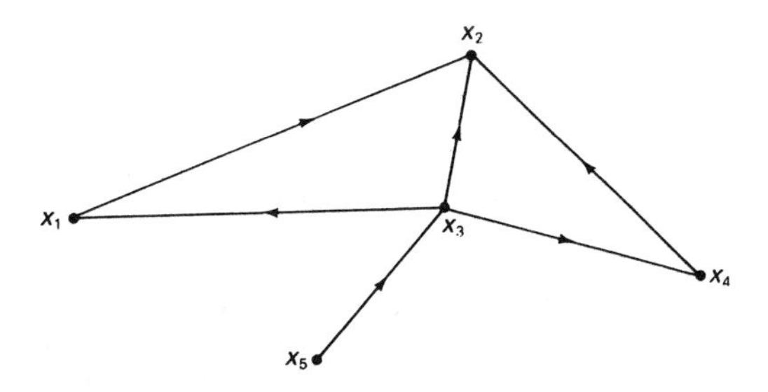

We then change the definition of R so that it consists of all pairs (x_i, x_j) for which there exists a one-way road from x_i to x_j. Then

$$R = \{(x_1, x_2), (x_3, x_1), (x_3, x_2), (x_3, x_4), (x_4, x_2), (x_5, x_3)\}.$$

Example 3.3 **(a)** Let X be the set of all people. Define a relation R on X to X as follows:

$$R = \{(x, y) \mid x \in X \wedge y \in X \wedge (y \text{ is the brother of } x)\}.$$

If $(x, y) \in R$, does it follow that $(y, x) \in R$? If $(x, y) \in R$ and $(y, z) \in R$, does it follow that $(x, z) \in R$? In answer to the first question, if x is the sister of y and y is the brother of x, then $(x, y) \in R$ but $(y, x) \notin R$. On the other hand if $(x, y) \in R$ and $(y, z) \in R$, then z is a male with the same parents as y, who in turn has the same parents as x. Therefore z is the brother of x. (It may very well be that x is a girl.)

(b) Let X be the set of all integers and define a relation R on X to X by

$$R = \{(x, y) \mid x \in X \wedge y \in X \wedge (y - x \text{ is divisible by } 3)\}.$$

It is obvious (see Exercise 13) that if x, y, and z are in X then:

(i) $(x, x) \in R$;
(ii) if $(x, y) \in R$ then $(y, x) \in R$;
(iii) if $(x, y) \in R$ and $(y, z) \in R$, then $(x, z) \in R$.

The preceding examples suggest the following definition.

Definition 3.3 ***Equivalence relation*** Let X be a set and let R be a relation on X to X satisfying the following three properties.

(i) (R is *reflexive*.) For every $x \in X$, $(x, x) \in R$, i.e.,

$$\{(x, x) \mid x \in X\} \subset R.$$

(ii) (R is *symmetric*.) If $(x, y) \in R$ then $(y, x) \in R$.

(iii) (R is *transitive*.) If $(x, y) \in R$ and $(y, z) \in R$, then

$$(x, z) \in R.$$

Then R is called an *equivalence relation*.

Example 3.4 Let X denote the set of all straight lines in the plane. Let

$$R = \{(x, y) \mid x \in X \wedge y \in X \wedge x \parallel y\}.$$

In other words, R consists of the totality of pairs of lines (x, y) for which x is parallel to y. As is customary in geometry, we consider every straight line to be parallel to itself and therefore R is reflexive. It is also clear that R is symmetric and transitive and hence is an equivalence relation.

Example 3.5 Let X denote the set of all men, and define R to be the totality of pairs (x, y) for which x and y have at least one parent in common. Is R an equivalence relation with the agreement that each person is considered to have at least one parent in common with himself? The relation R is clearly reflexive and symmetric, but it is certainly not transitive. For, y could be the half-brother of x, and z could be the half-brother of y, and yet x and z need not have any parent in common.

Suppose X and Y are two sets, each with a finite number of elements, say,

$$\nu(X) = n, \ \nu(Y) = m,$$

where $X = \{x_1, \ldots, x_n\}$ and $Y = \{y_1, \ldots, y_m\}$. Let R be any relation on X to Y. There is a very simple tabular or *matrix* form in which we can exhibit the relation R: construct a table with column headings $x_1, \ldots, x_n$ and row headings $y_1, \ldots, y_m$ and at the intersection of row i and column j of the matrix put 1 if $(x_j, y_i) \in R$ and put 0 if $(x_j, y_i) \notin R$.

Example 3.6 Let $X = \{1, 2, 3, 4\}$ and let

$$R = \{(i, j) \mid i \in X \wedge j \in X \wedge (i - j) \text{ is even}\}.$$

We easily check that R consists of the following set of pairs:

$$\{(1, 3), (3, 1), (2, 4), (4, 2), (1, 1), (2, 2), (3, 3), (4, 4)\}.$$

Then the matrix for the relation R is

$$\begin{array}{c} \\ 1 \\ 2 \\ 3 \\ 4 \end{array} \begin{array}{c} \begin{array}{cccc} 1 & 2 & 3 & 4 \end{array} \\ \begin{bmatrix} 1 & 0 & 1 & 0 \\ 0 & 1 & 0 & 1 \\ 1 & 0 & 1 & 0 \\ 0 & 1 & 0 & 1 \end{bmatrix} \end{array}.$$

Example 3.7 A pair of honest dice is thrown so that 36 possible ordered pairs of numbers can appear. Define a relation R on the set $\{1, \ldots, 6\}$ to consist of all pairs (i, j) for which $(i + j)$ is 7 or 11. Find the matrix for the relation R. The pairs in R are

$$(1, 6); \ (6, 1); \ (2, 5); \ (5, 2); \ (3, 4); \ (4, 3); \ (6, 5); \ (5, 6).$$

The matrix for R is

$$\begin{array}{c} \\ 1 \\ 2 \\ 3 \\ 4 \\ 5 \\ 6 \end{array} \begin{array}{c} \begin{array}{cccccc} 1 & 2 & 3 & 4 & 5 & 6 \end{array} \\ \begin{bmatrix} 0 & 0 & 0 & 0 & 0 & 1 \\ 0 & 0 & 0 & 0 & 1 & 0 \\ 0 & 0 & 0 & 1 & 0 & 0 \\ 0 & 0 & 1 & 0 & 0 & 0 \\ 0 & 1 & 0 & 0 & 0 & 1 \\ 1 & 0 & 0 & 0 & 1 & 0 \end{bmatrix} \end{array}.$$

Definition 3.4 ***Incidence matrix*** Let $X = \{x_1, \ldots, x_n\}$ and $Y = \{y_1, \ldots, y_m\}$. Let R be a relation on X to Y. Then the *incidence matrix* for R, denoted by $A(R)$, is the $m \times n$ array in which the entry in row i and column j is 1 or 0 according as $(x_j, y_i) \in R$ or $(x_j, y_i) \notin R$, respectively.

Thus, to form the incidence matrix for the relation R, write $x_1, \ldots, x_n$ as column headings and $y_1, \ldots, y_m$ as row headings. Then the entry opposite y_i and below x_j is 1 if $(x_j, y_i) \in R$ and is 0 if $(x_j, y_i) \notin R$. Observe that the incidence matrix $A(R)$ contains a complete description of the relation R and all properties of R appear as properties of $A(R)$.

In the preceding examples and definition, the matrices consisted only of the numbers 0 and 1. One can easily conceive of situations in which this "yes-no" approach is inadequate. For example, consider a set of four cities, $\{x_1, x_2, x_3, x_4\}$, which are connected by one-way roads according to the following map.

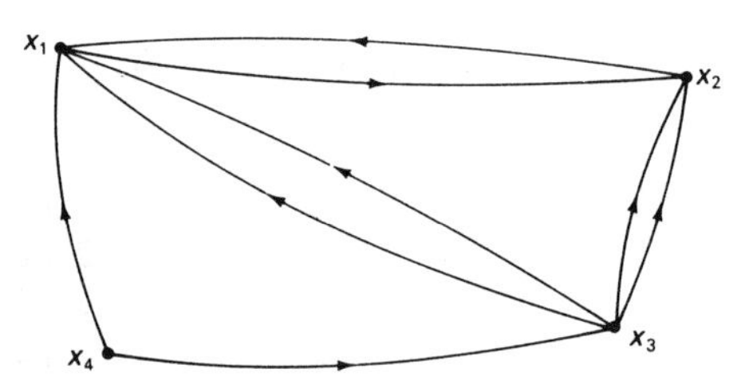

Suppose that we wish to construct a matrix which contains all of the information in the map, i.e., we want the matrix to include not only the information that there is a one-way road going directly from x_j to x_i, but we want to know precisely how many such roads there are. Thus, consider a matrix in which the entry in row i, column j is the number of one-way streets running from x_j to x_i.

We agree that any city is connected to itself once. The matrix is

$$\begin{array}{c} \\ 1 \\ 2 \\ 3 \\ 4 \end{array} \begin{array}{c} \begin{array}{cccc} 1 & 2 & 3 & 4 \end{array} \\ \begin{bmatrix} 1 & 1 & 2 & 1 \\ 1 & 1 & 2 & 0 \\ 0 & 0 & 1 & 1 \\ 0 & 0 & 0 & 1 \end{bmatrix} \end{array}.$$

Example 3.8 Consider the children's game of "rock, scissors, and paper": two players simultaneously exhibit a fist (rock), two fingers (scissors), or the flat of the hand (paper). Paper wins over rock (paper covers rock), scissors win over paper (scissors can cut paper), and rock wins over scissors (rock can break scissors). The winner collects a penny from the loser. The "payoff matrix" for this game can be described as follows: label the three possibilities rock, scissors, and paper by 1, 2, and 3, respectively, and construct a matrix in which the entry in row i, column j is the amount that the second player pays to the first for that particular combination:

$$\begin{array}{c} \\ \\ \\ \text{player I} \\ \\ \end{array} \begin{array}{c} \\ \\ 1 \\ 2 \\ 3 \end{array} \begin{array}{c} \text{player II} \\ \begin{array}{ccc} 1 & 2 & 3 \end{array} \\ \begin{bmatrix} 0 & 1 & -1 \\ -1 & 0 & 1 \\ 1 & -1 & 0 \end{bmatrix} \end{array}.$$

We have indicated that the entry is 1, 0, or -1 according as player II pays player I, there is no exchange of money, or player I pays player II, respectively.

A rather practical dietary problem can be solved by the use of incidence matrices.

Example 3.9 A vivarium houses four different species of animals. A total of seven different animal food mixtures are available on the market. Included among the constituents of these mixtures are five ingredients which certain of the species refuse to eat. Label the set of four species

$$S = \{s_1, s_2, s_3, s_4\},$$

the set of five ingredients

$$F = \{f_1, f_2, f_3, f_4, f_5\},$$

and the set of seven commercial mixtures

$$M = \{m_1, m_2, m_3, m_4, m_5, m_6, m_7\}.$$

Then species s_1 refuses to eat any mixture containing f_1 or f_2; species s_2 refuses to eat any mixture containing f_3; species s_3 refuses to eat any mixture containing either f_4 or f_5; species s_4 refuses to eat any mixture containing f_1, f_3, or f_4. Moreover the commercial mixtures contain the following combinations of ingredients of F: f_1, f_2, and f_5 are in m_1; f_2 is in m_2; f_1 and f_3 are in m_3; f_3 is in m_4; f_2, f_3, and f_4 are in m_5; f_1, f_4, and f_5 are in m_6; f_5 is in m_7. Problem: design all possible feedings using precisely four of the mixtures in M so that all of the species will have at least one acceptable mixture in any feeding.

We begin the solution by defining a relation R on S to F (i.e., $R \subset S \times F$) by

$$R = \{(s_i, f_j) \mid s_i \text{ will not eat any mixture containing } f_j\}.$$

Define another relation L on F to M ($L \subset F \times M$) by

$$L = \{(f_p, m_q) \mid f_p \text{ is a constituent of } m_q\}.$$

We construct the incidence matrices $A(R)$ and $A(L)$:

$$A(R) = \begin{array}{c} \\ f_1 \\ f_2 \\ f_3 \\ f_4 \\ f_5 \end{array} \begin{array}{c} \begin{array}{cccc} s_1 & s_2 & s_3 & s_4 \end{array} \\ \begin{bmatrix} 1 & 0 & 0 & 1 \\ 1 & 0 & 0 & 0 \\ 0 & 1 & 0 & 1 \\ 0 & 0 & 1 & 1 \\ 0 & 0 & 1 & 0 \end{bmatrix} \end{array}, \tag{5}$$

$$A(L) = \begin{array}{c} \\ m_1 \\ m_2 \\ m_3 \\ m_4 \\ m_5 \\ m_6 \\ m_7 \end{array} \begin{array}{c} \begin{array}{ccccc} f_1 & f_2 & f_3 & f_4 & f_5 \end{array} \\ \begin{bmatrix} 1 & 1 & 0 & 0 & 1 \\ 0 & 1 & 0 & 0 & 0 \\ 1 & 0 & 1 & 0 & 0 \\ 0 & 0 & 1 & 0 & 0 \\ 0 & 1 & 1 & 1 & 0 \\ 1 & 0 & 0 & 1 & 1 \\ 0 & 0 & 0 & 0 & 1 \end{bmatrix} \end{array}. \tag{6}$$

Consider the product of the entry in row i, column j of $A(L)$ and the entry in row j, column k of $A(R)$. Call these two entries ℓ_{ij} and r_{jk}, respectively. Then

$$\ell_{ij} r_{jk}$$

is 1 or 0 according as both $\ell_{ij} = 1$ and $r_{jk} = 1$ or at least one of ℓ_{ij} and r_{jk} is 0. But to say that $\ell_{ij} = r_{jk} = 1$ means that s_k won't eat f_j and f_j is in m_i. In other words, s_k won't eat m_i because it contains f_j. If for a fixed i and k we add the numbers $\ell_{ij} r_{jk}$ as j takes on the values 1, 2, 3, 4, 5, we will get as the sum the number of ingredients in m_i that s_k refuses to eat. This number is 0 if and only if s_k will eat m_i; otherwise this number is positive. Let us calculate each of these sums

$$\ell_{i1}r_{1k} + \ell_{i2}r_{2k} + \ell_{i3}r_{3k} + \ell_{i4}r_{4k} + \ell_{i5}r_{5k}, \qquad i = 1, \ldots, 7, \quad k = 1, \ldots, 4, \tag{7}$$

and arrange the answers in a 7×4 matrix in which the sum (7) appears in row i, column k. Thus, for example, the entry which appears in row 5, column 3 is

$$\ell_{51}r_{13} + \ell_{52}r_{23} + \ell_{53}r_{33} + \ell_{54}r_{43} + \ell_{55}r_{53}.$$

We will denote the resulting 7×4 matrix by simply juxtaposing $A(L)$ and $A(R)$:

$$A(L)A(R) = \begin{array}{c} \\ m_1 \\ m_2 \\ m_3 \\ m_4 \\ m_5 \\ m_6 \\ m_7 \end{array} \begin{array}{c} \begin{array}{cccc} s_1 & s_2 & s_3 & s_4 \end{array} \\ \begin{bmatrix} 2 & 0 & 1 & 1 \\ 1 & 0 & 0 & 0 \\ 1 & 1 & 0 & 2 \\ 0 & 1 & 0 & 1 \\ 1 & 1 & 1 & 2 \\ 1 & 0 & 2 & 2 \\ 0 & 0 & 1 & 0 \end{bmatrix} \end{array}. \tag{8}$$

Consider, for example, a choice of the four mixtures m_2, m_4, m_6, and m_7. According to (8), s_1 can eat m_4 and m_7; s_2 can eat m_2, m_6, m_7; s_3 can eat m_2 and m_4; and s_4 can eat m_2 and m_7. Is there another selection of four mixtures for which each species will find at least one acceptable mixture? What is required for such a selection? The answer to this second question is very simple. A selection of four mixtures, i.e., rows of the matrix (8), must have the property that at least one zero appears in each of the columns in these rows. We shall return in Chapter 2 to a systematic investigation of the composition of matrices of which $A(L)A(R)$ in (8) is a typical example.

A special class of relations are of central importance throughout mathematics.

Definition 3.5 ***Function*** If X and Y are sets, and f is a relation on X to Y such that $\operatorname{dmn} f = X$ and no two different ordered pairs in f have the same first member, then f is called a *function* on X to Y. This is written

$$f\colon X \to Y. \tag{9}$$

If $(x, y) \in f$, then y is called the *value* of the function f at x and is written

$$y = f(x). \tag{10}$$

Thus a function is a set of ordered pairs (x, y) in which $x \in X$, $y \in Y$ and y is uniquely determined once x is given. We can think of a function as a "machine" that cranks out a value $y \in Y$, given a value $x \in X$.

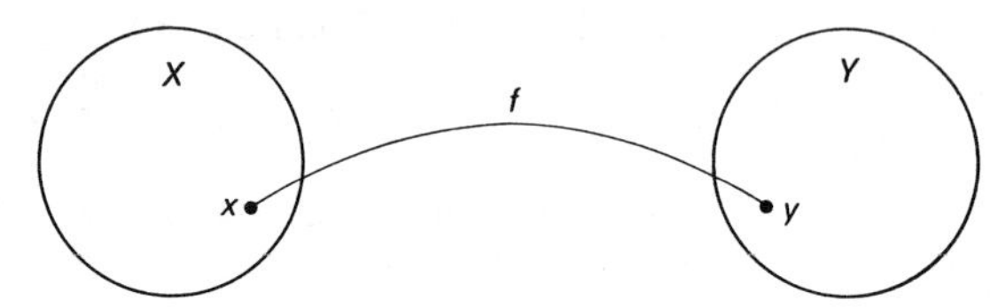

If $f(x) = y$, i.e., $(x, y) \in f$, we sometimes say that f *sends* x into y or f *maps* x into y or f *transforms* x into y. Functions are also called *mappings* or *transformations*.

If f is a relation on X to X and X is the set of all real numbers, there is a familiar pictorial representation for f called the *graph* of f. This is obtained by plotting the pairs $(x, y) \in f$ with respect to a pair of mutually perpendicular lines that depict X.

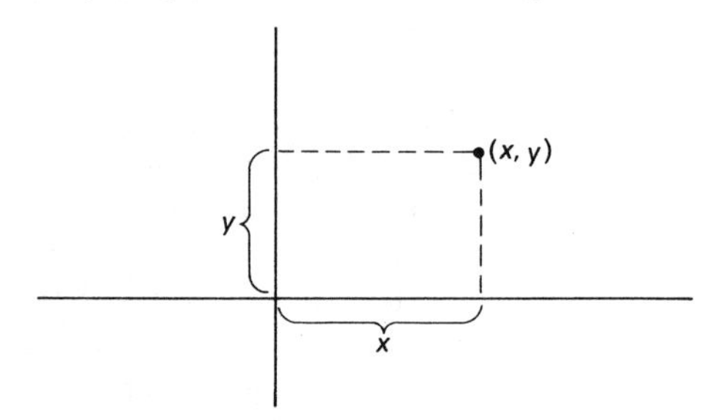

It is easy to recognize whether a relation is a function directly from the graph. For, no two different pairs in a function can have the same first member. This means that no vertical line can "hit" the graph more than once.

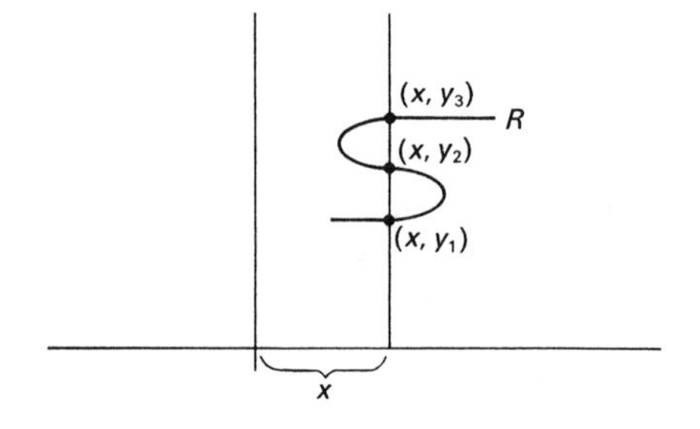

The relation R in the preceding diagram is definitely not a function. On the other hand if

$$f = \{(x, y) \mid x^2 + y^2 = 1 \wedge -1 \leq x \leq 1 \wedge y \geq 0\}$$

then f is indeed a function (recall that $a \leq b$ is read "a is less than or equal to b"; $a \geq b$ is read "a is greater than or equal to b"). Its graph is the following.

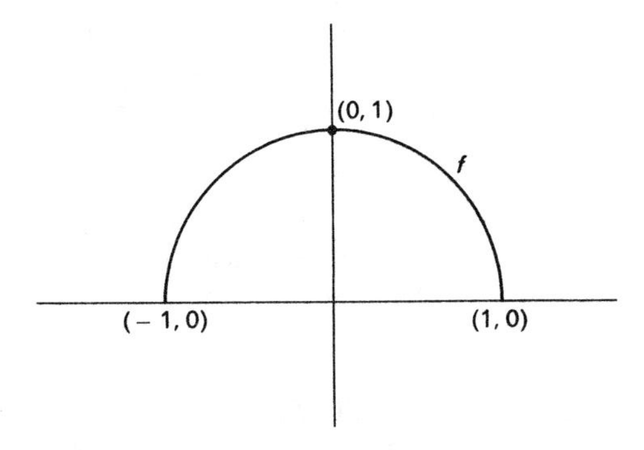

The domain of f, dmn f, is the set of all real numbers x in the interval $-1 \leq x \leq 1$. The range of f, rng f, is the set of all real numbers y in the interval $0 \leq y \leq 1$.

As another example consider the set of pairs

$$f = \{(0, 1), (1, 2), (2, 3)\}.$$

Clearly f satisfies the definition for a function: dmn $f = \{0, 1, 2\}$, rng $f = \{1, 2, 3\}$, $f(0) = 1$, $f(1) = 2$, $f(2) = 3$, and the graph of f consists of three pairs.

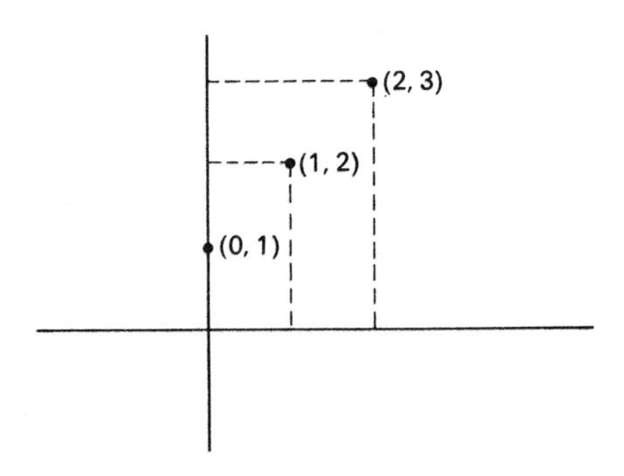

Observe that no vertical line intersects the graph of the function more than once.

Theorem 3.1 *Let f: $X \to Y$ and g: $X \to Y$ be two functions. Then $f = g$ if and only if $f(x) = g(x)$ for every $x \in X$. In logical notation,*

$$(f = g) \leftrightarrow \bigwedge_{x \in X} (f(x) = g(x)).$$

Proof Suppose that $f = g$ and $x \in X$. Then $(x, f(x)) \in f$ and hence $(x, f(x)) \in g$. However, $(x, g(x)) \in g$ and thus the two pairs in g, $(x, g(x))$ and $(x, f(x))$, have the same first member. According to the definition, the two pairs must be the same, and thus $f(x) = g(x)$. Conversely, suppose that

$$\bigwedge_{x \in X} (f(x) = g(x)).$$

An arbitrary element of f is of the form $(x, f(x)) = (x, g(x)) \in g$. Hence any pair in f is in g, i.e., $f \subset g$. Similarly $g \subset f$. It follows that $f = g$. ▮

The preceding theorem simply says that two functions with the same domains and ranges are equal if and only if they have the same function value at each element of their (common) domain.

The student has, of course, encountered functions in his previous training, but perhaps they were defined in a somewhat different way. In elementary mathematics the domain of a function f is usually a set of numbers and the range is another set of numbers produced by some formula. For example, consider the function

$$f = \{(x, y) \mid 0 \leq x \leq 1 \wedge y = x^2\}. \tag{11}$$

Formula (11) tells us that f is the set of pairs in which the first member of any pair in f is a real number between 0 and 1, and the second member is obtained from the first by squaring. A somewhat more familiar notation for f is

$$f(x) = x^2, \quad 0 \leq x \leq 1.$$

In other words, functions are usually given in a form in which the second member of the pair is expressed in terms of the first by some formula.

Example 3.10 Write the function

$$f(x) = \frac{1}{x}, \quad x > 0,$$

in a form analogous to (11). The answer is immediate:

$$f = \left\{(x, y) \mid x > 0 \wedge y = \frac{1}{x}\right\}.$$

We can ask for the range of this function f. Since $x > 0$, it is clear that $\frac{1}{x} > 0$ and thus the range consists of positive real numbers. The question is whether the range consists of *all* positive real numbers. In other words, can $\frac{1}{x}$ assume any positive value y for an appropriate $x > 0$? Of course, the answer is easy because given a positive number y, we simply choose x to be $\frac{1}{y}$.

Example 3.11 Let X be the set of all non-negative real numbers and let Y be the interval of all real numbers y satisfying $0 \leq y < 1$. Consider the function $f\colon X \to Y$ given by

$$f(x) = \frac{x}{x+1}, \quad x \in X.$$

Show that each number in Y occurs in precisely one pair of f. Since $x + 1$ exceeds x, we know that $\frac{x}{x+1}$ is less than 1, and hence $f(x) \in Y$ for any $x \in X$. Recall that the elements in f are just pairs of the form

$$(x, f(x)) = \left(x, \frac{x}{1+x}\right).$$

Suppose that two pairs $\left(x_1, \frac{x_1}{1+x_1}\right)$, $\left(x_2, \frac{x_2}{1+x_2}\right)$ in f have the same second members, that is,

$$\frac{x_1}{1+x_1} = \frac{x_2}{1+x_2}, \qquad x_1 \in X,\ x_2 \in X.$$

Then using a little elementary algebra (i.e., multiplying through the previous equation by the product $(1 + x_1)(1 + x_2)$, which is positive) we have

$$x_1(1 + x_2) = x_2(1 + x_1),$$

or

$$x_1 + x_1x_2 = x_2 + x_1x_2,$$

or

$$x_1 = x_2.$$

In other words, $f(x_1) = f(x_2)$ implies that $x_1 = x_2$. Next, let $y \in Y$. Then the question is whether there exists an $x \in X$ such that

$$\begin{aligned} f(x) &= \frac{x}{1+x} \\ &= y. \end{aligned}$$

Once again, we easily compute that

$$\begin{aligned} x &= (1 + x)y, \\ x &= y + xy, \\ x - xy &= y, \\ x(1 - y) &= y, \\ x &= \frac{y}{1 - y}. \end{aligned}$$

The last step is permissible since $(1 - y)$ is not zero (i.e., $0 \leq y < 1$). Moreover $y \geq 0$ and $(1 - y) > 0$ imply that $x \geq 0$ and thus $f(x) = y$.

These examples lead us to the following definition.

Definition 3.6 **1–1, *onto*, *inverse*** Let f be a function

$$f\colon X \to Y$$

whose domain is X. If rng $f = Y$, then f is said to be *onto* Y. If $f(x_1) = f(x_2)$ implies that $x_1 = x_2$, then f is said to be 1–1 (read "*one to one*"). If f is 1–1 and onto Y, then the relation $g \subset Y \times X$, defined by

$$g = \{(f(x), x) \mid x \in X\}, \tag{12}$$

is called the *inverse* of f and is written

$$g = f^{-1}. \tag{13}$$

Thus a function $f\colon X \to Y$ is *onto* Y if every $y \in Y$ is a function value of f, i.e., every $y \in Y$ is of the form $y = f(x)$ for some $x \in X$. Intuitively speaking, the values of f "cover" Y. To say that a function is 1–1 means that $x_1 \neq x_2$ implies $f(x_1) \neq f(x_2)$.

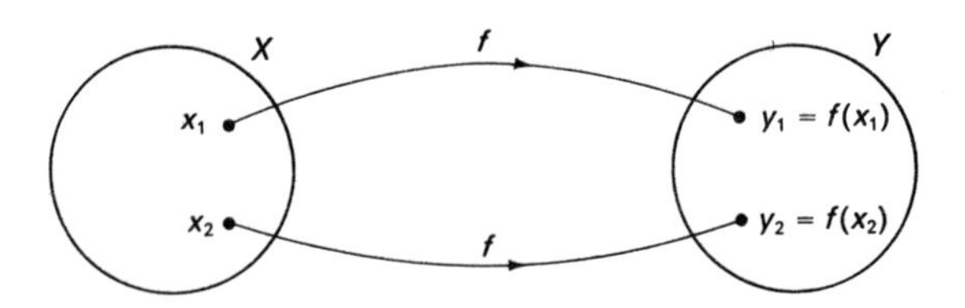

For, the contrapositive statement is simply the definition, i.e., if f is 1–1 and $f(x_1) = f(x_2)$, then x_1 cannot be different from x_2 (otherwise $f(x_1) \neq f(x_2)$), so that $x_1 = x_2$. The inverse of f is just the relation that goes from Y "back to" X, matching each point y in Y with that point x in X for which $f(x) = y$.

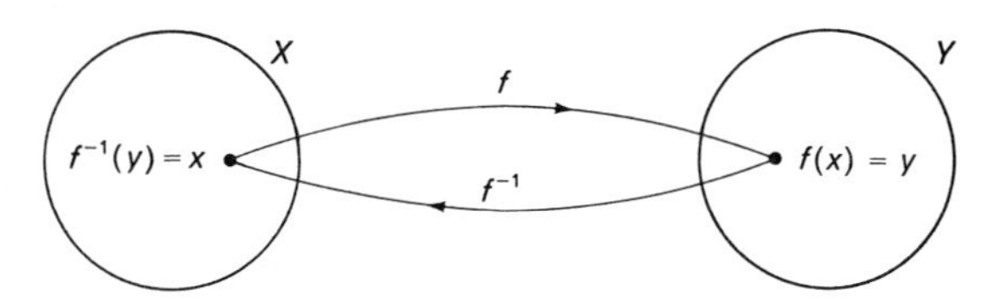

For example, suppose X is the set of working people in the United States and Y is the set of social security numbers of these people. Consider the function $f\colon X \to Y$ consisting of all ordered pairs (x, y) in which x is a worker and y is the social security number of x.

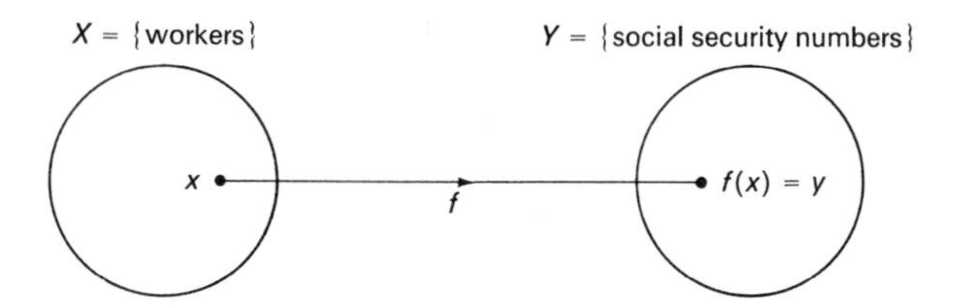

Clearly f is onto Y, for Y is defined as the set of all social security numbers for the members of X. Moreover f is 1–1. This simply means that different workers have different social security numbers. The inverse of f, f^{-1}, is the set of all pairs of the form $(y, x) = (f(x), x)$, where x is the worker whose social security number is y. Observe that $f^{-1}\colon Y \to X$ is also a 1–1 onto function. For, $y_1 = f(x_1) \neq f(x_2) = y_2$ means that x_1 and x_2 have different social security numbers and hence x_1 and x_2 must be different people. Moreover, f^{-1} is onto, since to each social security number there corresponds a worker.

Frequently it is required to find a "formula" for $g = f^{-1}$ given a formula for f. In Example 3.11 we saw that if X is the set of all nonnegative real numbers and Y is the interval of real numbers $0 \leq y < 1$, then $f\colon X \to Y$ defined by the formula

$$f(x) = \frac{x}{x+1}$$

is 1–1 and onto Y. We can ask for a similar formula for f^{-1}, i.e., given $y \in Y$ how do we find $f^{-1}(y)$? By definition $f^{-1}(y)$ is the element x of X for which $f(x) = y$. Hence, in the equation

$$\begin{aligned} y &= f(x) \\ &= \frac{x}{x+1}, \end{aligned}$$

we want to solve for x in terms of y and this was done in Example 3.11:

$$x = \frac{y}{1 - y}.$$

In other words,

$$f^{-1}(y) = \frac{y}{1 - y}.$$

This is the required formula.

Theorem 3.2 *If f: $X \to Y$ is a* 1–1 *onto function then the inverse of f is a function from Y to X. Moreover, f^{-1}: $Y \to X$ is a* 1–1 *onto function.*

Proof According to the definition (12), every member $f(x) \in Y$ is a first member in a pair of $g = f^{-1}$ and thus dmn $g = Y$. To see that g is a function, we must show that no two pairs in g have the same first member. Thus suppose $(f(x_1), x_1)$ and $(f(x_2), x_2)$ are in g and $f(x_1) = f(x_2)$. Then, since f is 1–1, it follows that $x_1 = x_2$. Finally, since g is a function, it is reasonable to ask whether it is 1–1. (It is clearly onto, since f is defined on all of X.) Two typical elements in g are, by definition, of the form $(f(x_1), x_1)$ and $(f(x_2), x_2)$ and g could fail to be 1–1 if and only if for some $x_1 \in X$ and $x_2 \in X$, $x_1 = x_2$ but $f(x_1) \neq f(x_2)$. However, f is a function and thus this situation can never occur. The proof is complete. ▮

Observe that if f: $X \to Y$ is 1–1 and onto, then

$$f^{-1}(f(x)) = x, \; x \in X, \tag{14}$$

and

$$f(f^{-1}(y)) = y, \; y \in Y. \tag{15}$$

The equality (14) is no mystery: it is just the statement that the pair in f^{-1} whose first member is $f(x)$ has x as a second member, i.e., $(f(x), x)$ is the pair in f^{-1} whose first member is $f(x)$. The reader should provide a similar argument for (15).

Example 3.12 Let X be the set of all isosceles triangles in the plane with vertices $(0, 0)$, $(1, 0)$, and $(\frac{1}{2}, t)$, as t varies over all positive real numbers.

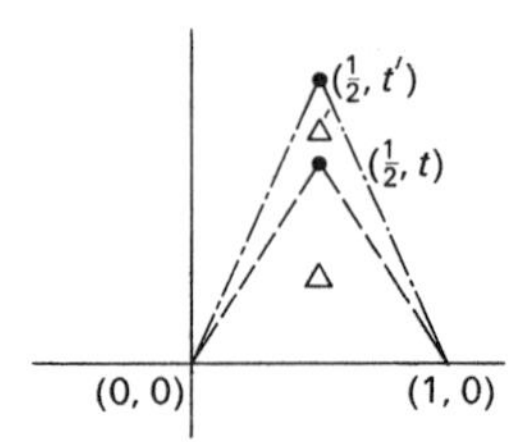

Let f: $X \to P$ be the function which associates with each triangle $\triangle \in X$ the area of $\triangle$, that is,

$$f(\triangle) = \text{area of } \triangle$$
$$= \tfrac{1}{2}t.$$

(The area of a triangle is $\frac{1}{2}$ the base times the altitude.) The set P is the set of positive real numbers. Observe first that f is 1–1 and onto P. For, let $\triangle$ and $\triangle'$ be two triangles in X ($\triangle'$ does not mean the complement of $\triangle$ here!) and assume that $f(\triangle) = f(\triangle')$. Then $\frac{1}{2}t = \frac{1}{2}t'$ and hence $t = t'$. Observe from the figure that $\triangle$ and $\triangle'$ then have the same vertices, so that $\triangle = \triangle'$. It is also obvious that f is onto P; for, if p is any positive number, then the triangle $\triangle \in X$ whose third vertex is $(\frac{1}{2}, 2p)$ has area p. Finally, f^{-1} is the function whose value for any positive number t is the isosceles triangle in X whose vertices are $(0, 0)$, $(1, 0)$, and $(\frac{1}{2}, 2t)$, that is,

$$f^{-1} = \{(t, \triangle) \mid t > 0 \wedge \triangle \in X \wedge (\triangle \text{ has vertices } (0,0), (1,0), \text{and } (\tfrac{1}{2}, 2t))\}.$$

Let f: $X \to Y$ and g: $Y \to Z$ be two functions.

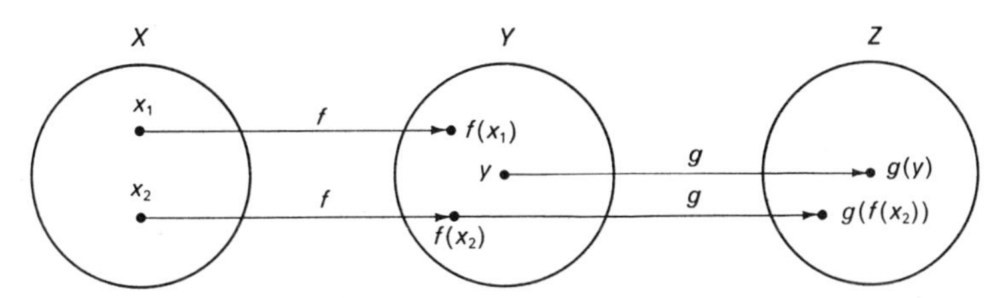

If f has domain X and range Y and g has domain Y and range Z it is evident that we may form a composite h: $X \to Z$ which associates with each element $x \in X$ the element $g(f(x)) \in Z$. In other words,

$$h(x) = g(f(x)).$$

Thus h maps x into the result of evaluating g at $f(x)$, i.e., first perform f and then perform g on this answer. For example, suppose f is given by the formula

$$f(x) = x^2$$

and g is given by the formula

$$g(x) = \frac{x}{1 + x^4}$$

(take the domain and range of both functions to be the set of non-negative real numbers). Then the value of the composite h of f

and g is found by computing the value of g at $f(x)$:

$$\begin{aligned} h(x) &= g(f(x)) \\ &= g(x^2) \\ &= \frac{x^2}{1 + (x^2)^4} \\ &= \frac{x^2}{1 + x^8}. \end{aligned}$$

In words: the value of h at x is

$$\frac{x^2}{1 + x^8};$$

or equivalently, h consists of the ordered pairs

$$\left(x, \frac{x^2}{1 + x^8}\right);$$

or finally, in set notation

$$h = \left\{(x, y) \mid x \geq 0 \wedge y = \frac{x^2}{1 + x^8}\right\}.$$

Definition 3.7 ***Composition of functions*** Let $f\colon X \to Y$ and $g\colon Y \to Z$ be two functions. Then the relation $h \subset X \times Z$,

$$h = \{(x, z) \mid x \in X \wedge z = g(f(x))\}, \tag{16}$$

is called the *composite* or *product* of f and g. We write

$$h = gf. \tag{17}$$

The notation (17) does *not* mean that the function values of g and f are to be multiplied together to obtain the function value of h, even when such an operation is possible.

Example 3.13 Let g be the function which associates with each positive number t the circle in the plane with center at the origin (0, 0) and area $2t$. Let f be the function in Example 3.12. Then the composition $h = gf$ of f and g is the function which associates with each triangle $\triangle$ with vertices (0, 0), (1, 0), and $(\frac{1}{2}, t)$, the circle with center at the origin whose area is t. Observe that $h(\triangle)$ is the circle of area t and therefore its radius must be $\sqrt{\frac{t}{\pi}}$.

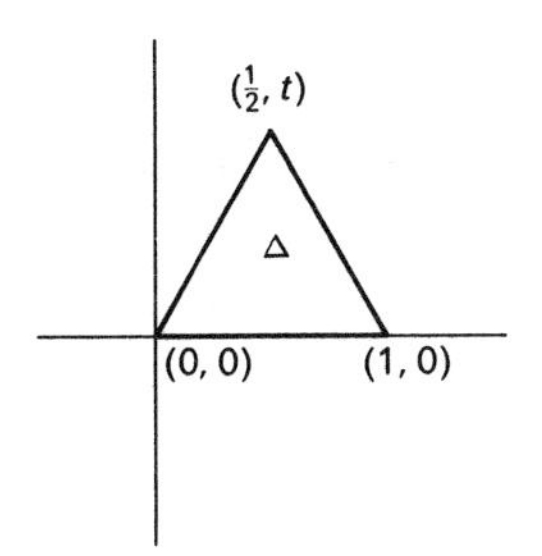

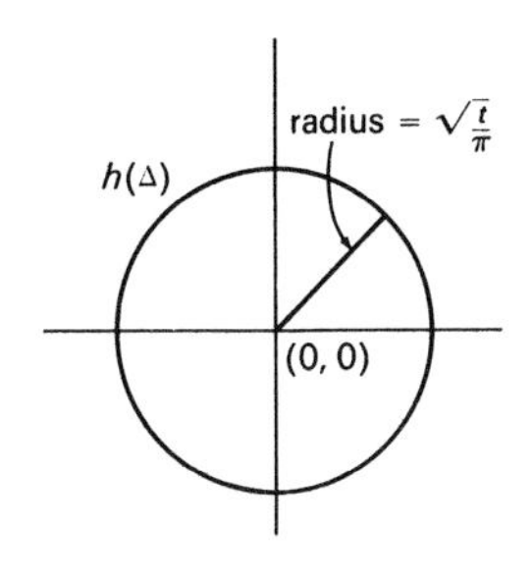

Theorem 3.3 *The composition of functions is an associative operation. That is, if* $f\colon X \to Y$, $g\colon Y \to Z$, *and* $k\colon Z \to W$ *are functions, then*

$$k(gf) = (kg)f. \tag{18}$$

Proof Let $x \in X$. The value of the left side of (18) at x is $(k(gf))(x)$ which, from the definition of function composition, is $k((gf)(x))$. Furthermore, $gf(x)$ is by definition $g(f(x))$. Thus

$$k((gf)(x)) = k(g(f(x)).$$

One can check similarly that

$$\begin{aligned}((kg)f)(x) &= (kg)(f(x)) \\ &= k(g(f(x))).\end{aligned}$$

Hence the two sides of (18) agree for each $x \in X$, and by Theorem 3.1, the two functions $k(gf)$ and $(kg)f$ are equal. ▮

As an example of the equality (18) let X, Y, and Z all be the set of positive real numbers and define f, g, and k by the formulas

$$\begin{aligned}f(x) &= \frac{x}{1+x}, \\ g(x) &= x^2, \\ k(x) &= \frac{1}{x}.\end{aligned}$$

Then

$$\begin{aligned}gf(x) &= g(f(x)) \\ &= g\left(\frac{x}{1+x}\right) \\ &= \left(\frac{x}{1+x}\right)^2 \\ &= \frac{x^2}{(1+x)^2}.\end{aligned}$$

Thus

$$\begin{aligned}(k(gf))(x) &= k(gf(x))\\ &= k\left(\frac{x^2}{(1+x)^2}\right)\\ &= \frac{1}{\left(\frac{x^2}{(1+x)^2}\right)}\\ &= \frac{(1+x)^2}{x^2}.\end{aligned}$$

Also

$$\begin{aligned}kg(x) &= k(g(x))\\ &= k(x^2)\\ &= \frac{1}{x^2}.\end{aligned}$$

Hence

$$\begin{aligned}((kg)f)(x) &= kg(f(x))\\ &= kg\left(\frac{x}{1+x}\right)\\ &= \frac{1}{\left(\frac{x}{1+x}\right)^2}\\ &= \frac{(1+x)^2}{x^2}.\end{aligned}$$

We see then that

$$(k(gf))(x) = ((kg)f)(x).$$

Example 3.14 Show that although function composition is associative (see Theorem 3.3), it is not *commutative*. This means that if $f\colon X \to X$ and $g\colon X \to X$, it is not necessarily the case that $fg = gf$. For example, let X be the totality of pairs of real numbers (x_1, x_2). Define

$$f((x_1, x_2)) = (x_1 + 2, x_2 + 3)$$

and

$$g((x_1, x_2)) = (x_2, x_1).$$

Then

$$\begin{aligned}fg((x_1, x_2)) &= f(g((x_1, x_2)))\\ &= f((x_2, x_1))\\ &= (x_2 + 2, x_1 + 3).\end{aligned}$$

However,

$$\begin{aligned} gf((x_1, x_2)) &= g(f((x_1, x_2))) \\ &= g((x_1 + 2, x_2 + 3)) \\ &= (x_2 + 3, x_1 + 2). \end{aligned}$$

A simple but important function defined on a set X is the *identity*, I_X. The identity function is defined as follows:

$$I_X = \{(x, x) \mid x \in X\}.$$

Thus

$$I_X(x) = x, \ x \in X.$$

Theorem 3.4 *Let f: $X \to Y$ be a* 1–1 *onto function. Then*

$$f^{-1}f = I_X,$$

and

$$ff^{-1} = I_Y.$$

This result is very easy to prove and will be left as an exercise (see (14) and (15)). ▮

As a mnemonic device, we can think of sets as "vertices" and functions as "arrows" connecting these vertices. There are some pertinent properties that these "vertices" and "arrows" must have.

(i) If f is an arrow from vertex X to vertex Y and g is an arrow from vertex Y to vertex Z, then a "composite" arrow, gf, from X to Z must exist.

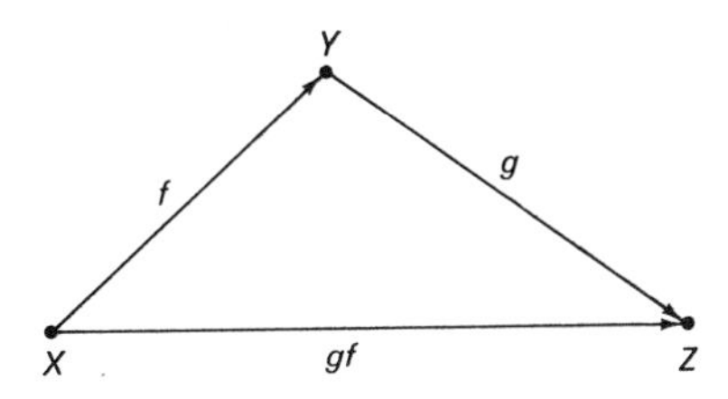

In other words, function composition is defined.

(ii) If X, Y, Z, W are vertices, f is an arrow from X to Y, g an arrow from Y to Z, and k an arrow from Z to W, then

$$k(gf) = (kg)f.$$

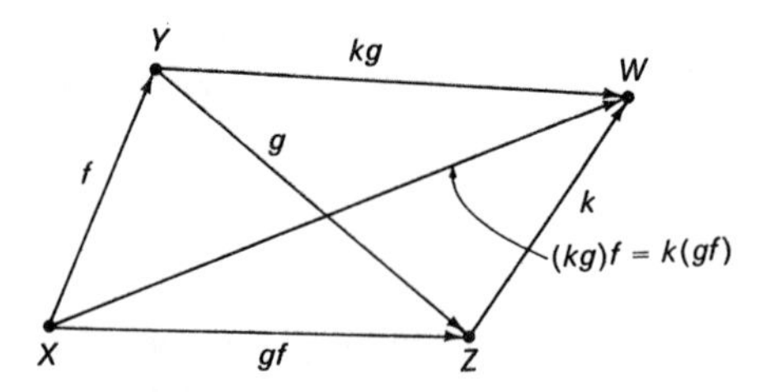

This is the associative law for function composition.

(iii) For vertices X and Y there exist identity arrows I_X and I_Y such that for any arrow f from X to Y,

$$\begin{aligned} I_Y f &= f \\ &= f I_X. \end{aligned}$$

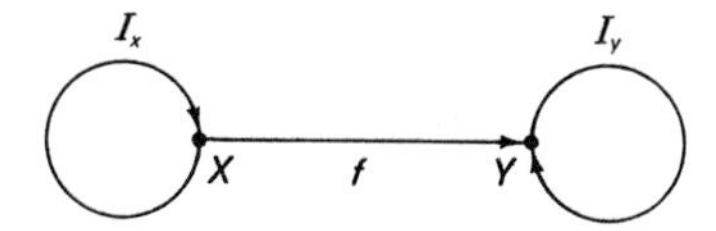

A collection of vertices $X, Y, Z, \ldots$, together with a collection of arrows f: $X \to Y$ is called a *category* of sets if the preceding three axioms are satisfied. Although we do not do so here, the usual axioms about sets can be replaced by axioms concerning categories. This was done by F. W. Lawvere in an article published in 1964.*

Some of the topics we now introduce should be of particular interest to those readers who require that abstract mathematics have practical and important applications. Certain classes of functions that we now discuss are of great utility in analyzing dice games and bridge hands, in doing cryptanalysis, and in dealing with many other manifestations of human gullibility. These functions will be introduced now and considered in detail in the next section.

Assume that Y is a set and let N_r be the set of integers

$$N_r = \{1, 2, 3, \ldots, r\}.$$

Then any function f: $N_r \to Y$ is called an *r-sample* of Y. For example, if $Y = \{-3, -5, -7\}$ and $r = 4$, then the function f defined by

$$\begin{aligned} f(1) &= -7, \\ f(2) &= -5, \\ f(3) &= -5, \\ f(4) &= -3 \end{aligned}$$

* An Elementary Theory of the Category of Sets, Proceedings of The National Academy of Sciences of the U.S.A., V. 52 (1964), pp. 1506–1511.

is a 4-sample of Y. Thus an r-sample can be thought of as a labelling of some of the elements of Y with the integers $1, \ldots, r$. That is, we label the element $y \in Y$ with the integer $k \in N_r$ if $f(k) = y$. Since no additional assumptions are made on f it may well be the case that a particular element $y \in Y$ is labelled by more than one integer (f is not necessarly 1–1) and some elements of Y are not labelled at all (f is not necessarily onto). If f is a 1–1 function, then f is called an *r-permutation* of Y. If f is 1–1 and onto, then Y is called an *r-set* or an *r-element set*. Stating this in another way, Y is an r-set if and only if $\nu(Y) = r$, since we can label each element of Y with precisely one of the r integers in N_r when and only when there is a 1–1 function from N_r onto Y.

Certain r-samples are of particular interest. Thus let the elements of Y be designated in some order,

$$Y = \{y_1, y_2, \ldots, y_n\},$$

and let $f\colon N_r \to Y$ be an r-sample of Y:

$$f(1) = y_{k_1}, f(2) = y_{k_2}, \ldots, f(r) = y_{k_r}, \tag{19}$$

where $k_1, \ldots, k_r$ are integers from the set $\{1, \ldots, n\}$. The subscript notation means simply that $f(1)$ is element number k_1 of Y, $f(2)$ is element number k_2 of Y, etc. For example, if $Y = \{y_1, y_2, y_3\}$ and f is the 4-sample defined by

$$f(1) = y_3, f(2) = y_2, f(3) = y_1, f(4) = y_3,$$

then

$$k_1 = 3, k_2 = 2, k_3 = 1, k_4 = 3.$$

The r-sample f is called an *r-selection* of Y if

$$k_1 \leq k_2 \leq \cdots \leq k_r. \tag{20}$$

If f is an r-selection of Y and f is 1–1, i.e., $k_1 < \cdots < k_r$, then f is called an *r-subset* of Y or an *r-combination* of Y. As an example, suppose $Y = \{1, 2, 3, 4, 5\}$, i.e., $y_1 = 1$, $y_2 = 2$, $y_3 = 3$, $y_4 = 4$, $y_5 = 5$. Let f be a 3-selection: $f(1) = y_{k_1}$, $f(2) = y_{k_2}$, $f(3) = y_{k_3}$, where

$$k_1 \leq k_2 \leq k_3.$$

Since $y_j = j$ in this case, we can associate with f the triple of numbers

$$(k_1, k_2, k_3).$$

A partial list of such 3-selections can easily be made: (1, 1, 1), (1, 1, 2), (1, 1, 3), (1, 1, 4), (1, 1, 5), (1, 2, 2), (1, 2, 3), (1, 2, 4), (1, 2, 5), (1, 3, 3), (1, 3, 4), (1, 3, 5), (1, 4, 4), (1, 4, 5), (1, 5, 5), (2, 2, 2), (2, 2, 3), etc. Certain of these 3-selections are 3-combinations, e.g., (1, 2, 3), (1, 2, 4), (1, 2, 5), (1, 3, 4), (1, 3, 5), (1, 4, 5), etc.

A further word of explanation is in order here. The inequalities (20) are not necessarily strict, i.e., some of the k_i may be equal. Thus the range of an r-selection is in effect a "choice" of r items from the ordered set Y, allowing repetitions. On the other hand, if f is an r-selection and f is 1–1 then (19) tells us that the k_i are distinct (e.g., $k_1 = k_2$ would imply $f(1) = y_{k_1} = y_{k_2} = f(2)$) and hence the inequalities (20) are strict:

$$k_1 < k_2 < \cdots < k_r.$$

This means that $\operatorname{rng} f$ contains precisely r elements of Y, i.e., $\nu(\operatorname{rng} f) = r$. This accounts for the name "r-subset". In addition, if we randomly select r different elements from Y, we can arrange them according to their subscripts:

$$y_{k_1}, \ldots, y_{k_r},$$

where

$$k_1 < k_2 < \cdots < k_r,$$

and then define a function $f\colon N_r \to Y$, by (19). Clearly f is then an r-combination of Y.

Example 3.15 We state some typical problems in which the above concepts appear. We are not yet in a position to solve these problems but they should be carefully read in order to provide motivation for some of the results obtained in the next section.

(a) Bridge is played by first distributing an entire deck of 52 cards among four players, 13 cards to a player. This situation can be described by a 52-sample of a set of four elements, $f\colon N_{52} \to P$, in which $P = \{P_1, P_2, P_3, P_4\}$ is the set of four players and

$$\nu(\{t \mid f(t) = P_i \wedge t \in N_{52}\}) = 13, \quad i = 1, 2, 3, 4. \tag{21}$$

For any such function f, one can identify the first hand as the set of 13 integers $t \in N_{52}$ for which $f(t) = P_1$ and a similar remark is true for each of the remaining three hands.

(b) Suppose we are interested in counting how many ways 5 boys and 3 girls can be arranged in a line so that no two girls stand next to each other. Label the boys $b_1, \ldots, b_5$ and the girls g_1, g_2, g_3. We can think of the line as consisting of a set of eight positions and any arrangement as an 8-permutation, $f\colon N_8 \to S$, where S is the set of eight people. It is further required that there exists no integer i

for which $f(i)$ and $f(i + 1)$ are in $\{g_1, g_2, g_3\}$.

(c) In how many ways can a class of 20 students be divided into three discussion sections consisting of 7, 7, and 6 students, respectively? Any such arrangement can be identified with an r-sample $f: N_{20} \to N_3$ in which

$$\nu(\{t \mid f(t) = i \wedge t \in N_{20}\}) = 7, \qquad i = 1, 2,$$

and

$$\nu(\{t \mid f(t) = 3 \wedge t \in N_{20}\}) = 6.$$

(d) In how many ways can 6 books be arranged on a shelf? Any such arrangements can be identified with a 6-permutation of N_6.

(e) The director of a vivarium is managing a population of p individual animals which exhibit a total of m behavior characteristics. Moreover, some of the animals exhibit more than one characteristic. For any k characteristics, $k = 1, \ldots, m$, there are at least k animals which taken together exhibit these k characteristics. Question: in how many ways, if any, can m distinct animals be chosen, each exhibiting a different characteristic? This problem is somewhat more difficult than the preceding examples. Let f be a relation on N_m to N_p defined by $(j, t) \in f$ if animal t has characteristic j. If f is an m-permutation of N_p, i.e., f is a 1–1 function, then the range of f consists of m individuals which exhibit the m different characteristics. The problem is to find how many such functions (if any) exist. (Note that f, as defined above, is not even a function if two animals exhibit the same characteristics.)

(f) Find the number of straight lines that can be drawn through pairs of points chosen from a set of five points if no three of the five points lie on the same straight line. If $f: N_2 \to N_5$ is a 2-combination of N_5, we can think of f as determining a straight line. For, if $f(1) = k_1$, $f(2) = k_2$, $k_1 < k_2$, then the pair of points (k_1, k_2) determine a line. Moreover, since the line determined by (k_2, k_1) is the same as the line determined by (k_1, k_2) we definitely do not want to count 2-permutations of N_5. Thus the problem is simply to count the number of 2-combinations of N_5.

(g) (Dance Problem) In a group of n boys and n girls, each boy has been introduced to precisely k girls and each girl has been introduced to precisely k boys, where k is a fixed integer, $1 \leq k \leq n$. Can the boys and girls pair off into dance partners who have been previously introduced and if so, in how many ways can this be done? Any such pairing into dance partners can be identified with an n-permutation, f, of N_n, i.e., we can think of boy t paired off with girl $f(t)$, $t = 1, \ldots, n$.

In general, problems in *combinatorial analysis*, of which the preceding are examples, all have a similar form. Namely, it is required to find the total number of functions from a set X to a set Y when certain conditions about these functions are stipulated. We shall soon consider methods of attack on combinatorial problems.

Quiz

Answer true or false.

1 $\nu(\emptyset) = 0$.

2 $\nu(\{\emptyset\}) = 1$, i.e., the number of elements in the set which consists of $\emptyset$ is 1.

3 If (x, y) is an ordered pair and $x \neq y$, then $\nu(\{(x, y)\}) = 1$.

4 Let $X = \{1, -1\}$ and let R be the relation on X to X defined by $R = \{(x, y) \mid y = x^2\}$. Then rng $R = X$.

5 Let X be the set of all people. Define a relation R on X to X by $R = \{(x, y) \mid x \in X \wedge y \in X \wedge (y \text{ is married to } x)\}$. Then R is an equivalence relation. (Polygamy is illegal.)

6 Let $X = \{1, 2, 3, 4\}$ and $Y = \{1, 2, 3, 4, 5\}$. Let R be the following relation on X to Y: $R = \{(x, y) \mid (y - x) \text{ is even}\}$. Then the incidence matrix for R is given by

$$A(R) = \begin{bmatrix} 1 & 0 & 1 & 0 \\ 0 & 1 & 0 & 1 \\ 1 & 0 & 1 & 0 \\ 0 & 1 & 0 & 1 \\ 1 & 0 & 1 & 0 \end{bmatrix}.$$

7 If $f\colon X \to X \times X$ is the function which associates with each element of X the ordered pair (x, x), i.e., $f(x) = (x, x)$, and g is the function whose domain is $X \times X$ and whose value at any ordered pair (x, y) is 1 or 0 according as $x = y$ or $x \neq y$, then the composite function satisfies $gf(x) = 1$ for all $x \in X$.

8 If $X = \{0, 1, 2\}$ and $g = \{(0, 1), (1, 2), (0, 0)\}$, then g is a function on X to X.

9 The number of 2-permutations of a two element set is 2.

10 The number of 3-samples of a two element set is 8.

Exercises

1 Identify each of the following sets as a function or a relation which is not a function, giving reasons:

(a) $\{(0, 1), (1, 2)\}$;
(b) $\{(0, 1), (0, -1)\}$;
(c) $\{(0, 1), (1, 2), (2, 3), (3, 2)\}$;
(d) $\{(x, y) \mid x^2 + y^2 = 1 \wedge -1 \leq x \leq 1 \wedge -1 \leq y \leq 1\}$;
(e) $\{(x, y) \mid (x = 1 \vee x = 2) \wedge (y = 3 \vee y = 5)\}$;
(f) $\{0, 1, 2\} \times \{2, 3\}$;
(g) $\{0, 1, 2\} \times \{2\}$;
(h) $\{2\} \times \{0, 1, 2\}$;

(i) $\{(\triangle, h) \mid (\triangle$ is a triangle in the plane$) \wedge (h$ is the perimeter of $\triangle)\}$;
(j) $\{(h, \triangle) \mid h$ is the perimeter of a triangle $\triangle$ in the plane$\}$;
(k) $\{\emptyset\} \times \{\emptyset\}$.

2 Write each of the following functions f with domain X as a set of ordered pairs:
(a) $f(x) = 2$, $X = \{0, 1, 3\}$ (Hint: $f = \{(0, 2), (1, 2), (3, 2)\}$);
(b) $f(x) = \dfrac{x}{1 + x}$, $X = \{3, -3, 0\}$;
(c) $f(x) = x^2$, $X = \{2^2, 3^2, 4^2\}$;
(d) $f(x) = x - 1$, $X = \{1, -1, 0, 2\}$;
(e) $f(x) = x(x - 1)$, $X = \{2, 4, -3\}$;
(f) $f(x) = x^x$, $X = \{1, 2, -1\}$.

3 Determine which of the following relations R is an equivalence relation on the set X.
(a) $R = X \times X$, $X = \{1, 2, 3\}$;
(b) $R = \{(a, b) \mid a$ is married to $b\}$, X is the set of all people;
(c) $R = \{(x, y) \mid x - y = 1\}$, $X = \{2, 3, 4, -2, -3, -4\}$;
(d) $R = \{(x, y) \mid x$ and y have a common parent$\}$, X is the set of all people;
(e) $R = \{(x, y) \mid x$ is the cousin of $y\}$, X is the set of all people;
(f) $R = \{(x, y) \mid x$ has the same name as $y\}$, X is the set of all people;
(g) $R = \{(x, y) \mid x$ and y have the same number of legs$\}$, X is the set of all animals.

4 Let $X = \{0, 1\}$. Write out all equivalence relations on X to X.

5 Show that the relation R in Example 3.6 is an equivalence relation.

6 Show that if R is an equivalence relation on X to X, then $\operatorname{dmn} R = \operatorname{rng} R = X$.

7 Each of the following matrices is an incidence matrix for a relation R on $N_3 = \{1, 2, 3\}$ to $N_4 = \{1, 2, 3, 4\}$. Explicitly write out the ordered pairs in R for each of the following:

$$\text{(a)}\ \begin{array}{c} \\ 1 \\ 2 \\ 3 \\ 4 \end{array}\!\!\begin{array}{c} \begin{array}{ccc} 1 & 2 & 3 \end{array} \\ \begin{bmatrix} 1 & 0 & 1 \\ 1 & 0 & 0 \\ 1 & 0 & 1 \\ 1 & 0 & 0 \end{bmatrix} \end{array}; \qquad \text{(b)}\ \begin{bmatrix} 0 & 0 & 0 \\ 0 & 0 & 0 \\ 0 & 0 & 0 \\ 1 & 1 & 1 \end{bmatrix}; \qquad \text{(c)}\ \begin{bmatrix} 1 & 0 & 1 \\ 1 & 1 & 1 \\ 0 & 0 & 0 \\ 1 & 1 & 1 \end{bmatrix};$$

$$\text{(d)}\ \begin{bmatrix} 0 & 1 & 0 \\ 0 & 0 & 0 \\ 1 & 0 & 1 \\ 0 & 0 & 0 \end{bmatrix}; \qquad \text{(e)}\ \begin{bmatrix} 0 & 0 & 1 \\ 1 & 0 & 0 \\ 0 & 1 & 0 \\ 0 & 0 & 0 \end{bmatrix}; \qquad \text{(f)}\ \begin{bmatrix} 0 & 0 & 0 \\ 0 & 0 & 1 \\ 1 & 0 & 1 \\ 0 & 1 & 1 \end{bmatrix};$$

$$\text{(g)}\ \begin{bmatrix} 1 & 0 & 0 \\ 0 & 1 & 0 \\ 0 & 0 & 1 \\ 0 & 0 & 0 \end{bmatrix}; \qquad \text{(h)}\ \begin{bmatrix} 0 & 0 & 1 \\ 0 & 1 & 0 \\ 0 & 0 & 0 \\ 1 & 1 & 0 \end{bmatrix}.$$

8 Identify those relations R in Exercise 7 that are functions. Which if any are 1–1 functions? Can any of these relations be 1–1 functions onto N_4? Why?

9 Let $X = \{0, 1\}$ and let $Y = X \times X$. Find $\nu(Y \times Y)$.

10 Find the domain of the relation $Y \times Y$ in Exercise 9.

11 Consider the following diagram of seven neurons in which the nerve impulses are assumed to travel from dendrite to axon in the direction indicated by the arrows. Let X be the set of seven neurons and let $R = \{(x, y) \mid x \in X \wedge y \in X \wedge (\text{the impulse travels from } x \text{ to } y)\}$. Construct the incidence matrix $A(R)$ for the relation R and verify that R is an equivalence relation on X. (Assume that R is transitive.)

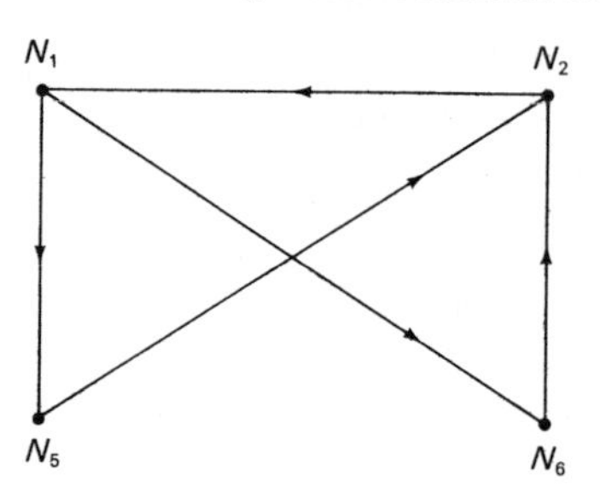

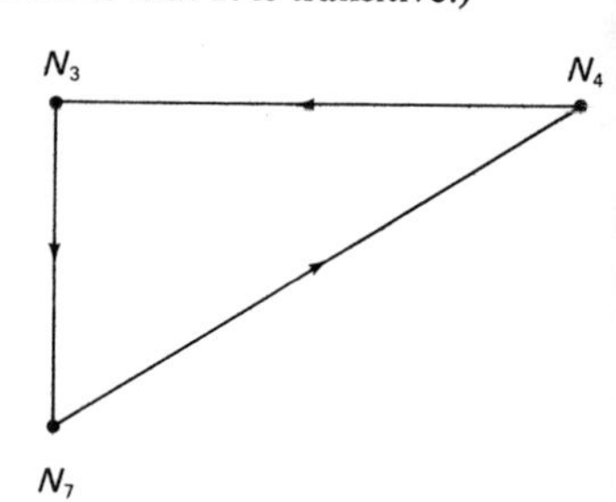

12 It is often argued that in the set of all people, "opposites" attract one another. Is the relation $R = \{(x, y) \mid (x \text{ attracts } y) \wedge (x \text{ is "opposite" to } y)\}$ an equivalence relation if it is assumed for convenience that $(x, x) \in R$ for all $x \in X$? Explain your answer.

†***13** Let X be the set of all integers (positive, negative, and zero).

(a) Define a relation R on X to X as follows:

$$R = \{(x, y) \mid (y - x) \text{ is divisible by } 3\}.$$

Prove that R is an equivalence relation.

(b) Let R be the relation in (a). For any $x \in X$, define $[x]$ to be the totality of integers $y \in X$ such that $(x, y) \in R$, i.e.,

$$[x] = \{y \mid (x, y) \in R \wedge y \in X\}.$$

Show that $[x_1] = [x_2]$ if and only if $(x_1, x_2) \in R$.

(c) Show that there are precisely three sets of the form $[x]$ in (b), namely $[0]$, $[1]$, $[2]$, and moreover $X = [0] \cup [1] \cup [2]$, and $[0] \cap [1] = [0] \cap [2] = [1] \cap [2] = \emptyset$.

(d) Show that $[x_1] = [x_2]$ if and only if x_1 and x_2 have the same remainder upon division by 3.

14 For each of the following relations R, construct the incidence matrix $A(R)$ as defined in Definition 3.4:

(a) $X = \{1, 2, 3\}$, $Y = \{1\}$,
$R = \{(x, y) \mid x \in X \wedge y \in Y \wedge y = x\}$;

(b) $X = \{0, 1, -1, 2, -2, 3, -3\}$, $Y = \{0, 1, 2, 3\}$,
$R = \{(x, y) \mid x \in X \wedge y \in Y \wedge y = -x\}$;

(c) X is a family consisting of two parents, two female children, and two male children,

$$R = \{(x, y) \mid x \in X \wedge y \in X \wedge (y \text{ is the brother of } x)\}.$$

15 Let X be a three element set. State which of the following 3×3 matrices are incidence matrices for functions $f: X \to X$. Further, which are incidence matrices for 1–1 functions? For functions onto X?

(a) $\begin{bmatrix} 1 & 0 & 0 \\ 0 & 1 & 0 \\ 0 & 0 & 1 \end{bmatrix}$; (b) $\begin{bmatrix} 0 & 0 & 1 \\ 0 & 1 & 0 \\ 1 & 0 & 0 \end{bmatrix}$;

(c) $\begin{bmatrix} 1 & 0 & 0 \\ 1 & 0 & 0 \\ 1 & 0 & 0 \end{bmatrix}$; (d) $\begin{bmatrix} 0 & 0 & 0 \\ 0 & 0 & 0 \\ 0 & 0 & 0 \end{bmatrix}$;

(e) $\begin{bmatrix} 0 & 0 & 0 \\ 0 & 0 & 0 \\ 1 & 1 & 1 \end{bmatrix}$; (f) $\begin{bmatrix} 1 & 1 & 1 \\ 1 & 1 & 1 \\ 1 & 1 & 1 \end{bmatrix}$;

(g) $\begin{bmatrix} 0 & 1 & 0 \\ 1 & 0 & 0 \\ 0 & 0 & 1 \end{bmatrix}$.

***16** Let X be a finite set, $X = \{x_1, \ldots, x_n\}$.

(a) Show that an $n \times n$ matrix, $A(f)$, is an incidence matrix for a function $f: X \to X$ if and only if each column of $A(f)$ has precisely one 1 and $n - 1$ zeros.

(b) Show that $A(f)$ is the incidence matrix for a 1–1 function f if and only if each row and each column of $A(f)$ contains precisely one 1 and $n - 1$ zeros.

17 Let R be a relation on a finite set X to X. What property must the incidence matrix $A(R)$ have in order that R be reflexive? (See Definition 3.3 (i).) Answer the same question for the symmetric property.

18 A physician finds that among three patients, p_1, p_2, and p_3, any k of the patients exhibit at least k of the symptoms of diseases d_1, d_2, d_3, and d_4 for $k = 1, 2$, or 3.

(a) Define a relation R as follows: $(p_j, d_i) \in R$ if and only if p_j exhibits the symptoms of the disease d_i. In the incidence matrix $A(R)$, how does the fact that any k of the patients exhibit at least k of the symptoms manifest itself?

(b) What is the interpretation of a set of three 1's, no two of which lie in either the same row or the same column?

19 In a university living group, four students came down with the measles; three other students were questioned as to whether they had had contacts with the four who were ill; then five more students were questioned as to whether they had had contacts with the preceding three. Let m_1, m_2, m_3, m_4 be the four students with measles; p_1, p_2, p_3 the students who were first questioned; and q_1, q_2, q_3, q_4, q_5 the third group of students.

We construct two incidence matrices,

$$\begin{array}{c} \\ m_1 \\ m_2 \\ m_3 \\ m_4 \end{array}\overset{\begin{array}{ccc} p_1 & p_2 & p_3 \end{array}}{\begin{bmatrix} 0 & 1 & 0 \\ 1 & 1 & 1 \\ 1 & 1 & 0 \\ 0 & 0 & 1 \end{bmatrix}} \quad \text{and} \quad \begin{array}{c} \\ p_1 \\ p_2 \\ p_3 \end{array}\overset{\begin{array}{ccccc} q_1 & q_2 & q_3 & q_4 & q_5 \end{array}}{\begin{bmatrix} 1 & 0 & 1 & 0 & 1 \\ 1 & 0 & 0 & 0 & 1 \\ 1 & 0 & 1 & 1 & 0 \end{bmatrix}},$$

in which an entry is 1 or 0 according as the person at the column heading has had contact with the person at the row heading or not. Find the 4×5 matrix in which the entry in column j and row i is the total number of exposures of student q_j which can be traced to student m_i, $i = 1, \ldots, 4$, $j = 1, \ldots, 5$.

20 With reference to Example 3.9, find another selection of four mixtures in which each species will find at least one acceptable mixture.

†21 Prove Theorem 3.4. (Hint: Refer to the equalities (14) and (15).)

22 Let X be the set of real numbers. Find the composite gf of the following pairs of functions. Express your answer in terms of a formula for $gf(x)$, $x \in X$.

(a) $f(x) = x$, $g(x) = -x$;
(b) $f(x) = 1$, $g(x) = 2$;
(c) $f(x) = x^2$, $g(x) = x + 1$;
(d) $f(x) = \dfrac{x}{1 + x^2}$, $g(x) = f(x)$;
(e) $f(x) = x - 1$, $g(x) = x + 1$;
(f) $f(x) = x + 1$, $g(x) = f(x)$;
(g) $f(x) = x^2$, $g(x) = -x$;
(h) $f(x) = (x - 1)(x + 1)$, $g(x) = x(x + 1)$;
(i) $f(x) = x^2 + 2x + 1$, $g(x) = 1$;
(j) $f(x) = x^3$, $g(x) = x^2$;
(k) $f(x) = x^2$, $g(x) = x^3$.

23 Let $Y = \{1, 2, 3\}$ and identify each of the following functions as an r-sample, r-permutation, r-selection, or r-combination of Y.

(a) $r = 1$, $f(1) = 3$;
(b) $r = 2$, $f = \{(1, 2), (2, 3)\}$;
(c) $r = 2$, $f(1) = 3$, $f(2) = 1$;
(d) $r = 3$, $f(1) = f(2) = f(3) = 2$;
(e) $r = 4$, $f(1) = 1$, $f(2) = f(3) = 2$, $f(4) = 3$;
(f) $r = 3$, $f = \{(1, 1), (2, 3), (3, 2)\}$;
(g) $r = 5$, $f = \{(1, 1), (2, 1), (3, 1), (4, 1), (5, 1)\}$;
(h) $r = 3$, $f = I_Y$;
(i) $r = 2$, $f(k) = k + 1$, $k \in N_2$;
(j) $r = 3$, $f(1) = 3$, $f(2) = 2$, $f(3) = 1$.

24 Write out all 3-combinations of the set $Y = \{1, 2, 3, 4\}$.

25 How many 2-combinations of a 2-element set are there?

26 How many 2-permutations of a 2-element set are there?

27 Find the number of 2-samples of a 3-set; of 2-permutations of a 3-set; of 2-selections of a 3-set; of 2-combinations of a 3-set.

†28 Let $f: X \to Y$ and $g: Y \to Z$ be 1–1, onto functions. Prove that $h = gf: X \to Z$ is 1–1, onto, and $h^{-1} = f^{-1}g^{-1}: Z \to X$.

1.4

Induction and Combinatorics

After repeated exposures to a certain event, rational people tend to believe that the event is inevitable. For example, in the spring of the first k years of a young man's life, his fancy turns to love. Hence it is plausible to assume that the same thing will happen in year $k + 1$ (precluding, of course, the onset of senility). Now consider the following series of similar events:

$$
\begin{aligned}
1 + 3 &= 2^2;\\
1 + 3 + 5 &= 3^2;\\
1 + 3 + 5 + 7 &= 4^2;\\
1 + 3 + 5 + 7 + 9 &= 5^2.
\end{aligned}
$$

It is not altogether nonsenical to guess that the sum of the first n odd natural numbers will be n^2. But this second example requires more of an argument than the first to be conclusive. Most of us are satisfied with the conclusion concerning a young man's fancy in the spring, based on preceding similar events. We can hardly argue, however, that the general proposition concerning the sum of the first n odd natural numbers can be established by an enumeration of a large number of special cases.

Again consider the following statement: for any natural number n, any n girls wear the same perfume. Of course, if $n = 1$, the statement is obvious. On the other hand, suppose that we assume any k girls wear the same perfume and on the basis of this assumption we prove that any $k + 1$ girls do. We can then go from 1 to 2, 2 to 3, 3 to 4, etc. Thus, suppose we have a set of $k + 1$ girls, $\{g_1, g_2, \ldots, g_k, g_{k+1}\}$, and consider the subset of k girls, $\{g_1, \ldots, g_k\}$. We are assuming that any k girls wear the same perfume and thus it must be true for $g_1, \ldots, g_k$. On the other hand, $\{g_2, g_3, \ldots, g_k, g_{k+1}\}$ is also a set of k girls, and hence $g_2, \ldots, g_{k+1}$ also wear the same perfume. But the two sets $\{g_1, \ldots, g_k\}$ and $\{g_2, \ldots, g_{k+1}\}$, intersect, and thus all $k + 1$ girls must wear the same perfume. Something is obviously wrong with this argument as one can easily detect by sniffing around. The difficulty, of course, is in going from the case $n = 1$ to the case $n = 2$. For, the preceding argument requires that the sets obtained by deleting one girl at a time must overlap, and this is not the case for $\{g_1, g_2\}$.

The *principle of mathematical induction* is the logical axiom used to establish a general proposition based on a series of repeated events. To make this statement more precise, consider the example we discussed before, namely, that the sum of the first n odd natural numbers is n^2. To formulate what we want to prove, let $p(n)$ be the statement, "For any natural number n,

$$1 + 3 + 5 + 7 + 9 + \cdots + (2n - 1) = n^2.\text{"} \tag{1}$$

(The number $2n - 1$ is the nth odd natural number.) For all we know at this stage of the argument, the statement $p(n)$ may have truth value T or F for any particular n. But in any event, $p(n)$ is a mathematically meaningful statement. We know that $p(1)$ has truth value T by direct verification, that is,

$$1 = 1^2.$$

Suppose we know that the following statement has truth value T:

$$\bigwedge_{k \in N} (p(k) \to p(k + 1)), \tag{2}$$

where N is the set of natural numbers, $N = \{1, 2, 3, \ldots\}$. In other words, assume that every one of the implications in the infinite list

$$\begin{aligned} p(1) &\to p(2), \\ p(2) &\to p(3), \\ &\vdots \\ p(k) &\to p(k + 1), \\ &\vdots \end{aligned}$$

has truth value T. Suppose, moreover, that we want to prove the truth of $p(n)$ for some prescribed n. Then the implication

$$p(1) \to p(2)$$

has truth value T by taking $k = 1$ in (2). Since $p(1)$ is true and $p(1) \to p(2)$ is true, it follows that $p(2)$ is true by the law of detachment. Next consider the case $k = 2$:

$$p(2) \to p(3)$$

has truth value T. Since $p(2)$ is true it again follows by the law of detachment that $p(3)$ is true. We simply work our way up to the statement that $p(n)$ has truth value T by employing the same reasoning one step at a time. Suppose, then, that we assume $p(k)$ is true, i.e., formula (1) holds for the case $n = k$:

$$1 + 3 + 5 + 7 + 9 + \cdots + (2k - 1) = k^2. \tag{3}$$

If we add the next odd natural number to both sides of (3), we obtain

$$1 + 3 + 5 + 7 + 9 + \cdots + (2k - 1) + (2k + 1) = k^2 + (2k + 1). \tag{4}$$

Now the left side of (4) is the sum of the first $(k + 1)$ odd natural numbers and, by a little high school algebra,

$$k^2 + (2k + 1) = (k + 1)^2.$$

In other words, the right-hand side of (4) is the square of $k + 1$ whereas the left-hand side of (4), as we have just remarked, is the sum of the first $k + 1$ odd natural numbers. The fact that

$$1 + 3 + 5 + 7 + 9 + \cdots + (2k + 1) = (k + 1)^2$$

is true is just the statement that $p(k + 1)$ has truth value T. We have proved that if $p(k)$ has truth value T then so does $p(k + 1)$, and thus we know that the implication

$$p(k) \rightarrow p(k + 1)$$

has truth value T.

It is essential for the reader to recognize that this argument is not circular: we have not proved that $p(k)$ has truth value T by assuming that $p(k)$ has truth value T. We have verified that $p(1)$ has truth value T and also that the *implication*

$$p(k) \rightarrow p(k + 1)$$

has truth value T for each $k \in N$. Our proof of the validity of the implication is based on our knowledge of truth tables. Namely, we know that the implication has truth value T if $p(k + 1)$ has truth value T whenever $p(k)$ has truth value T. That is, the case $p(k)$ true and $p(k + 1)$ false, which is the only case for which $p(k) \rightarrow p(k + 1)$ is false, cannot occur.

Example 4.1 For any natural number n let $p(n)$ denote the statement,

$$\text{"}n^2 + n + 41 \text{ is a prime number."}$$

(A natural number is prime if it is not divisible by any natural number except itself and 1.) Consider $p(1)$:

$$1 + 1 + 41 \text{ is a prime number.}$$

Surely $p(1)$ has truth value T since 43 is not divisible by any natural number except itself and 1. Once again, $p(2)$ is the statement

$$4 + 2 + 41 \text{ is a prime number,}$$

and since 47 is indeed prime, $p(2)$ has truth value T. If we proceed to replace n by 3, 4, 5, 6, 7, 8, 9, and 10, we obtain as values for

$n^2 + n + 41$ the numbers 53, 61, 71, 83, 97, 113, 131, and 151, all of which are prime. As a matter of fact, one can check by direct computation that $n^2 + n + 41$ is prime for all values of n not exceeding 39. Thus it is very tempting to try to prove that $p(n)$ has truth value T by establishing that

$$p(k) \rightarrow p(k+1)$$

has truth value T as we did in the example concerning the first n odd natural numbers. This approach, despite the 39 examples, is doomed to failure. For, if $n = 40$ we compute that

$$\begin{aligned} 40^2 + 40 + 41 &= 40(40 + 1) + 41 \\ &= 40 \cdot 41 + 41 \\ &= (40 + 1)41 \\ &= 41^2. \end{aligned}$$

Thus the statement,

$$\text{“}40^2 + 40 + 41 \text{ is a prime number”}$$

has truth value F.

We see then that to establish a proposition by the technique we have been discussing requires more than the verification of a large number of special cases.

We now state the *Principle of Mathematical Induction. Let $p(n)$ be a proposition which is either true or false for each natural number n. If*

(a) $p(1)$ *is true,*

and if

(b) *the implication*

$$p(k) \rightarrow p(k+1)$$

is true for each natural number k,
then $p(n)$ is true for every natural number n.

In somewhat more formal logical notation, the principle of mathematical induction states:

if

(a) $p(1)$ is true,

and if

(b) $$\bigwedge_{k \in N} (p(k) \rightarrow p(k+1))$$

is true, then

$$\bigwedge_{n \in N} (p(n) \text{ is true}).$$

The principle of mathematical induction is an axiom and we will not attempt to prove it in terms of more primitive notions. However, it is perhaps helpful to think of the induction principle in terms of a row of dominoes standing on end.

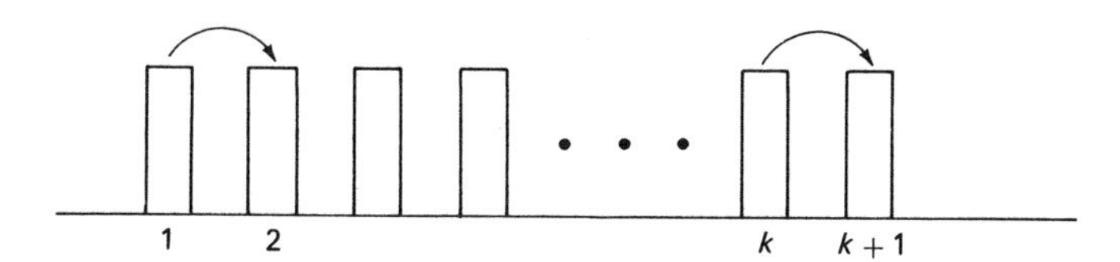

If

(a) the first domino is pushed over, and if successive dominoes are close enough so that

(b) whenever domino k falls over it pushes over domino $k + 1$ then

domino n will fall over for every n.

The statements (a) and (b) about the dominoes correspond to the statements (a) and (b) in the induction principle. The fact that all the dominoes fall over corresponds to the conclusion that $p(n)$ is true for every natural number n.

Example 4.2 Let a be a positive real number and consider the inequality

$$(1 + a)^n > 1 + na. \tag{5}$$

For $n = 1$, both sides of the proposed inequality (5) have value $1 + a$ and hence (5) is not true. In other words, if $p(n)$ is the statement that (5) holds for all $n \in N$, then $p(1)$ has truth value F. If we continue our investigation of (5) for $n = 2$, the left side has value

$$(1 + a)^2 = 1 + 2a + a^2,$$

and since a^2 is a positive number,

$$1 + 2a + a^2 > 1 + 2a,$$

that is,

$$(1 + a)^2 > 1 + 2a.$$

Again, if we take $n = 3$,

$$\begin{aligned}(1 + a)^3 &= 1 + 3a + 3a^2 + a^3\\ &> 1 + 3a,\end{aligned}$$

and we see that $p(3)$ is true. We would like to prove that (5) holds for every natural number n exceeding 1. But the principle of mathematical induction is not quite set up that way. It seems to require that

$p(1)$ be true. No doubt the reader has guessed by now what the appropriate modification of the principle should be. *Let n_0 be a natural number. Let $q(n)$ be a proposition which is either true or false for each natural number n, $n \geq n_0$. If*

(a) *$q(n_0)$ is true*

and if

(b) *the implication*

$$q(k) \rightarrow q(k + 1)$$

is true for each natural number k, $k \geq n_0$,
then $q(n)$ is true for all natural numbers n, $n \geq n_0$.

The fact that this modified statement of the induction principle holds is a direct consequence of the original statement. For, simply define $p(n)$ to be the proposition

$$q(n_0 + n - 1),$$

i.e., $p(1) = q(n_0)$, $p(2) = q(n_0 + 1)$, $p(3) = q(n_0 + 2), \ldots$. Then to prove by mathematical induction that $p(n)$ is true for all natural numbers n is precisely the same as the modified version of the principle as stated above. We leave this little detail to the reader. In terms of the "domino" model this modified version of the induction principle just amounts to pushing over domino n_0 first and forgetting about the first $n_0 - 1$ dominoes. Thus to prove the inequality (5), it suffices to show that $p(2)$ is true (as we have done) and to show, assuming $p(k)$ is true, that

$$p(k) \rightarrow p(k + 1)$$

is true for all natural numbers k, $k \geq 2$. Suppose, then, that

$$(1 + a)^k > 1 + ka.$$

This inequality is preserved if we multiply both sides by any positive number, in particular by the positive number $1 + a$:

$$\begin{aligned}(1 + a)(1 + a)^k &> (1 + a)(1 + ka),\\ (1 + a)^{k+1} &> 1 + ka + a + ka^2\\ &= 1 + (k + 1)a + ka^2.\end{aligned}$$

Now, ka^2 is a positive number and hence

$$(1 + a)^{k+1} > 1 + (k + 1)a,$$

and $p(k + 1)$ is true. We have proved: if $p(k)$ is true, then $p(k + 1)$ is true for any natural number k, $k \geq 2$. Hence the proposition $p(n)$ is true for any natural number n, $n \geq 2$.

The principle of mathematical induction is particularly useful in analyzing combinatorial problems, but before going into this interesting subject, we will introduce some classical language and notation.

Definition 4.1 ***Factorial, binomial coefficient*** Let n be a nonnegative integer. Then *factorial* n or n*-factorial*, written $n!$, is the product of the first n positive integers, if $n \geq 1$. That is,

$$n! = 1 \cdot 2 \cdot 3 \cdot 4 \cdots (n - 1) \cdot n.$$

The number 0-factorial is defined to be 1, i.e.,

$$0! = 1.$$

If r and n are integers, and $0 \leq r \leq n$, then the number

$$\frac{n!}{(n - r)!r!}$$

is called the *binomial coefficient* n *over* r and is written

$$\binom{n}{r}.$$

The main task in this section is to prove Theorem 4.2, which will show us how to count the number of samples, permutations, combinations, selections, and subsets of a given set. We briefly recapitulate the definitions of these items given at the end of Section 1.3, and add some remarks that should help to clarify their meaning.

Let X be an n-set. An r*-sample* of X is a function f: $N_r \to X$, $N_r = \{1, \ldots, r\}$. Thus an r-sample can be thought of as a labelling of the items in X with the integers $1, \ldots, r$ (i.e., if $f(k) = x$, then x is labelled with k), in which any item may have more than one label but it is not necessarily the case that every $x \in X$ have a label. In other words, an r-sample f is not necessarily 1–1 nor is it necessarily onto X. An r*-permutation* of X is just an r-sample, f: $N_r \to X$, that is 1–1. Thus we insist for an r-permutation that no item be labelled more than once.

Suppose the elements of X are designated in some order, say $X = \{x_1, \ldots, x_n\}$, and let f: $N_r \to X$ be an r-sample:

$$f(1) = x_{k_1}, f(2) = x_{k_2}, \ldots, f(r) = x_{k_r}.$$

If

$$k_1 \leq k_2 \leq \cdots \leq k_r$$

then f is called an *r-selection* of X. If f is also 1–1, so that

$$k_1 < k_2 < \cdots < k_r,$$

then f is called an *r-combination* or *r-subset* of X.

There is a very helpful and informative way of designating r-samples. If $X = \{x_1, \ldots, x_n\}$ and $f\colon N_r \to X$ is an r-sample we can write out the range of f as a sequence of length r:

$$(f(1), f(2), f(3), \ldots, f(r)).$$

But, after all, $f(1)$ is some element of X, say $f(1)$ is the element numbered k_1 of X; $f(2)$ is the element numbered k_2 of X; etc. The preceding sequence can thus be written

$$(x_{k_1}, x_{k_2}, x_{k_3}, \ldots, x_{k_r})$$

where $k_1, \ldots, k_r$ are just integers chosen from $1, \ldots, n$, allowing repetitions. Each r-sample f determines exactly one such sequence and each such sequence determines exactly one r-sample. If f is an r-permutation then the integers $k_1, \ldots, k_r$ must be distinct, i.e., f is 1–1. If f is an r-selection then $k_1 \leq \cdots \leq k_r$. If f is an r-combination then $k_1 < \cdots < k_r$.

We can further simplify and organize the preceding remarks. After all, given $k_1, \ldots, k_r$ from the set $\{1, \ldots, n\}$, the sequence $(x_{k_1}, x_{k_2}, x_{k_3}, \ldots, x_{k_r})$ is uniquely specified and hence the r-sample of X is uniquely specified. Thus to count the number of r-samples of X we need only count the number of sequences

$$(k_1, k_2, k_3, \ldots, k_r) \tag{6}$$

where $k_i \in \{1, \ldots, n\}$, $i = 1, \ldots, r$. If, for example, we want to count the number of r-selections of an n-set then we can simply count the number of sequences of integers (6) for which

$$k_1 \leq k_2 \leq \cdots \leq k_r.$$

We summarize these remarks in the following theorem.

Theorem 4.1 *Let X be the n-set*

$$X = \{x_1, x_2, x_3, \ldots, x_n\}.$$

Corresponding to any r-sample $f\colon N_r \to X$ there is a unique sequence of integers

$$(k_1, k_2, k_3, \ldots, k_r), \tag{7}$$

$1 \leq k_i \leq n$, defined by

$$f(i) = x_{k_i},\ i = 1, \ldots, r. \tag{8}$$

Conversely, for any such sequence (7), *there is precisely one r-sample f, defined by* (8). *Moreover,*

(a) *f is an r-permutation if and only if $k_1, \ldots, k_r$ are distinct;*

(b) *f is an r-combination if and only if $k_1 < k_2 < \cdots < k_r$;*

(c) *f is an r-selection if and only if $k_1 \leq k_2 \leq \cdots \leq k_r$.*

We can now go on to the main result of this section which tells us how to count these functions. In the proof of the following theorem and subsequently, we use a very elementary combinatorial principle for counting the total number of outcomes of two independent events. If the first event can happen in m ways and the second in n ways, then the two events can happen in mn ways. This statement is called the *multiplication principle*. For example, from five differently colored skirts and seven differently colored blouses, a total of $5 \cdot 7 = 35$ different outfits can be selected.

Theorem 4.2 *Let X be an n-set. Then:*

(a) *the number of r-samples of X is n^r;*

(b) *if $1 \leq r \leq n$, then the number of r-permutations of X is*

$$\frac{n!}{(n-r)!};$$

(c) *if $1 \leq r \leq n$, then the number of r-combinations or r-subsets of X is $\binom{n}{r}$;*

(d) *the number of r-selections of X is $\binom{n+r-1}{r}$;*

(e) *the number of subsets of X is 2^n.*

Proof

(a) Let $X = \{x_1, \ldots, x_n\}$ and let $s(r, n)$ denote the number of r-samples of X. According to Theorem 4.1, $s(r, n)$ is the number of sequences

$$(k_1, k_2, k_3, \ldots, k_r) \tag{9}$$

that can be formed in which $1 \leq k_i \leq n$, $i = 1, \ldots, r$. Now if $r = 1$, then (9) simply becomes a sequence of length 1,

$$(k_1),$$

and k_1 may be chosen in n ways from $\{1, \ldots, n\}$. Thus $s(1, n) = n$.

In other words, $s(r, n) = n^r$ in case $r = 1$ and we have the beginning of an induction argument on r. Suppose then that $s(r - 1, n) = n^{r-1}$. With any sequence

$$(k_1, k_2, \ldots, k_{r-1}) \tag{10}$$

we can form a sequence of length r by adjoining any one of the n integers $1, \ldots, n$ to (10). Thus each of the $s(r - 1, n) = n^{r-1}$ sequences (10) produces n sequences of the form (9) and it follows that

$$\begin{aligned} s(r, n) &= s(r - 1, n)n \\ &= n^{r-1}n \\ &= n^r. \end{aligned}$$

We have proved that there are n sequences (9) of length 1, and on the assumption that there are n^{r-1} sequences (10) of length $r - 1$ it follows that there are n^r sequences (9) of length r. This proves (a).

(b) Let $p(r, n)$ denote the number of r-permutations of an n-set. If $r = 1$, then $p(1, n)$ is just the number of sequences (9) of length 1 and, as above, this is obviously n, that is,

$$\begin{aligned} p(1, n) &= n \\ &= n \cdot \frac{(n - 1)!}{(n - 1)!} \\ &= \frac{n!}{(n - 1)!}. \end{aligned}$$

Thus, assume that $2 \leq r \leq n$ and

$$p(r - 1, n) = \frac{n!}{(n - (r - 1))!}. \tag{11}$$

According to Theorem 4.1(a) we want to count the number of sequences (9) in which $k_1, \ldots, k_r$ are distinct. We have verified that for $r = 1$ there are $p(1, n) = n$ of them, and we are assuming that there are $p(r - 1, n) = \dfrac{n!}{(n - (r - 1))!}$ of them of length $r - 1$.

Now, given a sequence (10) in which $k_1, \ldots, k_{r-1}$ are distinct, we can form a sequence for which $k_1, \ldots, k_{r-1}, k_r$ are distinct by simply choosing k_r to be any one of the $n - (r - 1) = n - r + 1$ integers in $\{1, 2, \ldots, n\}$ other than $k_1, \ldots, k_{r-1}$, and adjoining it to (10). Hence with each of the $p(r - 1, n)$ different sequences of the form (10) of distinct integers, we can construct $n - r + 1$ different sequences of the form (9) of distinct integers. From the induction assumption (11) we have

$$\begin{aligned}
p(r, n) &= (n - r + 1)p(r - 1, n) \\
&= (n - r + 1)\frac{n!}{(n - (r - 1))!} \\
&= (n - r + 1)\frac{n!}{1 \cdot 2 \cdot 3 \cdots (n - r)(n - r + 1)} \\
&= \frac{n!}{1 \cdot 2 \cdot 3 \cdots (n - r)} \\
&= \frac{n!}{(n - r)!}.
\end{aligned}$$

We have proved that if the formula in the statement of the theorem works for $r - 1$, then it works for r. However, this does not fully meet the requirements. After all, we have only the n propositions

$$p(r, n) = \frac{n!}{(n - r)!}, \qquad r = 1, 2, 3, \ldots, n,$$

but the induction principle requires that we have a set of propositions, one corresponding to each positive integer. This difficulty, however, is easily resolved, for, we can simply define a proposition for $n + 1, n + 2, \ldots$ to be any true statement, e.g., $1 = 1$. We will then have a list of propositions, one for each positive integer n, say $q(1), q(2), \ldots, q(n), q(n + 1), q(n + 2), \ldots$. The argument in the proof has the following form:

$$\begin{aligned}
&q(1) \text{ is true;} \\
&q(r - 1) \to q(r) \text{ is true,} \qquad r = 2, \ldots, n; \\
&q(k) \text{ is true,} \qquad k = n + 1,\ n + 2, \ldots.
\end{aligned}$$

It is obvious that for this set of propositions,

$$q(r - 1) \to q(r)$$

is true for all $r \geq 2$. Thus we are able to apply the induction principle. This technical device of simply adjoining a set of true propositions to a finite set of propositions allows us to use the principle of induction to establish a finite set of statements. This modification of the principle of induction is called *finite induction.*

(c) By Theorem 4.1(b) our task is to count the number of sequences (9) for which $k_1 < \cdots < k_r$. But this is easy now that we have proved part (b) of the present theorem. For, given any sequence (9) in which the integers are increasing, we can form $r!$ sequences in which the integers are distinct. The reason for this is that $\{k_1, \ldots, k_r\}$ is an r-element set, and according to Theorem 4.2(b) there are $r!$

(take $n = r$) different sequences of length r of distinct integers that can be formed from it. Also, different choices of $\{k_1, \ldots, k_r\}$ produce different sequences of length r, and every sequence (9) in which the integers are distinct arises in this way. Thus, if $c(r, n)$ denotes the number of r-combinations of X, we conclude that $c(r, n)r!$ is the number of r-permutations of X:

$$c(r, n)r! = \frac{n!}{(n-r)!},$$

or

$$\begin{aligned} c(r, n) &= \frac{n!}{(n-r)!r!} \\ &= \binom{n}{r}. \end{aligned}$$

(d) A tricky device is required to prove (d). According to Theorem 4.1(c) we must count the number of sequences (9) for which

$$1 \leq k_1 \leq k_2 \leq \cdots \leq k_r \leq n. \tag{12}$$

Now consider the sequence of integers obtained from $k_1, \ldots, k_r$ in (12) by adding 1 to k_2, 2 to $k_3, \ldots, r-1$ to k_r:

$$(k_1, k_2 + 1, k_3 + 2, \ldots, k_r + r - 1). \tag{13}$$

If we look at two successive integers in the sequence (13), say $k_3 + 2$ and $k_4 + 3$, we find that

$$(k_4 + 3) - (k_3 + 2) = k_4 - k_3 + 1 > 0,$$

and so it follows that

$$k_3 + 2 < k_4 + 3.$$

Clearly this argument works for any two successive integers in the sequence (13) and thus we may write

$$k_1 < k_2 + 1 < k_3 + 2 < \cdots < k_r + r - 1. \tag{14}$$

The integers in (14) are distinct and $k_r + r - 1 \leq n + r - 1$ (remember that $k_r \leq n$). In other words, given a sequence (9), i.e., $(k_1, \ldots, k_r)$, which satisfies (12) we can form a strictly increasing sequence (13) in which the integers are from the set $\{1, 2, \ldots, n + r - 1\}$. Conversely, suppose

$$(m_1, \ldots, m_r) \tag{15}$$

is a strictly increasing sequence of integers from

$$\{1, 2, \ldots, n + r - 1\};$$

this means that,

$$1 \le m_1 < \cdots < m_r \le n + r - 1. \tag{16}$$

Now, since $m_1 < m_2$, it follows that $m_1 \le m_2 - 1$. Similarly,

$$m_2 - 1 \le m_3 - 2 \le m_4 - 3 \le \cdots \le m_r - (r - 1). \tag{17}$$

Let $k_1 = m_1, k_2 = m_2 - 1, k_3 = m_3 - 2, \ldots, k_r = m_r - (r - 1)$. Then from (17) we have

$$k_1 \le k_2 \le k_3 \le \cdots \le k_r.$$

Moreover, since $m_r \le n + r - 1$ it follows that $k_r \le n$. Hence there is a 1–1 matching between sequences of the form (15) that satisfy (16) and sequences of the form (9) that satisfy (12). Thus we are able to say that the number of r-selections of the n-set X is the same as the number of r-combinations of an $(n + r - 1)$-set. But in part (c) we learned that the number of r-combinations of an $(n + r - 1)$-set is given by the binomial coefficient

$$\binom{n + r - 1}{r}. \tag{18}$$

Hence the number of r-selections of X is precisely the number given in (18).

(e) Probably the simplest way of seeing (e) is as follows. We can construct a subset of X by making a decision for each of the n elements of X as to whether or not it will be put in the subset. Such a sequence of n decisions clearly produces a subset of X, i.e., the subset consisting of precisely those elements for which the decision was "yes". The first decision can be made in two ways; the second decision in two ways; . . . ; the nth decision in two ways. Thus there is a total of

$$\overbrace{2 \cdot 2 \cdots 2}^{n} = 2^n$$

sequences of decisions, i.e., there are 2^n subsets of X. (A somewhat more pedestrian proof by mathematical induction may be given and will be left to the reader.) ∎

Example 4.3 In a class of ten boys, no two of whom are the same height, five are to be chosen and arranged in a row from left to right according to increasing height. In how many ways can this be done? In order to solve this problem, let $x_1, \ldots, x_{10}$ be the ten boys in the class ordered so that x_k is shorter than x_{k+1}, $k = 1, \ldots, 9$. Arranging 5 of the boys in a row according to increasing height now consists of selecting a 1–1 function $f\colon N_5 \to X$, where $X = \{x_1, x_2, \ldots, x_{10}\}$, such that if $f(1) = x_{k_1}$, $f(2) = x_{k_2}$, $f(3) = x_{k_3}$, $f(4) = x_{k_4}$, $f(5) = x_{k_5}$, then $k_1 < k_2 < k_3 < k_4 < k_5$. In other words, the number of ways of choosing 5 of the boys is exactly the number of 5-combinations of a 10-element set. By Theorem 4.2(c) this number is

$$\begin{aligned}\binom{10}{5} &= \frac{10!}{5!5!}\\ &= \frac{10 \cdot 9 \cdot 8 \cdot 7 \cdot 6 \cdot 5!}{1 \cdot 2 \cdot 3 \cdot 4 \cdot 5 \cdot 5!}\\ &= 252.\end{aligned}$$

Example 4.4 Using the digits 1, 2, 3, 4, how many 3-digit numbers can be formed in which the first digit is no bigger than the second and the second no bigger than the third? Choosing such a 3-digit integer amounts to defining a function $f\colon N_3 \to N_4$ in which $f(1)$ is the first digit, $f(2)$ is the second digit, and $f(3)$ is the third digit. It is required that

$$f(1) \le f(2) \le f(3).$$

Thus f is a 3-selection of the 4-element set N_4 and, according to Theorem 4.2(d), the number of such 3-selections is

$$\begin{aligned}\binom{4+3-1}{3} &= \frac{6!}{3!3!}\\ &= 20.\end{aligned}$$

Example 4.5 How many 3-digit numbers may be formed using the digits 1, 2, 3, 4, 5, such that no digit appears more than once? Let $f\colon N_3 \to N_5$, where $f(1)$, $f(2)$, $f(3)$ are the first, second, and third digits, respectively. Here it is required only that f be 1–1, i.e., that no digit appear more than once. Hence any such function is a 3-permutation of N_5 and, according to Theorem 4.2(b), there are precisely

$$\begin{aligned}\frac{5!}{(5-3)!} &= \frac{5!}{2!}\\ &= 60\end{aligned}$$

such permutations.

Quiz

Answer true or false.

1 $4! = 24$.

2 If n is a positive integer, then $\binom{n}{1} = \binom{n}{0}$.

3 $\binom{0}{0} = 1$.

4 The number of 3-samples of a 4-element set is 3^4.

5 All of the subsets of the 2-element set $X = \{a, b\}$ are $\{a\}$ and $\{b\}$.

6 The number of sequences of integers (k_1, k_2) satisfying $1 \leq k_1 < k_2 \leq 10$ is $\binom{10}{2}$.

7 The number of sequences of integers (k_1, k_2) satisfying $1 \leq k_1 < k_2 \leq 10$ is $\binom{11}{2}$.

8 If n is a nonnegative integer then $\binom{n}{n} = \binom{n+1}{0}$.

9 The number of r-selections of an n-element set is at least the number of r-combinations of an n-element set if $1 \leq r \leq n$.

10 The number of n-combinations of an n-element set is 1.

Exercises

1 Evaluate each of the following numbers:

(a) $\binom{7}{2}$; (b) $4!$; (c) $\frac{19!}{18!}$;

(d) $\frac{3!}{6!}$; (e) $\binom{4}{3}\binom{5}{4}$; (f) $\binom{7}{\binom{4}{2}}$;

(g) $\frac{1}{0!}$; (h) $\frac{0!}{1}$; (i) $\binom{253}{251}$;

(j) $\frac{1}{\binom{8}{3}}$.

2 Evaluate the number of
(a) 2-samples of a 5-element set;
(b) 5-samples of a 2-element set;
(c) 3-permutations of a 6-element set;
(d) 5-combinations of an 8-element set;
(e) 7-selections of a 2-element set;
(f) subsets of a 4-element set;
(g) 5-permutations of a 5-element set;
(h) 5-combinations of a 5-element set;
(i) 1-selections of a 10-element set;
(j) 10-selections of a 1-element set.

3 Let $X = \{x_1, x_2, x_3\}$. In accordance with Theorem 4.1, write down the sequence (7) corresponding to each of the following r-samples, $f\colon N_r \to X$, and identify combinations, permutations, and selections:

(a) $r = 1$, $f(1) = x_3$;
(b) $r = 2$, $f(1) = x_1$, $f(2) = x_3$;
(c) $r = 3$, $f(1) = x_3$, $f(2) = x_2$, $f(3) = x_1$;
(d) $r = 5$, $f(1) = x_1$, $f(2) = x_2$, $f(3) = x_2$, $f(4) = x_2$, $f(5) = x_3$;
(e) $r = 4$, $f(1) = x_1$, $f(2) = x_1$, $f(3) = x_2$, $f(4) = x_3$;
(f) $r = 2$, $f(1) = x_3$, $f(2) = x_2$;
(g) $r = 2$, $f(1) = x_2$, $f(2) = x_3$;
(h) $r = 3$, $f(1) = f(2) = f(3) = x_3$;
(i) $r = 4$, $f(1) = x_3$, $f(2) = x_2$, $f(3) = x_1$, $f(4) = x_1$;
(j) $r = 3$, $f(1) = x_1$, $f(2) = x_2$, $f(3) = x_3$.

4 Let $X = \{1, 2, 3, 4, 5\}$. Evaluate each of the following in which it is assumed that the digits come from X:

(a) the number of 2-digit integers;
(b) the number of 2-digit integers in which the digits are in increasing order;
(c) the number of 3-digit integers in which the digits are in nondecreasing order;
(d) the number of 3-digit integers not exceeding 299 in which the digits are in nondecreasing order;
(e) the number of 5-digit integers in which the digits are in decreasing order;
(f) the number of 5-digit integers in which the digits are in nonincreasing order.

5 Establish each of the following identities for positive integers using the principle of mathematical induction:

(a) $\dfrac{1}{1\cdot 2} + \dfrac{1}{2\cdot 3} + \dfrac{1}{3\cdot 4} + \cdots + \dfrac{1}{n(n+1)} = \dfrac{n}{n+1}$;

(b) $1\cdot 2 + 2\cdot 3 + \cdots + n\cdot(n+1) = \dfrac{n(n+1)(n+2)}{3}$;

(c) $1^2 - 2^2 + 3^2 - 4^2 + \cdots + (-1)^{n-1}n^2 = (-1)^{n-1}\dfrac{n(n+1)}{2}$;

(d) $1 + 2 + 3 + \cdots + n = \dfrac{n(n+1)}{2}$;

(e) $1\cdot(1!) + 2\cdot(2!) + 3\cdot(3!) + \cdots + n(n!) = (n+1)! - 1$.

6 By using the defining formulas for binomial coefficients, show that if r and n are positive integers, $1 \le r \le n$, then

(a) $\dbinom{n}{r} = \dbinom{n}{n-r}$;

(b) $\dbinom{n}{r} + \dbinom{n}{r+1} = \dbinom{n+1}{r+1}$;

(c) $\dbinom{n}{r} = \dbinom{n-2}{r} + 2\dbinom{n-2}{r-1} + \dbinom{n-2}{r-2}$, $(2 \le r \le n-2)$.

7 If $\dbinom{n}{3} = \dfrac{10}{21}\dbinom{n}{5}$, find n.

***8** Prove by the method of mathematical induction that if n is a positive integer then

$$3^{2n} - 1$$

is divisible by 8. (Hint: let $f(n)$ be the indicated expression and show that $f(n+1) - f(n)$ is divisible by 8.)

9 A committee of p people is to be chosen from g girls and b boys. In how many ways can this be done? By counting the number of committees in which there are precisely i girls and $p - i$ boys, $i = 0, \ldots, p$, show that

$$\binom{g}{0}\binom{b}{p} + \binom{g}{1}\binom{b}{p-1} + \binom{g}{2}\binom{b}{p-2} + \cdots + \binom{g}{p}\binom{b}{0} = \binom{b+g}{p}.$$

It is entirely possible that p is greater than g or b. Under these circumstances, how should $\binom{b}{p}$ or $\binom{g}{p}$, etc., be defined in order to make the preceding equality correct?

***10** In a class of 3 boys and 3 girls, each girl has been introduced to precisely 2 boys and each boy has been introduced to precisely 2 girls. In how many ways can they pair off into dance partners who have been previously introduced?

11 The Carcinoma Cigarette Company has devised the following quality control procedure. From a batch of 20 cigarettes, 6 of which are defective, a sample of 5 cigarettes is selected. The whole batch will be discarded if 2 or more cigarettes in the sample are defective. In how many ways can a sample be selected at random so that the entire batch is rejected? (Hint: there are $\binom{20}{5}$ samples altogether. Count the number of 5-element subsets containing 0 of the defective cigarettes and the number containing 1 of the defective cigarettes. Subtract.)

12 In how many ways can 8 people be seated at a round table if only their positions relative to one another matter?

***13** In how many ways can 4 girls and 4 boys be seated at a round table if boys and girls must alternate and only their positions relative to one another matter?

14 Show that if n, r, and s are nonnegative integers, $s < r \leq n$, satisfying

$$\binom{n}{r} = \binom{n}{s},$$

then $r + s = n$.

15 In how many ways can an 11-man football team be chosen from 15 players?

16 In how many ways can a committee of 5 people be chosen from a group of 8 people if:

(a) there are 3 people who dislike one another and refuse to serve with one another;

(b) there are 3 people who like one another and serve together.

17 Let $p(k)$ be the proposition, "If k is a nonnegative integer, then

$$2 + 2^2 + 2^3 + \cdots + 2^k = 2^{k+1}."$$

(a) Show that the implication

$$p(k) \rightarrow p(k + 1)$$

is true.

(b) For which nonnegative integers n is $p(n)$ true?

***18** Consider the inequality

$$2^k > k^2,$$

where k is a nonnegative integer. For what values of k does the inequality appear to be true? Formulate and prove by mathematical induction the appropriate theorem concerning the relationship between 2^k and k^2.

19 Prove the following inequality for all integers $n \geq 4$ by the method of mathematical induction:

$$2n + 2 < 2^n.$$

***20** Prove by the method of mathematical induction that for any nonnegative integer n,

$$1 \cdot 2 + 2 \cdot 3 + 3 \cdot 4 + \cdots + n(n + 1) > \frac{n^3}{3}.$$

1.5

Partitions

In this section we shall consider a class of combinatorial problems of which the following is typical.

Example 5.1 Suppose that a class S of 30 students is to be divided into 3 discussion sections consisting of 10 students each. In how many ways can this be done? The problem is to count the number of ways that a set of 30 items can be "partitioned" into 3 subsets (i.e., discussion sections) A_1, A_2, and A_3 such that $\nu(A_1) = \nu(A_2) = \nu(A_3) = 10$. There are 30 students from which to choose the 10 members of A_1. According to Theorem 4.2(c) there are $\binom{30}{10}$ 10-element subsets of a 30-element set. Once the assignments to A_1 have been made there are 20 students left and hence the assignments to A_2 can be made in $\binom{20}{10}$ ways.

Now, of course, the remaining 10 students must comprise A_3. Hence, by the multiplication principle (stated immediately before Theorem 4.2), there are

$$\binom{30}{10}\binom{20}{10} = \frac{30!}{10!20!}\cdot\frac{20!}{10!10!} = \frac{30!}{10!10!10!}$$

ways of assigning the students to the three sections.

Definition 5.1 **Partitions** Let S be a finite set and let $\mathfrak{A}$ be a family of r subsets of S satisfying the following two conditions:

(i) if A and B are in the family $\mathfrak{A}$, then $A \cap B = \emptyset$, i.e., the members of the family $\mathfrak{A}$ are pairwise disjoint;
(ii) the union of all the subsets in $\mathfrak{A}$ is S, i.e.,

$$\bigcup_{A \in \mathfrak{A}} A = S.$$

If f is an r-permutation of $\mathfrak{A}$ then f is called an *ordered partition* of S into r subsets. (Note that the function values of an r-permutation of $\mathfrak{A}$ are subsets.)

If $f\colon N_r \to \mathfrak{A}$ is an ordered partition of S into r subsets, and $f(1) = A_1, f(2) = A_2, \ldots, f(r) = A_r$, then it is easier just to think of f as the ordered sequence of r subsets:

$$(A_1, A_2, A_3, \ldots, A_r).$$

The analysis made in Example 5.1 is easily generalized.

Theorem 5.1 *Let n be a positive integer and let S be an n-element set. Let $n_1, n_2, \ldots, n_r$ be r nonnegative integers whose sum is $n_1 + n_2 + \cdots + n_r = n$. The number of ordered partitions of S into r subsets,*

$$(A_1, A_2, \ldots, A_r), \tag{1}$$

in which

$$\nu(A_1) = n_1,\ \nu(A_2) = n_2, \ldots, \nu(A_r) = n_r \tag{2}$$

is

$$\frac{n!}{n_1!n_2!\cdots n_r!}. \tag{3}$$

Proof In constructing the subset A_1 in (1), we can choose the n_1 elements in

(4) $$\binom{n}{n_1}$$

ways, for this is just the number of n_1-element subsets of the n-element set S. Next we can choose the n_2-element subset A_2 by selecting an n_2-element subset from the $n - n_1$ elements left in S (now that A_1 has been chosen), and this can be done in

(5) $$\binom{n - n_1}{n_2}$$

ways. Then we can choose the n_3-element subset A_3 by selecting n_3 of the elements of S from among the $n - (n_1 + n_2)$ elements left (now that A_1 and A_2 have been chosen). Thus there are

(6) $$\binom{n - (n_1 + n_2)}{n_3}$$

ways in which A_3 can be chosen. We continue in this fashion until all of the elements of S are used up. In other words, A_1 may be chosen in $\binom{n}{n_1}$ ways, A_2 in $\binom{n - n_1}{n_2}$ ways, A_3 in $\binom{n - (n_1 + n_2)}{n_3}$ ways, . . . , A_k in $\binom{n - (n_1 + n_2 + \cdots + n_{k-1})}{n_k}$ ways, ... Thus by the multiplication principle, the ordered partition (1) may be chosen in

(7)
$$\binom{n}{n_1}\binom{n - n_1}{n_2}\binom{n - (n_1 + n_2)}{n_3}\cdots\binom{n - (n_1 + \cdots + n_{k-1})}{n_k}\cdots\binom{n - (n_1 + \cdots + n_{r-1})}{n_r}$$

ways. The integer in (7) is equal to

(8)
$$\frac{n!}{n_1!(n - n_1)!}\cdot\frac{(n - n_1)!}{n_2!(n - (n_1 + n_2))!}\cdot\frac{(n - (n_1 + n_2))!}{n_3!(n - (n_1 + n_2 + n_3))!}\cdots\frac{(n - (n_1 + \cdots + n_{r-1}))!}{n_r!(n - (n_1 + \cdots + n_r))!}.$$

Observe that in (8), $(n - n_1)!$ cancels in the first two fractions, $(n - (n_1 + n_2))!$ cancels in the second and third fractions, . . . , $(n - (n_1 + n_2 + \cdots + n_{k-1}))!$ cancels in the $(k - 1)$st and kth fractions, etc. Thus all the numerators in (8) cancel except $n!$, and since $n = n_1 + \cdots + n_r$ and $1 = 0! = (n - (n_1 + \cdots + n_r))!$, the integer (8) collapses to

$$\frac{n!}{n_1!n_2!n_3!\cdots n_r!}. \quad \blacksquare \tag{9}$$

The integer in (3) is called a *multinomial coefficient* and is denoted by

$$\binom{n}{n_1\ n_2\cdots n_r}. \tag{10}$$

Hopefully there is some connection between multinomial and binomial coefficients, if the names have any significance. If we take $r = 2$ (i.e., "multi" becomes "bi") then $n = n_1 + n_2$ and (10) becomes

$$\begin{aligned}\binom{n}{n_1\ n_2} &= \frac{n!}{n_1!n_2!}\\ &= \frac{n!}{n_1!(n-n_1)!}.\end{aligned}$$

But this last expression is just $\binom{n}{n_1}$. Hence if $n_1 + n_2 = n$ then

$$\begin{aligned}\binom{n}{n_1\ n_2} &= \binom{n}{n_1}\\ &= \binom{n}{n_2}.\end{aligned}$$

Example 5.2 Find the coefficient of $x_1^2x_2x_3$ in the power $(x_1 + x_2 + x_3)^4$. To solve this problem, expand the expression $(x_1 + x_2 + x_3)^4$ by writing $x_1 + x_2 + x_3$ as a factor in the product four times:

$$(\cdots)(\cdots)(\cdots)(\cdots). \tag{11}$$

The terms in the product (11) are obtained by choosing an x_i, $i = 1, 2, 3$, from each of the four factors, and multiplying them together. Thus the terms are of the form

$$x_1^{n_1}x_2^{n_2}x_3^{n_3},$$

where $n_1 + n_2 + n_3 = 4$. That is, x_1 has been chosen from n_1 of the factors in (11), x_2 from n_2 factors, and x_3 from n_3 factors, and since there are four factors altogether, $n_1 + n_2 + n_3 = 4$. The problem is: find the number of different ways the term $x_1^2x_2x_3$ can be obtained. That is, in how many ways can two of the factors in (11) be used for the choice of the x_1 term, one of the factors for the choice of the x_2 term, and one of the factors for the choice of the x_3 term? This, however, is equal to the number of ordered partitions of 4 items into 3 subsets (A_1, A_2, A_3) in which $\nu(A_1) = 2, \nu(A_2) = 1$, and $\nu(A_3) = 1$. More explicitly, we can think of the set containing the four factors in (11) as partitioned into three subsets: A_1 is a subset consisting of two factors from which we choose x_1; A_2 is a subset consisting of one factor from which we choose x_2; and A_3 is the subset consisting of one factor from which we choose x_3. According to Theorem 5.1 the number of such ordered partitions is precisely

$$\binom{4}{2\ 1\ 1} = \frac{4!}{2!1!1!}$$
$$= 12.$$

Example 5.3 In how many ways can the white chess pieces be arranged in two lines on a chess board? A line is just a row of eight adjacent squares, and there are 8 pawns, 2 knights, 2 bishops, 2 rooks, 1 queen, and 1 king to be arranged on the two lines. Imagine that the squares in the two lines are numbered 1, 2, . . . , 16. We define an ordered partition of the set of integers $\{1, 2, \ldots, 16\}$ into six subsets,

$$(A_1, A_2, A_3, A_4, A_5, A_6),$$

where A_1 is the set of 8 integers representing the squares on which the pawns are to go, A_2 is the set of 2 integers representing the squares on which the knights are to go, A_3 is the set of 2 integers representing the squares on which the bishops are to go, A_4 is the set of 2 integers representing the squares on which the rooks are to go, A_5 is the set of 1 integer representing the square on which the queen is to go, and A_6 is the set of 1 integer representing the square on which the king is to go. The number of such ordered partitions is given by

$$\frac{16!}{8!2!2!2!1!1!} = 64{,}864{,}800.$$

Example 5.4 Two dice are rolled. Assuming the dice are honest, what are the chances that in four rolls, two sevens, a five, and a three will come up? We will give precise meaning to the question, "What are the chances . . . ?" and analyze this problem in detail. The possible sums that can appear on the dice are 2 through 12, and these sums can be achieved as the following combinations

$$\begin{gathered}(1, 1);\\(1, 2), (2, 1);\\(1, 3), (3, 1), (2, 2);\\(1, 4), (4, 1), (2, 3), (3, 2);\\(1, 5), (5, 1), (2, 4), (4, 2), (3, 3);\\(1, 6), (6, 1), (2, 5), (5, 2), (3, 4), (4, 3);\\(2, 6), (6, 2), (5, 3), (3, 5), (4, 4);\\(3, 6), (6, 3), (4, 5), (5, 4);\\(4, 6), (6, 4), (5, 5);\\(5, 6), (6, 5);\\(6, 6).\end{gathered}$$

We see that there are a total of 36 possible outcomes for any given throw and we partition the set of outcomes into 11 subsets according to the sum that appears. We must now be more explicit about what is meant by the phrase, "What are the chances . . .?" A 2 can be thrown in only one way, namely $1 + 1 = 2$. There are 36 possible throws that are equally likely so that it is reasonable to say that the chances are 1 in 36 of obtaining a 2, i.e., we can assign the number $\frac{1}{36}$ to the outcome that a 2 is thrown. Once again, a 3 can be thrown in two ways, namely $1 + 2 = 3$ and $2 + 1 = 3$. Hence it is sensible to assign the number $\frac{2}{36}$ to the outcome that a 3 is thrown. We proceed in this way to assign the number $\frac{3}{36}$ to the outcome that a 4 is thrown, $\frac{4}{36}$ that a 5 is thrown, $\frac{5}{36}$ that a 6 is thrown, $\frac{6}{36}$ that a 7 is thrown, $\frac{5}{36}$ that an 8 is thrown, $\frac{4}{36}$ that a 9 is thrown, $\frac{3}{36}$ that a 10 is thrown, $\frac{2}{36}$ that an 11 is thrown, and $\frac{1}{36}$ that a 12 is thrown. Let $p_2, p_3, \ldots, p_{12}$ denote these eleven numbers and suppose we throw the dice four times in succession. What number shall we assign to the outcome that 7, 5, 7, 3 appear, in precisely that order? Now, 7 can appear in 6 ways in the first throw, 5 can appear in 4 ways in the second throw, 7 in 6 ways in the third throw, and 3 in 2 ways in the fourth throw. Thus there are $6 \times 4 \times 6 \times 2$ ways in which the outcome 7, 5, 7, 3 can occur. There are a total of $36 \times 36 \times 36 \times 36$ ways in which the dice can be thrown four times and hence it is reasonable to assign the number

$$\begin{aligned}\frac{6 \times 4 \times 6 \times 2}{36 \times 36 \times 36 \times 36} &= \frac{6}{36} \cdot \frac{4}{36} \cdot \frac{6}{36} \cdot \frac{2}{36}\\&= p_7 p_5 p_7 p_3\\&= p_3 p_5 p_7^{\,2}\end{aligned}$$

to the indicated outcome. But two sevens, a five and a three can appear in other ways in four throws, e.g., 7, 7, 5, 3; 3, 7, 5, 7; etc. Of course, 7, 7, 5, 3 can appear in $6 \times 6 \times 4 \times 2$ ways as can

3, 7, 5, 7, etc. If c denotes the total number of sequences of length 4 consisting of two sevens, a five, and a three, then $c \times 6 \times 6 \times 4 \times 2$ is the total number of ways in which two sevens, a five and a three can appear in some order. Moreover, there are 36^4 possible ways of throwing the dice four times, and hence

$$c \cdot \frac{6 \times 6 \times 4 \times 2}{36 \times 36 \times 36 \times 36} = cp_3p_5p_7^2$$

is a reasonable number to assign to the cumulative event that two sevens, a five, and a three appear in some order in four throws of the dice. Consider the expression

$$(p_2 + p_3 + p_4 + \cdots + p_{12})^4. \tag{12}$$

The problem is to compute the coefficient c of $p_3p_5p_7^2$ in the expansion of (12). Now, a general term in the expansion of (12) is of the form

$$p_2^{n_1}p_3^{n_2}p_4^{n_3} \cdots p_{12}^{n_{11}}$$

in which $n_1 + n_2 + n_3 + \cdots + n_{11} = 4$. To find c we must determine how many ways we can use one of the 4 identical factors in (12) for the choice of the p_3 term, one of the factors for the choice of the p_5 term, and two for the p_7 term. This is clearly the number of ordered partitions of 4 items into 3 subsets, (A_3, A_5, A_7), in which $\nu(A_3) = 1$, $\nu(A_5) = 1$, $\nu(A_7) = 2$. That is, we can think of the four factors in (12) as partitioned into 3 subsets: A_3 is a subset consisting of one factor from which we choose p_3, A_5 is a subset consisting of one factor from which we choose p_5, and A_7 is a subset consisting of two factors from which we choose p_7. According to Theorem 5.1, the number of such ordered partitions is the multinomial coefficient

$$\begin{aligned}\binom{4}{1\ 1\ 2} &= \frac{4!}{2!1!1!} \\ &= 12.\end{aligned}$$

Hence the chance of obtaining two sevens, a five, and a three in some order in four throws of the dice is

$$\begin{aligned}12p_3p_5p_7^2 &= 12 \times \frac{6^2 \times 4 \times 2}{36^4} \\ &= \tfrac{1}{486}.\end{aligned}$$

We shall return to a more systematic investigation of this type of problem in the next section.

There are two standard notations which are useful in dealing with sums and products of more than two items.

Definition 5.2 ***Sigma and pi notation*** Let $\{a_1, \ldots, a_n\}$ be a set of mathematical items for which addition and multiplication are defined. Then

$$\sum_{i=1}^{n} a_i = a_1 + a_2 + a_3 + \cdots + a_n \tag{13}$$

and

$$\prod_{i=1}^{n} a_i = a_1 \cdot a_2 \cdot a_3 \cdots a_n. \tag{14}$$

The definitions (13) and (14) may be stated inductively as follows:

$$\sum_{i=1}^{1} a_i = a_1,$$

and for $n > 1$,

$$\sum_{i=1}^{n} a_i = \left(\sum_{i=1}^{n-1} a_i\right) + a_n. \tag{15}$$

Similarly,

$$\prod_{i=1}^{1} a_i = a_1,$$

and

$$\prod_{i=1}^{n} a_i = \left(\prod_{i=1}^{n-1} a_i\right) a_n. \tag{16}$$

The following is a somewhat more explicit definition of the inductive relations (15) and (16). Define

$$\sum_{i=1}^{n} a_i$$

to be a_1 for $n = 1$. Then, assuming that

$$\sum_{i=1}^{n-1} a_i$$

has been defined, define

$$\sum_{i=1}^{n} a_i$$

to be

$$\left(\sum_{i=1}^{n-1} a_i\right) + a_n.$$

Thus we can go from

$$\sum_{i=1}^{1} a_i$$

to

$$\sum_{i=1}^{2} a_i$$

to

$$\sum_{i=1}^{3} a_i,$$

etc. Similar remarks apply for the product notation. The symbols in (13) and (14) are read, "the sum as i runs from 1 to n of a_i," and "the product as i runs from 1 to n of a_i," respectively.

Example 5.5 (a) $\sum_{i=1}^{3} i^2 = 1^2 + 2^2 + 3^2$.

(b) $\prod_{j=1}^{4} (j + 1) = (1 + 1)(2 + 1)(3 + 1)(4 + 1)$.

(c) $\sum_{i=1}^{4} i^i = 1^1 + 2^2 + 3^3 + 4^4$.

(d) $\sum_{i=1}^{4} (a_i + b_i^2) = (a_1 + b_1^2) + (a_2 + b_2^2) + (a_3 + b_3^2) + (a_4 + b_4^2)$.

(e) $\prod_{i=1}^{5} a_i^s = a_1^s\, a_2^s\, a_3^s\, a_4^s\, a_5^s$.

(f) $\sum_{k=1}^{4} \frac{k}{k+1} = \frac{1}{1+1} + \frac{2}{2+1} + \frac{3}{3+1} + \frac{4}{4+1}$.

(g) The sigma and pi notations are quite flexible and can be used to simplify the writing of complicated sums and products. Suppose we want to add all products of the form $x_1^{k_1}x_2^{k_2}$ in which k_1 and k_2 are nonnegative integers whose sum is 4, i.e., $k_1 + k_2 = 4$. We would like to use the sigma notation to denote this sum:

$$x_1^4 + x_1^3x_2 + x_1^2x_2^2 + x_1x_2^3 + x_2^4. \tag{17}$$

We can incorporate the expression (17) into the sigma notation by writing

$$\sum_{k_1+k_2=4} x_1^{k_1}\, x_2^{k_2}. \tag{18}$$

Formula (18) is read, "the sum of the terms $x_1^{k_1}x_2^{k_2}$ over all pairs of nonnegative integers k_1, k_2 satisfying $k_1 + k_2 = 4$."

(h) Using the sigma notation, write the sum of all expressions of the form

$$x_1^{k_1}x_2^{k_2}x_3^{k_3}$$

in which $k_1 + k_2 + k_3 = 15$ and the k_i are nonnegative integers, $i = 1, 2, 3$. The required expression is

$$\sum_{k_1+k_2+k_3=15} x_1^{k_1}\, x_2^{k_2}\, x_3^{k_3}. \tag{19}$$

(i) Occasionally the sigma and pi notations are used together. Thus (19) can be written

$$\sum_{k_1+k_2+k_3=15} \prod_{i=1}^{3} x_i^{k_i}. \tag{20}$$

Each summand in (20) is a product, namely,

$$x_1^{k_1}\, x_2^{k_2}\, x_3^{k_3} = \prod_{i=1}^{3} x_i^{k_i}.$$

(j) Using the sigma and pi notations, find a formula for

$$(x_1 + x_2 + x_3 + x_4)^5. \tag{21}$$

To solve this problem (which will suggest a general theorem of this type), imagine the five factors in the expression (21) written out:

$$(\cdot\cdot\cdot\cdot)(\cdot\cdot\cdot\cdot)(\cdot\cdot\cdot\cdot)(\cdot\cdot\cdot\cdot)(\cdot\cdot\cdot\cdot). \tag{22}$$

To construct a term of the form $x_1^{k_1}x_2^{k_2}x_3^{k_3}x_4^{k_4}$ in the expansion of (21), we choose x_1 from k_1 of the factors in (22), x_2 from k_2 of the factors, x_3 from k_3 of the factors, and x_4 from k_4 of the factors. The problem, then, is to decide how many ways this can be done. Think of the five factors in (22) as being partitioned into four subsets, (A_1, A_2, A_3, A_4): A_1 is the subset consisting of the k_1 factors from which x_1 is chosen; A_2 is the subset consisting of the k_2 factors from which x_2 is chosen; A_3 is the subset consisting of the k_3 factors from which x_3 is chosen; and A_4 is the subset consisting of the k_4 factors from which x_4 is chosen. Since there are five factors altogether, $k_1 + k_2 + k_3 + k_4 = 5$. Thus the number of times $x_1^{k_1}x_2^{k_2}x_3^{k_3}x_4^{k_4}$ appears is just the number of ordered partitions of a 5-element set into four subsets, (A_1, A_2, A_3, A_4), such that $\nu(A_1) = k_1$, $\nu(A_2) = k_2$, $\nu(A_3) = k_3$, and $\nu(A_4) = k_4$. According to Theorem 5.1, this number is precisely the multinomial coefficient

$$\binom{5}{k_1\ k_2\ k_3\ k_4}.$$

Thus, (21) is equal to

(23) $$\sum_{k_1+k_2+k_3+k_4=5} \binom{5}{k_1\ k_2\ k_3\ k_4} \prod_{i=1}^{4} x_i^{k_i}.$$

(k) Find the coefficient of the term $x_1x_2^2x_3^2$ in (23). For this term $k_1 = 1$, $k_2 = 2$, $k_3 = 2$, and $k_4 = 0$. Then

$$\begin{aligned}\binom{5}{1\ 2\ 2\ 0} &= \frac{5!}{1!2!2!0!}\\ &= \frac{5!}{4}\\ &= 30.\end{aligned}$$

(l) Find the value of

$$\sum_{k_1+k_2+k_3+k_4=5} \binom{5}{k_1\ k_2\ k_3\ k_4}.$$

This expression is the sum of all the multinomial coefficients which appear in (23). We could solve this problem by writing out all the coefficients and then adding. This project would be tedious but fortunately another technique is available. If we set $x_1 = x_2 = x_3 = x_4 = 1$ in (23), then every term $\prod_{i=1}^{4} x_i^{k_i}$ is 1, and (23) becomes

(24) $$\sum_{k_1+k_2+k_3+k_4=5} \binom{5}{k_1\ k_2\ k_3\ k_4}.$$

On the other hand, we know that (23) is equal to

(25) $$(x_1 + x_2 + x_3 + x_4)^5,$$

so if we set $x_1 = x_2 = x_3 = x_4 = 1$ in (25), we obtain the value 4^5. It follows that (24) must be equal to $4^5 = 1{,}024$, and the solution is complete.

(m) Compute the total number of ordered partitions of a 5 element set S into four or fewer subsets. Let k_1, k_2, k_3, and k_4 be nonnegative integers satisfying $k_1 + k_2 + k_3 + k_4 = 5$. Imagine an ordered partition of S, (A_1, A_2, A_3, A_4), in which

(26) $$\nu(A_i) = k_i, \qquad i = 1, 2, 3, 4.$$

Since we are allowing the value 0 for some of the k_i, the preceding ordered partition consists, in fact, of four or fewer sets. The number of ordered partitions satisfying (26) is

(27) $$\binom{5}{k_1\ k_2\ k_3\ k_4},$$

and hence the total number of partitions is just the sum of the numbers in (27) over all k_i totaling 5, $i = 1, \ldots, 4$. As we saw in the preceding example, this is precisely 4^5.

As a final theorem in this section, we obtain a general formula for the expansion of a power of a multinomial. We have derived particular instances of this theorem in some of our examples.

Theorem 5.2 *Let r be a positive integer. Then*

$$\left(\sum_{i=1}^{r} x_i\right)^n = \sum_{k_1+k_2+\cdots+k_r=n} \binom{n}{k_1\ k_2 \cdots k_r} \prod_{i=1}^{r} x_i^{k_i}. \tag{28}$$

Proof Imagine the left side of (28) written out as n factors:

$$\overbrace{(\cdots)(\cdots)(\cdots)\cdots(\cdots)}^{n}. \tag{29}$$

To form a term in the product (29), we choose exactly one of the x_i's from each of the n factors. Suppose that we choose x_1 from k_1 of the factors, x_2 from k_2 factors, x_3 from k_3 factors, $\ldots$, x_r from k_r factors. The problem, then, is to compute in precisely how many ways this can be done. Think of the n factors in (29) as divided into r subsets, $(A_1, A_2, A_3, \ldots, A_r)$, where A_i is the subset consisting of the k_i factors from which x_i is chosen, $i = 1, \ldots, r$. Since there are n factors altogether, $k_1 + k_2 + k_3 + \cdots + k_r = n$. Thus the number of times $\prod_{i=1}^{r} x_i^{k_i}$ appears in the product (29) is just the number of ordered partitions of an n-element set into r subsets, $(A_1, A_2, A_3, \ldots, A_r)$, satisfying $\nu(A_i) = k_i$, $i = 1, \ldots, r$. (We are allowing the possibility that some of the k_i may be zero.) According to Theorem 5.1, the number of such ordered partitions is precisely the multinomial coefficient

$$\binom{n}{k_1\ k_2\ k_3 \cdots k_r}.$$

Hence the term

$$\prod_{i=1}^{r} x_i^{k_i}$$

appears precisely

$$\binom{n}{k_1\ k_2\ k_3 \cdots k_r}$$

times in the product (29), and the proof is complete. ▌

The result of Theorem 5.2 is useful in counting problems of a general type. Thus suppose that a certain event has a number of possible outcomes and suppose, moreover, that this event is performed a number of times. We often want to ask the question, "In how many ways is it possible for a certain collection of outcomes to occur?" Knowing the answer to such a question is, of course, of critical importance in analyzing games of chance.

Example 5.6 An unbiased penny is tossed five times in succession. In how many ways can two heads and three tails appear in some order? Consider the binomial power

$$(x_1 + x_2)^5 = (x_1 + x_2)(x_1 + x_2)(x_1 + x_2)(x_1 + x_2)(x_1 + x_2). \qquad (30)$$

In computing the product (30), we select x_1 from a certain subset of the five binomials and x_2 from the remaining binomials and we do this in all possible ways. According to Theorem 5.2, the number of ways in which x_1 may be chosen three times and x_2 twice is precisely the binomial coefficient $\binom{5}{3}$. In other words, in the expansion of $(x_1 + x_2)^5$, $x_1^3x_2^2$ occurs with the binomial coefficient $\binom{5}{3}$. If we associate with each toss of the coin one factor $(x_1 + x_2)$ in (30) and think of x_1 as representing the occurrence of a head and x_2 as representing the occurrence of a tail, then it is clear that the number of ways that three heads and two tails can come up in five tosses is just the number of terms in (30) in which x_1 appears 3 times and x_2 appears twice. That is, each of the $\binom{5}{3} = 10$ different combinations of three heads and two tails can be identified with precisely one of the following terms in the expansion of (30):

$$x_1x_1x_1x_2x_2;\ x_1x_1x_2x_1x_2;\ x_1x_2x_1x_1x_2;\ x_1x_1x_2x_2x_1;\ x_1x_2x_1x_2x_1;$$
$$x_1x_2x_2x_1x_1;\ x_2x_1x_2x_1x_1;\ x_2x_2x_1x_1x_1;\ x_2x_1x_1x_1x_2;\ x_2x_1x_1x_2x_1.$$

But all of these terms are the same, namely $x_1^3x_2^2$, and hence the required count is just the coefficient of this term in the expansion of (30).

Example 5.7 A cookie jar contains three vanilla cookies and four chocolate cookies, all of different sizes. A child grabs four cookies in succession, but each time he grabs a cookie his mother takes it away and puts it back in the jar. In how many ways can the child grab two vanilla and two chocolate cookies? Let x_1 be identified with the selection of a vanilla cookie; let x_2 be identified with the selection of a chocolate cookie. Consider the binomial power

(31) $$(3x_1 + 4x_2)^4 = (x_1 + x_1 + x_1 + x_2 + x_2 + x_2 + x_2)^4.$$

If we imagine the four factors on the right in (31) written out in succession, then a sequence of two vanilla cookie and two chocolate cookie grabs can obviously be identified with the choice of x_1 from two of the factors and x_2 from the remaining two factors. In other words, we want to count the total number of times that $x_1^2x_2^2$ occurs in the expansion of (31). We can apply Theorem 5.2 to compute that

(32) $$\begin{aligned}(3x_1 + 4x_2)^4 &= \sum_{k_1+k_2=4} \binom{4}{k_1\ k_2} (3x_1)^{k_1}(4x_2)^{k_2} \\ &= \sum_{k_1+k_2=4} \binom{4}{k_1\ k_2} 3^{k_1}4^{k_2}x_1^{k_1}\ x_2^{k_2}.\end{aligned}$$

We want the coefficient of $x_1^2x_2^2$; this corresponds to choosing $k_1 = k_2 = 2$ in (32). Thus we obtain

$$\begin{aligned}\binom{4}{2\ 2} 3^2 4^2 &= \binom{4}{2} 9 \cdot 16 \\ &= 6 \cdot 9 \cdot 16 \\ &= 864\end{aligned}$$

as the total number of ways.

Example 5.8 In the preceding example, in how many ways can the frustrated cookie snatcher grab vanilla, vanilla, chocolate, and chocolate cookies, in that order? If, once again, we imagine the four factors on the right in (31) written out in succession, then a sequence of two vanilla cookie and two chocolate cookie grabs, in that order, can be identified with the choice of x_1 from the first and second factors and x_2 from the third and fourth factors. It is clear that x_1 may be chosen in 3 ways from each of the first two factors and x_2 may be chosen in 4 ways from each of the third and fourth factors, so that there is a total of $3 \cdot 3 \cdot 4 \cdot 4 = 144$ ways in which the required sequence of cookie grabs can be accomplished.

Example 5.9 We modify Example 5.7 as follows. Mother leaves the house for a couple of hours to meet a friend for martinis. Junior returns to the cookie jar which still contains three vanilla and four chocolate cookies and proceeds to grab two cookies twice in succession. In how many ways can he get two vanilla and two chocolate cookies in that order? Here the problem is considerably different: we want to count the total number of subsets consisting of two vanilla and two chocolate cookies in a set of three vanilla and four chocolate cookies. There are a total of $\binom{3}{2}$ subsets of two vanilla cookies and similarly a total of $\binom{4}{2}$ subsets consisting of two chocolate cookies. A subset of two vanilla and two chocolate cookies can be obtained by putting together one each of these 2-element "vanilla and chocolate" subsets. This can be done in $\binom{3}{2}\binom{4}{2} = 18$ ways by the multiplication principle.

Example 5.10 In how many ways can the monster in Example 5.9 successively grab 2 vanilla and 2 chocolate cookies in that order if he grabs them one at a time? (His mother is still out.) He can grab the first vanilla cookie in any one of 3 ways, but then there are only 2 vanilla cookies left so that he can grab the second in any one of 2 ways. Similarly, he can grab two chocolate cookies in any one of 4 and 3 ways, respectively. Hence the required sequence can be achieved in $3 \cdot 2 \cdot 4 \cdot 3 = 72$ ways.

Example 5.11 Two dice are tossed three times in succession. In how many ways can two 4's and a 5 appear in some order? If we refer back to Example 5.4, we see that a 4 can come up in 3 different ways and a 5 can come up in 4 different ways. Identify x_i with the event that $i + 1$ comes up, $i = 1, \ldots, 11$. Then consider the expression

$$(x_1 + 2x_2 + 3x_3 + 4x_4 + 5x_5 + 6x_6 + 5x_7 + 4x_8 + 3x_9 + 2x_{10} + x_{11})^3. \tag{33}$$

The appearance of two 4's and a 5 in some order can be identified with an occurrence of an $x_3^2x_4$ term in the expansion of (33). By Theorem 5.2 the expression (33) is equal to

$$\sum_{k_1+\cdots+k_{11}=3} \binom{3}{k_1 \cdots k_{11}} x_1^{k_1}(2x_2)^{k_2}(3x_3)^{k_3}(4x_4)^{k_4} \cdots (2x_{10})^{k_{10}}x_{11}^{k_{11}}.$$

We are interested in computing the coefficient of $x_3^2x_4$, which amounts to choosing $k_3 = 2$, $k_4 = 1$, and the rest of the $k_i = 0$. We then obtain

$$\begin{aligned}\binom{3}{2\ 1}(3x_3)^2(4x_4) &= 3 \cdot 9 \cdot 4 \cdot x_3^2x_4 \\ &= 108x_3^2x_4.\end{aligned}$$

Thus there are 108 ways that two 4's and a 5 can appear in some order.

Quiz

Answer true or false.

†1 If n is a positive integer and p and q are nonnegative integers satisfying $p + q = n$, then $\binom{n}{p} = \binom{n}{p\ q}$.

2 $\binom{6}{3\ 2\ 1} = \binom{6}{1\ 2\ 3}$.

3 $\prod_{i=1}^{3} i^i = 108$.

4 $\sum_{i=1}^{3} i^i = 30$.

5 The number of ordered partitions of a 5-element set into 2 subsets consisting of 3 and 2 elements, respectively, is $\binom{5}{2}$.

6 $\binom{3}{1\ 1\ 1} = 3$.

7 $\binom{3}{1\ 1\ 1} = 1$.

8 $\sum_{i=1}^{17} 2 = 34$.

9 $\prod_{i=1}^{10} 2^3 = 2^{30}$.

10 $\sum_{i=1}^{3} \frac{i^2}{i} = 6$.

Exercises

1 Evaluate each of the following numbers:

(a) $\binom{7}{2}$; (b) $\binom{7}{2\ 5}$; (c) $\binom{7}{3\ 4}$;

(d) $\binom{4}{2\ 2\ 0}$; (e) $\binom{4}{1\ 2\ 1}$; (f) $\binom{6}{3\ 3\ 0}$;

(g) $\binom{6}{3\ 0\ 3}$; (h) $\binom{6}{0\ 3\ 3}$; (i) $\binom{5}{1\ 1\ 1\ 1\ 1}$;

(j) $\binom{2}{1}\binom{3}{2}\binom{4}{3}\binom{5}{4}\binom{6}{5}\binom{7}{6}$.

2 Find the coefficients of the indicated expressions in each of the following expansions:

(a) $(x + y)^7$, x^2y^5;

(b) $(x_1 - x_2 + x_3)^4$, $x_1^2x_2x_3$;

(c) $(2x_1 - 3x_2 + 4x_3 + 10)^3$, x_1;

(d) $(x_1 + x_2 + x_3 + 2x_4)^5$, $x_1x_2x_3x_4^2$;

(e) $\left(\sum_{i=1}^{3} ix_i\right)^5$, $x_1^2\,x_2^3$;

(f) $\left(\sum_{i=1}^{4} (-1)^i x_i\right)^3$, $x_1^2x_4$;

(g) $(x + y + z)^3$, y^2z;

(h) $\left(x + \frac{1}{x} + \frac{1}{y}\right)^3$, $\frac{1}{y}$;

(i) $\left(\sum_{i=1}^{n} \frac{1}{x_i}\right)^r$, $\prod_{i=1}^{r} \frac{1}{x_i}$;

(j) $\left(\binom{3}{2} x_1 + \binom{3}{1} x_2^2\right)^3$, $x_1^2\,x_2^2$;

(k) $(x_1 - x_2 + x_3 + 1)^{10}$, x_1^9;

(l) $(x_2 + x_3 + x_4)^5$, $x_2^2x_3x_4^2$;

(m) $(2x + 3y - 7z)^4$, xyz^2.

3 In how many different ways can 9 different toys be divided among Mary, Ann, and Sue if 4 are given to Mary, 2 to Ann, and 3 to Sue?

4 License plates in a certain state are made by using a 4-digit number followed by 1 letter. The letters O and I are omitted because they look like numbers. How many license plates can be made?

5 A shelf in a laboratory has space for at most 10 bottles. There are 5 bottles of chemical *A*, 6 bottles of chemical *B*, and 4 bottles of chemical *C*. It is required that at least 2 bottles of each of the 3 different chemicals be put on the shelf. In how many ways can this be done? (Order does not matter.)

6 In a standard deck of 52 cards, how many different 2-card hands are there that consist of:

(a) 2 face cards;

(b) no face cards;

(c) cards whose numerical values total less than 5;

(d) at least one ace;

(e) no kings?

7 A housewife wishes to set her 4 blue, 5 red, and 6 clear drinking glasses about a round table. In how many ways can this be done if only the relative positions of the glasses matter? (Assume that glasses of the same color are indistinguishable.)

8 A box contains 5 yellow, 4 green, and 6 blue candles. In how many ways can 5 candles be selected such that 2 are yellow, 2 are green, and 1 is blue?

9 A class of 16 students is to be divided into discussion sections A_1, A_2, and A_3 which are as nearly equal in number as possible. In how many ways can this be done?

10 Evaluate the following:

(a) $\sum_{i=1}^{3} \frac{i}{i+1}$;

(b) $\prod_{i=1}^{3} \frac{i}{i^2+1}$;

*(c) $\sum_{j=1}^{3} \left(\prod_{i=1}^{j} i\right)^j$;

*(d) $\prod_{j=1}^{3} \left(\sum_{i=1}^{j} i\right)^j$;

*(e) $\sum_{i=1}^{3} i\left(\prod_{j=1}^{i} j^2\right)$;

*(f) $\prod_{i=1}^{2} \left(\prod_{j=1}^{i} \left(\sum_{k=1}^{j} k\right)\right)$;

(g) $\sum_{i=1}^{5} (-1)^i$;

(h) $\prod_{i=1}^{5} (-1)^i$;

*(i) $\prod_{i=1}^{3} \left(\prod_{j=1}^{i} i\right)$;

*(j) $\prod_{i=1}^{3} \left(\prod_{j=1}^{4} \binom{3}{i}\binom{4}{j}\right)$;

(k) $\sum_{i=1}^{5} \binom{5}{i} + \sum_{j=1}^{5} \binom{5}{5-j}$;

(l) $\prod_{i=1}^{4} x^i$;

(m) $\prod_{i=1}^{4} x_i^i$;

(n) $\prod_{i=1}^{4} x$;

(o) $\prod_{i=1}^{4} x^{-i}$;

(p) $\prod_{i=1}^{4} x_i^{-i}$;

*(q) $\prod_{i=1}^{4} x_j^i$;

*(r) $\prod_{i=1}^{4} x_i^j$;

*(s) $\prod_{i=1}^{4} x^{ij}$;

(t) $\sum_{i=1}^{4} x^i$;

(u) $\sum_{i=1}^{4} x_i^i$;

(v) $\sum_{i=1}^{4} x$;

(w) $\sum_{i=1}^{4} x^{-i}$;

*(x) $\sum_{i=1}^{4} x_j^i$;

*(y) $\sum_{i=1}^{4} x_i^j$;

*(z) $\sum_{i=1}^{4} x^{ij}$.

11 Evaluate each of the following in which $k_1, k_2, \ldots, k_n$ range over all nonnegative integers satisfying the indicated equalities:

(a) $\sum_{k_1+k_2=3} 2^{k_1}3^{k_2}$;

(b) $\sum_{k_1+k_2+k_3=2} 2^{k_1}3^{k_2}4^{k_3}$;

(c) $\sum_{k_1+k_2+k_3=1} x^{k_1}$;

(d) $\sum_{k_1+k_2=5} \binom{5}{k_1\ k_2}$;

(e) $\prod_{k_1+k_2=4} x_1^{k_1} x_2^{k_2}$;

(f) $\prod_{k_1+k_2=1} \binom{1}{k_1\ k_2}$;

(g) $\sum_{k_1+k_2+k_3=3} \binom{3}{k_1\ k_2\ k_3} \prod_{i=1}^{3} x_i^{k_i}$;

(h) $\displaystyle\sum_{k_1+k_2+k_3=3} \binom{3}{k_1\ k_2\ k_3} \sum_{i=1}^{3} x_i^{k_i}.$

†**12** Show that Theorem 5.2 becomes the classical *binomial theorem*,

$$(x_1 + x_2)^n = \sum_{i=0}^{n} \binom{n}{i} x_1^{\,i} x_2^{\,n-i},$$

by taking $r = 2$ in (28).

13 Show that

$$\sum_{i=0}^{n} \binom{n}{i} = 2^n$$

by examining the expression $(x_1 + x_2)^n$ when $x_1 = x_2 = 1$.

14 Show that

$$\sum_{i=0}^{n} (-1)^i \binom{n}{i} = 0$$

by examining the expression $(x_1 + x_2)^n$ when $x_1 = -1$, $x_2 = 1$.

***15** Prove by the method of mathematical induction that for all positive integers n, $n \geq 3$,

$$n^n \geq (n + 1)!.$$

16 Find the coefficient of x^p in the product $(1 + x)^b(1 + x)^g$. Next find the coefficient of x^p in $(1 + x)^{b+g}$. Prove the identity

$$\sum_{i=0}^{p} \binom{g}{i} \binom{b}{p - i} = \binom{b + g}{p}$$

by comparing the two.

17 Using the digit 1 four times, 2 three times, and 6 five times, how many different 12-digit numbers can be written? Leave your answer in terms of factorials.

***18** An urn contains three red and five black balls. In how many ways can two red and three black balls be removed from the urn in succession (in any order) if after each removal, the ball is replaced? (Hint: Find the coefficient of $x_1^2x_2^3$ in $(3x_1 + 5x_2)^5$.)

19 In how many ways can a 4-letter word using the letters a, b, and c be formed in which a appears at most once, b appears at most twice, and c appears at most three times?

20 Two dice are tossed three times in succession. In how many ways can two 8's and a 5 occur in any order?

21 Two dice are tossed three times in succession. In how many ways can two fours and a five occur in that order? (Hint: Consider the expansion of the expression (33). There are three ways in which x_3 can be chosen from the first factor, three ways in which x_3 can be chosen from the second factor, and four ways in which x_4 can be chosen from the third factor.)

22 How many words can be formed using all of the letters in the word aardvark. (Hint: Number the positions in which letters are to be placed $1, \ldots, 8$. An ordered partition of these eight integers can be associated with the placement of the three a's, two r's, one d, one v, and one k in the numbered positions. Thus the answer is provided by a count of the number of ordered partitions $(A_1, A_2, A_3, A_4, A_5)$ of an 8-element set in which $\nu(A_1) = 3$, $\nu(A_2) = 2$, $\nu(A_3) = 1$, $\nu(A_4) = 1$, $\nu(A_5) = 1$.)

23 How many words can be formed by using all of the letters in each of the following words:

(a) eleemosynary;

(b) supercalifragilisticexpialidocious;

(c) antidisestablishmentarianism.

24 Ten differently colored marbles are divided among three boys named Tom, Dick, and Harry. In how many ways can this be done if two of the boys get three marbles each? In how many ways can this be done if Tom gets four marbles and Dick and Harry get three each?

25 A deck of 52 cards is dealt to four players. How many initial situations are possible for a bridge game? $\left(\text{Hint: The answer is } \dfrac{52!}{(13!)^4}.\right)$

***26** Show that if r is an even positive integer, $r = 2p$, then

$$0 = \sum_{k_1+k_2+\cdots+k_r=n} \binom{n}{k_1\ k_2 \cdots k_r} (-1)^{k_{p+1}+\cdots+k_r}.$$

(Hint: Set $x_1 = x_2 = \cdots = x_p = 1$, $x_{p+1} = x_{p+2} = \cdots = x_r = -1$ in (28).)

27 Show that

$$\sum_{k_1+\cdots+k_r=n} \binom{n}{k_1 \cdots k_r} = r^n.$$

28 The following game is played. Three white balls and four black balls are placed in an urn. A player rolls two dice and then draws a ball from the urn. He wins if and only if he throws a seven or an eleven and draws a black ball. How many outcomes are possible in this game? How many of these are winning outcomes? If there are two people playing, how much money should a player demand to be put up against his own ten dollars?

29 Two dice are rolled twice. In how many ways can a 4 and a 5 come up in some order?

1.6

Introduction to Probability

Suppose that in flipping a coin n times we observe that it comes up heads f times and tails $n - f$ times. The *relative frequency* of the occurrence of heads is defined to be $\frac{f}{n}$. If n is "large enough", we believe that $\frac{f}{n}$ is a reliable estimate upon which to base a prediction of what will happen at any given toss. For example, if it is observed that this ratio tends to be near $\frac{1}{2}$ for a large number of trials, then it would be sensible to bet even money, e.g., a dollar against a dollar, that it will come up heads on any given trial. Of course, before we are prepared to make such a bet, it is necessary to satisfy ourselves that $\frac{f}{n}$ is indeed close to $\frac{1}{2}$. For, after all, the coin could be defective so that $\frac{f}{n}$ might be close to some number other than $\frac{1}{2}$ after a large number of trials. We have an intuitive feeling, however, that under normal circumstances, whatever defects the coin may have will be discovered by flipping it a large number of times.

Consider another example: rolling two dice. As we have computed a number of times, a seven can come up in six ways. On the other hand, if the dice are loaded, then there is no reason to think that seven must come up at all. Thus, anyone participating in a game of dice without first observing the behavior of the dice over a relatively large number of throws will rapidly be separated from his money. In order to arrive at a rational basis for making bets, we require that the dice be thrown n times and that the numbers $f_2, f_3, \ldots, f_{12}$ be recorded, where f_i denotes the number of occurrences of the sum i on the two dice. Our next step is to compute the relative frequencies $\frac{f_i}{n}$, $i = 2, \ldots, 12$. If these numbers seem reasonable for large n, we can at least attempt to make rational bets. Observe that

$$\sum_{i=2}^{12} \frac{f_i}{n} = 1, \tag{1}$$

since $f_2 + f_3 + \cdots + f_{12} = n$ and there are only 11 sums that can occur (if one of the dice happens to land so as to balance on either an edge or a corner, it is best to pick up your money and go home, anyway).

Probability theory is concerned with the problem of making rational decisions about events in which the outcome cannot be predicted in advance with any certainty. As with any mathematical theory which purports to have applications in the real world, the validity of the theory depends on its success. However, the theory of probability can be put on a completely mathematical basis, and the theorems which result will be like those in any other mathematical discipline. The relationship of a mathematical theory to physical reality (whatever this may mean) is not a new idea. The points and lines that one studies in elementary Euclidean geometry are intellectual constructs. After all, no one has ever held a point in his hand, nor has anyone ever constructed a perfect right triangle, or precisely measured the distance between two points. Nevertheless, we apply the theorems of plane geometry daily. It is a successful theory because it works. We will develop the theory of probability and see that it, too, works in this sense.

However, before launching into definitions and theorems, it is instructive to consider some familiar dice games that contain most of the ingredients of the formal theory. Suppose that we wish to make bets on the appearance of certain totals in the throw of two "honest" dice. We recall from Example 5.4 that totals 2 through 12 can appear on any given throw and that these can be achieved as follows:

$$\begin{gathered}(1, 1);\\ (1, 2), (2, 1);\\ (1, 3), (3, 1), (2, 2);\\ (1, 4), (4, 1), (2, 3), (3, 2);\\ (1, 5), (5, 1), (2, 4), (4, 2), (3, 3);\\ (1, 6), (6, 1), (2, 5), (5, 2), (3, 4), (4, 3);\\ (2, 6), (6, 2), (5, 3), (3, 5), (4, 4);\\ (3, 6), (6, 3), (4, 5), (5, 4);\\ (4, 6), (6, 4), (5, 5);\\ (5, 6), (6, 5);\\ (6, 6).\end{gathered}$$

Any individual throw of the dice is obviously identifiable with precisely one of the preceding pairs of numbers. Thus the "universe" U of elementary outcomes is just the set of 36 pairs (i, j), $1 \leq i \leq 6$, $1 \leq j \leq 6$.

Suppose we want to bet on the occurrence of a total of 4 in a single throw. Corresponding to this event is the subset $X \subset U$ consisting of (1, 3), (3, 1) and (2, 2), i.e.,

$$X = \{(1, 3), (3, 1), (2, 2)\}.$$

If, however, we want to bet on the occurrence of a total of 3 or 4 in a single throw, then the appropriate subset of U corresponding to this outcome is

$$X \cup Y,$$

where

$$Y = \{(1, 2), (2, 1)\}.$$

Once again, suppose we are interested in the occurrence of a total not exceeding 5. Then the appropriate subset of U corresponding to this event is

$$Z = \{(1, 1), (1, 2), (2, 1), (1, 3), (3, 1), (2, 2), (1, 4), (4, 1), (2, 3), (3, 2)\}.$$

Next, suppose a bet is made that an odd number will appear on exactly one of the two dice. Then, of course, we are interested in the following subset of U:

$$\begin{aligned} W = \{&(1, 2), (2, 1), (1, 4), (4, 1), (2, 3), (3, 2), \\ &(1, 6), (6, 1), (2, 5), (5, 2), (3, 4), (4, 3), \\ &(3, 6), (6, 3), (4, 5), (5, 4), (5, 6), (6, 5)\}. \end{aligned}$$

In case we want to bet on the occurrence of a total not exceeding 5 in which precisely one of the dice shows an odd number, then obviously

$$Z \cap W = \{(1, 2), (2, 1), (1, 4), (4, 1), (2, 3), (3, 2)\}$$

is the pertinent subset of U. Again, if we are interested in the occurrence of an odd number on either both or neither of the dice then the appropriate subset is the complement of W.

We see, then, that in order to begin the analysis of a dice game we require a universe U of "elementary" or "atomistic" outcomes and a collection or family of subsets of U that correspond to the possible outcomes of interest. If we call this family of subsets $\mathfrak{A}$ then in order to consider *compound* events (i.e., those consisting of several elementary events), it is desirable for $\mathfrak{A}$ to have the following properties: if X and Y are any subsets of U in $\mathfrak{A}$ then:

(i) $X \cup Y$ is in $\mathfrak{A}$;
(ii) X' is in $\mathfrak{A}$;
(iii) $X \cap Y$ is in $\mathfrak{A}$.

But of course this is not enough. The really pertinent question is: what are the chances that a particular outcome will occur? For example, what are the chances that a total not exceeding 5 will come up in a single throw of the two dice? In order to answer this question, recall that we assume the dice are "honest" in the sense that any particular pair (i, j) is as likely to come up as any other. This means we expect that in a large number of throws the relative frequency of occurrence of a particular pair (i, j) will be close to $\frac{1}{36}$. Thus we can set up a "measure" to be assigned to the elementary outcome that a particular pair (i, j) comes up on one throw. We can think of this measure in terms of a function p that assigns the number $\frac{1}{36}$ to the elementary outcome $\{(i, j)\}$, i.e.,

$$p(\{(i, j)\}) = \tfrac{1}{36}.$$

The set Z that corresponds to the event that a total not exceeding 5 comes up consists of 10 pairs:

$$Z = \{(1, 1), (1, 2), (2, 1), (1, 3), (3, 1), (2, 2), (1, 4), (4, 1), (2, 3), (3, 2)\}.$$

Our intuition tells us that in a large number of throws each of the pairs in Z will occur with a relative frequency of $\frac{1}{36}$. Thus it is sensible to expect that the relative frequency of occurrence of at least one of the 10 pairs in Z in a large number of throws will be $\frac{10}{36}$. In other words, a reasonable measure to assign to the outcome corresponding to Z is

$$p(Z) = \tfrac{10}{36}.$$

Now consider $X = \{(1, 3), (3, 1), (2, 2)\}$ and $Y = \{(1, 2), (2, 1)\}$. By similar arguments we assign measures to these outcomes: $p(X) = \frac{3}{36}$ and $p(Y) = \frac{2}{36}$. Moreover since $X \cup Y$ contains 5 elements, and $X \cap Y = \emptyset$, we assign the measure $\frac{5}{36}$ to $X \cup Y$, i.e.,

$$p(X \cup Y) = p(X) + p(Y).$$

What measure should be assigned to U, the entire universe? Obviously, the appropriate number is $\frac{36}{36} = 1$:

$$p(U) = 1.$$

These considerations lead us to the next two definitions.

Definition 6.1 ***Sigma field*** Let U be a universe and let $\mathfrak{A}$ be a family of subsets of U such that:

(i) the union of any sub-collection of sets in $\mathfrak{A}$ is also a set in $\mathfrak{A}$;
(ii) if X is a set in $\mathfrak{A}$, then its complement X' is also in $\mathfrak{A}$;
(iii) the intersection of any sub-collection of sets in $\mathfrak{A}$ is also a set in $\mathfrak{A}$.

Then $\mathfrak{A}$ is called a *sigma field* of subsets of U.

In our treatment of probability theory, we shall limit ourselves to a universe U consisting of only a finite number of elements and to the "trivial" sigma field $\mathfrak{A}$ consisting of *all* subsets of U. One might wonder, then, why the notion of a sigma field is pertinent. At the end of the next section we shall examine a situation in which finite sets will not suffice for the analysis and moreover, not all subsets of the universe U can be allowed as members of $\mathfrak{A}$.

Definition 6.2

Probability space Let U be a finite set and let $\mathfrak{A}$ be the sigma field of all subsets of U. Let $p\colon \mathfrak{A} \to R_+$ be a function, where R_+ denotes the set of nonnegative real numbers. If p satisfies

(i) $p(U) = 1$,

and

(ii) $p(X \cup Y) = p(X) + p(Y)$ for any two disjoint subsets X and Y of U,

then p is called a *probability measure* on U, and the space U together with p, i.e., the pair (U, p), is called a *finite probability space.*

Example 6.1

Suppose we toss a coin. Let $U = \{H, T\}$ be the set consisting of the two letters H (heads) and T (tails). Let $\mathfrak{A}$ be the sigma field of all subsets of U:

$$\mathfrak{A} = \{\emptyset, \{H\}, \{T\}, \{H, T\}\}.$$

Define a function $p\colon \mathfrak{A} \to R_+$ by $p(\emptyset) = 0$, $p(\{H\}) = \frac{1}{2}$, $p(\{T\}) = \frac{1}{2}$, $p(\{H, T\}) = 1$. Then p is a probability measure and the pair (U, p) is a finite probability space. For, $p(U) = p(\{H, T\}) = 1$, and $p(\{H\} \cup \{T\}) = p(\{H, T\}) = 1 = p(\{H\}) + p(\{T\})$. Except for the empty set, $\{H\}$ and $\{T\}$ are the only subsets that need be considered in verifying Definition 6.2 (ii). There is nothing sacred about the definition of this probability measure p. The one given above corresponds to our belief that for an "unbiased" coin, heads and tails are equally likely alternatives. If we were to discover after tossing the coin a large number of times that the relative frequency of occurrence of heads seemed to stabilize at $\frac{2}{3}$, then a more appropriate probability measure for an experiment involving this particular coin would be $p(\emptyset) = 0$, $p(\{H\}) = \frac{2}{3}$, $p(\{T\}) = \frac{1}{3}$, $p(\{H, T\}) = 1$.

The set U is sometimes referred to as a *sample space* and a subset in $\mathfrak{A}$ is called an *event.* If $x \in U$, then $\{x\}$, the set which consists of x, is called an *elementary event.* For finite probability spaces, any subset of U is an event.

If we are performing an experiment, the choice of an appropriate sample space U can vary depending on the kind of information we seek. We do require, though, that the sample space satisfy one fundamental criterion in relation to the experiment, namely, there must be a 1–1 correspondence between the elements of U and the possible elementary outcomes of the experiment.

Example 6.2 Two pennies are tossed. Let $U = \{0, 1\}$ where 0 corresponds to no heads appearing and 1 corresponds to at least one head appearing. This is surely a suitable sample space for an experiment in which we are interested only in the appearance or non-appearance of a head. That is, the correspondence

$$\text{(no head)} \leftrightarrow 0$$
$$\text{(at least one head)} \leftrightarrow 1$$

is a 1–1 correspondence between the sample space and the possible outcomes of the experiment. On the other hand, if we want to bet on the number of possible ways in which heads or tails can occur, then a more appropriate sample space is

$$U = \{(H, H), (H, T), (T, H), (T, T)\}.$$

Example 6.3 Three cookie jars contain two chocolate and no vanilla cookies, one chocolate and one vanilla cookie, and two vanilla and no chocolate cookies, respectively. Junior performs the following experiment. He reaches up and takes down a cookie jar. He then reaches into this cookie jar and takes a cookie without looking. If we are interested in analyzing the probability that he gets a chocolate (or a vanilla) cookie, what would an appropriate sample space be? To answer this question, suppose we label the jars 0, 1, and 2, respectively, where jar i contains i vanilla cookies and $2 - i$ chocolate cookies, $i = 0, 1, 2$. Consider the set

$$U = \{(0, c), (1, c), (1, v), (2, v)\}.$$

The following indicates a 1–1 correspondence between the possible elementary outcomes of this experiment and the sample space U:

(Junior takes down jar 0 and grabs a chocolate cookie) $\leftrightarrow (0, c)$,
(Junior takes down jar 1 and grabs a chocolate cookie) $\leftrightarrow (1, c)$,
(Junior takes down jar 1 and grabs a vanilla cookie) $\leftrightarrow (1, v)$,
(Junior takes down jar 2 and grabs a vanilla cookie) $\leftrightarrow (2, v)$.

(Why don't $(2, c)$ and $(0, v)$ appear?) This compound experiment can also be summarized by a helpful device known as a *tree diagram*.

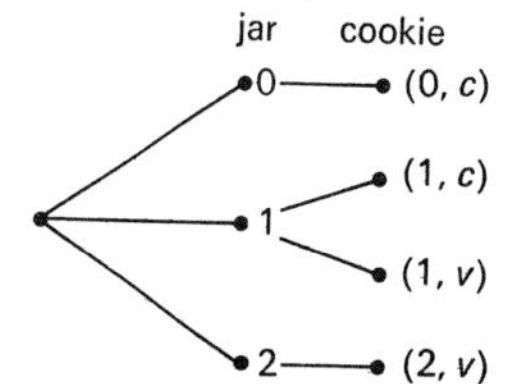

The initial branches of the tree to the left indicate the choices of jars and the secondary branches indicate the choices of cookies. Suppose we assume that the probabilities of the four possible outcomes are:

$$p(\{(0, c)\}) = \tfrac{1}{3},\ p(\{(1, c)\}) = \tfrac{1}{6},\ p(\{(1, v)\}) = \tfrac{1}{6},\ p(\{(2, v)\}) = \tfrac{1}{3}.$$

By Definition 6.2 (ii), the probability of the union of disjoint sets must be the sum of the individual probabilities, and the four sets $\{(0, c)\}$, $\{(1, c)\}$, $\{(1, v)\}$, $\{(2, v)\}$ are disjoint. Thus

$$\begin{aligned} p(\{(0, c), (1, c)\}) &= p(\{(0, c)\}) + p(\{(1, c)\}) \\ &= \tfrac{1}{3} + \tfrac{1}{6} \\ &= \tfrac{1}{2}. \end{aligned}$$

In other words, the probability that Junior will get a chocolate (or vanilla) cookie is $\frac{1}{2}$. Observe that the probability measure p does not have the same value for every one of the elementary events $\{(0, c)\}$, $\{(1, c)\}$, $\{(1, v)\}$, $\{(2, v)\}$. However the values of p are perfectly reasonable in terms of the possible outcomes of the experiment. For, once jar 0 has been chosen there are in fact two chocolate cookies available for Junior's grab, say c_1 and c_2. Both of these are lumped together in the single symbol $(0, c)$. We can further identify the separate cookies by: $(1, c_3)$, $(1, v_1)$, $(2, v_2)$, $(2, v_3)$. Thus

$$U_1 = \{(0, c_1), (0, c_2), (1, c_3), (1, v_1), (2, v_2), (2, v_3)\}$$

is a somewhat more refined sample space than U. Now if the jars and cookies are chosen at random, then any one of these 6 pairs is as likely to occur as any other, so that the assignment of a measure of $\frac{1}{6}$ to each elementary event consisting of one of these pairs is reasonable. But in our original sample space the pair $(0, c)$ accounts for both $(0, c_1)$ and $(0, c_2)$ and thus the value $\frac{1}{3} = \frac{1}{6} + \frac{1}{6}$ is assigned to $\{(0, c)\}$. Similar remarks can be made for each of the other elementary events in U. We did not start with U_1 because the question asked only for Junior's chances of getting a chocolate cookie and not his chances of getting any particular cookie.

Definition 6.3 ***Equi-probable space*** Let (U, p) be a finite probability space. Then (U, p) is an *equi-probable space* if the probability of every elementary event is the same. In other words, if x and y are any two elements of U then $p(\{x\}) = p(\{y\})$.

As we remarked before, the definition of the probability measure p on the space U can be arbitrary as long as the conditions of Definition 6.2, (i) and (ii), are satisfied. It should be emphasized that a space is equi-probable only by assumption. For example, in flipping a coin to decide who pays for the drinks, most friendly people would hesitate before demanding a long series of trials to estimate the relative frequencies of heads and tails; instead, they would assume the sample space to be equi-probable.

The main result in this section is the following sequence of elementary facts about finite probability spaces.

Theorem 6.1 *Let (U, p) be a finite probability space. Then:*

(a) $p(\emptyset) = 0$;

(b) *if X is an event, then*

$$p(X') = 1 - p(X);$$

(c) *if X and Y are events and $X \subset Y$, then*

$$p(X) \leq p(Y);$$

(d) *if $X_1, \ldots, X_r$ are pairwise disjoint events* (i.e., $X_i \cap X_j = \emptyset$ *for $i \neq j$*), *then*

$$p\left(\bigcup_{i=1}^{r} X_i\right) = \sum_{i=1}^{r} p(X_i);$$

(e) *if $X_1, \ldots, X_r$ are arbitrary events, then*

$$p\left(\bigcup_{i=1}^{r} X_i\right) \leq \sum_{i=1}^{r} p(X_i);$$

(f) *if X and Y are events, then*

$$p(X \cap Y') = p(X) - p(X \cap Y);$$

(g) *if $X = \{x_1, \ldots, x_r\}$ is an event, then*

$$p(X) = \sum_{i=1}^{r} p(\{x_i\});$$

(h) *if (U, p) is an equi-probable space and X is an event then*

$$p(X) = \frac{\nu(X)}{\nu(U)}.$$

Proof

(**a**) Since $p(U) = 1$, $U \cup \emptyset = U$, and $U \cap \emptyset = \emptyset$, we have

$$\begin{aligned} 1 &= p(U) \\ &= p(U \cup \emptyset) \\ &= p(U) + p(\emptyset) \\ &= 1 + p(\emptyset). \end{aligned}$$

Hence $p(\emptyset) = 0$.

(**b**) We know that X and X' are disjoint and that $X \cup X' = U$.

Hence

$$\begin{aligned} 1 &= p(U) \\ &= p(X \cup X') \\ &= p(X) + p(X'). \end{aligned}$$

Thus

$$p(X') = 1 - p(X).$$

(c) Since $X \subset Y$, we can write

$$Y = X \cup (X' \cap Y).$$

Moreover,

$$\begin{aligned} X \cap (X' \cap Y) &= (X \cap X') \cap Y \\ &= \emptyset. \end{aligned}$$

Hence X and $X' \cap Y$ are disjoint, and we may apply Definition 6.2 (ii) to obtain

$$\begin{aligned} p(Y) &= p(X \cup (X' \cap Y)) \\ &= p(X) + p(X' \cap Y). \end{aligned} \tag{2}$$

Now,

$$p(X' \cap Y) \geq 0$$

and hence from (2) we conclude that

$$p(X) \leq p(Y).$$

(d) We know from Definition 6.2 that in case $r = 2$, the required equality holds. The proof of the theorem for arbitrary r is a very easy induction. For, write

$$\bigcup_{i=1}^{r} X_i = X_1 \cup \bigcup_{i=2}^{r} X_i;$$

the fact that the X_i are pairwise disjoint implies that

$$\begin{aligned} X_1 \cap \bigcup_{i=2}^{r} X_i &= \bigcup_{i=2}^{r} (X_1 \cap X_i) \\ &= \bigcup_{i=2}^{r} \emptyset \\ &= \emptyset. \end{aligned}$$

Hence

$$\begin{aligned} p\left(\bigcup_{i=1}^{r} X_i\right) &= p\left(X_1 \cup \bigcup_{i=2}^{r} X_i\right) \\ &= p(X_1) + p\left(\bigcup_{i=2}^{r} X_i\right). \end{aligned} \tag{3}$$

Now, the induction assumption states that for $r - 1$ pairwise disjoint events,

$$p\left(\bigcup_{i=2}^{r} X_i\right) = \sum_{i=2}^{r} p(X_i).$$

Combining this with (3) establishes the required equality for r events.

(e) Let X_1 and X_2 be arbitrary events. Then

$$X_1 \cup X_2 = (X_1 \cap X_2') \cup X_2$$

and

$$(X_1 \cap X_2') \cap X_2 = \emptyset.$$

Hence, by Definition 6.2 (ii) and part (c) of this theorem,

$$\begin{aligned} p(X_1 \cup X_2) &= p(X_1 \cap X_2') + p(X_2) \\ &\leq p(X_1) + p(X_2), \end{aligned} \tag{4}$$

since $X_1 \cap X_2' \subset X_1$. This proves (e) for $r = 2$. The general case is proved by induction in the same way as part (d) and is left as an exercise.

(f) We write

$$\begin{aligned} X &= X \cap U \\ &= X \cap (Y \cup Y') \\ &= (X \cap Y) \cup (X \cap Y'). \end{aligned}$$

Moreover, $X \cap Y$ and $X \cap Y'$ are disjoint. Hence

$$p(X) = p(X \cap Y) + p(X \cap Y'),$$

the required equality.

(g) Observe that X is the union of the pairwise disjoint sets $\{x_i\}$, $i = 1, \ldots, r$, and hence (g) follows immediately by applying part (d).

(h) Let $U = \{x_1, \ldots, x_n\}$ and let $X = \{x_1, \ldots, x_r\}$, $r \leq n$, so that $x_{r+1}, \ldots, x_n$ are the elements in U which are not in X. By applying (g) to U, we have

$$\begin{aligned} 1 &= p(U) \\ &= \sum_{i=1}^{n} p(\{x_i\}). \end{aligned} \tag{5}$$

By the definition of an equi-probable space, $p(\{x_i\})$ is the same for each i, $i = 1, \ldots, n$, and hence from (5),

$$p(\{x_i\}) = \frac{1}{n}\cdot$$

Similarly, by again applying (g), we have

$$\begin{aligned} p(X) &= p\left(\bigcup_{i=1}^{r} \{x_i\}\right) \\ &= \sum_{i=1}^{r} p(\{x_i\}) \\ &= \sum_{i=1}^{r} \frac{1}{n} \\ &= \overbrace{\frac{1}{n} + \cdots + \frac{1}{n}}^{r} \\ &= \frac{r}{n} \\ &= \frac{\nu(X)}{\nu(U)}. \quad \blacksquare \end{aligned}$$

Example 6.4 In a group of 5 people, what is the probability that at least two of them were born on the same day of the year? (For simplicity, we assume every year has 365 days.) To analyze this problem, we let the sample space U be the $(365)^5$ 5-tuples of dates, $(d_1, d_2, d_3, d_4, d_5)$, and we assume that the space is equi-probable, i.e.,

$$p(\{(d_1, d_2, d_3, d_4, d_5)\}) = (\tfrac{1}{365})^5,$$

whatever the choice of d_i, $i = 1, \ldots, 5$. In other words, we assume that five people are equally likely to be born on any choice of 5 days. The event X which interests us is the totality of 5-tuples in which $d_i = d_j$ for at least one pair $i \neq j$, i.e.,

$$X = \{(d_1, \ldots, d_5) \mid \exists_{i,j} (i \neq j \wedge d_i = d_j)\}.$$

Rather than compute $p(X)$, we compute $p(X')$ and then use Theorem 6.1 (b). The set X' is the totality of 5-tuples $(d_1, \ldots, d_5)$ in which the d_i, $i = 1, \ldots, 5$, are all different. Counting the total number of such 5-tuples is equivalent to counting the number of 5-permutations of a 365-element set. According to Theorem 4.2 (b), this number is

$$365 \cdot 364 \cdot 363 \cdot 362 \cdot 361.$$

Thus by Theorem 6.1 (h),

$$p(X') = \frac{365 \cdot 364 \cdot 363 \cdot 362 \cdot 361}{(365)^5},$$

and applying Theorem 6.1 (b), we know that

$$\begin{aligned} p(X) &= 1 - p(X') \\ &= 1 - \frac{365 \cdot 364 \cdot 363 \cdot 362 \cdot 361}{(365)^5} \\ &= 1 - .973 \\ &= .027 \text{ (approximately).} \end{aligned}$$

The probability is rather small that at least two people have the same birth date in a group of five. The same argument will show, however, that in a group of twenty-three or more people, the probability of at least two having the same birthday exceeds $\frac{1}{2}$.

It is frequently the case that in considering two events X and Y, the knowledge of Y affects the probability of X.

Example 6.5 A pair of fair dice is rolled. As we have computed many times, the sample space U can be chosen to consist of the 36 ordered pairs (i, j), $1 \leq i \leq 6$, $1 \leq j \leq 6$. The assumption that the dice are fair is equivalent to saying that we have an equi-probable space (U, p), where

$$p(\{(i, j)\}) = \tfrac{1}{36}$$

for all i and j. We also know that 7 can come up in six ways and 11 in two ways. Thus, if X is the event consisting of all pairs (i, j) for which $i + j$ is 7 or 11, we can compute that

$$\begin{aligned} p(X) &= \frac{\nu(X)}{36} \\ &= \tfrac{8}{36} \\ &= \tfrac{2}{9}. \end{aligned}$$

So far we have said nothing new. But suppose we play a game in which it is required that we compute the probability that a given pair of dice comes up 7 or 11, having been told that the sum appearing on the dice is odd. Then we can no longer say that $p(X) = \frac{2}{9}$. In other words, if Y is the set of pairs (i, j) for which $i + j$ is odd, the knowledge of Y affects the probability of X. More precisely, the pertinent elementary events (in this case, pairs totaling 7 or 11) are no longer chosen from all possible pairs in U, but rather from those which are known to have an odd sum. There are 18 such pairs, i.e., $\nu(Y) = 18$. Thus it is reasonable to say that the probability that the dice show 7 or 11, knowing in advance that the sum is odd, is

$$\frac{8}{18} = \frac{\nu(X \cap Y)}{\nu(Y)} = \frac{\dfrac{\nu(X \cap Y)}{36}}{\dfrac{\nu(Y)}{36}} = \frac{p(X \cap Y)}{p(Y)}.$$

The preceding example suggests the following definition.

Definition 6.4 ***Conditional probability*** Let (U, p) be a finite probability space and let X and Y be events for which $p(Y) \neq 0$. Then the *conditional probability of X, given Y*, denoted by $p(X \mid Y)$, is defined by

$$p(X \mid Y) = \frac{p(X \cap Y)}{p(Y)}.$$

Example 6.6 There are three cookie jars containing chocolate and vanilla cookies. Jar I contains 3 chocolate and 4 vanilla cookies; jar II contains 2 chocolate and 3 vanilla cookies; and jar III contains 5 chocolate and 2 vanilla cookies. Although Nancy is on a diet, she steals a cookie at midnight. After biting into it, she discovers it is chocolate. What is the probability that it came from jar II? We define the sample space U as follows: U consists of the pairs

$$\begin{array}{ll} (\text{I}, c_i), & i = 1, 2, 3; \\ (\text{I}, v_i), & i = 1, 2, 3, 4; \\ (\text{II}, c_i), & i = 4, 5; \\ (\text{II}, v_i), & i = 5, 6, 7; \\ (\text{III}, c_i), & i = 6, 7, 8, 9, 10; \\ (\text{III}, v_i), & i = 8, 9; \end{array}$$

the first member of the pair indicates the jar, and the c_i and v_i indicate the 10 chocolate and 9 vanilla cookies. Let the event Y denote the set of all pairs in U in which the second element of the pair is a c_i and let X denote the set of all pairs in which the first element of the pair is a II. We are interested in the probability that the cookie came from jar II, knowing that it is chocolate, i.e., we want to compute $p(X \mid Y)$. According to Definition 6.4,

$$p(X \mid Y) = \frac{p(X \cap Y)}{p(Y)}.$$

In order to compute $p(X \mid Y)$, we must define the probability measure p. We assign probabilities to the elementary events in U based on the following argument. Nancy is equally likely to steal a cookie out of any of the three jars. However, once she has her hand in a jar, the probability of choosing a cookie of any given flavor depends on the number of chocolate and vanilla cookies in that jar. She can choose jar I with probability $\frac{1}{3}$, and once her hand is in jar I, she can choose a chocolate cookie with probability $\frac{3}{7}$ and a vanilla cookie with probability $\frac{4}{7}$. Thus it is reasonable to assign the probability $\frac{1}{3} \cdot \frac{1}{7}$ to each of the elementary events $\{(\mathrm{I}, c_i)\}$, $i = 1, 2, 3$, and to assign the probability $\frac{1}{3} \cdot \frac{1}{7}$ to each of the elementary events $\{(\mathrm{I}, v_i)\}$, $i = 1, 2, 3, 4$. Similarly, we define

$$\begin{aligned} p(\{(\mathrm{II}, c_i)\}) &= \tfrac{1}{3} \cdot \tfrac{1}{5}, & i &= 4, 5; \\ p(\{(\mathrm{II}, v_i)\}) &= \tfrac{1}{3} \cdot \tfrac{1}{5}, & i &= 5, 6, 7; \\ p(\{(\mathrm{III}, c_i)\}) &= \tfrac{1}{3} \cdot \tfrac{1}{7}, & i &= 6, 7, 8, 9, 10; \\ p(\{(\mathrm{III}, v_i)\}) &= \tfrac{1}{3} \cdot \tfrac{1}{7}, & i &= 8, 9. \end{aligned}$$

We compute immediately that

$$\begin{aligned} p(Y) &= 3(\tfrac{1}{3} \cdot \tfrac{1}{7}) + 2(\tfrac{1}{3} \cdot \tfrac{1}{5}) + 5(\tfrac{1}{3} \cdot \tfrac{1}{7}) \\ &= \tfrac{18}{35}. \end{aligned}$$

The event $X \cap Y$ is the set of pairs in which the first element of the pair is a II and the second is a c_i, i.e.,

$$X \cap Y = \{(\mathrm{II}, c_4), (\mathrm{II}, c_5)\}.$$

It follows that

$$\begin{aligned} p(X \cap Y) &= 2(\tfrac{1}{3} \cdot \tfrac{1}{5}) \\ &= \tfrac{2}{15}. \end{aligned}$$

Hence

$$\begin{aligned} p(X \mid Y) &= \frac{\frac{2}{15}}{\frac{18}{35}} \\ &= \tfrac{7}{27}. \end{aligned}$$

Example 6.7 Three fair dice are thrown and a different number appears on each die. What is the probability that the sum is 6? Here the sample space U is the set of 216 triples (i, j, k), $i = 1, \ldots, 6$, $j = 1, \ldots, 6$, $k = 1, \ldots, 6$. Let Y denote the event consisting of all triples (i, j, k) in which i, j, and k are distinct, and let X denote the event consisting of all triples (i, j, k) for which $i + j + k = 6$. Now we must compute

$$p(X \mid Y) = \frac{p(X \cap Y)}{p(Y)}.$$

The set $X \cap Y$ is the totality of triples (i, j, k) in which i, j, and k

are distinct and $i + j + k = 6$. These are

$$(1, 2, 3), (1, 3, 2), (2, 1, 3), (2, 3, 1), (3, 1, 2), (3, 2, 1);$$

hence $\nu(X \cap Y) = 6$. On the other hand, $\nu(Y) = 120$ since Y, in fact, consists of all 3-permutations of a 6-element set. Thus

$$\begin{aligned} p(X \mid Y) &= \frac{\dfrac{\nu(X \cap Y)}{216}}{\dfrac{\nu(Y)}{216}} \\ &= \frac{\frac{6}{216}}{\frac{120}{216}} \\ &= \tfrac{1}{20}. \end{aligned}$$

The following result concerning conditional probability is easy to prove by mathematical induction. The proof will be followed by an example of the use of the theorem and then we will go on to devise two formulas which will equip us with the tools to analyze a wide variety of interesting problems.

Theorem 6.2 ***Multiplication Theorem*** Let $X_1, X_2, \ldots, X_n$ *be events in a finite probability space* (U, p). *Then*

$$(6) \qquad \begin{aligned} p\left(\bigcap_{i=1}^{n} X_i\right) &= p(X_1)p(X_2 \mid X_1)p(X_3 \mid X_1 \cap X_2) \cdots \\ &\qquad p(X_n \mid X_1 \cap X_2 \cap \cdots \cap X_{n-1}) \\ &= \prod_{i=1}^{n} p\left(X_i \,\middle|\, \bigcap_{j=1}^{i-1} X_j\right), \end{aligned}$$

assuming that each conditional probability is defined.

Proof We initially remark that the first factor in the product (6) is just $p(X_1)$. (The symbol $\bigcap_{j=1}^{0} X_j$ does not make sense without some explanation.) We prove (6) by induction. If $n = 2$, (6) becomes

$$p(X_1 \cap X_2) = p(X_1)p(X_2 \mid X_1),$$

which follows immediately from the definition of $p(X_2 \mid X_1)$. Now, assume that the formula (6) holds for k events, $X_1, \ldots, X_k$, i.e.,

$$(7) \qquad p\left(\bigcap_{i=1}^{k} X_i\right) = \prod_{i=1}^{k} p\left(X_i \,\middle|\, \bigcap_{j=1}^{i-1} X_j\right).$$

Multiply both sides of (7) by

$$p\left(X_{k+1} \,\middle|\, \bigcap_{j=1}^{k} X_j\right),$$

to obtain

$$p\left(X_{k+1} \,\middle|\, \bigcap_{j=1}^{k} X_j\right) p\left(\bigcap_{i=1}^{k} X_i\right) = \prod_{i=1}^{k+1} p\left(X_i \,\middle|\, \bigcap_{j=1}^{i-1} X_j\right). \tag{8}$$

If we let $Y = \bigcap_{j=1}^{k} X_j$, then the left side of (8) is

$$\begin{aligned} p(X_{k+1} \mid Y)p(Y) &= p(X_{k+1} \cap Y) \\ &= p\left(X_{k+1} \cap \bigcap_{j=1}^{k} X_j\right) \\ &= p\left(\bigcap_{j=1}^{k+1} X_j\right). \end{aligned} \tag{9}$$

Combining (8) and (9), we see that the equality (6) holds for $k + 1$ events, $X_1, \ldots, X_{k+1}$. Therefore, assuming that (6) holds for $n = k$, we have proved that it holds for $n = k + 1$. ▌

Theorem 6.2 is called the multiplication theorem for fairly obvious reasons. Heuristically, Theorem 6.2 can be stated as follows: the probability that events $X_1, X_2, \ldots, X_n$ all happen is equal to the probability that X_1 happens, times the probability that X_2 happens given X_1, times the probability that X_3 happens given both X_1 and X_2, etc.

We next state and prove two important consequences of Theorem 6.1.

Theorem 6.3 *Let (U, p) be a finite probability space, and let $X_1, \ldots, X_n$ be a collection of pairwise disjoint sets whose union is U, $U = \bigcup_{i=1}^{n} X_i$. If X is any event, then*

$$p(X) = \sum_{i=1}^{n} p(X_i)p(X \mid X_i). \tag{10}$$

Moreover,

$$p(X_i \mid X) = \frac{p(X_i)p(X \mid X_i)}{\sum_{j=1}^{n} p(X_j)p(X \mid X_j)}. \tag{11}$$

We assume that all conditional probabilities that appear are defined.

(The formula (11) is called Bayes' Theorem, after the Reverend Thomas Bayes, and was published posthumously in 1763.)

Proof To prove (10), first observe that

$$\begin{aligned} X &= X \cap U \\ &= X \cap \bigcup_{i=1}^{n} X_i \\ &= \bigcup_{i=1}^{n} X \cap X_i, \end{aligned}$$

and $(X \cap X_i) \cap (X \cap X_j) = X \cap (X_i \cap X_j) = \emptyset$ whenever $i \neq j$, i.e., the sets $X \cap X_i$ are pairwise disjoint and their union is X. Hence by Theorem 6.1 (d),

$$\begin{aligned} p(X) &= p\left(\bigcup_{i=1}^{n} (X \cap X_i)\right) \\ &= \sum_{i=1}^{n} p(X \cap X_i). \qquad (12) \end{aligned}$$

However, by Definition 6.4,

$$p(X \cap X_i) = p(X_i)p(X \mid X_i), \qquad i = 1, \ldots, n,$$

and substituting this in (12) we have

$$p(X) = \sum_{i=1}^{n} p(X_i)p(X \mid X_i),$$

the required formula (10). To prove (11), we observe that

$$\begin{aligned} p(X_i \mid X) &= \frac{p(X_i \cap X)}{p(X)} \\ &= \frac{p(X_i)p(X \mid X_i)}{\sum_{j=1}^{n} p(X_j)p(X \mid X_j)}, \end{aligned}$$

where we have used (10) to replace $p(X)$. ▌

Example 6.8 Suppose that there are n automobile factories in the United States, and the ith factory produces a total of a_i autos in a given year. Assume that d_i of the autos produced by the ith factory have defects which constitute safety hazards. Moreover, assume that in purchasing a car, no irrelevant prejudices are involved (e.g., it is not necessarily the case that there is a Ford in your future). What is the probability of obtaining a defective car in such a random purchase? Let U be the totality of cars produced, so that U has $\sum_{i=1}^{n} a_i = a$ elements. Let X_i be the set of a_i cars produced by the ith factory. Let X be the event consisting of $\sum_{i=1}^{n} d_i = d$ defective cars. Observe first that

$$\begin{aligned} p(X \mid X_i) &= \frac{p(X \cap X_i)}{p(X_i)} \\ &= \frac{\nu(X \cap X_i)}{\nu(X_i)} \\ &= \frac{d_i}{a_i}, \qquad i = 1, \ldots, n. \end{aligned}$$

In other words, the probability that a car is defective, given the knowledge that it came from the ith factory, is $\frac{d_i}{a_i}$. On the other hand,

$$\begin{aligned} p(X_i) &= \frac{\nu(X_i)}{\nu(U)} \\ &= \frac{a_i}{a}. \end{aligned}$$

Substituting these quantities in (10), we have

$$p(X) = \sum_{i=1}^{n} \left(\frac{a_i}{a}\right)\left(\frac{d_i}{a_i}\right). \tag{13}$$

Of course, (13) simplifies immediately to $\frac{d}{a}$, namely, the ratio of the total number of defective cars produced to the total number of cars produced. But the numbers $p(X_i) = \frac{a_i}{a}$ and $p(X \mid X_i) = \frac{d_i}{a_i}$ can be estimated by sampling without knowing the numbers d_i or a_i. This means that we can take a sample of buyers from the public and tabulate the numbers of each of the brands of cars purchased as well as the numbers of defective cars of each make. This would allow us to estimate $p(X_i)$ and $p(X \mid X_i)$, provided the sample is of significant size. This is a realistic consideration, since it is unlikely that an automobile manufacturer will willingly release information either about the number of cars produced or the number of defective cars produced.

Example 6.9 An animal house in Australia maintains a population of three distinct strains of rabbits for use in experiments devoted to eradicating the rabbit population. Myxamatosis is introduced into a sample of laboratory rabbits with the following effects:

80% of strain I die after exposure;
90% of strain II die after exposure;
86% of strain III die after exposure.

Further, 30% of the rabbits in the sample come from strain I and 40% from strain II. If a rabbit that died after exposure is chosen at random, what is the probability that it came from strain I? Let U be the totality of laboratory rabbits and let X_1, X_2, and X_3 denote strains I, II, and III, respectively. Let X denote the set of rabbits that die after exposure to the disease. We want to compute the probability that a rabbit that died after exposure came from strain I, i.e., $p(X_1 \mid X)$. According to Bayes' Theorem, (11),

$$p(X_1 \mid X) = \frac{p(X_1)p(X \mid X_1)}{p(X_1)p(X \mid X_1) + p(X_2)p(X \mid X_2) + p(X_3)p(X \mid X_3)}.$$

The number $p(X_1)$ is just the probability that a rabbit in U is of strain I, while $p(X \mid X_1)$ is the probability that a rabbit succumbed, given that he came from strain I. The given information states that

$$\begin{aligned} p(X_1) &= 30\% \\ &= .3, \end{aligned}$$

and

$$\begin{aligned} p(X \mid X_1) &= 80\% \\ &= .8. \end{aligned}$$

Similarly,

$$\begin{aligned} p(X_2) &= 40\% \\ &= .4, \end{aligned}$$

$$\begin{aligned} p(X \mid X_2) &= 90\% \\ &= .9, \end{aligned}$$

$$\begin{aligned} p(X_3) &= 100\% - (30\% + 40\%) \\ &= 30\% \\ &= .3, \end{aligned}$$

and

$$\begin{aligned} p(X \mid X_3) &= 86\% \\ &= .86. \end{aligned}$$

Substituting, we have

$$\begin{aligned} p(X_1 \mid X) &= \frac{(.3)(.8)}{(.3)(.8) + (.4)(.9) + (.3)(.86)} \\ &= \frac{.24}{.858} \\ &= .2797. \end{aligned}$$

Quiz

Answer true or false.

1 In an experiment with a finite number of outcomes, the sum of the relative frequencies of the different outcomes is strictly less than 1.

2 If $\mathfrak{A}$ is a sigma field of a finite universe U, $\nu(U) = n$, then $\nu(\mathfrak{A}) = 2^n$.

3 The empty set $\emptyset$ is an elementary event in any finite probability space.

4 The fundamental criterion that must be satisfied for a sample space of an experiment is that there must be a 1–1 correspondence between the elements of U and the possible elementary outcomes of the experiment.

5 If X and Y are events in a finite probability space (U, p), then $p(X \cap Y) = p(X) - p(X \cap Y')$.

6 The conditional probability of X given Y, $p(X \mid Y)$, is defined for any non-empty event Y.

7 If X_1, X_2, and X_3 are events in a finite probability space (U, p), then

$$p(X_1 \cap X_2 \cap X_3) = p(X_1)p(X_2 \mid X_1)p(X_3 \mid X_1 \cap X_2),$$

assuming that each conditional probability is defined.

8 If X and Y are events and $p(X)p(Y) \neq 0$, then both $p(X \mid Y)$ and $p(Y \mid X)$ are defined.

9 If X is any event in the finite probability space (U, p) then $p(X)^2 \leq p(X)$.

10 If (U, p) is a finite probability space and q is defined by $q(X) = 1 - p(X)$ for any $X \subset U$, then (U, q) is a finite probability space.

Exercises

1 In each of the following experiments, decide upon an appropriate probability space and probability measure.

(a) A fair coin is tossed twice.
(b) A two-headed coin and a fair coin are tossed three times.
(c) A pair of fair dice is rolled 5 times.
(d) In the draw of two cards from a deck, it is required to find the probability of obtaining two face cards.
(e) A committee of 4 people is chosen at random from a group of 5 men and 4 women and it is required to find the probability that the committee is equally balanced between the sexes.

2 In each of the following questions, set up an appropriate finite probability space, (U, p), and solve the problem.

(a) What is the probability that a random day of the week has more than 6 letters in its spelling?
(b) What is the probability that a random card drawn from a standard deck of 52 cards is a face card?
(c) Two cards are drawn at random from a standard deck of 52 cards. What is the probability that they are both of the same suit?
(d) An urn contains 2 red, 2 white, and 4 black marbles. If a marble is drawn at random, what is the probability that it is black or white?

3 Following the method used in Example 6.4, find the probability that in a group of 5 people, at least 2 of them were born in the same week of the year. (Assume that the year has 52 weeks.)

***4** In a group of five people, find the probability that at least two people were born in the same month.

5 With reference to Definition 6.1, verify that the family of all subsets of the universe U satisfies the definition of a sigma field.

6 Using the de Morgan formulas (Theorem 2.1), show that condition (iii) of Definition 6.1 is superfluous if conditions (i) and (ii) are assumed. Similarly, show that (i) follows from (ii) and (iii).

7 Illustrate Theorem 6.1, parts (b), (c), (d), and (f), by means of appropriate Venn diagrams. Take $r = 3$ in (d).

8 The four faces of an equilateral pyramidal die are numbered 1, 2, 3, and 4. (Recall that an equilateral pyramid has 4 faces, each in the shape of an equilateral triangle.)

(a) For an experiment consisting of one roll of two such fair dice, define an appropriate probability space and a probability measure. The score is the number on the face upon which the die rests.

(b) Given that an odd sum appears on the two dice, what is the probability that it is a five?

9 Five students are assigned five numbered seats. Before knowing the numbers of their seats, they all enter the classroom and distribute themselves at random in the five places.

(a) Find the probability that each student takes his assigned seat.

(b) Find the probability that no student takes his assigned seat.

(c) Find the probability that at least one student takes his assigned seat.

(d) Find the probability that exactly two students take their assigned seats.

†10 Let (U, p) be a finite probability space and let X and Y be two events. Show that

$$p(X \cup Y) = p(X) + p(Y) - p(X \cap Y).$$

11 A pair of fair dice is rolled and we are told that the total appearing on the dice is even. Knowing this, what is the probability that the total exceeds 5?

12 A die is weighted so that the even faces appear twice as often as the odd faces. The die is rolled twice and it is known that the sum of the numbers appearing in the 2 throws is at least 7. What is the probability that it is precisely 7?

13 Three fair dice are thrown and the same number appears on each die. What is the probability that the sum is not 6? (Hint: See Example 6.7.)

14 A box contains 12 light bulbs of which 2 are known to be defective. Four light bulbs are selected in succession from the box. What is the probability that none of the 4 is defective? (Hint: This is an application of Theorem 6.2.)

15 It is known that 70% of the men and 90% of the women are overweight in a certain ethnic group. Moreover, 55% of the population of this group are women. If a person is selected at random from the group and is found to be overweight, what is the probability that this person is a woman? (Hint: This is an example of Theorem 6.3, in particular, formula (11).)

16 In a seminar consisting of 5 students, $s_1, \ldots, s_5$, it is known that in compiling a bibliography for a joint research project, s_1 does 30% of the work, s_2 and s_3 each do 25% of the work, and s_4 and s_5 each do 10% of the work. It is also known that s_1 makes mistakes 20% of the time, s_2 and s_3 make mistakes 15% of the time, and s_4 and s_5 each make mistakes 5% of the time. If an entry from the bibliography is chosen at random, what is the probability that it is wrong? (Hint: Apply formula (10), Theorem 6.3.)

17 Three cards are selected in succession from a standard deck of 52 cards. What is the probability that all three are face cards? (Hint: Use Theorem 6.2.)

†18 Let X and X_1 be events in a finite probability space (U, p). Assume that $p(X_1)p(X \mid X_1) + p(X_1')p(X \mid X_1') > 0$. Show that

$$p(X_1 \mid X) = \frac{p(X_1)p(X \mid X_1)}{p(X_1)p(X \mid X_1) + p(X_1')p(X \mid X_1')}.$$

(Hint: Take $n = 2$ and $U = X_1 \cup X_1'$ in Theorem 6.3, Formula (11).)

19 A recently devised test for the detection of cancer has been found to have the following reliability. The test detects 75% of those people who have cancer and does not detect the disease in 25% of this group. Among those people who do not have cancer, the test detects 85% as not having cancer, but erroneously detects 15% of this group as having the disease. It is known that in a large sample of the population, 2% have cancer. Suppose that a random individual is given the test for cancer and registers as having the disease. What is the probability that he actually has cancer? (Hint: Let X_1 denote the event consisting of those people having cancer, and let X denote the event consisting of those people whom the test shows as having the disease. We are given that $p(X_1) = .02$, $p(X_1') = .98$, $p(X \mid X_1) = .75$, $p(X \mid X_1') = .15$. Apply the result of the preceding exercise.)

20 An automobile rental company orders three cars from a lot of five cars of the following colors: red, green, blue, yellow, black. What is the probability that the company will receive red and green cars; blue and yellow cars; or blue, yellow, and black cars? (Hint: Let X_1 be the set of 3-element subsets of the set of five cars which contain a red and a green car; let X_2 be the set of 3-element subsets of the set of five cars which contain a blue and a yellow car; let X_3 be the set of 3-element subsets of the set of five cars which contain a blue, a yellow, and a black car. Of course, X_3 is a 1-element subset. The conditions of the problem imply that

$X_1 \cap (X_2 \cup X_3) = \emptyset$. Use the equi-probable measure and the result of Exercise 10 to compute

$$\begin{aligned} p(X_1 \cup X_2 \cup X_3) &= p(X_1 \cup (X_2 \cup X_3)) \\ &= p(X_1) + p(X_2 \cup X_3) \\ &= p(X_1) + p(X_2) + p(X_3) - p(X_2 \cap X_3) \\ &= \frac{1}{\binom{5}{3}} [\nu(X_1) + \nu(X_2) + \nu(X_3) - \nu(X_2 \cap X_3)]. \end{aligned}$$
)

21 In a sample of people it is found that 40% of the men and 75% of the women are under 5′5″ tall. Assume that 50% of the sample are women. If a person is selected at random, and this person is under 5′5″ tall, what is the probability that this person is a woman? (Hint: let X_1 be the set of women in the sample, X_2 the set of men in the sample, and X the set of people under 5′5″ tall in the sample. Then $p(X_1 \mid X)$ is the probability that the person chosen is a woman, given that the person is under 5′5″. The probability $p(X \mid X_1)$ is the probability that the person is under 5′5″, given that the person is a woman, and $p(X \mid X_2)$ is the probability that the person is under 5′5″, given that the person is a man. The conditions of the problem state that

$$p(X_1) = .5,\ p(X_2) = .5,\ p(X \mid X_1) = .75,\ p(X \mid X_2) = .4.$$

Bayes' Theorem states that

$$p(X_1 \mid X) = \frac{p(X_1)p(X \mid X_1)}{p(X_1)p(X \mid X_1) + p(X_2)p(X \mid X_2)}.$$
)

***22** A grocery chain purchases crates each containing ten boxes of a dozen eggs from an egg supplier. The chain has found over a long period of time that 80% of such crates contain all fresh eggs, 15% contain one box with rotten eggs in it, and 5% contain two boxes of rotten eggs. A housewife selects a box of eggs from a crate and finds that it contains rotten eggs. What is the probability that the crate contains two boxes of rotten eggs? (Hint: Let X_1 be the set of all boxes of eggs coming from crates containing all fresh eggs, X_2 the set of all boxes of eggs which come from crates containing one box of rotten eggs, and X_3 the set of all boxes of eggs which come from crates containing two boxes of rotten eggs. Let X be the totality of boxes of rotten eggs. We are given that $p(X_1) = .8$, $p(X_2) = .15$, and $p(X_3) = .05$. Now $p(X \mid X_1)$ is the probability that the housewife obtained a box of rotten eggs from a crate of all good eggs, i.e., $p(X \mid X_1) = 0$. Also, $p(X \mid X_2)$ is the probability that the housewife obtained a box of rotten eggs from a crate containing one box of rotten eggs, i.e., $p(X \mid X_2) = .1$. Similarly, $p(X \mid X_3) = \frac{2}{10} = .2$. Then $p(X_3 \mid X)$ is the probability that the housewife obtained her box of eggs from a crate containing two boxes of rotten eggs, given that the box she chose was rotten. From Bayes' Theorem, we have

$$p(X_3 \mid X) = \frac{p(X_3)p(X \mid X_3)}{p(X_1)p(X \mid X_1) + p(X_2)p(X \mid X_2) + p(X_3)p(X \mid X_3)}.$$
)

†**23** Complete the induction argument indicated in Theorem 6.1(e).

24 In a computer laboratory, 60% of the computation is done by computer I, 25% is done by computer B, and 15% by computer M. These machines make computational errors in 6%, 7%, and 10% of their computations, respectively. What is the probability that any random computation is wrong? (Hint: This is an instance of the multiplication theorem.)

1.7

Finite Stochastic Processes

A sequence of experiments, finite in length, in which each stage has a finite number of outcomes is called a *finite stochastic process*. The word "stochastic" is derived from the Greek word "stochos" meaning "guess". It is often not at all clear what the probability space and probability measure should be for analyzing any particular stochastic process. Recall that in Example 6.3 we introduced a tree diagram to describe Junior's cookie proclivities. Such diagrams are sometimes helpful in setting up an appropriate probability space and measure. Our first example is a typical case in point.

Example 7.1 Jars I and II contain black (b) and white (w) marbles as follows:

Jar I —3 white, 4 black;
Jar II—2 white, 2 black.

The following sequence of experiments is performed. A jar is chosen at random. Then a marble is selected from it at random and placed in the other jar. Finally a marble is selected at random from this latter jar. What is the probability that the two marbles taken from the jars are of opposite colors? We construct a tree to indicate the initial part of this experiment.

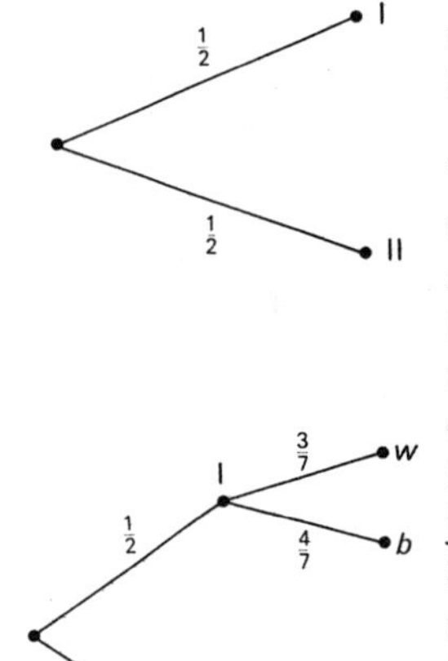

The number $\frac{1}{2}$ that appears on each branch is a translation of the statement that a jar is chosen "at random", i.e., the number $\frac{1}{2}$ represents the probability of picking jar I or jar II. To construct the tree representation of the second stage of the experiment, we must account for all possibilities. Suppose then that jar II has been chosen. Then either a white marble or a black marble may be selected from it to be placed in jar I. There are 2 black and 2 white marbles in jar II and hence the probability of selecting either w or b is $\frac{1}{2}$. Similarly, there are 7 marbles in jar I, so the probability of selecting a white marble is $\frac{3}{7}$ and the probability of selecting a black marble is $\frac{4}{7}$. The tree grows four new branches.

The selection at this stage will affect the outcome of the next stage of the experiment, namely, the probability of obtaining a marble of a given color from the other jar. For example, suppose we are on the top branch of the preceding tree, i.e., a white marble has been chosen from jar I. According to the description of the experiment, this marble is now placed in jar II and then a marble is randomly selected from jar II. Putting w in jar II results in jar II having $3w$ and $2b$. Thus the probability of obtaining a w is $\frac{3}{5}$ and the probability of obtaining a b is $\frac{2}{5}$. We can add two branches to the top branch of the preceding tree.

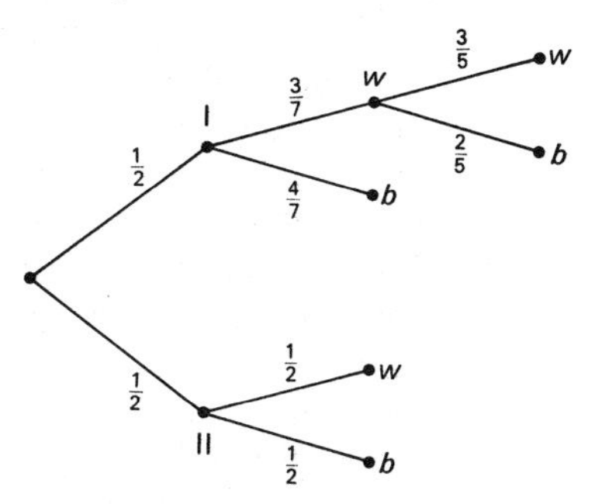

We can complete each of the remaining three branches by precisely the same kind of analysis. For example, if we are at the lowest branch at the second stage, it means we have chosen a black marble from jar II. We are to place it in jar I and then select a marble from jar I at random. But then jar I will have $3w$ and $5b$, so that the probability of selecting a w is $\frac{3}{8}$ and the probability of selecting a b is $\frac{5}{8}$. Continuing in this way, we complete the tree.

(1)

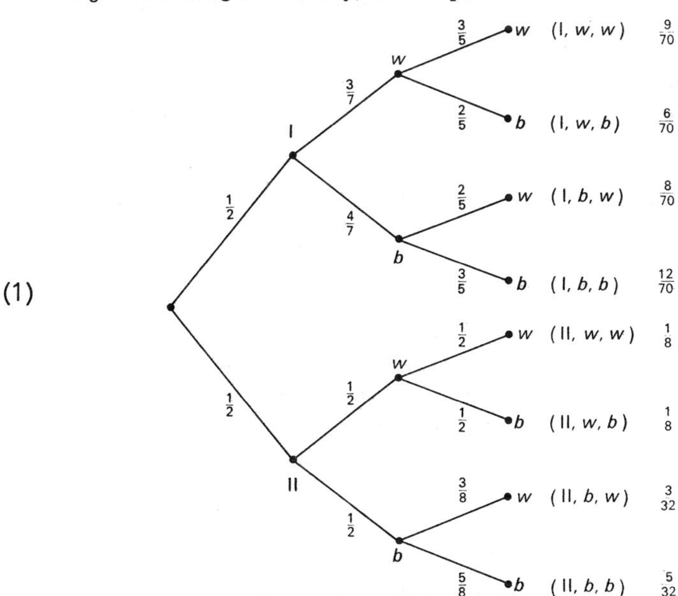

We can now define the sample space for this experiment and, using Theorem 6.2, we can determine an appropriate probability measure p. The sample space consists of all ordered triples of items in which the first item of the triple is the number of the jar chosen at the first stage, the second item is the color of the marble selected at the second stage, and the third item is the color of the marble at the third stage. These triples may be determined directly from diagram (1) and are written immediately to the right of each of the final branches. This is certainly a reasonable choice for a sample space because it takes into account all possible outcomes at each stage of the experiment. Now, the next problem is to define the probability measure on this space of triples and this can be done as follows. The probability of obtaining the triple (I, w, w) is $(\frac{1}{2}) \cdot (\frac{3}{7}) \cdot (\frac{3}{5})$. For, the probability of obtaining jar I is $\frac{1}{2}$; the probability of choosing w, given that jar I has been selected, is $\frac{3}{7}$; the probability of then selecting a white marble from jar II after putting the chosen white marble in jar II is $\frac{3}{5}$. Thus, by Theorem 6.2, the probability to be assigned to (I, w, w) is $\frac{9}{70}$. By similarly applying Theorem 6.2 to each of the remaining branches, we obtain the set of numbers indicated on the far right of diagram (1). Observe that the sum of the probabilities assigned to the elementary events is

$$\tfrac{9}{70} + \tfrac{6}{70} + \tfrac{8}{70} + \tfrac{12}{70} + \tfrac{1}{8} + \tfrac{1}{8} + \tfrac{3}{32} + \tfrac{5}{32} = 1.$$

(Question: Is this just a fortuitous accident?) The original problem was to determine the probability that the two marbles selected were of opposite color. This defines an event X in the sample space:

$$X = \{(\mathrm{I}, w, b), (\mathrm{I}, b, w), (\mathrm{II}, w, b), (\mathrm{II}, b, w)\}.$$

This event X is the disjoint union of the four indicated elementary events, and thus we have:

$$\begin{aligned} p(X) &= p(\{(\mathrm{I}, w, b)\}) + p(\{(\mathrm{I}, b, w)\}) + p(\{(\mathrm{II}, w, b)\}) + p(\{(\mathrm{II}, b, w)\}) \\ &= \tfrac{6}{70} + \tfrac{8}{70} + \tfrac{1}{8} + \tfrac{3}{32} \\ &= \tfrac{67}{160}. \end{aligned}$$

Example 7.1 is typical of many problems in probability in which a sequence of experiments is performed. These problems often require the computation of the probability of certain sets of ordered sequences of outcomes. For example, the set of outcomes pertinent to Example 7.1 was

$$X = \{(\mathrm{I}, w, b), (\mathrm{I}, b, w), (\mathrm{II}, w, b), (\mathrm{II}, b, w)\}.$$

The original problem was to determine the probability that the two marbles selected were of opposite color. In the construction of the sample space and the probability measure in Example 7.1, we assigned probabilities to every branch extending from a given node. Moreover, the sum of these probabilities is always 1. In the tree (1), two branches extend from each node. Of course, there can be many branches extending from a node, depending on the number of possible outcomes of the experiment at that stage. For example, suppose that we consider tosses of a single fair die: there are six possible outcomes for each toss (trial). The tree for this sequence of experiments has the following form.

(2)

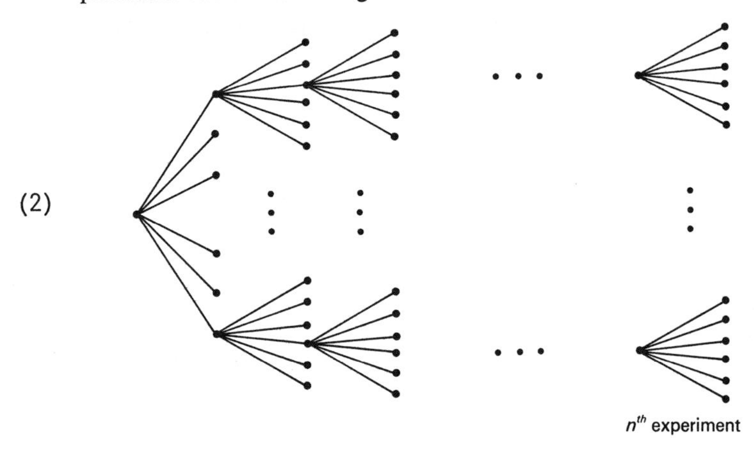

In performing such a sequence of n experiments, one is frequently interested in a sequence of n events, one from each trial. Thus we can think of a sequence of n experiments as a single experiment in which an outcome, i.e., an elementary event, is an n-tuple of outcomes corresponding to the n stages in the experiment. For example, the sequence of three experiments described in Example 7.1 can be thought of as a composite experiment in which the elementary events are (I, w, w), (I, w, b), . . . , (II, b, b). If we generalize what we did in Example 7.1, then the probability measure assigned to each of the elementary events in the composite experiment is the product of the probabilities assigned to each component. This definition is motivated by the multiplication theorem (Theorem 6.2). The situation is precisely the same for the composite experiment described by the tree (2). An elementary event in the composite experiment is just the set consisting of an n-tuple

$$(i_1, i_2, \ldots, i_n): \tag{3}$$

i_1 is the number which comes up on the first toss, i_2 is the number which comes up on the second toss, . . . , i_n is the number which comes up on the nth toss. Moreover, the probability that i_1 comes up on the first toss is $\frac{1}{6}$, and similarly for $i_2, \ldots, i_n$. Thus the multiplication theorem tells us that we should assign the probability

$(\frac{1}{6})^n$ to the event consisting of the n-tuple (3).

It is not necessarily the case that the number of branches extending from a given node is the same for every node at a given trial. Moreover, the probability assigned to the branches may vary from one node to the other, much as in the tree (1). It is not absolutely obvious that this way of assigning probabilities to the elementary events in the composite experiment results in a probability measure. The general theorem to this effect can be proved by an inductive procedure, but the notational difficulties of the proof are rather awkward. We can make the idea perfectly clear by an example.

(4)

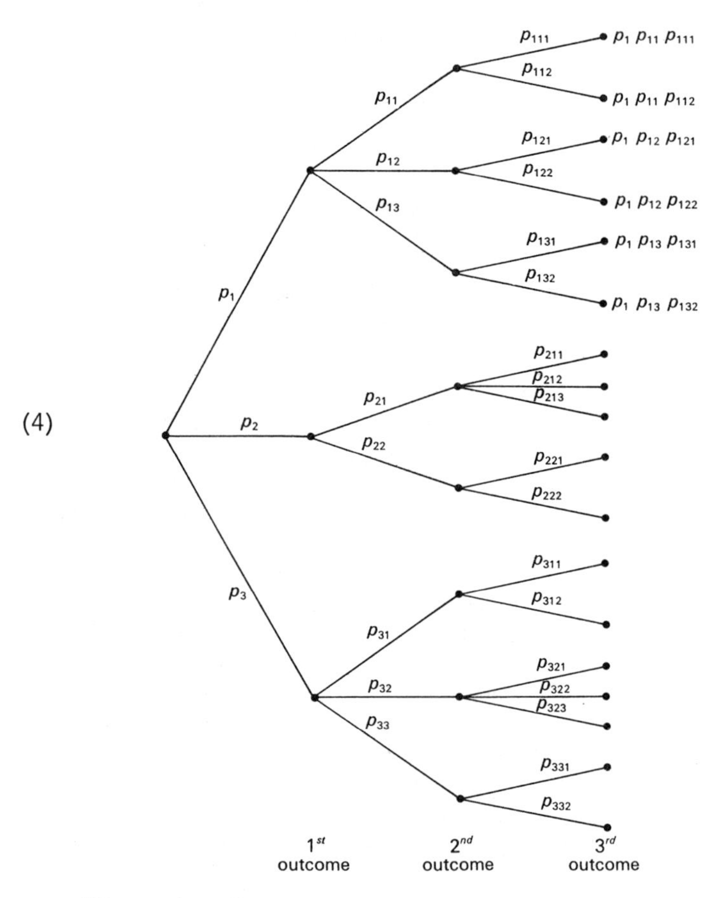

We are given that

$$
\begin{aligned}
p_1 + p_2 + p_3 &= 1, \\
p_{11} + p_{12} + p_{13} &= 1, \\
p_{21} + p_{22} &= 1, \\
p_{31} + p_{32} + p_{33} &= 1, \\
p_{111} + p_{112} &= 1,
\end{aligned}
$$

$$\begin{aligned} p_{121} + p_{122} &= 1, \\ p_{131} + p_{132} &= 1, \\ p_{211} + p_{212} + p_{213} &= 1, \\ p_{221} + p_{222} &= 1, \\ p_{311} + p_{312} &= 1, \\ p_{321} + p_{322} + p_{323} &= 1, \\ p_{331} + p_{332} &= 1. \end{aligned}$$

Consider all the nodes at the end of the tree that are connected by an unbroken line to the uppermost node at the first outcome. Forming the sum of the probabilities which are assigned to these nodes yields the result:

$$\begin{aligned} p_1p_{11}p_{111} + p_1p_{11}p_{112} &+ p_1p_{12}p_{121} + p_1p_{12}p_{122} + p_1p_{13}p_{131} + p_1p_{13}p_{132} \\ &= p_1p_{11}(p_{111} + p_{112}) + p_1p_{12}(p_{121} + p_{122}) + p_1p_{13}(p_{131} + p_{132}) \\ &= p_1p_{11} + p_1p_{12} + p_1p_{13} \\ &= p_1(p_{11} + p_{12} + p_{13}) \\ &= p_1. \end{aligned}$$

Similarly, forming the sums of the probabilities that are assigned to nodes that are connected to the second and third nodes at the first outcome yields p_2 and p_3, respectively. Adding these sums we get

$$p_1 + p_2 + p_3 = 1.$$

Putting this in a different manner, the assignment of probabilities in the tree (4) does in fact, result in a probability measure. The probability space in this case will again be a set of 3-tuples:

$$\{(1, 1, 1), (1, 1, 2), (1, 2, 1), (1, 2, 2), (1, 3, 1), (1, 3, 2), (2, 1, 1), (2, 1, 2), (2, 1, 3), (2, 2, 1), (2, 2, 2), (3, 1, 1), (3, 1, 2), (3, 2, 1), (3, 2, 2), (3, 2, 3), (3, 3, 1), (3, 3, 2)\}.$$

That is, $\{(i, j, k)\}$ is the elementary event in which i is the outcome of the first trial; j is the outcome of the second trial, given that i has occurred at the first stage; and k is the outcome of the third trial, given that i and j have occurred at the first and second stages, respectively.

It is frequently the case that the outcome of a given trial of an experiment is "independent" of what has occurred at previous stages. In Example 7.1 this was definitely not the situation: the various outcomes possible for choosing the first marble affect the possible outcomes for choosing the second marble. But it is easy to conceive of a series of trials which we would call independent, e.g., n successive tosses of a coin. At each trial the probability of heads (or tails) coming up does not depend in any way on what has happened before. The tree for this simple experiment has two branches emanating from every node.

(5)

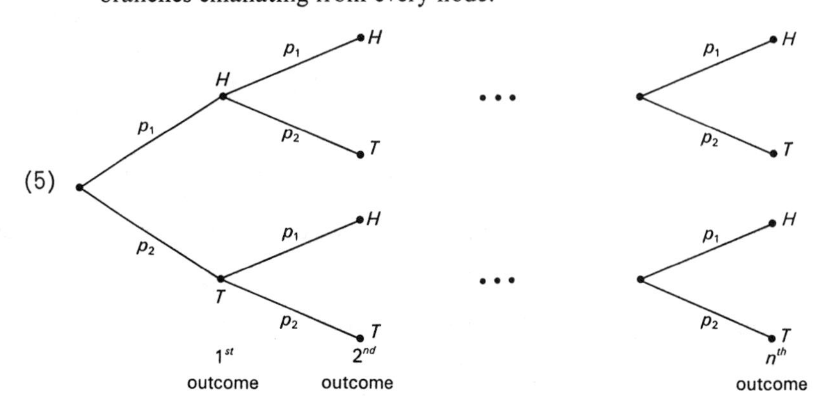

The number p_1 is the probability that H comes up and p_2 is the probability that T comes up. An n-tuple of H's and T's designating a sequence of n successive outcomes,

$$(H, T, \ldots, H), \tag{6}$$

is assigned the probability $p_1^{k_1}p_2^{k_2}$, $k_1 + k_2 = n$, where k_1 is the number of occurrences of H in (6) and k_2 is the number of occurrences of T. There are $\binom{n}{k_1\ k_2}$ different sequences of the form (6) in which there are k_1 H's and k_2 T's. Note that the sample space U is the totality of n-tuples of the form (6) so that $\nu(U) = 2^n$, and observe that the sum of the probabilities for all elementary events in U is

$$\begin{aligned}\sum_{k_1+k_2=n} \binom{n}{k_1\ k_2} p_1^{k_1}p_2^{k_2} &= (p_1 + p_2)^n \\ &= 1.\end{aligned}$$

Example 7.2 It is observed that in the toss of a defective coin, the probability of heads coming up is $\frac{1}{3}$ and the probability of tails coming up is $\frac{2}{3}$. In 10 tosses of the coin, find the probability that at least as many heads as tails come up. The sample space U is the set of 2^{10} 10-tuples of H's and T's. Let X denote the event consisting of all 10-tuples in which the number of H's is at least 5. The probability assigned to each elementary event contained in X is of the form

$$p_1^{k_1}p_2^{k_2}$$

in which $k_1 \geq 5$ and $k_1 + k_2 = 10$. For a fixed k_1, there are precisely $\binom{10}{k_1}$ 10-tuples containing precisely k_1 H's. Thus we can split the set X into a number of disjoint subsets depending on the number of H's involved. It follows that the probability of X is given by the sum

$$\begin{aligned}\sum_{\substack{k_1+k_2=10\\k_1\geq 5}} \binom{10}{k_1} p_1^{k_1} p_2^{k_2} &= \binom{10}{5}\left(\frac{1}{3}\right)^5\left(\frac{2}{3}\right)^5 + \binom{10}{6}\left(\frac{1}{3}\right)^6\left(\frac{2}{3}\right)^4 \\ &+ \binom{10}{7}\left(\frac{1}{3}\right)^7\left(\frac{2}{3}\right)^3 + \binom{10}{8}\left(\frac{1}{3}\right)^8\left(\frac{2}{3}\right)^2 \\ &+ \binom{10}{9}\left(\frac{1}{3}\right)^9\left(\frac{2}{3}\right) + \binom{10}{10}\left(\frac{1}{3}\right)^{10} \\ &= \left(\frac{1}{3}\right)^{10}\left[\binom{10}{5}2^5 + \binom{10}{6}2^4 + \binom{10}{7}2^3 + \binom{10}{8}2^2 + \binom{10}{9}2 + \binom{10}{10}\right].\end{aligned}$$

It is just a matter of straightforward computation to evaluate this latter number and we leave this to the reader.

At this point we formulate precisely what we mean by independent events.

Definition 7.1 ***Independent events*** Let (U, p) be a finite probability space and let $X_1, \ldots, X_n$ be a collection of events. Then $X_1, \ldots, X_n$ are *independent* if

$$p\left(\bigcap_{i=1}^{n} Y_i\right) = \prod_{i=1}^{n} p(Y_i), \tag{7}$$

for all choices of Y_i which are either X_i or X_i', $i = 1, \ldots, n$.

For $n = 2$, Definition 7.1 states that the following conditions must hold if X_1 and X_2 are to be independent events:

$$\begin{aligned} p(X_1 \cap X_2) &= p(X_1)p(X_2); \\ p(X_1 \cap X_2') &= p(X_1)p(X_2'); \\ p(X_1' \cap X_2) &= p(X_1')p(X_2); \\ p(X_1' \cap X_2') &= p(X_1')p(X_2'). \end{aligned} \tag{8}$$

We can relate the notion of independent events to the idea of conditional probability. Suppose we want to compute the conditional probability of X_1 given X_2, i.e., $p(X_1 \mid X_2)$, where X_1 and X_2 are independent. Assume that $p(X_1) \neq 0$ and $p(X_2) \neq 0$, so that

$$\begin{aligned} p(X_1 \mid X_2) &= \frac{p(X_1 \cap X_2)}{p(X_2)} \\ &= \frac{p(X_1)p(X_2)}{p(X_2)} \\ &= p(X_1), \end{aligned}$$

and similarly,

$$\begin{aligned} p(X_2 \mid X_1) &= \frac{p(X_1 \cap X_2)}{p(X_1)} \\ &= \frac{p(X_1)p(X_2)}{p(X_1)} \\ &= p(X_2). \end{aligned}$$

Hence we see that the conditional probability of X_1 given X_2 is the "unconditional" probability of X_1. The outcome X_1 does not depend on X_2, and similarly, the outcome X_2 does not depend on X_1. Observe that in concluding this, we did not make use of the equations in (8) which involve complements. As a matter of fact, these can be concluded directly from the first equation in (8). For example,

$$\begin{aligned} p(X_1) &= p(X_1 \cap U) \\ &= p(X_1 \cap (X_2 \cup X_2')) \\ &= p((X_1 \cap X_2) \cup (X_1 \cap X_2')) \\ &= p(X_1 \cap X_2) + p(X_1 \cap X_2') \\ &= p(X_1)p(X_2) + p(X_1 \cap X_2'). \end{aligned}$$

Thus,

$$\begin{aligned} p(X_1 \cap X_2') &= p(X_1) - p(X_1)p(X_2) \\ &= p(X_1)(1 - p(X_2)) \\ &= p(X_1)p(X_2'). \end{aligned}$$

The reader should verify that the last two equations in (8) follow by similar computations.

In view of the preceding discussion in the case $n = 2$, one might wonder whether it is generally the case that if $X_1, \ldots, X_n$ are pairwise independent, i.e., if X_i and X_j are independent, $1 \leq i < j \leq n$, then $X_1, \ldots, X_n$ are in fact independent in the sense of Definition 7.1. This does not always happen.

Example 7.3 Let U be the sample space for a toss of 2 unbiased coins:

$$U = \{(H, H), (H, T), (T, H), (T, T)\}.$$

Each of the elementary events is to be assigned probability $\frac{1}{4}$. Let X_1 be the event

$$X_1 = \{(H, H), (H, T)\},$$

that is, a head comes up on the first coin. Also, let

$$X_2 = \{(H, H), (T, H)\}$$

and let

$$X_3 = \{(H, T), (T, H)\}.$$

Then $p(X_1) = p(X_2) = p(X_3) = \frac{1}{2}$. Moreover, $X_1 \cap X_2 = \{(H, H)\}$, $X_1 \cap X_3 = \{(H, T)\}$, $X_2 \cap X_3 = \{(T, H)\}$, and hence $p(X_1 \cap X_2) = p(X_1 \cap X_3) = p(X_2 \cap X_3) = \frac{1}{4} = \frac{1}{2} \cdot \frac{1}{2} = p(X_1)p(X_2) = p(X_1)p(X_3) = p(X_2)p(X_3)$. Thus the events X_1, X_2, and X_3 are pairwise independent according to Definition 7.1. On the other hand, $X_1 \cap X_2 \cap X_3 = \emptyset$ and hence $p(\emptyset) = 0 = p(X_1 \cap X_2 \cap X_3) \neq \frac{1}{8} = p(X_1)p(X_2)p(X_3)$.

As we have seen in a number of examples, we are often confronted with a sequence of experiments (e.g., tossing coins, throwing dice) in which the information we require can be formulated as a question concerning a set of n-tuples in which the ith element of the n-tuple is the ith outcome of the experiment. We can formalize this situation as follows.

Definition 7.2 ***Product space*** Let $\{(U_k, p_k), k = 1, \ldots, n\}$ be a set of finite probability spaces. Let U be the space of all ordered sequences of length n,

$$u = (u_1, u_2, \ldots, u_n),$$

where $u_k \in U_k$, $k = 1, \ldots, n$. Define a function p on U by

$$p(\{u\}) = \prod_{k=1}^{n} p_k(\{u_k\}). \tag{9}$$

Then the pair (U, p) will be called the *product space* and p will be called the *product probability measure.*

Theorem 7.1 *The space (U, p) in Definition 7.2 is a finite probability space.*

Proof It follows immediately from Definition 7.2 that the product has only a finite number of elements in it and further,

$$\nu(U) = \prod_{k=1}^{n} \nu(U_k).$$

Also, the function p can take on only nonnegative values, since it is a product of probability measures. We need only check that the value of $p(U)$ is 1. But

$$\begin{aligned} p(U) &= \sum_{u \in U} p(\{u\}) \\ &= \sum_{\substack{u_k \in U_k \\ k=1,\ldots,n}} p_1(\{u_1\})p_2(\{u_2\}) \cdots p_n(\{u_n\}) \\ &= \Big(\sum_{u_1 \in U_1} p_1(\{u_1\})\Big)\Big(\sum_{u_2 \in U_2} p_2(\{u_2\})\Big) \cdots \Big(\sum_{u_n \in U_n} p_n(\{u_n\})\Big) \\ &= 1. \end{aligned}$$

This calculation is just a formalization of the fact that the sum of the probabilities of all the elementary events in U is the sum of all possible products of probabilities of n elementary events, the first out of U_1, the second out of $U_2, \ldots,$ the nth out of U_n. (The notation $\sum_{u \in U} p(\{u\})$ denotes the sum of all values of $p(\{u\})$ for all elements u in U.) ▌

Example 7.4 Find the probability of rolling precisely m 7's in n throws of a pair of fair dice. To solve this problem, we let (U_k, p_k) be a finite probability space, $k = 1, \ldots, n$, where U_k is the set of 11 possible sums obtainable on any roll of the dice and p_k is the following probability measure:

$$\begin{aligned} p_k(\{2\}) &= \tfrac{1}{36}; \\ p_k(\{3\}) &= \tfrac{2}{36}; \\ p_k(\{4\}) &= \tfrac{3}{36}; \\ p_k(\{5\}) &= \tfrac{4}{36}; \\ p_k(\{6\}) &= \tfrac{5}{36}; \\ p_k(\{7\}) &= \tfrac{6}{36}; \\ p_k(\{8\}) &= \tfrac{5}{36}; \\ p_k(\{9\}) &= \tfrac{4}{36}; \\ p_k(\{10\}) &= \tfrac{3}{36}; \\ p_k(\{11\}) &= \tfrac{2}{36}; \\ p_k(\{12\}) &= \tfrac{1}{36}. \end{aligned}$$

The event X in the product space concerning us is the set of n-tuples

$$u = (u_1, \ldots, u_n)$$

in which precisely m of the u_i equal 7. The probability measure p on the product space is the product of the probability measures on the individual spaces. Thus $p(X)$ is the sum of all possible products

$$p_1(\{u_1\})p_2(\{u_2\}) \cdots p_n(\{u_n\}) \tag{10}$$

in which precisely m of the u_i's are 7. Let x_k be the probability of obtaining $k + 1$ on any given roll of the dice. Then, according to (10), $p(X)$ is the sum of all products of the form

$$x_1^{k_1} x_2^{k_2} \cdots x_{11}^{k_{11}}$$

in which $k_1 + \cdots + k_{11} = n$ and $k_6 = m$. In other words, each of the factors which appears in (10) must be one of $x_1, \ldots, x_{11}$, and we coalesce identical x_i's. Thus k_i is the number of times x_i occurs among the n factors in (10), $i = 1, \ldots, 11$. To evaluate this, note that $x_1 + \cdots + x_6 + x_7 + x_8 + \cdots + x_{11} = \frac{1}{36} + \cdots + \frac{5}{36} + \frac{6}{36} + \frac{5}{36} + \cdots + \frac{1}{36} = 1$ and consider

$$(11) \qquad 1 = \left(\sum_{i=1}^{11} x_i\right)^n = \sum_{k_1+k_2+\cdots+k_{11}=n} \binom{n}{k_1\ k_2 \ldots k_{11}} x_1^{k_1} \cdots x_6^{k_6} \cdots x_{11}^{k_{11}}.$$

Then $p(X)$ is the sum of all terms in the sum (11) for which $k_6 = m$. Since we know the value of each x_i, we can compute this explicitly:

$$p(X) = \sum_{\substack{k_1+k_2+\cdots+k_{11}=n \\ k_6=m}} \binom{n}{k_1\ k_2 \ldots k_{11}} x_1^{k_1} \cdots x_6^{m} \cdots x_{11}^{k_{11}}.$$

For example, if $n = 3$ and $m = 2$, then the probability of obtaining precisely two 7's in 3 rolls of the dice is

$$(12) \qquad \sum_{\substack{k_1+\cdots+k_{11}=3 \\ k_6=2}} \binom{3}{k_1 \ldots k_{11}} x_1^{k_1} \cdots x_6^{2} \cdots x_{11}^{k_{11}}.$$

The sum (12) looks worse than it is. For, $k_6 = 2$ implies that the following are the only possibilities for the k_i:

$$k_6 = 2,\ k_t = 1,\ k_j = 0\ (j \neq t, 6), \qquad t = 1, \ldots, 5, 7, \ldots, 11.$$

Hence (12) becomes

$$\begin{aligned} \binom{3}{1\ 2} x_6^2[x_1 + x_2 + x_3 + x_4 + x_5 + x_7 + x_8 + x_9 + x_{10} + x_{11}] &= \binom{3}{2} x_6^2(1 - x_6) \\ &= 3(\tfrac{6}{36})^2(1 - \tfrac{6}{36}) \\ &= \tfrac{5}{72}. \end{aligned}$$

The preceding considerations lead us to the following result.

Theorem 7.2 *Let (U, p) be a finite probability space with n possible elementary events,* i.e., *$\nu(U) = n$. Let x_i be the probability of the* ith *elementary event u_i in U, $i = 1, \ldots, n$. In the product space of U with itself m times, the probability of obtaining the* ith *outcome exactly k_i times, $i = 1, \ldots, n$, where $k_1 + \cdots + k_n = m$, is precisely*

$$\binom{m}{k_1\ k_2 \ldots k_n} x_1^{k_1} \cdots x_n^{k_n}.$$

Proof Let X denote the event in the product space consisting of precisely those m-tuples of elements of U in which u_i appears exactly k_i times, $i = 1, \ldots, n$. Then the probability of any elementary event contained in X is

$$x_1^{k_1} x_2^{k_2} \cdots x_n^{k_n} \tag{13}$$

(by Definition 7.2). But we know that there are precisely $\binom{m}{k_1\ k_2 \ldots k_n}$ elements in X. Thus

$$p(X) = \binom{m}{k_1\ k_2 \ldots k_n} x_1^{k_1} x_2^{k_2} \cdots x_n^{k_n}. \blacksquare$$

Example 7.5 In tossing a fair coin 4 times, find the probability of obtaining exactly 3 heads. Let $x_1 = \frac{1}{2}$ be the probability of obtaining heads on any trial and similarly, let $x_2 = \frac{1}{2}$ be the probability of obtaining tails. In this case, $n = 4$, $k_1 = 3$, $k_2 = 1$, and, according to Theorem 7.2, the required probability is

$$\binom{4}{3\ 1}\left(\frac{1}{2}\right)^3\left(\frac{1}{2}\right) = \frac{1}{4}.$$

Example 7.6 In seven rolls of a fair die, what is the probability of getting two 7's, three 3's, and two 4's? Let x_i be the probability of obtaining $i + 1$, $i = 1, \ldots, 11$. Here $n = 7$, $k_6 = 2$, $k_2 = 3$, $k_3 = 2$. Hence the required probability is

$$\binom{7}{2\ 3\ 2} x_2^3 x_3^2 x_6^2 = (210)\left(\frac{2}{36}\right)^3\left(\frac{3}{36}\right)^2\left(\frac{6}{36}\right)^2$$
$$= \frac{35}{3(36)^4}.$$

As a final example in this section, we consider a problem in which the probability space can no longer be assumed finite. Although the techniques we have developed are not adequate to deal with such a problem, they do, nevertheless, suggest a method of attack.

Example 7.7 A matchstick is broken at two places. What is the probability that the three pieces so obtained can be formed into a triangle? To answer this question, we should first observe that it is not the case that *any* two breaks will result in a triangle. For example, if the breaks are too close to the ends of the match, the segments will not form a triangle. (Try it!) The problem here is to decide on a probability space and an appropriate probability measure. To formulate the problem mathematically, we may begin by assuming that the match is one unit long. We designate the break-points by x_1 and x_2.

(14)

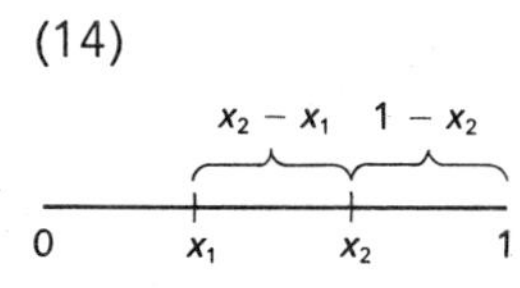

As we see from figure (14), the three segments have lengths x_1, $x_2 - x_1$, and $1 - x_2$. Moreover, $0 \leq x_1 \leq 1$ and $0 \leq x_2 \leq 1$. We define the probability space U to be the totality of points (x_1, x_2), $x_2 > x_1$, $0 \leq x_i \leq 1$, $i = 1, 2$. An elementary event, then, will consist of a point in this triangle. Now, necessary and sufficient conditions for the three segments to form a triangle, possibly degenerate, are the following:

(15)
$$\begin{aligned} x_1 &\leq (x_2 - x_1) + (1 - x_2), \\ x_2 - x_1 &\leq x_1 + (1 - x_2), \\ 1 - x_2 &\leq x_1 + (x_2 - x_1). \end{aligned}$$

The three inequalities in (15) are simply statements of the fact that three line segments can form a triangle if and only if the sum of the lengths of any two of the segments is at least as great as the third. Thus the "event" that a triangle can be formed from the three segments is precisely the set of points in the unit square which satisfy the inequalities (15). We first determine this set of points graphically and then decide upon the sigma field of subsets of the unit square and the appropriate probability measure. If we simplify the inequalities (15), we obtain

(16)
$$\begin{aligned} x_1 &\leq \tfrac{1}{2}, \\ x_2 &\leq x_1 + \tfrac{1}{2}, \\ x_2 &\geq \tfrac{1}{2}. \end{aligned}$$

The first and third of the inequalities (16) imply that the event that a triangle can be formed is in the upper left quarter of the unit square. If we graph the line $x_2 = x_1 + \frac{1}{2}$, we obtain the following diagram.

(17)

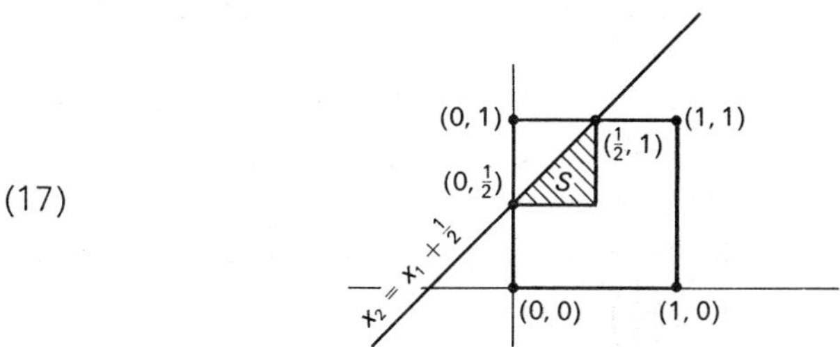

The second inequality in (16) is satisfied by precisely those points which lie below the line $x_2 = x_1 + \frac{1}{2}$. For the other two inequalities to be satisfied as well, it is necessary and sufficient that the point

(x_1, x_2) lie in the shaded region S in diagram (17). It seems reasonable then that we should choose our sigma field of sets $\mathfrak{A}$ so that it contains S and, moreover, is the smallest sigma field of subsets of U such that each subset has the property that its area can be defined. The probability measure p should be chosen to be the area of any set in $\mathfrak{A}$. It is clear from the diagram that if X denotes the event that a triangle can be formed from the three segments, then

$$p(X) = \frac{\text{area of } S}{\text{area of } U} = \frac{1}{4}.$$

There are severe and profound technical difficulties that we have glossed over rather glibly in this example. These have to do with the definition of the sigma field $\mathfrak{A}$. For it is not clear precisely what one means by area and by sets which have an area. It is this problem which prevents us from systematically investigating probability measures on spaces which are not finite.

Quiz

Answer true or false.

1 A coin is tossed 3 times. The number of branches in the tree diagram for this experiment is 6.

2 In a tree diagram, the number of branches extending from a given node is the same for every node at a given trial.

3 If X_1 and X_2 are events in a finite probability space (U, p), then X_1 and X_2 are independent if and only if $p(X_1 \cap X_2) = p(X_1)p(X_2)$.

4 If X_1, X_2, and X_3 are events in (U, p), then X_1, X_2, and X_3 are independent if and only if $p(X_1 \cap X_2 \cap X_3) = p(X_1)p(X_2)p(X_3)$.

5 If X_1, X_2, and X_3 are independent events in (U, p), then $X_1 \cup X_2$ and X_3 are independent events.

6 If X_1 and X_2 are independent events in (U, p), then X_1' and X_2' are independent events.

7 If X_1 and X_2 are independent events in (U, p), then $X_1 \cap X_2 = \emptyset$.

8 If X_1 and X_2 are events in (U, p) and $X_1 \cap X_2 = \emptyset$, then X_1 and X_2 are independent.

9 The probability of obtaining exactly 2 heads in 3 tosses of a fair coin is $\binom{3}{2}\left(\frac{1}{2}\right)^3$.

10 The probability of obtaining no heads in 3 tosses of a fair coin is $\frac{7}{8}$.

Exercises

1 For each of the following finite stochastic processes draw an appropriate tree diagram, label each branch with the appropriate probability, and define a sample space and probability measure.

(a) A fair coin is tossed 3 times.

(b) A biased coin for which heads appears twice as frequently as tails is tossed 4 times.

(c) Three marbles are successively drawn at random from an urn containing 3 white and 4 black marbles, where no marbles are replaced once they are removed.

(d) In the preceding problem, the marbles are replaced after each draw.

(e) Three pairs of marbles are successively drawn at random from an urn containing 5 black and 4 white marbles, without replacements.

(f) In the preceding problem, the marbles are replaced after each draw.

(g) Three marbles are successively drawn at random from an urn containing 2 white, 2 black, and 3 red marbles, without replacements.

(h) A loaded die for which the even faces appear twice as often as the odd faces, is thrown 3 times.

2 One fair die is tossed 3 times. Find the probability that a 3 or a 4 appears exactly twice. (Hint: Let $X = \{3, 4\}$. Then $x_1 = p(X) = \frac{1}{3}$ (why?) and $x_2 = p(X') = \frac{2}{3}$. Define a two element space $U = \{S, F\}$, where S means success in obtaining a 3 or a 4 and F means failure. Define a probability measure q by $q(\{S\}) = x_1 = \frac{1}{3}$, $q(\{F\}) = x_2 = \frac{2}{3}$. Then use Theorem 7.2 with $k_1 = 2$, $k_2 = 1$, $n = 3$.)

3 In Example 7.1, find the probability that the two marbles taken from the jars are of the same color.

4 Jar I contains 3 white, 4 black, and 2 red marbles; Jar II contains 4 white, 2 black, and 3 red marbles. A jar is selected at random, two marbles are drawn from it at random and placed in the other jar. Then two marbles are drawn at random from this second jar. What is the probability that all four marbles drawn are of the same color? Construct the appropriate tree diagram.

5 A loaded die for which the even faces appear three times as often as the odd faces, is tossed 4 times. Find the probability that a 4 or a 5 comes up exactly twice.

6 Find the probability of obtaining precisely 3 sevens in 4 throws of a pair of fair dice. (Hint: See Example 7.4.)

7 Find the probability of obtaining at least 1 seven in 3 throws of a pair of fair dice.

8 Suppose that an experiment has two outcomes: success and failure. Also assume that the probability of success is x_1 and the probability of failure is $x_2 = 1 - x_1$. Find the probability of precisely k successes in n trials of the experiment. (Hint: Let $U = \{S, F\}$, where S and F denote the two possible outcomes on any trial. Apply Theorem 7.2.)

9 It is estimated that the chances of survival on any given mission in Vietnam is 0.8. What are the chances of surviving 5 missions?

10 Find the probability of obtaining more heads than tails in 5 tosses of a fair coin. (Hint: Find the probability of obtaining 3, 4, or 5 heads.)

11 Let X_1 and X_2 be events in a finite probability space (U, p). Assume that

$$p(X_1 \cap X_2) = p(X_1)p(X_2),$$

and show that X_1 and X_2 are independent events.

12 Three fair coins are tossed. Let X_1 be the event that the first coin comes up heads and let X_2 be the event that all three coins show the same face. Are these events independent? (Hint: The sample space consists of eight 3-tuples: (H, H, H), (H, H, T), etc. Find events X_1 and X_2 and check if $p(X_1 \cap X_2) = p(X_1)p(X_2)$.)

13 The four faces of an equilateral pyramidal die are numbered 1, 2, 3, 4. (Recall that an equilateral pyramid has 4 faces, each in the shape of an equilateral triangle.) An experiment consists of one throw of two such fair dice. The total obtained is the sum of the numbers on the faces upon which the dice rest. If the dice are thrown 4 times, what is the probability of obtaining two 5's, one 7, and one 3, in that order? What is the probability of obtaining different totals in 3 throws of the dice?

14 A publisher observes that among five prolific authors, the probability of signing a best-seller is .2, .4, .6, .8, and .5, respectively. The publisher decides to sign all five authors for one book each. What is the probability that the company will publish at least two best sellers from among these five? (Hint: With each of the five authors associate a 2-element sample space $\{0, 1\}$ and the following probability measures: $p_1(\{1\}) = .2$, $p_1(\{0\}) = .8$; $p_2(\{1\}) = .4$, $p_2(\{0\}) = .6$; etc. On the product space consisting of all 5-tuples with 0 or 1 in each component, let p be the product measure, i.e.,

$$p(\{(u_1, \ldots, u_5)\}) = p_1(\{u_1\}) \ldots p_5(\{u_5\}).$$

Then compute $p(X)$ where X is the set of all 5-tuples in which at least two of the u_i are 1's.)

15 Five fair pennies are tossed 4 times. What is the probability of obtaining exactly 2 heads in exactly 3 of the tosses? (Hint: First compute the probability of obtaining exactly 2 heads on a given toss, call it x_1. Let $x_2 = 1 - x_1$. Then define a 2-element sample space $U = \{S, F\}$, where S means that exactly 2 heads come up on a given toss of the five coins. Let $p(\{S\}) = x_1$, $p(\{F\}) = x_2$. Then compute the probability of obtaining S exactly 3 times in 5 trials by using Theorem 7.2.)

16 A fair penny is tossed 4 times. What is the probability of obtaining at least one 2-run? (A 2-run is the occurrence of either 2 heads or 2 tails in succession.)

17 Assume that it is equally probable that a boy or a girl will be born to a family. What is the probability that in a family of 7 children there are at least 2 girls but not more than 5 girls?

18 With the same assumption as in the preceding exercise, find the number of children that a family must plan on having in order that the probability of having at least 2 boys is at least .75.

19 Five pennies are thrown at 3 boxes and we assume that each penny must land in a box. Let x_j denote the probability of a tossed penny landing in box j, $j = 1, 2, 3$. In terms of x_1, x_2, and x_3, find the probability of 2 pennies landing in box 1, 1 penny landing in box 2, and 2 pennies landing in box 3. (Hint: This can be thought of as 1 penny being tossed 5 times at the boxes with 3 outcomes of probabilities x_1, x_2, and x_3 at each trial. Apply Theorem 7.2.)

20 In the preceding exercise find the probability that all five pennies land in box 1.

21 An archery target is divided into 4 non-overlapping concentric rings numbered 1, 2, 3, and 4. The probability of hitting ring 1 is $\frac{1}{5}$, ring 2 is $\frac{1}{4}$, ring 3 is $\frac{1}{4}$, and ring 4 is $\frac{3}{10}$. In 6 trials, what is the probability of hitting ring 1 twice, ring 2 once, ring 3 once, and ring 4 twice?

†22 Let $U = \{u_1, \ldots, u_n\}$ be an n-element sample space and let f be a function from U to the real numbers. Then f is called a *random variable*. If p is a probability measure defined on U then the number

$$E(f) = \sum_{i=1}^{n} p(\{u_i\})f(u_i)$$

is called the *expectation* of the random variable f.

(a) Show that

$$E(1 - f) = 1 - E(f),$$

where 1 denotes the function whose value at each u_i is 1.

(b) Show that if g is another random variable, then

$$E(f + g) = E(f) + E(g).$$

(c) Two fair coins are tossed and the following betting game is played by players I and II. If 2 heads come up, player II pays player I one dollar. Otherwise player I pays player II half a dollar. What can each player expect to win or lose in five trials? (Hint: A 2-element sample space U can be defined as follows. Let $U = \{S, F\}$, where S means that 2 heads come up and F means that 2 heads do not come up on a toss of the coins. Now it is a familiar computation to see that the probability of obtaining 2 heads is $\frac{1}{4}$. Thus we can define a probability measure p on U by $p(\{S\}) = \frac{1}{4}$, $p(\{F\}) = \frac{3}{4}$. Let f be a random variable defined by $f(S) = 1$, $f(F) = -\frac{1}{2}$, i.e., the random variable f corresponds to the amount won or lost by player I. Then

$$\begin{aligned} E(f) &= p(\{S\})f(S) + p(\{F\})f(F) \\ &= \tfrac{1}{4}\cdot 1 + \tfrac{3}{4}\cdot(-\tfrac{1}{2}) \\ &= \tfrac{1}{4} - \tfrac{3}{8} \\ &= -\tfrac{1}{8}. \end{aligned}$$

In other words, player I can expect to lose $\frac{1}{8}$ of a dollar on a given trial. In 5 trials player I can expect to lose $\frac{5}{8}$ of a dollar. The same kind of computation can be made for player II.)

(d) Two fair dice are thrown 10 times. Player II pays player I three dollars if a 3 or a 4 comes up and otherwise player II pays player I two dollars. Find the amount that each player can expect to win or lose. (Study the hint for part (c) carefully!)

linear algebra

chapter 2

2.1

Vectors and Matrices

In this section we begin the study of linear algebra and its applications to the social, biological, and physical sciences. Linear algebra is a highly developed and remarkably successful branch of mathematics. However, it has only been in recent times (about the last twenty years) that its techniques and results have been applied in the social sciences.

In this first section we shall be concerned mainly with the most elementary definitions and results, and our emphasis will be primarily on developing facility with matrix and vector operations. In succeeding sections we will obtain some additional results, and then apply the entire theory to a multitude of interesting problems from many fields.

The reader will recall that any real number r can be associated with a unique point on a number line once the positions of 0 and 1 have been specified.

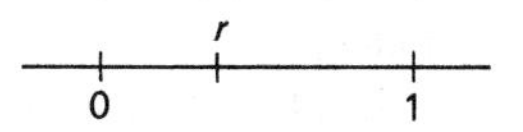

Similarly, corresponding to any ordered pair of numbers (a_1, a_2), there is a unique point P in the Cartesian plane. The number a_1

is the projection of P on the horizontal axis and a_2 is the projection of P on the vertical axis.

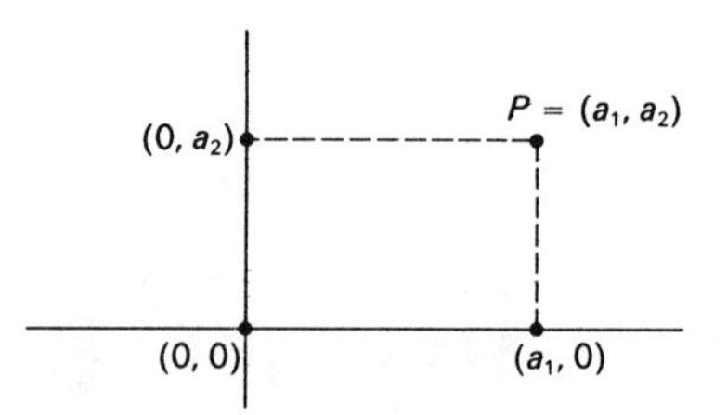

In fact, it is customary to refer to the *point* (a_1, a_2) once a pair of mutually perpendicular axes have been chosen. By exactly the same kind of argument, if three mutually perpendicular axes are chosen then a point P in 3-dimensional space can be uniquely associated with an ordered triple of real numbers, (a_1, a_2, a_3).

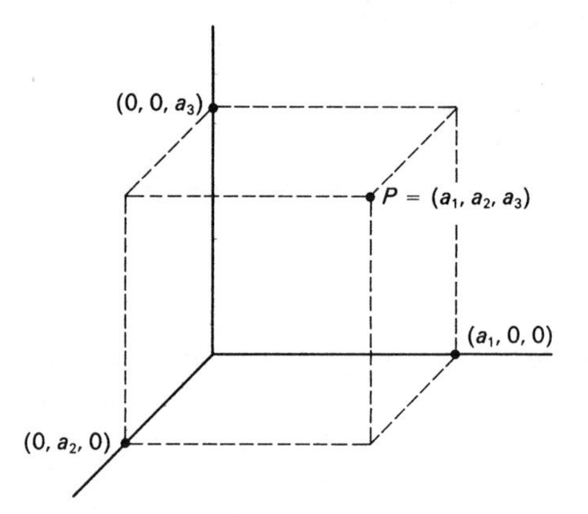

In many practical situations, sequences of 1, 2, or 3 numbers are inadequate. For example, suppose it is required to locate a moving body in space. This can be done with respect to some fixed rectangular frame of reference, but it is also necessary to have a method of measuring the time elapsed from a given initial time. Thus, in a 4-tuple of numbers

$$(a_1, a_2, a_3, a_4),$$

the first three coordinates can specify the position of the point with respect to some frame of reference and the last coordinate can be the number of time units, e.g., minutes, which have elapsed on a clock.

As another example in which a sequence of real numbers is used, consider a model of a segment of the economy devoted to the manufacture of n different items. Suppose that there are m factories engaged in the production of these items. Let a_{ik} denote the number of units of the kth item produced by the ith factory. Then the productivity of the ith factory is completely described by the ordered "n-tuple" of numbers

$$(a_{i1}, a_{i2}, \dots, a_{in}).$$

Definition 1.1 ***Vector*** An n-tuple of real numbers is called an *n-vector*, or simply a *vector* if n is understood. More precisely, an n-vector is a function $v\colon N_n \to R$, where $N_n = \{1, \dots, n\}$ and R is the set of real numbers. Rather than write out the set of ordered pairs $\{(1, v(1)), \dots, (n, v(n))\}$, it is customary to write out the range of v in order as follows: if $v(i) = a_i$, $i = 1, \dots, n$, then

$$v = (a_1, a_2, \dots, a_n). \tag{1}$$

The number a_k is called the kth *component* of the vector, $k = 1, \dots, n$.

Example 1.1 Two factories each produce the same three items. Let a_{ik} denote the number of units of item k produced by factory i, $k = 1, 2, 3$; $i = 1, 2$. Factory i increases its production by a factor λ_i, $i = 1, 2$. Find a 3-vector whose kth component is the total number of units produced by both factories. The original productivity of the two factories is represented by the vectors

$$(a_{11}, a_{12}, a_{13})$$

and

$$(a_{21}, a_{22}, a_{23}).$$

After expansion, the first factory will produce λ_1 times as many items of each kind as before expansion, i.e., the new productivity vector for the first factory is

$$(\lambda_1 a_{11}, \lambda_1 a_{12}, \lambda_1 a_{13}).$$

Similarly, the productivity vector for the second factory after expansion is

$$(\lambda_2 a_{21}, \lambda_2 a_{22}, \lambda_2 a_{23}).$$

After expansion, the two factories together produce $\lambda_1 a_{11} + \lambda_2 a_{21}$ units of item 1, $\lambda_1 a_{12} + \lambda_2 a_{22}$ units of item 2, and $\lambda_1 a_{13} + \lambda_2 a_{23}$ units of item 3. Thus the vector describing the combined productivity of both expanded factories is

$$(\lambda_1 a_{11} + \lambda_2 a_{21}, \lambda_1 a_{12} + \lambda_2 a_{22}, \lambda_1 a_{13} + \lambda_2 a_{23}). \tag{2}$$

If we set

$$v_1 = (a_{11}, a_{12}, a_{13}),$$
$$v_2 = (a_{21}, a_{22}, a_{23}),$$

and if we define $\lambda_i v_i$ by

$$\lambda_i v_i = (\lambda_i a_{i1}, \lambda_i a_{i2}, \lambda_i a_{i3}),$$

and if the addition sign between two vectors designates componentwise addition (e.g., $(a_1, a_2, a_3) + (b_1, b_2, b_3) = (a_1 + b_1, a_2 + b_2, a_3 + b_3)$), then the vector (2) can be more compactly written as

$$\lambda_1 v_1 + \lambda_2 v_2. \tag{3}$$

Definition 1.2 ***Vector operations*** Let $v = (a_1, \ldots, a_n)$ be an n-vector and let λ be a number. Then the vector λv, called the *scalar product* of the vector v and the scalar (i.e., number) λ, is defined by

$$\lambda v = (\lambda a_1, \lambda a_2, \ldots, \lambda a_n).$$

We sometimes write $v\lambda$ for λv. If u is another n-vector, $u = (b_1, b_2, \ldots, b_n)$, then the *sum* of the vectors u and v, denoted $u + v$, is defined by

$$u + v = (a_1 + b_1, a_2 + b_2, \ldots, a_n + b_n).$$

The two vector operations in Definition 1.2 have interesting geometric interpretations. If $n = 1$, i.e., if we are talking about 1-tuples, then we can identify any such 1-vector with a point on the line.

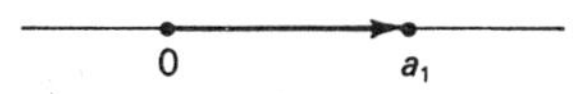

If λ is a real number, then the various possibilities for the point λa_1 look as follows.

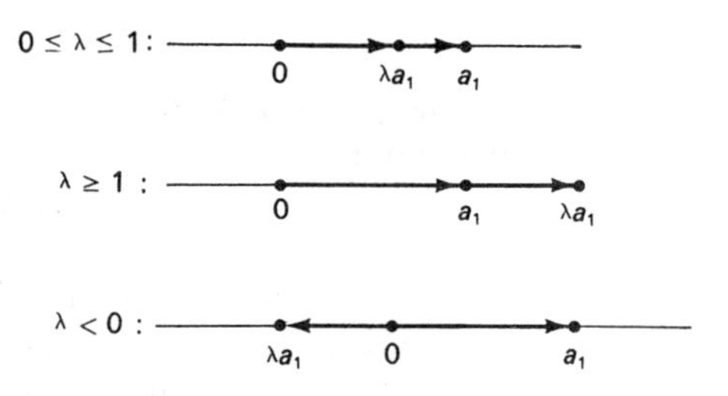

These diagrams pertain to the case in which a_1 is positive. For a_1 negative or 0, the modifications will be supplied by the student.

Addition of 1-vectors a_1 and b_1 is just ordinary addition.

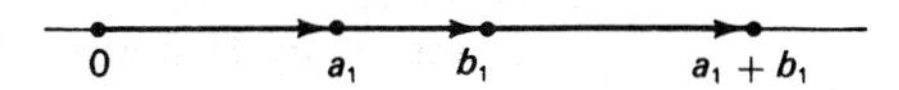

A 2-vector (a_1, a_2) can be represented geometrically as a directed line segment from the origin to the point (a_1, a_2).

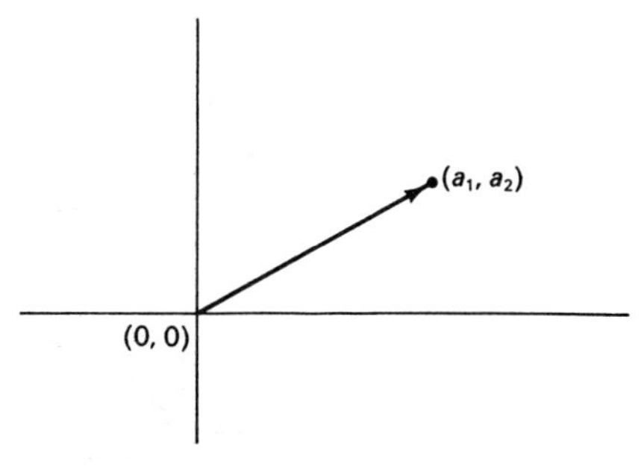

If λ is a number, then $\lambda(a_1, a_2)$ is represented by the following three diagrams for various choices of λ.

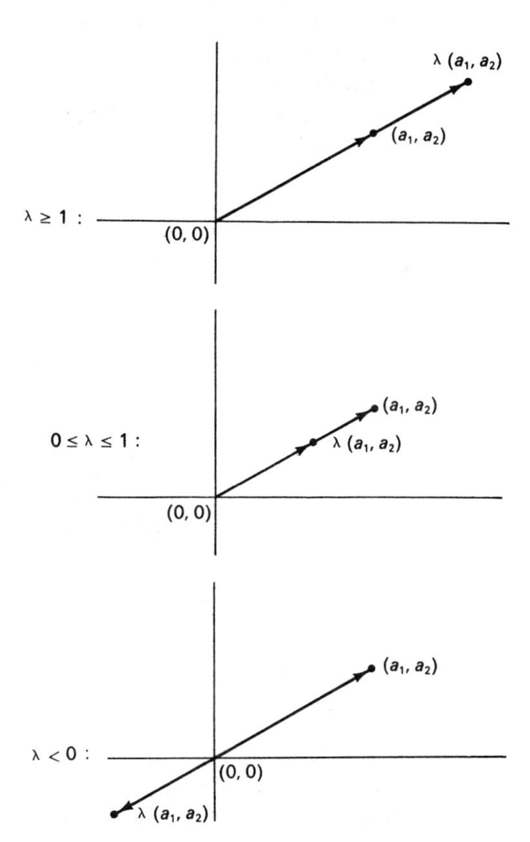

The addition of 2-vectors has a somewhat more interesting geometric interpretation. Suppose $a = (a_1, a_2)$ and $b = (b_1, b_2)$ are two 2-vectors. By Definition 1.2,

$$(a_1, a_2) + (b_1, b_2) = (a_1 + b_1, a_2 + b_2).$$

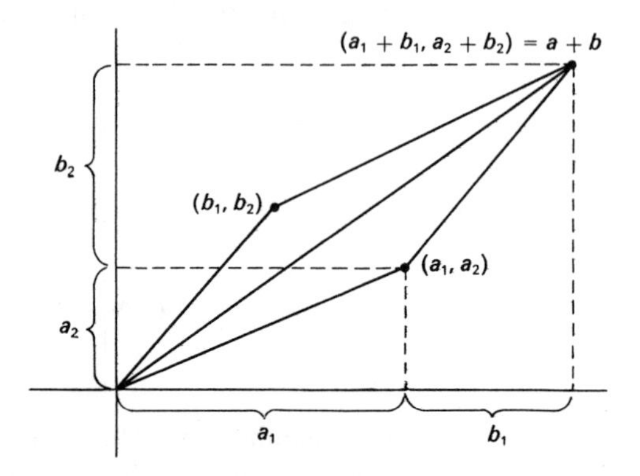

We see from the preceding diagram that the sum of the two vectors is constructed in the following way. Translate the line segment denoting b parallel to itself until the initial point (i.e., the point previously at the origin) coincides with the point (a_1, a_2). Then the diagonal of the parallelogram so formed is the vector $a + b$.

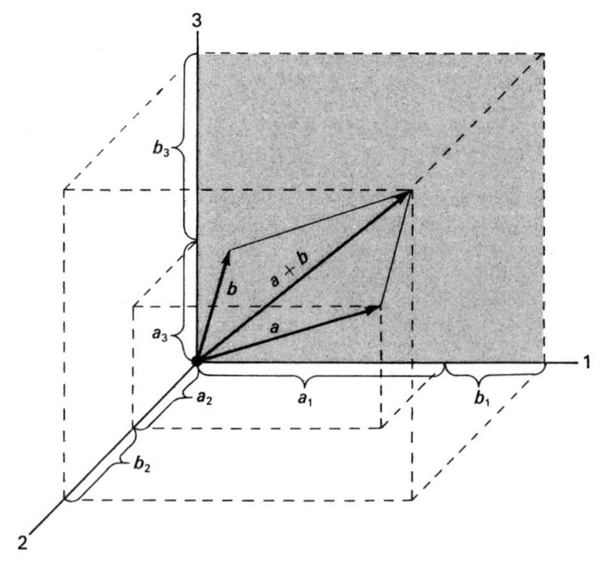

Similarly, in three dimensions a scalar multiple of a vector expands, contracts, or changes direction according to the value of the scalar. The sum of two vectors $a = (a_1, a_2, a_3)$ and $b = (b_1, b_2, b_3)$

can be obtained by translating the line segment denoting b parallel to itself until the initial point coincides with the point (a_1, a_2, a_3). Then the sum $a + b$ is represented by the line segment from the origin to the terminal end of the translated vector.

Geometric representations for vectors are helpful in predicting what is likely to be true. However, even in three dimensions, the geometric interpretation is often awkward to draw. Of course, if we are confronted with n-vectors, $n > 3$, then no suitable geometric interpretation is available. Nevertheless, n-vectors occur in very practical situations, e.g., the example of m factories producing n different items. But vector operations are defined for n-vectors and from their definitions we can easily prove a number of purely algebraic results which have significant applications. We first define what is meant by the "vector space" of n-tuples.

Definition 1.3 ***Vector space of n-tuples*** Let n be a fixed positive integer and let R^n denote the totality of n-vectors with real components. Then R^n together with the operations of vector addition and scalar multiplication is called the *vector space of real n-tuples*. The vector whose components are all zero is called the *zero vector* and is denoted by 0_n (or simply by 0 if n is understood). If $u \in R^n$, then $(-1)u$ is denoted by $-u$.

A number of very elementary results are contained in the following theorem.

Theorem 1.1 *Let u, v, and w be any vectors in R^n and let λ and μ be any two real numbers. Then*

(a) $\lambda u \in R^n$ and $(u + v) \in R^n$;
(b) $u + (v + w) = (u + v) + w$ (*associative law for addition*);
(c) $u + v = v + u$ (*commutative law for addition*);
(d) $u + (-u) = 0_n$;
(e) $1u = u$;
(f) $0u = 0_n$;
(g) $(\lambda + \mu)u = \lambda u + \mu u$;
(h) $\lambda(u + v) = \lambda u + \lambda v$;
(i) $(\lambda\mu)u = \lambda(\mu u)$.

Proof We prove (a), (b), and (i) and leave the rest of the statements as very simple exercises.

(a) Let $u = (u_1, \ldots, u_n)$. Then by definition, $\lambda u = (\lambda u_1, \ldots, \lambda u_n)$ which is obviously an n-vector. Similarly, if $v = (v_1, \ldots, v_n)$, then

$u + v = (u_1, \ldots, u_n) + (v_1, \ldots, v_n) = (u_1 + v_1, \ldots, u_n + v_n)$, which is also an n-vector.

(b) Let $u = (u_1, \ldots, u_n), v = (v_1, \ldots, v_n)$, and $w = (w_1, \ldots, w_n)$. Then $v + w = (v_1 + w_1, \ldots, v_n + w_n)$ and hence $u + (v + w) = (u_1, \ldots, u_n) + (v_1 + w_1, \ldots, v_n + w_n) = (u_1 + v_1 + w_1, \ldots, u_n + v_n + w_n)$. A similar calculation shows that $(u + v) + w = (u_1 + v_1 + w_1, \ldots, u_n + v_n + w_n)$. In other words, the associativity of ordinary addition of numbers implies the associativity of vector addition.

(i) Observe that $\mu u = (\mu u_1, \ldots, \mu u_n)$ and hence $\lambda(\mu u) = (\lambda(\mu u_1), \ldots, \lambda(\mu \mu_n)) = ((\lambda\mu)u_1, \ldots, (\lambda\mu)u_n) = (\lambda\mu)u$. Here the associativity of ordinary real number multiplication implies (i). ▮

A word of caution in the notation of the following definition is necessary: u_t is not the tth component of a vector u. It is the tth vector in a certain set of m n-vectors.

Definition 1.4 ***Linear combination*** Let $u_t \in R^n$, $t = 1, \ldots, m$, and let λ_t be a real number for each $t = 1, \ldots, m$. Then the n-vector

$$\lambda_1 u_1 + \lambda_2 u_2 + \cdots + \lambda_m u_m \tag{4}$$

is called a *linear combination* of the vectors u_t, $t = 1, \ldots, m$, and the scalars λ_t are called the *coefficients*. It is often convenient to use the sigma notation to denote the linear combination (4):

$$\sum_{t=1}^{m} \lambda_t u_t. \tag{5}$$

Example 1.2 It is observed in a nursery that 60% of the seeds from yellow poppies yield yellow poppies in the next generation, and the remaining 40% yield white poppies. Moreover, 30% of the seeds from white poppies yield yellow poppies in the next generation, and the remaining 70% yield white poppies. Suppose that a sample of λ_1 seeds from yellow poppies and λ_2 seeds from white poppies is selected at random. Express as a 2-vector the expected distribution of poppies after one generation. We can restate the information in terms of probabilities as follows. Let p_{11} denote the probability that a seed from a yellow poppy produces a yellow poppy in the next generation: $p_{11} = .6$. Similarly, let $p_{21} = .4$ denote the probability that a yellow poppy produces a white poppy in the next generation. Again, $p_{12} = .3$

denotes the probability that a white poppy produces a yellow poppy in the next generation. Finally, $p_{22} = .7$ is the probability that a white poppy produces a white poppy in the next generation. Consider the 2-vector (p_{11}, p_{21}). If we multiply this vector by λ_1, we obtain

$$\lambda_1(p_{11}, p_{21}) = (\lambda_1 p_{11}, \lambda_1 p_{21}). \tag{6}$$

Consider the first component, $\lambda_1 p_{11}$. The number λ_1 is the initial number of yellow poppy seeds and $p_{11} = .6$ is the probability that a yellow poppy seed produces a yellow poppy in the next generation. Hence, $\lambda_1 p_{11}$ is the expected number of yellow poppies in the next generation which will come from yellow poppies. Similarly, $\lambda_1 p_{21}$ is the expected number of white poppies in the next generation which will come from yellow poppies. Thus the vector (6) gives the distribution of white and yellow poppies in the next generation which will come from yellow poppies. Similarly, the first and second components of

$$\lambda_2(p_{12}, p_{22}) = (\lambda_2 p_{12}, \lambda_2 p_{22}) \tag{7}$$

give the expected distribution of yellow and white poppies in the next generation which will come from white poppies. Now consider the vector

$$\begin{aligned} \lambda_1(p_{11}, p_{21}) + \lambda_2(p_{12}, p_{22}) &= (\lambda_1 p_{11} + \lambda_2 p_{12}, \lambda_1 p_{21} + \lambda_2 p_{22}) \\ &= (\lambda_1(.6) + \lambda_2(.3), \lambda_1(.4) + \lambda_2(.7)). \end{aligned} \tag{8}$$

The first component of the vector (8) is the expected number of yellow poppies in the next generation, and the second component is the expected number of white poppies.

Example 1.3 With the same data as in Example 1.2, express as a 2-vector the expected distribution of yellow and white poppies after two generations. (For the second generation, choose one seed from each poppy.) We observe that an initial distribution vector, (λ_1, λ_2) resulted in a new distribution vector (μ_1, μ_2) after one generation:

$$\mu_1 = \lambda_1 p_{11} + \lambda_2 p_{12}, \quad \mu_2 = \lambda_1 p_{21} + \lambda_2 p_{22}.$$

Thus, applying what we have learned to the distribution vector (μ_1, μ_2), we see that (μ_1, μ_2) results in a new distribution vector

$$(\mu_1 p_{11} + \mu_2 p_{12}, \mu_1 p_{21} + \mu_2 p_{22}). \tag{9}$$

Substituting the values of μ_1 and μ_2, we can express (9) directly in terms of λ_1 and λ_2:

$$\begin{aligned}
&((\lambda_1 p_{11} + \lambda_2 p_{12})p_{11} + (\lambda_1 p_{21} + \lambda_2 p_{22})p_{12}, (\lambda_1 p_{11} + \lambda_2 p_{12})p_{21} + (\lambda_1 p_{21} + \lambda_2 p_{22})p_{22})\\
&\quad = (\lambda_1(p_{11}^2 + p_{12}p_{21}) + \lambda_2(p_{11}p_{12} + p_{12}p_{22}), \lambda_1(p_{21}p_{11} + p_{22}p_{21})\\
&\qquad + \lambda_2(p_{21}p_{12} + p_{22}^2))\\
&\quad = \lambda_1(p_{11}^2 + p_{12}p_{21}, p_{21}p_{11} + p_{22}p_{21}) + \lambda_2(p_{11}p_{12} + p_{12}p_{22}, p_{21}p_{12} + p_{22}^2)\\
&\quad = \lambda_1(.36 + .12, .24 + .28) + \lambda_2(.18 + .21, .12 + .49)\\
&\quad = \lambda_1(.48, .52) + \lambda_2(.39, .61).
\end{aligned}$$

The probabilities p_{ij}, $i = 1, 2$, $j = 1, 2$, in Examples 1.2 and 1.3 can be summarized in a 2×2 matrix:

$$P = \begin{bmatrix} p_{11} & p_{12} \\ p_{21} & p_{22} \end{bmatrix}. \tag{10}$$

We saw that given a color distribution of poppies, (λ_1, λ_2), the expected color distribution in the next generation is obtained by forming a linear combination of the columns of P with coefficient λ_1 for the first column and coefficient λ_2 for the second column. That is, the color distribution after one generation is

$$\lambda_1(p_{11}, p_{21}) + \lambda_2(p_{12}, p_{22}). \tag{11}$$

We can look at (11) in a different way. Let $\lambda = (\lambda_1, \lambda_2)$ be a 2-vector and define the "product" $P\lambda$ of the 2×2 matrix P with the vector λ to be the vector (11). That is, the matrix-vector product $P\lambda$ is the vector obtained by forming a linear combination of the columns of P, using the components of λ as coefficients. This method is described in general in the following definition.

Definition 1.5 **Matrix-vector product** Let A be an $m \times n$ matrix,

$$A = \begin{bmatrix} a_{11} & a_{12} & \cdots & a_{1n} \\ a_{21} & a_{22} & \cdots & a_{2n} \\ \vdots & \vdots & & \vdots \\ a_{m1} & a_{m2} & \cdots & a_{mn} \end{bmatrix}, \tag{12}$$

that is, A is an $m \times n$ rectangular array of real numbers in which the entry in row i, column j, is designated by a_{ij}. Let $v = (v_1, v_2, \ldots, v_n)$ be an n-vector in R^n. Then the *matrix-vector product* Av is the m-vector obtained by forming the linear combination of the columns of A using v_k as the coefficient multiplying the kth column of A, $k = 1, \ldots, n$.

If we denote the columns of the matrix A in (12) by $A^{(1)}, A^{(2)}, \ldots, A^{(n)}$, then the matrix-vector product Av is

$$Av = A^{(1)}v_1 + A^{(2)}v_2 + \cdots + A^{(n)}v_n = \sum_{j=1}^{n} A^{(j)}v_j, \tag{13}$$

which is an m-vector.

Example 1.4 Three manufacturing plants produce four different items as follows: the ith plant produces a_{ik} units of the kth item, $i = 1, 2, 3$, $k = 1, 2, 3, 4$. Moreover, the sale of the kth item results in a profit of v_k dollars per unit, $k = 1, 2, 3, 4$. Express the profit earned by the three plants as a matrix-vector product. Consider the 3×4 matrix whose entry in row i, column k, is a_{ik}:

$$A = \begin{bmatrix} a_{11} & a_{12} & a_{13} & a_{14} \\ a_{21} & a_{22} & a_{23} & a_{24} \\ a_{31} & a_{32} & a_{33} & a_{34} \end{bmatrix}. \tag{14}$$

Consider the scalar multiple

$$A^{(1)}v_1 = (a_{11}v_1, a_{21}v_1, a_{31}v_1).$$

The first component, $a_{11}v_1$, is the number of units of item 1 produced by plant 1 multiplied by the profit per unit, that is, the product, $a_{11}v_1$ is the profit earned by plant 1 on the sale of item 1. Similarly, $a_{21}v_1$ is the profit earned by plant 2 through the sale of item 1, and $a_{31}v_1$ is the profit earned by plant 3 through the sale of item 1.

By identical reasoning we see that $A^{(2)}v_2$, $A^{(3)}v_3$, and $A^{(4)}v_4$ have as ith components the profit earned by plant i through the sale of items 2, 3, and 4, respectively. If we add these four vectors,

$$A^{(1)}v_1 + A^{(2)}v_2 + A^{(3)}v_3 + A^{(4)}v_4$$

we obtain a 3-vector whose first component is the total profit earned by plant 1 through the sale of all four items; the second component is the total profit earned by plant 2 through the sale of all four items, etc. In other words, the matrix-vector product Av is a 3-vector whose ith component is the total profit earned by plant i, $i = 1, 2, 3$.

Example 1.5 Suppose that a fluctuation in demand for the items produced in the plants in Example 1.4 results in a change in the profit earned per unit as follows. In the first half of the year, the vector $v_1 = (v_{11}, v_{21}, v_{31}, v_{41})$ represents the profit per unit on each of the four items. Similarly, $v_2 = (v_{12}, v_{22}, v_{32}, v_{42})$ represents the profit per unit on each of the four items during the second half of the year. Then the two matrix-vector products, Av_1 and Av_2, represent the profit earned by each of the three plants for each of the two six-month periods. We can write the two vectors Av_1 and Av_2 as the two columns of a 3×2 matrix. This provides a neat summary of the profits earned by the three plants semiannually. Now,

$$\begin{aligned} Av_1 &= A^{(1)}v_{11} + A^{(2)}v_{21} + A^{(3)}v_{31} + A^{(4)}v_{41} \\ &= \sum_{j=1}^{4} A^{(j)}v_{j1} \\ &= \sum_{j=1}^{4} (a_{1j}, a_{2j}, a_{3j})v_{j1} \\ &= \sum_{j=1}^{4} (a_{1j}v_{j1}, a_{2j}v_{j1}, a_{3j}v_{j1}) \\ &= \left(\sum_{j=1}^{4} a_{1j}v_{j1}, \sum_{j=1}^{4} a_{2j}v_{j1}, \sum_{j=1}^{4} a_{3j}v_{j1}\right). \end{aligned}$$

Similarly,

$$Av_2 = \left(\sum_{j=1}^{4} a_{1j}v_{j2}, \sum_{j=1}^{4} a_{2j}v_{j2}, \sum_{j=1}^{4} a_{3j}v_{j2}\right).$$

Thus the matrix whose columns are Av_1 and Av_2 is

$$\begin{bmatrix} \sum_{j=1}^{4} a_{1j}v_{j1} & \sum_{j=1}^{4} a_{1j}v_{j2} \\ \sum_{j=1}^{4} a_{2j}v_{j1} & \sum_{j=1}^{4} a_{2j}v_{j2} \\ \sum_{j=1}^{4} a_{3j}v_{j1} & \sum_{j=1}^{4} a_{3j}v_{j2} \end{bmatrix}. \tag{15}$$

If we let V be the 4×2 matrix whose columns are v_1 and v_2,

$$V = \begin{bmatrix} v_{11} & v_{12} \\ v_{21} & v_{22} \\ v_{31} & v_{32} \\ v_{41} & v_{42} \end{bmatrix},$$

then in analogy to the notation for matrix-vector products, we write AV for the matrix (15). The matrix AV is composed of the two matrices A and V and it is therefore called the *product* of the matrices A and V.

The discussion in Example 1.5 leads us to the following definition.

Definition 1.6 **Product of matrices** Let A be an $m \times n$ matrix and let B be an $n \times p$ matrix. Then the *product* of A and B, denoted by

$$AB,$$

is the matrix whose kth column is the matrix-vector product

$$AB^{(k)}, \qquad k = 1, \ldots, p,$$

where $B^{(k)}$ is the kth column of B regarded as an n-vector, i.e., $B^{(k)}$ is the n-vector

$$B^{(k)} = (b_{1k}, b_{2k}, \ldots, b_{nk}).$$

Thus the product of the two matrices A and B is the $m \times p$ matrix

$$\begin{bmatrix} AB^{(1)} & AB^{(2)} & \ldots & AB^{(p)} \\ \vdots & \vdots & & \vdots \end{bmatrix}, \tag{16}$$

where the dots indicate that for $k = 1, \ldots, p$, the kth column in (16) is the m-vector $AB^{(k)}$ written vertically.

Observe that the product AB is defined if and only if the number of columns of A is the same as the number of rows of B. Moreover, it is easy to see that the (i, j) entry of the product AB in (16) is the sum

$$\sum_{k=1}^{n} a_{ik}b_{kj}$$

(see Exercise 4). We can also think of the jth column of B, $B^{(j)}$, as an $n \times 1$ matrix,

$$B^{(j)} = \begin{bmatrix} b_{1j} \\ b_{2j} \\ \vdots \\ b_{nj} \end{bmatrix}.$$

Similarly, the n-vector which is the ith row of A, denoted by $A_{(i)} = (a_{i1}, \ldots, a_{in})$, can also be thought of as a matrix, namely, the $1 \times n$ matrix

$$A_{(i)} = [a_{i1}\ a_{i2} \ldots a_{in}].$$

Then the (i, j) entry of AB is also the single entry in the 1×1 matrix $A_{(i)}B^{(j)}$ (see Exercise 10).

Example 1.6 For each of the following pairs of matrices find the product AB.

(a) $A = \begin{bmatrix} 1 & 1 \\ -1 & 1 \end{bmatrix}$, $B = \begin{bmatrix} 1 & 2 & 0 \\ -1 & 5 & 3 \end{bmatrix}$.

We compute that

$$\begin{aligned} AB^{(1)} &= \begin{bmatrix} 1 & 1 \\ -1 & 1 \end{bmatrix}\begin{pmatrix} 1 \\ -1 \end{pmatrix} \\ &= \begin{pmatrix} 1 \\ -1 \end{pmatrix} 1 + \begin{pmatrix} 1 \\ 1 \end{pmatrix} (-1) \\ &= \begin{pmatrix} 1 \\ -1 \end{pmatrix} + \begin{pmatrix} -1 \\ -1 \end{pmatrix} \\ &= \begin{pmatrix} 1 + (-1) \\ -1 + (-1) \end{pmatrix} \\ &= \begin{pmatrix} 0 \\ -2 \end{pmatrix}. \end{aligned}$$

Similarly,

$$\begin{aligned} AB^{(2)} &= \begin{bmatrix} 1 & 1 \\ -1 & 1 \end{bmatrix}\begin{pmatrix} 2 \\ 5 \end{pmatrix} \\ &= \begin{pmatrix} 1 \\ -1 \end{pmatrix} 2 + \begin{pmatrix} 1 \\ 1 \end{pmatrix} 5 \\ &= \begin{pmatrix} 7 \\ 3 \end{pmatrix}, \end{aligned}$$

and

$$\begin{aligned} AB^{(3)} &= \begin{bmatrix} 1 & 1 \\ -1 & 1 \end{bmatrix}\begin{pmatrix} 0 \\ 3 \end{pmatrix} \\ &= \begin{pmatrix} 1 \\ -1 \end{pmatrix} 0 + \begin{pmatrix} 1 \\ 1 \end{pmatrix} 3 \\ &= \begin{pmatrix} 3 \\ 3 \end{pmatrix}. \end{aligned}$$

Thus

$$\begin{aligned} AB &= \begin{bmatrix} AB^{(1)} & AB^{(2)} & AB^{(3)} \\ \vdots & \vdots & \vdots \end{bmatrix} \\ &= \begin{bmatrix} 0 & 7 & 3 \\ -2 & 3 & 3 \end{bmatrix}. \end{aligned}$$

Observe that we have written the columns $B^{(k)}$ vertically because this notation suggests that they are columns. After all, a vector is simply an m-tuple, and its meaning is unchanged by the way it is written.

(b) $A = B = \begin{bmatrix} 0 & 1 \\ 0 & 0 \end{bmatrix}$. The product in this case is computed as follows:

$$AB^{(1)} = A\begin{pmatrix} 0 \\ 0 \end{pmatrix} = \begin{pmatrix} 0 \\ 0 \end{pmatrix}$$

and

$$\begin{aligned} AB^{(2)} &= \begin{bmatrix} 0 & 1 \\ 0 & 0 \end{bmatrix}\begin{pmatrix} 1 \\ 0 \end{pmatrix} \\ &= \begin{pmatrix} 0 \\ 0 \end{pmatrix} 1 + \begin{pmatrix} 1 \\ 0 \end{pmatrix} 0 \\ &= \begin{pmatrix} 0 \\ 0 \end{pmatrix}. \end{aligned}$$

Thus

$$\begin{aligned} AB &= AA \\ &= \begin{bmatrix} 0 & 0 \\ 0 & 0 \end{bmatrix}. \end{aligned}$$

This example shows that the product of two matrices can have every entry zero although neither factor has this property.

(c) Let A be any $m \times n$ matrix and let $B = I_n$ be the $n \times n$ matrix defined as follows: for $k = 1, \ldots, n$, the kth column is the n-tuple with its kth component equal to 1 and all other components equal to 0, $k = 1, \ldots, n$. The kth column of the product AI_n is by definition:

$$\begin{aligned} A\begin{pmatrix} 0 \\ \vdots \\ 0 \\ 1 \\ 0 \\ \vdots \\ 0 \end{pmatrix} \begin{matrix} \\ \\ \\ \leftarrow k \\ \\ \\ \\ \end{matrix} &= A^{(1)} \cdot 0 + A^{(2)} \cdot 0 + \cdots + A^{(k-1)} \cdot 0 + A^{(k)} \cdot 1 + A^{(k+1)} \cdot 0 + \cdots + A^{(n)} \cdot 0 \\ &= A^{(k)}, \ k = 1, \ldots, n. \end{aligned}$$

In other words, the kth column of AI_n is precisely $A^{(k)}$, so that

$$AI_n = A.$$

The matrix I_n is called the $n \times n$ *identity matrix*.

(d) Let I_m be the $m \times m$ identity matrix. We compute I_mA for an arbitrary $m \times n$ matrix A. According to the definition of matrix multiplication, the kth column of $I_m A$ is

$$I_m A^{(k)} = I_m \begin{pmatrix} a_{1k} \\ a_{2k} \\ \vdots \\ a_{mk} \end{pmatrix}$$

$$= \begin{pmatrix} 1 \\ 0 \\ 0 \\ \vdots \\ 0 \end{pmatrix} a_{1k} + \begin{pmatrix} 0 \\ 1 \\ 0 \\ \vdots \\ 0 \end{pmatrix} a_{2k} + \cdots + \begin{pmatrix} 0 \\ 0 \\ \vdots \\ 0 \\ 1 \end{pmatrix} a_{mk}$$

$$= A^{(k)}.$$

In other words, the kth column of $I_m A$ is $A^{(k)}$, so that

$$I_m A = A.$$

Corresponding to scalar multiplication and vector addition, we can define similar operations for matrices.

Definition 1.7 ***Addition and scalar multiplication of matrices***

(a) Let A be an $m \times n$ matrix and let c be a number. Then the *scalar product*

$$cA = Ac$$

is the matrix obtained from A by multiplying every entry of A by the number c. In other words, the kth column of cA is the scalar multiple of the m-tuple $A^{(k)}$,

$$(cA)^{(k)} = cA^{(k)}.$$

(b) If A and B are $m \times n$ matrices, then the *sum* of A and B is the $m \times n$ matrix whose kth column is the sum $A^{(k)} + B^{(k)}$, $k = 1, \ldots, n$. The sum is denoted by

$$A + B.$$

Thus the entry in row i column k of the matrix $A + B$ is the sum

$$a_{ik} + b_{ik},$$

where a_{ik} and b_{ik} are the entries in row i, column k of A and B, respectively, $i = 1, \ldots, m$ and $k = 1, \ldots, n$.

Theorem 1.2 (a) *Let A be an $m \times n$ matrix and let B and C be $n \times p$ matrices. Then*

$$A(B + C) = AB + AC. \tag{17}$$

(b) *If A, B, and C are $m \times n$, $n \times p$, and $p \times q$ matrices, respectively, then*

$$A(BC) = (AB)C. \tag{18}$$

Proof

(a) By the definition of the sum of two matrices, the kth column of $B + C$ is the n-vector

$$(B + C)^{(k)} = B^{(k)} + C^{(k)} = \begin{pmatrix} b_{1k} + c_{1k} \\ \vdots \\ b_{nk} + c_{nk} \end{pmatrix}.$$

Hence the kth column of the matrix $A(B + C)$ is

$$\begin{aligned} A(B^{(k)} + C^{(k)}) &= A^{(1)}(b_{1k} + c_{1k}) + \cdots + A^{(n)}(b_{nk} + c_{nk}) \\ &= \sum_{j=1}^{n} A^{(j)} b_{jk} + \sum_{j=1}^{n} A^{(j)} c_{jk} \\ &= AB^{(k)} + AC^{(k)}. \end{aligned} \tag{19}$$

By the definition of matrix product, $AB^{(k)}$ is the kth column of the product AB, and $AC^{(k)}$ is the kth column of the product AC. But then, by the definition of matrix addition, the right hand side of (19) is the kth column of the sum $AB + AC$. It should be observed that the second equality in (19) is an application of Theorem 1.1 (h) and Theorem 1.1 (c), extended to several vectors (see Exercise 24). Equation (17) is appropriately called the *distributive law* for matrices.

(b) By definition, the kth column of the matrix $A(BC)$ is

$$\begin{aligned} A(BC)^{(k)} &= A \sum_{j=1}^{p} B^{(j)} c_{jk} \\ &= \sum_{j=1}^{p} A(B^{(j)} c_{jk}) \\ &= \sum_{j=1}^{p} (AB)^{(j)} c_{jk}. \end{aligned} \tag{20}$$

The second equality in (20) follows from the equality

$$A(v_1 + v_2 + \cdots + v_p) = Av_1 + Av_2 + \cdots + Av_p$$

where $v_1, \ldots, v_p$ are any n-vectors and A is an $m \times n$ matrix (see Exercise 5). The third equality in (20) follows from the fact that $A(vc) = (Av)c$ when v is any n-vector and c is a number (see Exercise 6). But the last expression in (20) is by definition the matrix-vector product $(AB)C^{(k)}$. Conversely, the kth column of $(AB)C$ is, by definition of matrix multiplication, $(AB)C^{(k)}$. Therefore the kth column of $A(BC)$ is the same as the kth column of $(AB)C$. In view of this, we write ABC for the triple matrix product

$$A(BC) = (AB)C. \qquad (21)$$ ▮

The equality (21) is the *associative law* of matrix multiplication. The associative law allows us to compute powers of an *n-square matrix* (i.e., an $n \times n$ matrix) precisely as we do with ordinary numbers. Thus in the product

$$AAA$$

it makes no difference where we insert the parentheses:

$$(AA)A = A(AA).$$

Similarly, the numerous ways (How many?) in which we can convert

$$AAAA$$

into a product of two matrices, e.g.,

$$(AA)(AA) = A(A(AA)),$$

must all have the same value. In general, if we want to form a product using A as a factor p times, we will denote the answer by

$$A^p. \qquad (22)$$

When p is a positive integer, A^p is called the pth *power* of A. All the usual laws of exponents must hold:

$$\begin{aligned} A^pA^q &= \overbrace{A \cdots A}^{p} \cdot \overbrace{A \cdots A}^{q} \\ &= \overbrace{A \cdots A}^{p+q} \\ &= A^{p+q}. \end{aligned}$$

In case $p = 0$, we define

$$A^0 = I_n.$$

Several other items of notation are convenient. If A is an $n \times n$ matrix with $a_{ij} = 0$ whenever $i \neq j$, then A is called a *diagonal* matrix:

$$A = \begin{bmatrix} a_{11} & 0 & \cdots & 0 \\ 0 & a_{22} & & \vdots \\ \vdots & 0 & \ddots & \vdots \\ \vdots & \vdots & \ddots & 0 \\ 0 & 0 & \cdots\ 0 & a_{nn} \end{bmatrix}.$$

We observe that although the associative and distributive laws hold for matrix multiplication, the *commutative law*, $AB = BA$, definitely is not valid for arbitrary matrices A and B, even when both are square matrices. It is easy to construct 2×2 examples for which commutativity fails. (See, e.g., Exercise 1(d), (e).) The matrix A is sometimes denoted by

$$A = \text{diag}(a_{11}, \ldots, a_{nn}).$$

In general, if A is any $n \times n$ matrix, then the n-tuple of numbers $(a_{11}, \ldots, a_{nn})$ is called the *main diagonal* of A. The $m \times n$ matrix each of whose entries is zero is called a *zero matrix* and is denoted by

$$0_{mn}.$$

Often, if the meaning is clear, we simply write 0 instead of 0_{mn}.

If A is an $m \times n$ matrix, then we can construct a related $n \times m$ matrix called the *transpose* of A by defining the (i, j) entry of the transpose to be the (j, i) entry of A. The usual notation for the transpose of A is

$$A^T.$$

Thus if

$$A = \begin{bmatrix} a_{11} & a_{12} & \cdots & a_{1n} \\ a_{21} & a_{22} & \cdots & a_{2n} \\ \vdots & & & \vdots \\ a_{m1} & a_{m2} & \cdots & a_{mn} \end{bmatrix},$$

then

$$A^T = \begin{bmatrix} a_{11} & a_{21} & \cdots & a_{m1} \\ a_{12} & a_{22} & \cdots & a_{m2} \\ \vdots & & & \vdots \\ a_{1n} & a_{2n} & \cdots & a_{mn} \end{bmatrix}.$$

The first column of A^T is the first row of A; the second column of A^T is the second row of A; . . . ; the mth column of A^T is the mth row of A. For example, if

$$A = \begin{bmatrix} 1 & 4 & 5 \\ 3 & 6 & 2 \end{bmatrix},$$

then

$$A^T = \begin{bmatrix} 1 & 3 \\ 4 & 6 \\ 5 & 2 \end{bmatrix}.$$

Quiz

Answer true or false.

1 If u is an m-vector, then $0u = 0_m$. (Here 0 is the scalar 0 and 0_m is the m-vector with each component 0.)

2 If u and v are n-vectors, then $u = v$ if and only if the kth component of u is equal to the kth component of v, $k = 1, \ldots, n$. (Hint: What is the requirement for the equality of two functions? Think in particular about functions with domain $\{1, 2, \ldots, n\}$.)

3 If c and d are numbers and v is an n-vector, then $(cd)v = c(dv)$.

4 If v is a 2-vector both of whose components are positive, then $-v$ has the following geometric representation.

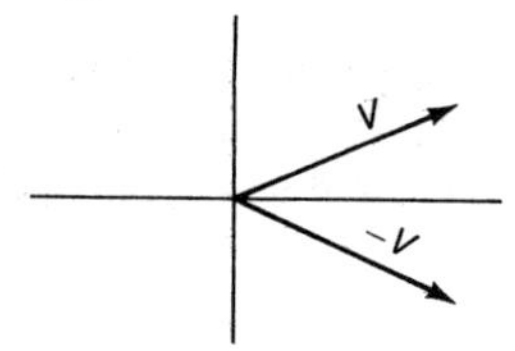

5 If every entry of a 2×2 matrix A is non-zero, then A^2 cannot have every entry equal to zero.

6 The product of a 2×3 and a 3×2 matrix is a 2×3 matrix.

7 If the product of two matrices is an n-square matrix, then both matrices in the product must be square.

†8 If A is an $m \times n$ matrix, and the *negative* of A (denoted by $-A$) is the $m \times n$ matrix whose entries are the negatives of the entries of A, then $A + (-A)$ has every entry equal to zero.

†9 The following equality is correct:

$$B - A = -(A - B).$$

(The notation is that of the previous question: A and B are $m \times n$ matrices, and $B - A$ denotes the matrix $B + (-A)$, called the *difference* of B and A.)

10 If I_m denotes the m-square identity matrix, then $I_m I_m = I_m$.

Exercises

1 Compute the following matrix products:

(a) $\begin{bmatrix} 1 & 5 \\ 4 & 3 \\ 6 & 1 \end{bmatrix} \begin{bmatrix} 1 & 3 & 7 \\ 0 & 5 & -2 \end{bmatrix}$;

(b) $[1 \quad 0 \quad -1] \begin{bmatrix} 5 & 2 \\ 8 & 2 \\ 6 & -7 \end{bmatrix}$;

(c) $\begin{bmatrix} 0 & 1 & 3 & 2 \\ 1 & 1 & 0 & 3 \\ 2 & 0 & 3 & 0 \end{bmatrix} \begin{bmatrix} 0 & 3 & 2 & 3 & 0 \\ 1 & 2 & 2 & 3 & 3 \\ 3 & 1 & 2 & 1 & 0 \\ 0 & 3 & 3 & 0 & 0 \end{bmatrix}$;

(d) $\begin{bmatrix} 0 & 5 & 0 \\ 2 & 7 & 6 \\ 3 & -3 & 5 \end{bmatrix} \begin{bmatrix} 1 & 4 & 5 \\ 5 & 3 & -1 \\ 0 & -2 & 0 \end{bmatrix}$;

(e) $\begin{bmatrix} 1 & 4 & 5 \\ 5 & 3 & -1 \\ 0 & -2 & 0 \end{bmatrix} \begin{bmatrix} 0 & 5 & 0 \\ 2 & 7 & 6 \\ 3 & -3 & 5 \end{bmatrix}$

(compare with (d));

(f) $[2][3]$;

(g) $\begin{bmatrix} 1 & 0 & 0 \\ 0 & 1 & 0 \\ 0 & 0 & 1 \end{bmatrix} \begin{bmatrix} 0 & 0 & 1 \\ 0 & 1 & 0 \\ 1 & 0 & 0 \end{bmatrix}$;

(h) $\begin{bmatrix} -1 & 2 & 3 \\ 2 & -1 & 4 \\ 0 & 0 & 0 \end{bmatrix} \begin{bmatrix} 0 & 5 & 1 & 4 \\ 1 & 0 & 2 & 1 \\ -2 & 3 & 0 & 5 \end{bmatrix} \begin{bmatrix} 3 & 0 \\ 3 & 1 \\ 4 & 2 \\ 4 & 5 \end{bmatrix}$;

(i) $\begin{bmatrix} 0 & 1 & 0 \\ 1 & 0 & 0 \\ 0 & 0 & 1 \end{bmatrix} \begin{bmatrix} 2 & 5 & 7 \\ 3 & 2 & 1 \\ 4 & 3 & 2 \end{bmatrix}$;

(j) $\begin{bmatrix} 1 & 0 & 0 & 0 \\ 0 & 0 & 1 & 0 \\ 0 & 0 & 0 & 1 \\ 0 & 1 & 0 & 0 \end{bmatrix} \begin{bmatrix} 4 & 3 \\ 1 & 2 \\ 2 & 2 \\ 3 & 7 \end{bmatrix}$;

(k) $\begin{bmatrix} 1 & -1 & 0 \\ 0 & 0 & -1 \\ 0 & 0 & 0 \end{bmatrix}^3$;

(l) $\begin{bmatrix} 1 & 2 & 3 \\ 0 & -1 & 0 \\ 1 & 1 & 1 \end{bmatrix} \begin{bmatrix} 1 & 0 & 1 \\ 2 & -1 & 1 \\ 3 & 0 & 1 \end{bmatrix}$;

(m) $\begin{bmatrix} 1 & 0 & 1 \\ 2 & -1 & 1 \\ 3 & 0 & 1 \end{bmatrix} \begin{bmatrix} 1 & 2 & 3 \\ 0 & -1 & 0 \\ 1 & 1 & 1 \end{bmatrix}$;

(n) $\begin{bmatrix} 1 & 1 \\ 1 & -1 \\ 1 & 1 \end{bmatrix}\begin{bmatrix} 2 & -1 & 0 \\ 1 & 0 & 0 \end{bmatrix}$;

(o) $\begin{bmatrix} 0 & 1 & 0 & 0 \\ 0 & 0 & 1 & 0 \\ 0 & 0 & 0 & 1 \\ 0 & 0 & 0 & 0 \end{bmatrix}^4$.

†2 Show that if c and d are numbers, A and B are $m \times n$ matrices, and D is an $n \times p$ matrix, then:

(a) $(c + d)A = cA + dA$;
(b) $(cd)A = c(dA)$;
(c) $c(A + B) = cA + cB$;
(d) $c(A - B) = cA - cB$ (see Quiz Question 9);
(e) $(cA)D = c(AD)$;
(f) $(A - B)D = AD - BD$.

†3 If A is an $m \times n$ matrix, show that

$$A0_{np} = 0_{mp}$$

and

$$0_{pm}A = 0_{pn}.$$

†4 Let A be an $m \times n$ matrix and B be an $n \times p$ matrix. Let $C = AB$. If c_{ij} is the (i, j) entry of C, show that

$$c_{ij} = \sum_{k=1}^{n} a_{ik}b_{kj}.$$

(Hint: As in the matrix (16), the jth column of C is

$$AB^{(j)} = \begin{bmatrix} A^{(1)} & \cdots & A^{(n)} \\ \cdot & & \cdot \\ \cdot & & \cdot \\ \cdot & & \cdot \end{bmatrix}\begin{pmatrix} b_{1j} \\ b_{2j} \\ \vdots \\ b_{nj} \end{pmatrix}$$

$$= A^{(1)}b_{1j} + A^{(2)}b_{2j} + \cdots + A^{(n)}b_{nj}.$$

We want to compute the ith component of this linear combination of column vectors, for it is precisely the entry in the ith row and jth column of C, i.e., it is c_{ij}.)

†5 Prove that if A is an $m \times n$ matrix and $v_1, \ldots, v_p$ are n-vectors, then

$$A\left(\sum_{j=1}^{p} v_j\right) = \sum_{j=1}^{p} Av_j.$$

†6 Prove that if A is an $m \times n$ matrix, v is an n-vector, and c is a number, then

$$c(Av) = A(cv) = (Av)c = A(vc) = (Ac)v.$$

7 Show by induction that if A is an n-square matrix with nonnegative entries, then A^p has nonnegative entries for any positive integer p.

(Hint: For $p = 1$, $A^1 = A$ has nonnegative entries. Thus suppose that A^{p-1} has nonnegative entries. Then $A^p = AA^{p-1}$, so that any column of A^p is a linear combination of columns of A using entries from A^{p-1} as scalar coefficients.)

8 If A is an $m \times n$ matrix, show that

$$(A^T)^T = A.$$

9 Find the transposes of the following matrices:

(a) $\begin{bmatrix} 1 \\ 2 \\ 3 \end{bmatrix}$;

(b) $[4 \quad 5 \quad 6]$;

(c) $\begin{bmatrix} 1 & 5 & -2 & -3 \\ 2 & -4 & 3 & -1 \\ 0 & 0 & 0 & 0 \end{bmatrix}$;

(d) $\begin{bmatrix} 1 & 2 & 0 \\ 1 & 0 & 2 \end{bmatrix} \begin{bmatrix} 1 & 1 \\ 2 & 0 \\ 0 & 2 \end{bmatrix}$;

(e) $\begin{bmatrix} 1 & 1 & 1 & 1 \\ 1 & 1 & 1 & 1 \\ 1 & 1 & 1 & 1 \end{bmatrix}$.

†10 Let A be an $m \times n$ matrix and B be an $n \times p$ matrix. Let $A_{(i)}$ be the ith row of A regarded as a $1 \times n$ matrix and let $B^{(j)}$ be the jth column of B regarded as an $n \times 1$ matrix. Show that the product $A_{(i)}B^{(j)}$ is a 1×1 matrix whose single entry is the (i, j) entry of AB. (Hint:

$$\begin{aligned} A_{(i)}B^{(j)} &= [a_{i1}a_{i2} \dots a_{in}] \begin{bmatrix} b_{1j} \\ b_{2j} \\ \vdots \\ b_{nj} \end{bmatrix} \\ &= [a_{i1}b_{1j} + a_{i2}b_{2j} + \cdots + a_{in}b_{nj}]. \end{aligned}$$

Compare this with Exercise 4.)

†11 Let A be an $m \times n$ matrix and B be an $n \times p$ matrix. Show that

$$(AB)^T = B^T A^T.$$

(Hint: The (s, t) entry of $(AB)^T$ is by definition the (t, s) entry of AB. According to the preceding exercise, this is just the single entry in the 1×1 matrix $A_{(t)}B^{(s)}$. On the other hand, the (s, t) entry of $B^T A^T$ is the single entry in the 1×1 matrix $(B^T)_{(s)}(A^T)^{(t)}$. But the sth row of B^T is by definition the sth column of B and the tth column of A^T is the tth row of A, i.e., $(B^T)_{(s)} = B^{(s)}$ and $(A^T)^{(t)} = A_{(t)}$. Hence $(B^T)_{(s)}(A^T)^{(t)} = A_{(t)}B^{(s)}$. But we saw that $A_{(t)}B^{(s)}$ is the 1×1 matrix whose single entry is the (s, t) entry of $(AB)^T$. We have proved that the (s, t) entry of $(AB)^T$ is the same as the (s, t) entry of $B^T A^T$.)

12 Let

$$A = \begin{bmatrix} a_{11} & a_{12} & a_{13} & a_{14} \\ a_{21} & a_{22} & a_{23} & a_{24} \\ a_{31} & a_{32} & a_{33} & a_{34} \end{bmatrix}$$

and

$$B = \begin{bmatrix} b_{11} & b_{12} \\ b_{21} & b_{22} \\ b_{31} & b_{32} \\ b_{41} & b_{42} \end{bmatrix}.$$

Write out the following matrices explicitly for this choice of A and B:

(a) $A_{(1)}$, $A_{(2)}$, $A_{(3)}$, as 1×4 matrices;
(b) $A^{(1)}$, $A^{(2)}$, $A^{(3)}$, $A^{(4)}$, as 3×1 matrices;
(c) $(A^T)_{(1)}$;
(d) $(A_{(1)})^T$;
(e) $(B^T)_{(1)}$;
(f) $(B_{(1)})^T$;
(g) $(AB)_{(2)}$;
(h) $(AB)^T$;
(i) $B^T A^T$;
(j) $A_{(1)} B^{(2)}$;
(k) $(B^T)_{(2)}(A^T)^{(1)}$;
(l) $((A^T)^T)_{(1)}$;
(m) $((A^T)_{(1)})^T$;
(n) $((B^T)^{(2)})^T$;
(o) $(((B^T)_{(1)})^T)_{(2)}$.

13 If

$$A = \begin{bmatrix} 1 & 1 \\ 1 & 1 \end{bmatrix},$$

find A^2, A^3, A^4. Do you see a pattern emerging? Write down a formula for A^p, and show that your formula is valid for all positive integers p by using mathematical induction.

†**14** Let p and q be nonnegative integers and let A be an n-square matrix. Show that $A^p A^q = A^q A^p$.

†**15** Let A be an $m \times n$ matrix and B and C be $p \times m$ matrices. Show that

$$(B + C)A = BA + CA.$$

16 Show that for any $n \times m$ matrix A, it is always possible to form both products AA^T and $A^T A$. Show that if A is not a square matrix, then

$$AA^T \neq A^T A.$$

Show by constructing an example that even if A is a square matrix, it is not necessarily the case that $AA^T = A^T A$.

17 Let

$$A = \begin{bmatrix} 1 & -1 & 0 & -4 \\ 3 & -2 & 4 & 0 \end{bmatrix}.$$

Compute AA^T and A^TA.

†**18** Let A and B be two $m \times n$ matrices. Show that

$$(A + B)^T = A^T + B^T,$$

and

$$(cA)^T = cA^T$$

for any number c.

19 If

$$A = \begin{bmatrix} 1 & -1 & 0 \\ 0 & 0 & 0 \\ 1 & 2 & 3 \end{bmatrix},$$

find a non-zero 3-vector $v = (v_1, v_2, v_3)$ such that $Av = 0_3$.

20 Find a linear combination of the 4-vectors $(1, 0, 0, 0)$, $(0, 1, 0, 0)$, $(0, 0, 1, 0)$, and $(0, 0, 0, 1)$ equal to the vector (v_1, v_2, v_3, v_4). Is there more than one solution to this problem? Why?

21 Let A be an $m \times n$ matrix. Let E^j be the $n \times 1$ matrix whose $(j, 1)$ entry is 1 and whose other entries are 0. Show that

$$AE^j = A^{(j)},$$

regarded as an $m \times 1$ matrix.

22 Let E_i be the $1 \times m$ matrix whose $(1, i)$ entry is 1 and whose other entries are 0 and let A be an $m \times n$ matrix. Show that

$$E_iA = A_{(i)},$$

regarded as a $1 \times n$ matrix.

23 In the notation of the preceding two exercises, show that

$$E_iAE^j = [a_{ij}].$$

24 State and prove Theorem 1.1(b), (c), and (h) extended to more than three vectors. (Hint: Use induction.)

2.2

Incidence Matrices and Applications

In the preceding section, we introduced and developed the elementary properties of matrix addition and multiplication. Our task now is to consider some applications of matrix theory to a number of interesting combinatorial problems.

We remind the reader of the definition of an incidence matrix of a relation (Chapter 1, Definition 3.4): if $X = \{x_1, \ldots, x_n\}$ and $Y = \{y_1, \ldots, y_m\}$, and if R is a relation on X to Y, then the incidence matrix for R is the $m \times n$ matrix $A(R)$ whose (i, j) entry is 1 or 0 according as $(x_j, y_i) \in R$ or $(x_j, y_i) \notin R$, respectively.

An important question that immediately arises is this: What is the interpretation, in terms of relations, of the algebraic operations of addition and multiplication of matrices?

Example 2.1 Let $X = \{x_1, x_2, x_3, x_4, x_5, x_6, x_7, x_8, x_9\}$ be a set of nine geographical points. Consider the road map below which connects the points in X. The arrows indicate the directions in which the roads may be traversed.

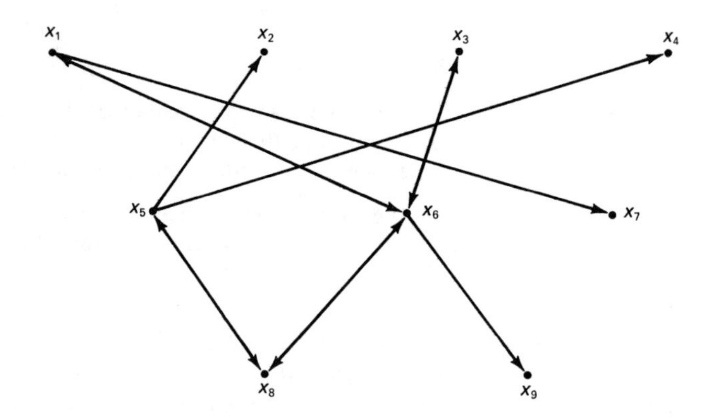

Define a relation R on X to X as follows: a pair (x_i, x_j) is an element of R if and only if there is a road from x_i to x_j. We omit the pairs (x_i, x_i) since if someone is already at x_i, there is no problem in getting there. If $A = A(R)$ denotes the incidence matrix for R then, for example, $a_{17} = 0$ and $a_{71} = 1$ because there is only a one-way road from x_1 to x_7. On the other hand, $a_{68} = a_{86} = 1$, since the road connecting x_6 and x_8 runs in both directions. Suppose we consider the product $a_{ik}a_{kj}$ of two entries in $A = A(R)$. Then $a_{ik}a_{kj}$ is 1 when

$$a_{ik} = a_{kj} = 1 \tag{1}$$

and is 0 when either

$$a_{ik} = 0 \quad \text{or} \quad a_{kj} = 0. \tag{2}$$

For the conditions (1) to hold there must be a road connecting x_j to x_k and a road connecting x_k to x_i. In other words, there must be a road from x_j to x_i via x_k. On the other hand, if either or both of the conditions (2) hold, it means that there is no road connecting x_j to x_k or there is no road connecting x_k to x_i, i.e., it is not possible to go from x_j to x_i via x_k. We can summarize this discussion: the product

$$a_{ik}a_{kj} \tag{3}$$

is 1 or 0 according as x_j is connected to x_i via x_k or not. If we sum the numbers in (3) over $k = 1, \ldots, 9$, we obtain a count of the total number of roads connecting x_j to x_i via intermediate points. But the sum of the numbers in (3) is

$$\sum_{k=1}^{9} a_{ik}a_{kj},$$

and this is just the (i, j) entry of the matrix A^2. In other words, the (i, j) entry of the matrix A^2 is the total number of ways in which the trip from x_j to x_i can be made via any intermediate point.

Next, consider the sum of the two matrices A and A^2:

$$S = A + A^2. \tag{4}$$

The (i, j) entry of S is the sum of the (i, j) entry of A and the (i, j) entry of A^2. Now the (i, j) entry of A is 1 if x_j is connected directly to x_i and 0 otherwise; the (i, j) entry of A^2 is the number of routes from x_j to x_i via an intermediate x_k $(k \neq i, k \neq j)$. Thus the (i, j) entry of S is the number of routes from x_j to x_i which are either direct or else go through precisely one intermediate point. We compute that

$$A = \begin{array}{c} \\ x_1 \\ x_2 \\ x_3 \\ x_4 \\ x_5 \\ x_6 \\ x_7 \\ x_8 \\ x_9 \end{array} \begin{array}{c} \begin{array}{ccccccccc} x_1 & x_2 & x_3 & x_4 & x_5 & x_6 & x_7 & x_8 & x_9 \end{array} \\ \begin{bmatrix} 0 & 0 & 0 & 0 & 0 & 1 & 0 & 0 & 0 \\ 0 & 0 & 0 & 0 & 1 & 0 & 0 & 0 & 0 \\ 0 & 0 & 0 & 0 & 0 & 1 & 0 & 0 & 0 \\ 0 & 0 & 0 & 0 & 1 & 0 & 0 & 0 & 0 \\ 0 & 0 & 0 & 0 & 0 & 0 & 0 & 1 & 0 \\ 1 & 0 & 1 & 0 & 0 & 0 & 0 & 1 & 0 \\ 1 & 0 & 0 & 0 & 0 & 0 & 0 & 0 & 0 \\ 0 & 0 & 0 & 0 & 1 & 1 & 0 & 0 & 0 \\ 0 & 0 & 0 & 0 & 0 & 1 & 0 & 0 & 0 \end{bmatrix} \end{array}.$$

Then

$$A^2 = \begin{bmatrix} 1 & 0 & 1 & 0 & 0 & 0 & 0 & 1 & 0 \\ 0 & 0 & 0 & 0 & 0 & 0 & 0 & 1 & 0 \\ 1 & 0 & 1 & 0 & 0 & 0 & 0 & 1 & 0 \\ 0 & 0 & 0 & 0 & 0 & 0 & 0 & 1 & 0 \\ 0 & 0 & 0 & 0 & 1 & 1 & 0 & 0 & 0 \\ 0 & 0 & 0 & 0 & 1 & 3 & 0 & 0 & 0 \\ 0 & 0 & 0 & 0 & 0 & 1 & 0 & 0 & 0 \\ 1 & 0 & 1 & 0 & 0 & 0 & 0 & 2 & 0 \\ 1 & 0 & 1 & 0 & 0 & 0 & 0 & 1 & 0 \end{bmatrix}$$

and

$$S = A + A^2$$

$$= \begin{bmatrix} 1 & 0 & 1 & 0 & 0 & 1 & 0 & 1 & 0 \\ 0 & 0 & 0 & 0 & 1 & 0 & 0 & 1 & 0 \\ 1 & 0 & 1 & 0 & 0 & 1 & 0 & 1 & 0 \\ 0 & 0 & 0 & 0 & 1 & 0 & 0 & 1 & 0 \\ 0 & 0 & 0 & 0 & 1 & 1 & 0 & 1 & 0 \\ 1 & 0 & 1 & 0 & 1 & 3 & 0 & 1 & 0 \\ 1 & 0 & 0 & 0 & 0 & 1 & 0 & 0 & 0 \\ 1 & 0 & 1 & 0 & 1 & 1 & 0 & 2 & 0 \\ 1 & 0 & 1 & 0 & 0 & 1 & 0 & 1 & 0 \end{bmatrix}.$$

Example 2.1 tells us that the square of the incidence matrix A is significant in that if b_{ij} is the (i, j) element of A^2, then b_{ij} is precisely the number of elements x_k such that $(x_j, x_k) \in R$ and $(x_k, x_i) \in R$. We prove a theorem about the elements in a product of matrices and then show how this can be interpreted for powers of incidence matrices.

Theorem 2.1 *Let $A_1, A_2, \ldots, A_p$ be p matrices for which the product*

$$C = A_1 A_2 \cdots A_p \tag{5}$$

is defined, i.e., A_1 is $m \times n_1$, A_2 is $n_1 \times n_2$, A_3 is $n_2 \times n_3, \ldots,$ A_{p-1} is $n_{p-2} \times n_{p-1}$, A_p is $n_{p-1} \times n$. Let $a_{ij}^{(t)}$ denote the (i, j) entry of A_t, $t = 1, \ldots, p$. Then the (i, j) entry of C is the sum of all products

$$a_{ik_1}^{(1)} a_{k_1k_2}^{(2)} a_{k_2k_3}^{(3)} \cdots a_{k_{p-2}\,k_{p-1}}^{(p-1)} a_{k_{p-1}j}^{(p)} \tag{6}$$

as $k_1, \ldots, k_p$ take on all values, $k_1 = 1, \ldots, n_1$; $k_2 = 1, \ldots, n_2$; $k_3 = 1, \ldots, n_3; \ldots; k_{p-1} = 1, \ldots, n_{p-1}$.

Proof The proof is by induction on p, the number of matrices. By the definition of matrix multiplication, we know that the (s, t) entry in a product of any two matrices XY is the sum of all products of the form

$$x_{sk} y_{kt} \tag{7}$$

in which we sum over $k = 1, \ldots, r$, r being the common number of columns of X and rows of Y. Now let the role of X be played by $A_1 A_2 \cdots A_{p-1}$, the product of the first $p - 1$ of the A_i's, and let the role of Y be played by A_p. Then the (i, j) entry of XY is the sum of all terms

(8) $$x_{ik_{p-1}}a^{(p)}_{k_{p-1}j}$$

as k_{p-1} takes on the values $1, 2, \ldots, n_{p-1}$. By the induction hypothesis, the (i, k_{p-1}) entry of the product $X = A_1A_2 \cdots A_{p-1}$ is the sum of all products of the form

(9) $$a^{(1)}_{ik_1}a^{(2)}_{k_1k_2}a^{(3)}_{k_2k_3} \cdots a^{(p-1)}_{k_{p-2}k_{p-1}}$$

in which k_1 takes on the values $1, \ldots, n_1$; k_2 takes on the values $1, \ldots, n_2; \ldots ; k_{p-2}$ takes on the values $1, \ldots, n_{p-2}$. If we substitute (9) into (8), we see that the (i, j) entry of

$$XY = (A_1A_2 \cdots A_{p-1})A_p = A_1 \cdots A_p$$

is just the sum of all products of the form

(10) $$(a^{(1)}_{ik_1}a^{(2)}_{k_1k_2}a^{(3)}_{k_2k_3} \cdots a^{(p-1)}_{k_{p-2}k_{p-1}})a^{(p)}_{k_{p-1}j}$$

as k_1 takes on the values $1, \ldots, n_1; \ldots ; k_{p-1}$ takes on the values $1, \ldots, n_{p-1}$. But (10) is precisely the same as (6) and this completes the proof. ▮

Example 2.2 Let A_1, A_2, and A_3 be three matrices for which the product $C = A_1A_2A_3$ is defined. Show that if the ith row of A_1 consists entirely of zeros, then so does the ith row of C. This is quite easy if we use the result of Theorem 2.1, for we know that the (i, j) entry of C is a sum of products of the form

(11) $$a^{(1)}_{ik_1}a^{(2)}_{k_1k_2}a^{(3)}_{k_2j}$$

as k_1 and k_2 vary. But $a^{(1)}_{ik_1} = 0$ for any value of k_1 since we are assuming that the ith row of A_1 consists of zeros. Thus each of the products in (11) is zero and hence their sum is zero.

Definition 2.1 ***p-step connection*** Let $X_1, \ldots, X_{p+1}$ be $p + 1$ sets. Let R_1 be a relation on X_p to X_{p+1}, R_2 a relation on X_{p-1} to $X_p, \ldots, R_p$ a relation on X_1 to X_2. If $x_1 \in X_1$ and $x_{p+1} \in X_{p+1}$, then we say that there exists a *p-step connection* from x_1 to x_{p+1} if there are elements $x_2 \in X_2$, $x_3 \in X_3, \ldots, x_p \in X_p$ such that

$$(x_p, x_{p+1}) \in R_1,\ (x_{p-1}, x_p) \in R_2, \ldots,\ (x_1, x_2) \in R_p.$$

A p-step connection is sometimes written

$$x_1 \to x_2 \to x_3 \to \cdots \to x_{p-1} \to x_p \to x_{p+1}.$$

Our next result provides an interpretation of Theorem 2.1 in terms of relations.

Theorem 2.2 *Let $X_1, X_2, \ldots, X_{p+1}$ be $p + 1$ sets. Let R_1 be a relation on X_p to X_{p+1}, R_2 a relation on X_{p-1} to $X_p, \ldots, R_p$ a relation on X_1 to X_2. Let $A_k = A(R_k)$ be the incidence matrix for R_k, $k = 1, \ldots, p$. Then the product*

$$C = A_1 A_2 \cdots A_p \tag{12}$$

is well-defined and the (i, j) entry of C is equal to the number of p-step connections from the jth element of X_1 to the ith element of X_{p+1}.

Proof Suppose $\nu(X_{p+1}) = m$, $\nu(X_p) = n_1$, $\nu(X_{p-1}) = n_2, \ldots$, $\nu(X_2) = n_{p-1}$, $\nu(X_1) = n$ are the numbers of elements in each of the sets. Then A_1 is $m \times n_1$, A_2 is $n_1 \times n_2, \ldots, A_{p-1}$ is $n_{p-2} \times n_{p-1}$, A_p is $n_{p-1} \times n$, and hence the product

$$C = A_1 A_2 \cdots A_p$$

is well-defined and is an $m \times n$ matrix. Now according to Theorem 2.1, the (i, j) entry of C is the sum of all products

$$a_{ik_1}^{(1)} a_{k_1k_2}^{(2)} a_{k_2k_3}^{(3)} \cdots a_{k_{p-2}k_{p-1}}^{(p-1)} a_{k_{p-1}j}^{(p)}. \tag{13}$$

The factors in the product in (13) are all 1 or 0 and thus the product in (13) is 1 if and only if

$$a_{k_{p-1}j}^{(p)} = a_{k_{p-2}k_{p-1}}^{(p-1)} = \cdots = a_{k_1k_2}^{(2)} = a_{ik_1}^{(1)} = 1.$$

Now let x_s^r be the sth element of X_r. Then

$$\begin{aligned}
a_{ik_1}^{(1)} &= 1 \quad \text{means that} \quad (x_{k_1}^p, x_i^{p+1}) \in R_1; \\
a_{k_1k_2}^{(2)} &= 1 \quad \text{means that} \quad (x_{k_2}^{p-1}, x_{k_1}^p) \in R_2; \\
&\vdots \\
a_{k_{p-2}k_{p-1}}^{(p-1)} &= 1 \quad \text{means that} \quad (x_{k_{p-1}}^2, x_{k_{p-2}}^3) \in R_{p-1}; \\
a_{k_{p-1}j}^{(p)} &= 1 \quad \text{means that} \quad (x_j^1, x_{k_{p-1}}^2) \in R_p.
\end{aligned}$$

In other words,

$$x_j^1 \to x_{k_{p-1}}^2 \to x_{k_{p-2}}^3 \to \cdots \to x_{k_2}^{p-1} \to x_{k_1}^p \to x_i^{p+1},$$

so that there is a p-step connection from the jth element of X_1, x_j^1, to the ith element of X_{p+1}, x_i^{p+1}. Hence for each product in (13) having value 1 there is a p-step connection from x_j^1 to x_i^{p+1}. It follows that the sum of the terms in (13) is the number of such p-step connections; but by Theorem 2.1 this is precisely the (i, j) entry of C. This completes the proof. ▮

Example 2.3 Denote a group of five neighborhood housewives by $H = \{h_1, h_2, h_3, h_4, h_5\}$. The incidence matrix describing who speaks to whom is

$$A = \begin{array}{c} \\ h_1 \\ h_2 \\ h_3 \\ h_4 \\ h_5 \end{array} \begin{array}{c} \begin{array}{ccccc} h_1 & h_2 & h_3 & h_4 & h_5 \end{array} \\ \begin{bmatrix} 0 & 1 & 0 & 1 & 1 \\ 1 & 0 & 0 & 1 & 0 \\ 0 & 0 & 0 & 1 & 1 \\ 1 & 1 & 1 & 0 & 1 \\ 1 & 0 & 1 & 1 & 0 \end{bmatrix} \end{array}. \tag{14}$$

The (i, j) entry of A is 1 or 0 according as h_j speaks to h_i or not. Housewife h_5 develops laryngitis and can no longer spend the afternoon gossiping. Is it possible for a rumor to spread among the remaining four ladies? To answer this question, we must decide precisely what it means to remove h_5 from the communication network. Clearly the 4×4 matrix lying in the first four rows and columns of the matrix in (14) describes the network when h_5 is out of the picture:

$$B = \begin{bmatrix} 0 & 1 & 0 & 1 \\ 1 & 0 & 0 & 1 \\ 0 & 0 & 0 & 1 \\ 1 & 1 & 1 & 0 \end{bmatrix}. \tag{15}$$

According to Theorem 2.2, the (i, j) entry of B^p (B multiplied times itself p times) is the number of p-step connections from h_j to h_i. Thus the question reduces to asking whether there exists a positive integer p such that

$$B + B^2 + B^3 + \cdots + B^p$$

has only positive entries except possibly along the main diagonal. For, the sum of entries (i, j) in $B, B^2, B^3, \ldots, B^p$ is the total number of k-step connections, $k = 1, \ldots, p$, between h_j and h_i. We compute that

$$B^2 = \begin{bmatrix} 0 & 1 & 0 & 1 \\ 1 & 0 & 0 & 1 \\ 0 & 0 & 0 & 1 \\ 1 & 1 & 1 & 0 \end{bmatrix} \begin{bmatrix} 0 & 1 & 0 & 1 \\ 1 & 0 & 0 & 1 \\ 0 & 0 & 0 & 1 \\ 1 & 1 & 1 & 0 \end{bmatrix},$$

which will then yield

$$B^2 = \begin{bmatrix} 2 & 1 & 1 & 1 \\ 1 & 2 & 1 & 1 \\ 1 & 1 & 1 & 0 \\ 1 & 1 & 0 & 3 \end{bmatrix}.$$

We see that not every entry in B^2 is positive, but every entry in $B + B^2$ is positive, that is, there is a 2-step or a 1-step connection between every pair of housewives. A rumor can spread despite the departure of h_5.

Example 2.4 A set of five countries, $C = \{c_1, c_2, c_3, c_4, c_5\}$, maintains the following diplomatic relations: c_1 recognizes c_2 and c_2 recognizes c_1; c_3 recognizes c_4, c_4 recognizes c_5, and c_5 recognizes c_3. Is it possible for these countries to communicate with each other directly or through several intermediaries? Let A be the matrix whose (i, j) entry is 1 or 0 according as c_j does or does not recognize c_i (we take $a_{11} = \cdots = a_{55} = 0$, since it is communication between *different* countries that interests us):

$$A = \begin{matrix} & c_1 & c_2 & c_3 & c_4 & c_5 \\ c_1 & 0 & 1 & 0 & 0 & 0 \\ c_2 & 1 & 0 & 0 & 0 & 0 \\ c_3 & 0 & 0 & 0 & 0 & 1 \\ c_4 & 0 & 0 & 1 & 0 & 0 \\ c_5 & 0 & 0 & 0 & 1 & 0 \end{matrix}.$$

We easily compute that I_5, A, A^2, A^3, A^4, and A^5 are the only distinct powers of A, and each of these matrices has zero entries in the intersection of columns 1 and 2 and rows 3, 4, and 5. It follows that every sum of powers of A will have zeros in these positions. Hence, neither c_1 nor c_2 can communicate with any of c_3, c_4, and c_5 either directly or through intermediaries.

Quiz

Answer true or false.

1 Let $X = \{1, 2\}$ and let R be the following relation on X to X: $R = \{(i, j) \mid i^2 + j^2 = 2\}$. Then the incidence matrix $A(R)$ is $\begin{bmatrix} 1 & 0 \\ 0 & 0 \end{bmatrix}$.

2 The matrix

$$\begin{bmatrix} 1 & 1 & 1 \\ 0 & 0 & 0 \end{bmatrix}$$

is the incidence matrix of a function.

3 The matrix

$$\begin{bmatrix} 1 & 0 \\ 1 & 0 \\ 1 & 0 \end{bmatrix}$$

is the incidence matrix of a function.

4 The matrix

$$\begin{bmatrix} 1 & 0 & 0 \\ 0 & 0 & 1 \end{bmatrix}$$

is the incidence matrix of an onto function from a 3-element set to a 2-element set.

5 If A is the incidence matrix of a 1–1 function f: $X \to X$, X a finite set, then A is a square matrix with precisely one 1 in each row and each column.

6 If $X = \{x_1, x_2, x_3\}$, then the matrix

$$\begin{bmatrix} 1 & 1 \\ 0 & 0 \\ 0 & 0 \end{bmatrix}$$

is the incidence matrix for a 2-selection of X.

7 Let $X = \{x_1, x_2, x_3\}$ and let

$$A(f) = \begin{bmatrix} 0 & 0 & 1 \\ 1 & 0 & 0 \\ 0 & 1 & 0 \end{bmatrix}$$

be the incidence matrix for a function f: $X \to X$. Then f is a 1–1 function and the incidence matrix for f^{-1} is

$$A(f^{-1}) = \begin{bmatrix} 0 & 1 & 0 \\ 0 & 0 & 1 \\ 1 & 0 & 0 \end{bmatrix}.$$

8 If A_1 and A_2 are matrices for which the product $C = A_1A_2$ is defined and if the last row of C consists entirely of zeros, then the last row of A_1 consists entirely of zeros.

9 If A is a square matrix with only positive entries, then any power A^p, $p \geq 1$, has only positive entries.

10 Let A be the incidence matrix for a relation R on an n-set X to itself. If the (i, i) entry of A^p is positive, $p \geq 1$, then there is at least one p-step connection from an element of X to itself.

Exercises

1 Consider the following two-way road map connecting three sets of cities denoted by $X = \{x_1, x_2, x_3\}$, $Y = \{y_1, y_2, y_3\}$ and $Z = \{z_1, z_2\}$.

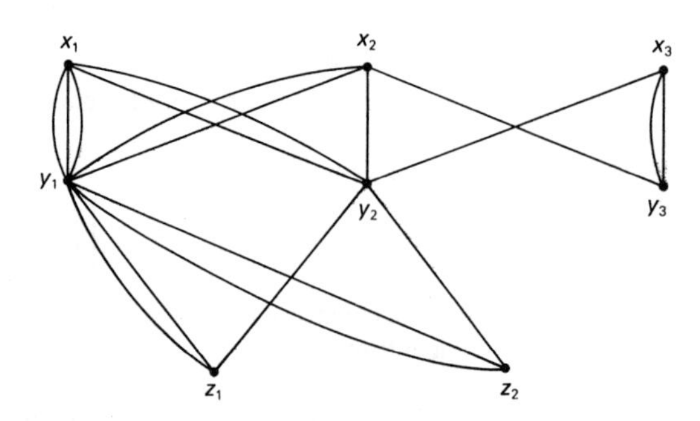

Write down the matrix A whose (i, j) entry is the number of roads connecting x_j to y_i. Similarly, write down the matrix B whose (i, j) entry is the number of roads between y_j and z_i. Compute the product BA and interpret the entries in terms of the roadmap.

†2 Let $A_1, \ldots, A_p$ be p matrices for which the product $C = A_1A_2 \cdots A_p$ is defined. Show that if the jth column of A_p consists entirely of 0's, then the jth column of C consists entirely of 0's. Interpret this result if $A_1, \ldots, A_p$ are incidence matrices for relations $R_1, \ldots, R_p$.

†3 Let $X = \{x_1, \ldots, x_n\}$ be an n-element set and let R be a relation on X to X. Let A be the incidence matrix for R. Using Theorem 2.2 show that the (i, j) element of A^p is the number of p-step connections from x_j to x_i.

4 Four countries, c_1, c_2, c_3, c_4, recognize each other as follows: c_1 recognizes c_2, c_3, and c_4; c_2 recognizes c_3 and c_4; and c_3 recognizes c_4.

(a) Show that c_2 cannot communicate with c_1 directly or indirectly if it is assumed that c_i communicates with c_j if and only if c_i recognizes c_j.

(b) Show that if c_4 recognizes c_1, then it is still not possible for every country to communicate with all other countries either directly or through one intermediary.

(c) Show that if c_4 recognizes c_1, then any country can communicate with any other through a sufficient number of intermediaries.

(d) Find the least integer k such that any country can communicate with any other through exactly k intermediaries, if c_4 recognizes c_1.

†5 Let f be a 1–1 function on an n-element set to itself, f: $X \to X$. Show that the incidence matrix for f^{-1} is the transpose of the incidence matrix for f, i.e.,

$$A(f^{-1}) = (A(f))^T.$$

(Hint: $f^{-1}(x_j) = x_i$ if and only if $f(x_i) = x_j$. In other words, the (i, j) entry of $A(f^{-1})$ is 1 if and only if the (j, i) entry of $A(f)$ is 1.)

†6 Let S be an $n \times n$ matrix. A "diagonal" of S can be thought of as a set of n positions in the matrix S, precisely one from each row and each column of S.

(a) Prove that there is a 1–1 function from the set of n-permutations of $N_n = \{1, \ldots, n\}$ onto the set of all diagonals of S.

(b) Show that S has precisely $n!$ diagonals.

(c) Let

$$S = \begin{bmatrix} 1 & 2 & 3 \\ 4 & 5 & 6 \\ 7 & 8 & 9 \end{bmatrix}.$$

Write down all six diagonals of S by expressing each diagonal as an ordered triple of entries from S.

†7 Let $X = \{x_1, \ldots, x_n\}$ be an n-element set and let R be a relation on X to X satisfying the following two properties: for no i is it true that $(x_i, x_i) \in R$ and for all $i \neq j$, if $(x_i, x_j) \in R$, then $(x_j, x_i) \in R$.

(a) What properties does the incidence matrix $A(R)$ have that reflect these two conditions on R?

(b) The relation R is said to be *connected* if for any i and j there is a positive integer p such that there exists a p-step connection

$$x_i \to x_{k_1} \to x_{k_2} \to \cdots \to x_{k_{p-1}} \to x_j$$

in which $k_1, k_2, \ldots, k_{p-1}$ are distinct integers. Show that R is not connected if and only if $S = A + A^2 + A^3 + \cdots + A^{n-1}$ has a zero entry ($n > 2$).

(Hint: (a) The answer is $a_{ii} = 0$, $i = 1, \ldots, n$, and $a_{ij} = a_{ji}$ for all $i \neq j$. (b) Suppose that S has a zero entry, say $s_{ij} = 0$. Since A has only 0's and 1's as entries, it follows that the (i, j) entry of A^p is 0, $p = 1, 2, \ldots, n - 1$. Now, according to Theorem 2.1, the (i, j) entry of A^p is the sum of all products of the form

$$a_{ik_1}a_{k_1k_2}a_{k_2k_3} \cdots a_{k_{p-2}k_{p-1}}a_{k_{p-1}j}. \tag{16}$$

Hence there can be no p-step connection between x_j and x_i, $p \leq n - 1$. Suppose that there is a p-step connection between x_j and x_i in which $p \geq n$, where p is the least integer for which such a p-step connection exists. This means that there exist integers $k_1, k_2, \ldots, k_{p-1}$ such that the product (16) is 1. Now, there are a total of at least $n + 1$ integers (counting multiple occurences) among the integers $i, k_1, k_2, \ldots, k_{p-1}, j$, i.e., $p \geq n$ and hence there are at least $n + 1$ items in this list of integers chosen from $1, \ldots, n$. Thus there must be repetitions. To complete the solution, the reader will show that a repetition contradicts the minimality condition in the choice of p. Conversely, assume R is not connected and suppose that every entry of S is positive. This means that for any i and j, there exist integers $k_1, \ldots, k_{p-1}$ such that the product (16) is 1, where $p \leq n - 1$. If there are any repetitions among $i, k_1, \ldots, k_{p-1}, j$, then we can find a non-zero product of the form (16) involving fewer factors.)

8 Four people, x_1, x_2, x_3, x_4, speak to one another as follows: x_1 speaks to x_2, x_3, x_4; x_2 speaks to x_4; x_3 speaks to x_1 and x_4; and x_4 speaks to x_2 and x_3. Is it possible for a rumor to spread from any person to any other person without having the rumor repeated twice by someone?

(a) Assume that when x_i speaks to x_j, x_j speaks to x_i, and no x_i speaks to himself.

(b) Next answer the question by assuming that if x_i speaks to x_j, then x_j does not speak to x_i unless specified above.

(Hint: Define a relation R as follows: $(x_i, x_j) \in R$ if and only if x_i speaks to x_j. Then the incidence matrix A for R in (a) is

$$A = \begin{bmatrix} 0 & 1 & 1 & 1 \\ 1 & 0 & 0 & 1 \\ 1 & 0 & 0 & 1 \\ 1 & 1 & 1 & 0 \end{bmatrix}.$$

Compute $A + A^2 + A^3$ and use the preceding exercise. To answer (b) observe that the incidence matrix becomes

$$A = \begin{bmatrix} 0 & 0 & 1 & 0 \\ 1 & 0 & 0 & 1 \\ 1 & 0 & 0 & 1 \\ 1 & 1 & 1 & 0 \end{bmatrix}.)$$

9 Suppose that in the preceding problem (part (a)), x_3 departs from the scene. We still assume that whenever x_i speaks to x_j, x_j speaks to x_i and no x_i speaks to himself, $i = 1, 2, 4$. Is it true that a rumor can go from any one of the remaining three people to any other? (Hint: For x_3 to depart means that the incidence matrix is altered by having both its third row and third column deleted. (Why?) The resulting matrix is

$$B = \begin{bmatrix} 0 & 1 & 1 \\ 1 & 0 & 1 \\ 1 & 1 & 0 \end{bmatrix}.)$$

10 Let f be an n-permutation of the set N_n, i.e., f is a 1–1 function from $N_n = \{1, 2, \ldots, n\}$ to itself. Show that by choosing n and the n-permutation f appropriately, it is not necessarily the case that an integer k exists satisfying $1 \leq k \leq n$ such that the incidence matrix $A = A(f)$ satisfies

$$A^k = I_n.$$

(Hint: Try $n = 5$.)

11 Show that no example of the kind required in Exercise 10 exists for $n \leq 4$.

12 Let f be a 3-permutation of a 3-set, and let A be the incidence matrix for f. Show that $A(f)(A(f))^T = I_3$. (Hint: Consider all possible f and use Exercise 5.)

13 Let A be the incidence matrix for a function f on an n-set to itself. Show that if $\sum_{i=1}^{n} a_{ii} = n$, then $f(x) = x$, for every $x \in X$, i.e., if the sum of all entries on the main diagonal of A is n, then f must be the identity function that sends each element of X into itself.

14 A pen contains 3 chickens. It is observed that for any two chickens c and c', either c always pecks c' or c' always pecks c, but not both. Moreover, no chicken ever pecks itself. Show that there is a dominant chicken, C, in the following sense. Given any other chicken c, either C pecks c, or C pecks d who in turn pecks c. (Hint: Define an incidence matrix A in which the (i, j) entry is 1 if the jth chicken pecks the ith chicken and is 0 otherwise. What properties does A have? Show that $A + A^2$ has every entry positive except those along the main diagonal by considering all possible choices for A.)

2.3

Powers of Incidence Matrices

As we saw in the previous section, the product of incidence matrices for relations can provide useful information about these relations. If we are interested in a single relation, then the properties of powers and sums of powers of the incidence matrix yield detailed insights into the corresponding relation.

Definition 3.1 ***Asymmetric relation*** A relation R on a set $X = \{x_1, \ldots, x_n\}$ to itself is called *asymmetric* if $(x_j, x_i) \in R$ if and only if $(x_i, x_j) \notin R$, and if $(x_i, x_i) \notin R$, $i, j = 1, \ldots, n$.

It is easy to think of examples of asymmetric relations. Suppose that $\{x_1, \ldots, x_n\}$ are participants in a tennis tournament and R is the set of pairs (x_j, x_i) in which x_j beats x_i. Then $(x_j, x_i) \in R$ and $(x_i, x_j) \notin R$. There is an additional property that this relation has which is not necessarily the case for all relations: for any two players x_i and x_j, either x_j beats x_i or x_i beats x_j.

We can very neatly summarize the information about such relations in a single matrix equation. To do this, let J_n be the n-square matrix each of whose entries is 1:

$$J_n = \underbrace{\begin{bmatrix} 1 & \cdots & 1 \\ \vdots & & \vdots \\ 1 & \cdots & 1 \end{bmatrix}}_{n} \Bigg\} n. \tag{1}$$

Let A be the incidence matrix for the relation R. For any two elements x_j and x_i in X, we are assuming that precisely one of $(x_j, x_i) \in R$ or $(x_i, x_j) \in R$ is true and it is never the case that $(x_i, x_i) \in R$. Thus the matrix A satisfies

(2) $$A + A^T = J_n - I_n;$$

for, the (i, j) entry of $A + A^T$ is $a_{ij} + a_{ji}$ and if $i \neq j$, then precisely one of $(x_j, x_i) \in R$ or $(x_i, x_j) \in R$ holds. In terms of the matrix A, this means that precisely one of a_{ij} and a_{ji} is 1 and the other is 0. Hence $a_{ij} + a_{ji} = 1$ if $i \neq j$. Conversely, if $i = j$, then since $(x_i, x_i) \notin R$ it follows that $a_{ii} = 0$ and hence the (i, i) entry of $A + A^T$ is 0. A moment's reflection will convince the reader that the (i, j) entry of $J_n - I_n$ is 1 if $i \neq j$ and 0 if $i = j$.

We are often interested in the existence of p-step connections. In particular, certain information concerning 2-step connections can be deduced for a matrix satifying (2).

Theorem 3.1 *Let A be an n-square 0,1 matrix (i.e., the entries of A are 0 or 1). Furthermore, suppose A satisfies (2):*

$$A + A^T = J_n - I_n.$$

Then there exists a row and a column of $A + A^2$, each of which contains exactly $n - 1$ positive entries.

Proof Multiply both sides of the equation (2) by A on the left:

$$\begin{aligned} A(A + A^T) &= A(J_n - I_n) \\ &= AJ_n - AI_n. \end{aligned}$$

Thus

$$A^2 + AA^T = AJ_n - A$$

or equivalently

(3) $$A + A^2 = AJ_n - AA^T.$$

First, we compute the product AJ_n:

$$AJ_n = \begin{bmatrix} a_{11} & \cdots & a_{1n} \\ \vdots & & \vdots \\ a_{n1} & \cdots & a_{nn} \end{bmatrix} \begin{bmatrix} 1 & \cdots & 1 \\ \vdots & & \vdots \\ 1 & \cdots & 1 \end{bmatrix}$$

$$= \begin{bmatrix} r_1 & r_1 & \cdots & r_1 \\ r_2 & r_2 & \cdots & r_2 \\ \vdots & \vdots & & \vdots \\ r_n & r_n & \cdots & r_n \end{bmatrix},$$

where r_i is the sum of the entries in the ith row of A. Next consider the (i, j) entry of the matrix AA^T:

$$a_{i1}a_{j1} + a_{i2}a_{j2} + \cdots + a_{in}a_{jn}. \tag{4}$$

In other words, (4) is just the sum of the products of corresponding entries in the ith and jth rows of A. Every entry of A is 0 or 1 and hence it follows that for $i = j$, (4) just becomes r_i. Thus we see from (3) that every main diagonal entry of $A + A^2$ must be 0 (i.e., the (i, i) entry is $r_i - r_i$, $i = 1, \ldots, n$). Now suppose that row i of A has at least as many 1's as any other row of A, and suppose that row i of $A + A^2$ has fewer than $n - 1$ positive entries. Then from (3) it follows that for some $j \neq i$, the (i, j) entry of AJ_n, r_i, must be the same as the (i, j) entry of AA^T, the sum (4):

$$\begin{aligned} r_i &= a_{i1} + a_{i2} + \cdots + a_{in} \\ &= a_{i1}a_{j1} + a_{i2}a_{j2} + \cdots + a_{in}a_{jn}. \end{aligned} \tag{5}$$

From (5) we have

$$a_{i1}(1 - a_{j1}) + a_{i2}(1 - a_{j2}) + \cdots + a_{in}(1 - a_{jn}) = 0. \tag{6}$$

It follows from (6) that whenever $a_{ik} = 1$ then $a_{jk} = 1$ in order that each nonnegative term be zero. Hence row j of A must contain 1's in those columns in which row i contains 1's. But our choice of row i stipulates that it has at least as many 1's as row j does and it follows that row i and row j are identical. But since $a_{ii} = 0$ it must be the case that

$$a_{ji} = 0$$

(since rows i and j are identical) and hence that $a_{ij} = 1$ (recall (2)). But rows i and j are identical and hence $a_{jj} = 1$, which conflicts with (2). This contradiction allows us to conclude that row i of $A + A^2$ has exactly $n - 1$ positive entries. To prove that there is a column of $A + A^2$ with $n - 1$ positive entries we use the fact that

$$A^T + (A^T)^2 = (A + A^2)^T \tag{7}$$

together with what we know about the rows of $A + A^2$. The remainder of the proof is left as Exercise 12 (with copious hints). ▌

Example 3.1 Suppose that in a tennis club with n members it is known to each member precisely whom he can beat and who can beat him. Show that there is a member of the club who is "best" in the following sense: given any other member, either he can beat that other member or he can beat someone who can. Moreover, there is a player who is "worst" in the sense that either every other member can defeat him or every other member can defeat someone who can defeat him. Let $X = \{x_1, \ldots, x_n\}$ denote the n players. Define a relation R on X to X as follows. A pair $(x_s, x_t) \in R$ if and only if x_s can defeat x_t, i.e.,

$$R = \{(x_s, x_t) \mid (x_s \in X) \wedge (x_t \in X) \wedge (x_s \text{ can defeat } x_t)\}.$$

Let $A = A(R)$ be the incidence matrix for the relation R, that is, A is an n-square matrix whose (i, j) entry, a_{ij}, is 1 or 0 according as $(x_j, x_i) \in R$ or $(x_j, x_i) \notin R$, respectively. The conditions of the problem state that each player knows precisely who are his inferior and superior opponents. In terms of the matrix A, this can be interpreted as the condition that $a_{ij} = 1$ or $a_{ji} = 1$, but not both. Since the main diagonal of A consists of 0's (tennis requires two participants!) it follows that the (i, j) entry of $A + A^T$ is 1 if $i \neq j$ and 0 if $i = j$. In other words, A satisfies the matrix equation

$$A + A^T = J_n - I_n.$$

According to Theorem 3.1 the matrix $A + A^2$ has the following property: some row, say row s, has $n - 1$ positive entries, and some column, say column t, also has $n - 1$ positive entries (we saw that the main diagonal entries of $A + A^2$ are all 0 in the proof of Theorem 3.1 so that no row or column of $A + A^2$ can have n positive entries). Let $D = A + A^2$, so that $d_{1t} > 0$, $d_{2t} > 0, \ldots,$ $d_{t-1,t} > 0, d_{t+1,t} > 0, \ldots, d_{nt} > 0$. Moreover d_{it} is the sum of the (i, t) entry of A and the (i, t) entry of A^2. Hence $d_{it} > 0$ means that either A or A^2 has a positive (i, t) entry (perhaps both). According to Theorem 2.2, Section 2.2, for a given i, $i \neq t$, either $(x_t, x_i) \in R$ or there is a 2-step connection from x_t to x_i, that is, there exists an x_k such that $(x_t, x_k) \in R$ and $(x_k, x_i) \in R$. In other words, either x_t defeats x_i or x_t defeats x_k who in turn defeats x_i. Thus x_t is the best player. Once again, to say that $d_{s1} > 0$, $d_{s2} > 0, \ldots, d_{s,s-1} > 0$, $d_{s,s+1} > 0, \ldots, d_{sn} > 0$ means that the (s, j) entry of A is positive or the (s, j) entry of A^2 is positive, perhaps both, $j \neq s$, $j = 1,$ $2, \ldots, n$. That is, $(x_j, x_s) \in R$ or there exists a player x_k such that $(x_k, x_s) \in R$ and $(x_j, x_k) \in R$. Hence, either x_j defeats x_s or x_j defeats x_k who in turn defeats x_s. Thus x_s is the worst player.

Suppose that $X = \{x_1, \ldots, x_n\}$ and $Y = \{y_1, \ldots, y_m\}$ and f is a function from X to Y, $f\colon X \rightarrow Y$. The function f is a particular kind of relation on X to Y and hence the incidence matrix $A(f)$ can be defined. Now for a pair (x_j, y_i) to be in f means that $f(x_j) = y_i$. Since there is only one pair in f whose first member is x_j, it follows that the (i, j) entry of A, a_{ij}, is 1 and moreover if $k \neq i$, then $a_{kj} = 0$. Thus an incidence matrix for a function can be recognized by observing that each column has exactly one 1 in it. Now let $Z = \{z_1, \ldots, z_q\}$ and suppose that $g: Y \to Z$ is a function. We can also construct the incidence matrix for g, $A(g)$. The composition of the functions f and g, $gf: X \to Z$, is again a function and we can construct the incidence matrix $A(gf)$. Now $A(g)$ is $q \times m$, $A(f)$ is $m \times n$ and $A(gf)$ is $q \times n$. Thus it is possible to form the matrix product $A(g)A(f)$ to obtain a $q \times n$ matrix. This suggests the possibility of a relationship between the matrices $A(g)A(f)$ and $A(gf)$.

Example 3.2 Let $X = \{x_1, x_2, x_3\}$, $Y = \{y_1, y_2, y_3, y_4\}$, and $Z = \{z_1, z_2\}$. Let $f: X \to Y$ be defined by $f(x_1) = y_3, f(x_2) = y_4$, and $f(x_3) = y_1$, and let $g: Y \to Z$ be defined by $g(y_1) = z_1, g(y_2) = z_1, g(y_3) = z_1$, and $g(y_4) = z_2$. Then $(gf)(x_1) = g(f(x_1)) = g(y_3) = z_1$ and similarly $(gf)(x_2) = z_2$, $(gf)(x_3) = z_1$. We construct the incidence matrices $A(g)$, $A(f)$ and $A(gf)$:

$$A(g) = \begin{array}{c} \\ z_1 \\ z_2 \end{array} \begin{array}{c} \begin{array}{cccc} y_1 & y_2 & y_3 & y_4 \end{array} \\ \begin{bmatrix} 1 & 1 & 1 & 0 \\ 0 & 0 & 0 & 1 \end{bmatrix} \end{array};$$

$$A(f) = \begin{array}{c} \\ y_1 \\ y_2 \\ y_3 \\ y_4 \end{array} \begin{array}{c} \begin{array}{ccc} x_1 & x_2 & x_3 \end{array} \\ \begin{bmatrix} 0 & 0 & 1 \\ 0 & 0 & 0 \\ 1 & 0 & 0 \\ 0 & 1 & 0 \end{bmatrix} \end{array};$$

$$A(gf) = \begin{array}{c} \\ z_1 \\ z_2 \end{array} \begin{array}{c} \begin{array}{ccc} x_1 & x_2 & x_3 \end{array} \\ \begin{bmatrix} 1 & 0 & 1 \\ 0 & 1 & 0 \end{bmatrix} \end{array}.$$

Computing the matrix product $A(g)A(f)$, we have

$$\begin{aligned} A(g)A(f) &= \begin{bmatrix} 1 & 1 & 1 & 0 \\ 0 & 0 & 0 & 1 \end{bmatrix} \begin{bmatrix} 0 & 0 & 1 \\ 0 & 0 & 0 \\ 1 & 0 & 0 \\ 0 & 1 & 0 \end{bmatrix} \\ &= \begin{bmatrix} 1 & 0 & 1 \\ 0 & 1 & 0 \end{bmatrix} \\ &= A(gf). \end{aligned}$$

This example suggests the following result.

Theorem 3.2 *Let $X = \{x_1, \ldots, x_n\}$, $Y = \{y_1, \ldots, y_m\}$, and $Z = \{z_1, \ldots, z_q\}$. Let f: $X \to Y$ and g: $Y \to Z$. Then*

$$A(gf) = A(g)A(f). \tag{8}$$

Proof We can make the proof of (8) depend on Theorem 2.2. In this case we have three sets, so that, in the notation of the statement of Theorem 2.2, the role of X_1 is taken by X, X_2 by Y, and X_3 by Z. The role of the relation R_1 is played by g and the role of the relation R_2 is played by f. In the present theorem, $A_1 = A(g)$ and $A_2 = A(f)$. Theorem 2.2 states that the (i, j) entry of the product

$$A_1A_2 = A(g)A(f)$$

is equal to the number of 2-step connections from the jth element of $X_1 = X$ to the ith element of $X_3 = Z$. In other words, the (i, j) element of $A(g)A(f)$ is the number of elements y_k of $Y = X_2$ for which

$$(y_k, z_i) \in g \tag{9}$$

and

$$(x_j, y_k) \in f. \tag{10}$$

There is exactly one ordered pair in f whose first member is x_j. Thus (10) is satisfied by precisely one element of Y. Now (9) means that $g(y_k) = z_i$ and (10) means that $f(x_j) = y_k$. Hence (9) and (10) together imply that

$$\begin{aligned}(gf)(x_j) &= g(f(x_j)) \\ &= g(y_k) \\ &= z_i.\end{aligned} \tag{11}$$

Summarizing, we can say that for a fixed i and j, either there is precisely one element $y_k \in Y$ such that $f(x_j) = y_k$ and $g(y_k) = z_i$, or no such element exists. In the first instance, the (i, j) element of $A(g)A(f)$ is 1, and by (11) the (i, j) element of $A(gf)$ is also 1. In the second alternative, there is no element $y_k \in Y$ for which $g(y_k) = z_i$ and hence $(gf)(x_j) = g(f(x_j))$ cannot be z_i, for otherwise $f(x_j)$ would be an element y_k in Y for which $g(y_k) = z_i$. Hence in this case, the (i, j) entries of both $A(g)A(f)$ and $A(gf)$ are 0. We have proved that the (i, j) entry of the two matrices in (8) are the same, $i = 1, \ldots, q$, $j = 1, \ldots, n$, and this completes the argument. ▌

Example 3.3 Let f be an r-permutation of an n-set $X = \{x_1, \ldots, x_n\}$. Discuss the incidence matrix for f. According to the definition, an r-permutation f is a 1–1 function f: $N_r \to X$ where $N_r = \{1, \ldots, r\}$. Thus the incidence matrix $A(f)$ is an $n \times r$ matrix in which the (i, j) entry is 1 or 0 according as $f(j) = x_i$ or not, respectively. But if $f(j) = f(t)$, then since f is a permutation, it follows that $j = t$. Stated differently, if $j \neq t$, then $f(j) \neq f(t)$. Hence the incidence matrix $A(f)$ has the following property: no two 1's in the matrix appear in the same row and, since f is a function, we already know that each column has precisely one 1 in it. Conversely, if A is an $n \times r$ matrix with these two properties, it is clear that we may define an r-permutation of X: for any $j \in N_r$, let $f(j) = x_i$, where i is the number of the row in which the 1 in column j of A appears. For example, if $r = 3$ and $X = \{x_1, \ldots, x_4\}$ and

$$A = \begin{bmatrix} 0 & 0 & 1 \\ 1 & 0 & 0 \\ 0 & 0 & 0 \\ 0 & 1 & 0 \end{bmatrix},$$

then A is the incidence matrix for the 3-permutation of X defined by $f(1) = x_2, f(2) = x_4, f(3) = x_1$.

Example 3.4 Let f be an r-selection of $X = \{x_1, \ldots, x_n\}$. Discuss the incidence matrix for f. By definition, $f\colon N_r \to X$ must have the property that if $f(j) = x_{k_j}$, $j = 1, \ldots, r$, then $k_1 \leq k_2 \leq \cdots \leq k_r$. Now the only positive entry in column j of $A(f)$ is a 1 in row k_j. Hence if $j > t$, the 1 in column j must be in a row which is not above the row with the 1 in column t. Moreover, since f is a function, every column contains precisely one 1. Conversely, any matrix which contains precisely r ones and satisfies the above properties is an incidence matrix for an r-selection of X. For example, take $r = 3$, $n = 4$, and

$$A = \begin{bmatrix} 0 & 0 & 0 \\ 1 & 1 & 0 \\ 0 & 0 & 0 \\ 0 & 0 & 1 \end{bmatrix}.$$

Then A is the incidence matrix for the 3-selection of $X = \{x_1, x_2, x_3, x_4\}$ defined by $f(1) = x_2$, $f(2) = x_2$, $f(3) = x_4$. Observe that $k_1 = 2$, $k_2 = 2$, $k_3 = 4$, and $k_1 \leq k_2 \leq k_3$ (in fact, $k_1 = k_2 < k_3$).

Example 3.5 We analyze Exercise 18 in Section 1.3. Recall that occurring among the three patients p_1, p_2, and p_3 are four diseases, d_1, d_2, d_3, and d_4. The conditions are such that if k is 1, 2, or 3, then any k of the patients exhibit symptoms of at least k of the diseases. If we define a relation R by the condition that $(p_j, d_i) \in R$ if p_j exhibits the symptoms of the disease d_i, then we can analyze this situation in terms of the incidence matrix $A = A(R)$. The matrix A is a 4×3 matrix:

$$A = \begin{matrix} \\ d_1 \\ d_2 \\ d_3 \\ d_4 \end{matrix} \begin{matrix} p_1 & p_2 & p_3 \\ \left[\begin{matrix} & \cdot & \\ \cdot & \cdot & \cdot \\ & & \\ & \cdot & \end{matrix} \right] \end{matrix}.$$

The conditions state that in any k columns of A, 1's must appear in at least k different rows of these columns. Finding a set of three 1's in A, no two of which lie in either the same row or the same column, is equivalent to finding three different diseases, d_{i_1}, d_{i_2}, d_{i_3}, such that p_j exhibits the symptoms of d_{i_j}, $j = 1, 2, 3$. The question of whether such a distribution of symptoms exists amounts to asking whether the conditions on the matrix A imply that A possesses three 1's in the three positions which correspond to the 1's in the incidence matrix of some 3-permutation of the set $\{d_1, d_2, d_3, d_4\}$. To find out if A has this property, we first add a 4th column of 1's to A to obtain a 4×4 matrix S with the following appearance:

(12) $$S = \left[\begin{array}{c|c} & 1 \\ A & 1 \\ & 1 \\ & 1 \end{array} \right].$$

Now, a "diagonal" of S will be a set of 4 positions in S, no two in any one row or column. It is easy to confirm that there are 4! diagonals of S, i.e., we can choose the element of a diagonal from the first row in any one of four ways, we can choose the element from the second row in any one of three ways, etc. For, elements from different rows must come from different columns. Suppose that the conditions on A are enough to conclude that S possesses a set of 1's in positions $\{(i_1, 1), (i_2, 2), (i_3, 3), (i_4, 4)\}$. Then it is clear from (12) that there are three different rows of A, namely rows i_1, i_2, i_3, such that in column j of A there is a 1 in row i_j, $j = 1, 2, 3$. Thus the question becomes: Does S possess a diagonal consisting entirely of 1's? We shall show in the next section that if every diagonal of an $n \times n$ matrix of 0's and 1's contains a 0, then there must exist s rows and t columns, $s + t = n + 1$, such that each of the st entries lying at the intersection of these rows and columns is 0 (see Theorem 4.2, Section 2.4). If we anticipate this result now we can argue that unless S has a diagonal of 1's, every diagonal must contain a 0. Then we can conclude that there exist s rows and t columns of S, $s + t = n + 1 = 4 + 1 = 5$, such that S has 0's at the intersections of these s rows and t columns. Since the last column of S consists entirely of 1's, it follows that $t < 4$. The possibilities are $s = 2$, $t = 3$; $s = 3$, $t = 2$; $s = 4$, $t = 1$. The conditions that any k patients exhibit the symptoms of k diseases among them, $k = 1, 2, 3$, are unaltered by a renumbering of patients or diseases. Thus, by a relabelling of the rows and columns of S, the $s \times t$ box of 0's in S may be assumed to be in one of the following three standard positions:

$$\begin{bmatrix} 0 & * & * & 1 \\ 0 & * & * & 1 \\ 0 & * & * & 1 \\ 0 & * & * & 1 \end{bmatrix} \quad (s = 4, t = 1); \tag{13}$$

$$\begin{bmatrix} 0 & 0 & * & 1 \\ 0 & 0 & * & 1 \\ 0 & 0 & * & 1 \\ * & * & * & 1 \end{bmatrix} \quad (s = 3, t = 2); \tag{14}$$

$$\begin{bmatrix} 0 & 0 & 0 & 1 \\ 0 & 0 & 0 & 1 \\ * & * & * & 1 \\ * & * & * & 1 \end{bmatrix} \quad (s = 2, t = 3). \tag{15}$$

Now (13) implies that p_1 does not exhibit any symptoms; (14) implies that p_1 and p_2 can exhibit at most the symptoms of d_4; and (15) implies that p_1, p_2, and p_3 can exhibit among them at most the symptoms of d_3 and d_4. In every case we have contradicted the fact that any k of the patients exhibit among them the symptoms of k of the diseases, $k = 1, 2, 3$. Thus there must be three diseases $d_{i_1}, d_{i_2}, d_{i_3}$, such that p_j exhibits the symptoms of d_{i_j}, $j = 1, 2, 3$.

Quiz

Answer true or false.

1 The matrix

$$A = \begin{bmatrix} 0 & 0 & 0 \\ 1 & 0 & 0 \\ 1 & 1 & 0 \end{bmatrix}$$

satisfies $A + A^T = J_3 - I_3$ (see formula (1)).

2 If A is the matrix in the preceding question, then the main diagonal entries of $A + A^2$ are positive numbers.

3 The matrix

$$\begin{bmatrix} 1 & 1 & 1 \\ 1 & 1 & 1 \\ 1 & 0 & 1 \end{bmatrix}$$

is the sum of three incidence matrices for functions.

4 The matrix

$$\begin{bmatrix} 0 & 0 & 0 \\ 1 & 0 & 0 \\ 0 & 1 & 0 \end{bmatrix}$$

is an incidence matrix for an asymmetric relation.

5 Define a relation R on $N_3 = \{1, 2, 3\}$ to itself by $R = \{(i, j) \mid i - j$ is even$\}$. Then R is an asymmetric relation.

6 Define a relation R on $N_4 = \{1, 2, 3, 4\}$ to itself by $R = \{(i, j) \mid i > j\}$. Then R is an asymmetric relation.

7 Let X be the set of all people and define a relation R on X to itself by $R = \{(x_1, x_2) \mid x_1$ is the mother of $x_2\}$. Then R is an asymmetric relation.

8 Any matrix which satisfies the equation $A + A^T = J_n - I_n$ is an incidence matrix for an asymmetric relation.

9 $J_3^4 = 27J_3$.

10 $(J_3 - I_3)^3 = 3J_3 - I_3$.

Exercises

1 Compute each of the following matrices explicitly:

(a) J_4^2;
(b) $(\frac{1}{3}J_3)^3$;
(c) $(2J_3 - I_3)^2$;
(d) $(I_4 - J_4)(I_4 + J_4)$;
(e) J_2^{15}.

2 Write out explicitly all 2×2 and 3×3 matrices that are incidence matrices for asymmetric relations.

3 For each of the following $n \times n$ matrices find a row and a column of the matrix $A + A^2$ each of which contains exactly $n - 1$ positive entries:

(a) $A = \begin{bmatrix} 0 & 0 & 1 & 0 \\ 1 & 0 & 0 & 1 \\ 0 & 1 & 0 & 1 \\ 1 & 0 & 0 & 0 \end{bmatrix}$; (b) $A = \begin{bmatrix} 0 & 1 & 1 \\ 0 & 0 & 1 \\ 0 & 0 & 0 \end{bmatrix}$;

(c) $A = \begin{bmatrix} 0 & 0 & 1 & 1 & 1 \\ 1 & 0 & 0 & 0 & 1 \\ 0 & 1 & 0 & 0 & 1 \\ 0 & 1 & 1 & 0 & 0 \\ 0 & 0 & 0 & 1 & 0 \end{bmatrix}$; (d) $A = \begin{bmatrix} 0 & 1 & 1 & 0 \\ 0 & 0 & 0 & 0 \\ 0 & 1 & 0 & 1 \\ 1 & 1 & 0 & 0 \end{bmatrix}$.

4 Let X be an n-element set and let $R \subset X \times X$ be a relation which satisfies the following conditions: given any x and y in X, either $(x, y) \in R$ or $(y, x) \in R$, but not both, and moreover, $(x, x) \notin R$ for any $x \in X$. Prove that, for any $y \in X$, there exists an $x \in X$, $x \neq y$, such that either $(x, y) \in R$ or, for some $z \in X$, $(x, z) \in R$ and $(z, y) \in R$. (Hint: The incidence matrix for R satisfies $A + A^T = J_n - I_n$. Apply Theorem 3.1.)

5 In *The Admirable Crichton*, J. M. Barrie establishes that in a given environment the following interrelations must hold: for any two people, one must dominate the other and moreover, no person dominates himself. Prove that a "dominant" character must emerge, i.e., given any other character, the dominant character either dominates this character, or dominates someone who does.

6 For each of the following sets X and pairs of functions f and g on X to X construct the corresponding incidence matrices and verify the equality in (8):

(a) $X = \{1, 2, 3\}$,
$f(1) = 2,\ f(2) = 3,\ f(3) = 1,$
$g(1) = 1,\ g(2) = 1,\ g(3) = 2;$

(b) $X = \{x, y, z, u, v\}$,
$f(x) = y,\ f(y) = x,\ f(z) = v,\ f(u) = v,\ f(v) = x,$
$g(x) = y,\ g(y) = z,\ g(z) = u,\ g(u) = v,\ g(v) = z;$

(c) $X = \{x_1, x_2, x_3, x_4\}$,
$f(x_1) = x_1,\ f(x_2) = x_1,\ f(x_3) = x_4,\ f(x_4) = x_3,$
$g(x_1) = x_4,\ g(x_2) = x_4,\ g(x_3) = x_1,\ g(x_4) = x_2.$

***7** Show that if f is an n-permutation of $N_n = \{1, \ldots, n\}$, then there exists a positive integer p such that $(A(f))^p = I_n$. (Hint: From Theorem 3.2, $(A(f))^p = A(f^p)$, where f^p denotes the composition of f with itself p times. Now we know that at least one of $f(j), f(f(j)), \ldots, f^n(j)$ must be equal to j, say $f^{p_j}(j) = j$, $j = 1, \ldots, n$. (Why?) Then let $p = \prod_{k=1}^{n} p_k$ and show that $f^p(j) = j$, $j = 1, \ldots, n$.)

8 It has been established among the countries in the Middle East that the following relationships exist. In a war between any two, say X and Y, either X defeats Y or Y defeats X. Prove that there is a strongest country, I, in the following sense: either I can defeat any other country X or I can defeat a country Y which can defeat X. Also show that there is a weakest country J in the following sense: if X is any other country, then either X can defeat J or X can defeat a country Y which can defeat J.

9 In a closed community of neurotics, it is found that:

(i) no one loves himself; and
(ii) given any two people x and y, either x loves y or y loves x but not both.

Prove that there is a person p in this community such that if x is any other person, then either x loves p, or x loves y who loves p.

†10 Let A be an $m \times n$ matrix.

(a) Show that the (i, j) entry of $J_m A$ is c_j, where c_j is the sum of the entries in column j of A.

(b) Show that $J_m^p = m^{p-1} J_m$.

(c) Show that every entry of $J_m A J_n$ is s, where s is the sum of all the entries in the matrix A.

(d) Let e_n be the n-vector each of whose components is 1. Show that the matrix-vector product Ae_n is equal to e_m if and only if the sum of the entries in each row of A is 1.

(e) Show that $A^T e_m = e_n$ if and only if the sum of the entries in each column of A is 1.

11 Write J_3 as a sum of incidence matrices of 1–1 functions. In how many ways can this be done?

†12 Let A be an n-square 0,1 matrix which satisfies the equation $A + A^T = J_n - I_n$. Show that there exists a column of $A + A^2$ which contains exactly $n - 1$ positive entries. (Hint: Use Theorem 3.1 in the following way. Let $B = A^T$ so that $B + B^T = A^T + (A^T)^T = A^T + A = A + A^T = J_n - I_n$. Then according to Theorem 3.1, there exists a row of $B + B^2$ which contains exactly $n - 1$ positive entries. But $B + B^2 = A^T + (A^T)^2 = A^T + (A^2)^T = (A + A^2)^T$. Hence the columns of $A + A^2$ are the rows of $B + B^2$.)

2.4

Introduction to Combinatorial Matrix Theory

We briefly recapitulate Example 3.15 (g) in Section 1.3. In a group of n boys and n girls, each boy has been previously introduced to precisely k girls and each girl has been introduced to precisely k boys. Can the boys and girls pair off into dance partners with no further introductions and if so, in how many ways can this be done?

At the end of the preceding section in Example 3.5 we saw that problems of this type can be solved if we know an appropriate fact about 0,1 matrices (i.e., matrices with only 0's and 1's as entries) which we now state in detail. Let A be an $n \times n$ matrix and suppose a 0 is in every set of n entries obtained by choosing exactly one entry from each row and each column. Then there must exist s rows and t columns in A such that $s + t = n + 1$ and, moreover, each of the st entries lying at the intersections of these rows and columns is 0. This fact is of crucial importance in analyzing a variety of combinatorial problems. We shall examine its proof and a number of interesting applications in this section.

Example 4.1 Let $S = \{1, 2, 3, 4, 5\}$ and define four subsets of S as follows: $S_1 = \{1, 2, 3\}$; $S_2 = \{1, 2\}$; $S_3 = \{1, 3\}$; $S_4 = \{2, 4\}$. Consider the 4-permutation of S defined by $d = (1, 2, 3, 4)$, i.e., $d(t) = t$, $t = 1, 2, 3, 4$. (This notation is discussed in the remarks immediately preceding Theorem 4.1 in Chapter 1.) Then observe that $d(t) \in S_t$, $t = 1, 2, 3, 4$. The sequence d consists of distinct integers and the integers appear in different sets S_t. Moreover, each S_t has its own "representative" in d. Suppose we replace the set $S_4 = \{2, 4\}$ by $\bar{S}_4 = \{2, 3\}$. We can then ask if there exists a system of distinct representatives, $\bar{d}$, such that $\bar{d}(1) \in S_1$, $\bar{d}(2) \in S_2$, $\bar{d}(3) \in S_3$, and $\bar{d}(4) \in \bar{S}_4$. The original d will not do, for $d(4) = 4 \notin \bar{S}_4$. As a matter of fact, $\bar{d}(4)$ must be 2 or 3, $\bar{d}(3)$ must be 1 or 3, $\bar{d}(2)$ must be 1 or 2, and $\bar{d}(1)$ must be one of 1, 2, or 3. Now consider the following 4×5 incidence matrix:

$$A = \begin{matrix} & \begin{matrix} 1 & 2 & 3 & 4 & 5 \end{matrix} \\ \begin{matrix} S_1 \\ S_2 \\ S_3 \\ \bar{S}_4 \end{matrix} & \begin{bmatrix} 1 & 1 & 1 & 0 & 0 \\ 1 & 1 & 0 & 0 & 0 \\ 1 & 0 & 1 & 0 & 0 \\ 0 & 1 & 1 & 0 & 0 \end{bmatrix} \end{matrix}.$$

Opposite each subset we enter 1 or 0 according as the integer at the column heading belongs to the subset or not. In terms of the matrix A, the existence of $\bar{d}$ is equivalent to the appearance of four 1's, no two of which lie in the same row and no two of which lie in the same column of A. This is clearly impossible, for only the first three columns of A contain 1's. On the other hand, if we construct the incidence matrix for the original subsets S_1, S_2, S_3, and S_4, we obtain

$$A = \begin{matrix} & \begin{matrix} 1 & 2 & 3 & 4 & 5 \end{matrix} \\ \begin{matrix} S_1 \\ S_2 \\ S_3 \\ S_4 \end{matrix} & \begin{bmatrix} 1 & 1 & 1 & 0 & 0 \\ 1 & 1 & 0 & 0 & 0 \\ 1 & 0 & 1 & 0 & 0 \\ 0 & 1 & 0 & 1 & 0 \end{bmatrix} \end{matrix}.$$

The sequence d corresponds to the choice of entries a_{ii}, $i = 1, 2, 3, 4$. There are other such systems of representatives that we can compute. For example, consider the following set of entries in the matrix A: a_{12}, a_{21}, a_{33}, and a_{44}. This corresponds to a sequence $d(1) = 2$, $d(2) = 1$, $d(3) = 3$, $d(4) = 4$. Another such sequence corresponds to the set of entries $a_{13}, a_{22}, a_{31}, a_{44}$: $d(1) = 3, d(2) = 2, d(3) = 1$, $d(4) = 4$. These are the only three systems which satisfy $d(t) \in S_t$, $t = 1, 2, 3, 4$. (Why?)

Such examples lead us to make the following general definition.

Definition 4.1 **System of distinct representatives** Let S be a set and let $S_1, \ldots, S_m$ be m subsets of S. An m-permutation of S, $d = (r_1, r_2, \ldots, r_m)$, is called a *system of distinct representatives* for the sets $S_1, \ldots, S_m$ if

$$d(i) = r_i \in S_i, \qquad i = 1, \ldots, m. \tag{1}$$

Thus a system of distinct representatives for the m sets $S_1, \ldots, S_m$ is simply a choice of m distinct elements, one from each set. For example, the question of the existence of a pairing into dance partners can be formulated in terms of systems of distinct representatives.

Example 4.2 Let S_j denote the set of boys who have been introduced to girl j, $j = 1, \ldots, n$. (We have labelled the girls in some order.) Suppose that d is a system of distinct representatives for the sets $S_1, \ldots, S_n$. Then according to Definition 4.1, $d(j) \in S_j, j = 1, \ldots, n$, and d is an n-permutation of the set S of all n boys. That is, $d(j)$ is a boy who has been introduced to girl $j, j = 1, \ldots, n$, and moreover, the boys $d(1), \ldots, d(n)$ are all different. Thus we see that a count of the total number of pairings into dance partners is precisely equal to the total number of systems of distinct representatives for the sets $S_1, \ldots, S_n$.

There is a standard abbreviation which we shall use. A system of distinct representatives is referred to as an SDR. Example 4.1 suggests that there is an intimate relationship between the distribution of 1's in an incidence matrix and the existence of an SDR.

Definition 4.2 **Diagonal of a matrix** Let A be an $m \times n$ matrix. If $m \leq n$, then a *diagonal of a matrix A* is just a sequence of m entries of A of the form

$$(a_{1\sigma(1)}, a_{2\sigma(2)}, \ldots, a_{m\sigma(m)}), \tag{2}$$

in which σ is an m-permutation of the set $N_n = \{1, \ldots, n\}$. If $n \leq m$, then a diagonal of A is a sequence of n entries of A of the form

$$(a_{\varphi(1)1}, a_{\varphi(2)2}, \ldots, a_{\varphi(n)n}) \tag{3}$$

where φ is an n-permutation of N_m.

Thus a diagonal of an $m \times n$ matrix, $m \leq n$, is a choice of m entries of A lying in all m rows and in m distinct columns. In case $n \leq m$, a diagonal is a choice of n entries of A lying in all n columns and in n different rows. In the event that $n = m$, then a diagonal is just a choice of n entries, one from precisely each row and each column.

Example 4.3 List all diagonals of the following matrices:

$$\text{(a) } A = \begin{bmatrix} x_1 & x_2 & x_3 \\ y_1 & y_2 & y_3 \end{bmatrix}; \qquad \text{(b) } A = \begin{bmatrix} x_1 & y_1 \\ x_2 & y_2 \\ x_3 & y_3 \end{bmatrix};$$

$$\text{(c) } A = \begin{bmatrix} x_1 & x_2 & x_3 \\ y_1 & y_2 & y_3 \\ z_1 & z_2 & z_3 \end{bmatrix}; \qquad \text{(d) } A = \begin{bmatrix} x_1 & y_1 & z_1 \\ x_2 & y_2 & z_2 \\ x_3 & y_3 & z_3 \end{bmatrix}.$$

(a) (x_1, y_2); (x_2, y_1); (x_1, y_3); (x_3, y_1); (x_2, y_3); (x_3, y_2).
(b) (x_1, y_2); (x_1, y_3); (x_2, y_1); (x_2, y_3); (x_3, y_1); (x_3, y_2).
(c) We first list the diagonals of this matrix by choosing one entry from each of *rows* 1, 2, and 3 in succession. We then list the diagonals by choosing one entry from each of *columns* 1, 2, and 3 in succession.

By rows: (x_1, y_2, z_3); (x_1, y_3, z_2); (x_2, y_1, z_3); (x_2, y_3, z_1); (x_3, y_1, z_2); (x_3, y_2, z_1).

By columns: (x_1, y_2, z_3); (x_1, z_2, y_3); (y_1, x_2, z_3); (y_1, z_2, x_3); (z_1, x_2, y_3); (z_1, y_2, x_3).

Observe that, except for the order in which the elements appear, these two lists of diagonals are identical. This is to be expected since, after all, we are selecting one element from each row and each column. In the first instance the ith element of the diagonal is from the ith row, $i = 1, 2, 3$, and in the second instance, the jth element of the diagonal is chosen from the jth column, $j = 1, 2, 3$.

(d) *By columns:* (x_1, y_2, z_3); (x_1, y_3, z_2); (x_2, y_1, z_3); (x_2, y_3, z_1); (x_3, y_1, z_2); (x_3, y_2, z_1).

By rows: (x_1, y_2, z_3); (x_1, z_2, y_3); (y_1, x_2, z_3); (y_1, z_2, x_3); (z_1, x_2, y_3); (z_1, y_2, x_3).

Notice that the list by row (column) choice for the matrix in (d) is the same as the list by column (row) choice for the matrix in (c). Again this is to be expected since the matrix in (d) is simply the transpose of the matrix in (c) and the rows and columns interchange roles.

Our next result formally relates the existence of an SDR for a collection of sets to the existence of a diagonal consisting entirely of 1's in an appropriate incidence matrix.

Theorem 4.1 *Let $S = \{x_1, \ldots, x_n\}$ be an n-set and let $S_1, \ldots, S_m$ be m subsets of S, $m \leq n$. Let R be the following relation: R is the set of pairs of integers (i, j) for which $x_j \in S_i$. Let A be the $m \times n$ incidence matrix for R:*

$$A = \begin{array}{c} \\ S_1 \\ S_2 \\ \vdots \\ S_m \end{array} \overset{\begin{array}{cccc} x_1 & x_2 & \cdots & x_n \end{array}}{\begin{bmatrix} & & & \\ & & a_{ij} & \\ & & & \end{bmatrix}}$$

where a_{ij} is 1 if $x_j \in S_i$ and 0 otherwise. The number of SDR's for the sets $S_1, \ldots, S_m$ is equal to the number of diagonals of A consisting entirely of 1's.

Proof A diagonal of A is a sequence of m entries of A of the form

$$(a_{1\sigma(1)}, a_{2\sigma(2)}, \ldots, a_{m\sigma(m)}) \tag{4}$$

where σ is an m-permutation of $N_n = \{1, \ldots, n\}$. In other words, a diagonal (4) is just a sequence of m entries in A, no two of which lie in the same row or the same column. If each element in the sequence in (4) is 1, then

$$x_{\sigma(i)} \in S_i, \qquad i = 1, \ldots, m. \tag{5}$$

The statement (5) implies that

$$(x_{\sigma(1)}, x_{\sigma(2)}, \ldots, x_{\sigma(m)}) \tag{6}$$

is an SDR for the sets $S_1, \ldots, S_m$. Moreover, a different σ in the sequence (4) will yield a different SDR in (6) if the corresponding elements in A are all 1. Conversely, if (6) is an SDR for the sets $S_1, \ldots, S_m$, then σ must be an m-permutation of $\{1, \ldots, n\}$, because the elements $x_{\sigma(i)}$, $i = 1, \ldots, m$, must be distinct, by definition of an SDR. But then, since (5) holds, it follows that $a_{i\sigma(i)} = 1$, $i = 1, \ldots, m$, and hence (4) must be a diagonal of A consisting entirely of 1's. ▮

Example 4.4 Suppose that three different hospital patients, p_1, p_2, p_3, exhibit the symptoms of four diseases, d_1, d_2, d_3, d_4, as follows: p_1 exhibits the symptoms of d_1, d_3, and d_4; p_2 exhibits the symptoms of d_1, d_2, and d_4; and p_3 exhibits the symptoms of d_2 and d_4. In how many ways is it possible to choose three out of the four diseases so that each patient exhibits the symptoms of a different one of these three diseases. We analyze this problem by defining subsets of the set of diseases $\{d_1, \ldots, d_4\}$ as follows: S_i will be the set of diseases exhibited by p_i, $i = 1, 2, 3$. Then an SDR for the sets S_1, S_2, S_3 is a sequence of three different diseases, i.e., it is a sequence of the form

$$(d_{\sigma(1)}, d_{\sigma(2)}, d_{\sigma(3)})$$

in which σ is a 3-permutation of N_4 and

$$d_{\sigma(i)} \in S_i, \qquad i = 1, 2, 3. \tag{7}$$

Statement (7) says that p_i exhibits the symptoms of the disease $d_{\sigma(i)}$, $i = 1, 2, 3$. We construct a 3×4 incidence matrix A in which $a_{ij} = 1$ whenever $d_j \in S_i$ and $a_{ij} = 0$ otherwise:

$$A = \begin{array}{c} \\ S_1 \\ S_2 \\ S_3 \end{array} \begin{array}{c} \begin{array}{cccc} d_1 & d_2 & d_3 & d_4 \end{array} \\ \begin{bmatrix} 1 & 0 & 1 & 1 \\ 1 & 1 & 0 & 1 \\ 0 & 1 & 0 & 1 \end{bmatrix} \end{array}.$$

According to Theorem 4.1, the answer to the problem is just a count of the total number of diagonals of A which consist entirely of 1's. We can count these by omitting the four columns of A one at a time and then counting diagonals of 1's in each of the four resulting 3×3 matrices. The list of diagonals of A consisting of 1's is:

$$(a_{13}, a_{21}, a_{32}),$$
$$(a_{11}, a_{22}, a_{34}),$$
$$(a_{11}, a_{24}, a_{32}),$$
$$(a_{14}, a_{21}, a_{32}),$$
$$(a_{13}, a_{21}, a_{34}),$$
$$(a_{13}, a_{24}, a_{32}),$$
$$(a_{13}, a_{22}, a_{34}).$$

There are a total of seven ways to choose three out of the four diseases so that each patient exhibits the symptoms of a different one of these three diseases.

If $S_1, \ldots, S_m$ is a collection of sets and $d = (r_1, \ldots, r_m)$ is an SDR for these sets, then $r_i \in S_i$, $i = 1, \ldots, m$, i.e., $\nu(S_i) \geq 1$. Moreover, the union of any two of the sets $S_{i_1} \cup S_{i_2}$ contains r_{i_1} and r_{i_2} and hence, $\nu(S_{i_1} \cup S_{i_2}) \geq 2$. In general, the existence of an SDR obviously ensures that the union of any k of the sets $S_{i_1} \cup S_{i_2} \cup \cdots \cup S_{i_k}$ has at least k elements in it:

$$\nu(S_{i_1} \cup S_{i_2} \cup \cdots \cup S_{i_k}) \geq k, \tag{8}$$

for any $1 \leq i_1 < i_2 < \cdots < i_k \leq m$, $k \leq m$.

We shall subsequently prove that the inequalities (8) are in fact enough by themselves to guarantee that the sets $S_1, \ldots, S_m$ have an SDR. Indeed, we shall obtain an estimate of the number of SDR's. We make our result depend on a simple and elegant combinatorial theorem concerning matrices. This result is known as the Frobenius-König theorem, named after its discoverers.

In order to make our statements somewhat briefer, we first introduce some standard language. If S is a matrix, then a *submatrix* of A is just the matrix that can be formed using the entries that are situated at the intersections of certain specified rows and columns of A. For example, if

$$A = \begin{bmatrix} 1 & 2 & 3 \\ 4 & 5 & 6 \end{bmatrix}$$

then

$$B = \begin{bmatrix} 1 & 3 \\ 4 & 6 \end{bmatrix}$$

is the submatrix of A lying in rows 1 and 2 and columns 1 and 3. Similarly,

$$C = [2 \quad 3]$$

is the submatrix of A lying in row 1 and columns 2 and 3. On the other hand

$$D = \begin{bmatrix} 1 & 2 \\ 5 & 6 \end{bmatrix}$$

is not a submatrix of A because D is not situated at the intersections of any sets of rows and columns of A. We can now state and prove our central theorem.

Theorem 4.2 **Frobenius-König** *Let A be an $n \times n$ 0,1 matrix. Then every diagonal of A has a 0 in it if and only if A contains an $s \times t$ submatrix whose entries are all 0 and for which $s + t = n + 1$.*

Proof Before we embark on a formal proof of this result it is instructive to consider some examples. If $n = 3$, then every diagonal of

$$A = \begin{bmatrix} 0 & 0 & 0 \\ 1 & 1 & 1 \\ 1 & 1 & 1 \end{bmatrix}$$

has a 0 in it and A contains the 1×3 submatrix

$$[0 \quad 0 \quad 0]$$

of 0's. Moreover $1 + 3 = 4 = 3 + 1 = n + 1$. As another example take

$$A = \begin{bmatrix} 0 & 1 & 0 \\ 1 & 1 & 1 \\ 0 & 1 & 0 \end{bmatrix}.$$

Then every diagonal of A has a 0 in it and once again A contains the 2×2 submatrix

$$\begin{bmatrix} 0 & 0 \\ 0 & 0 \end{bmatrix}$$

lying at the intersections of rows 1 and 3 and columns 1 and 3. Also, $2 + 2 = 4 = 3 + 1 = n + 1$. We remark further that if we interchange columns in A then the diagonals of A are unaltered except for the order in which the elements in the diagonals occur. (Recall that in a square matrix, a diagonal is just a choice of n entries, precisely one from each row and each column of A.) It is also clear that interchanging rows or interchanging columns will alter a submatrix of 0's by at most changing its location in the matrix. We will not hesitate to rearrange a matrix to bring it into a more convenient form. Thus although the conclusion may be proved for a matrix different from A, it will also apply to A.

To proceed to the proof, first suppose that A does contain an $s \times t$ submatrix of 0's, $s + t = n + 1$. We prove that any diagonal of A must contain a 0. First bring the submatrix of zeros into the first s rows and first t columns of A by appropriate row interchanges and column interchanges. We can then assume that the matrix has the following appearance:

(9) 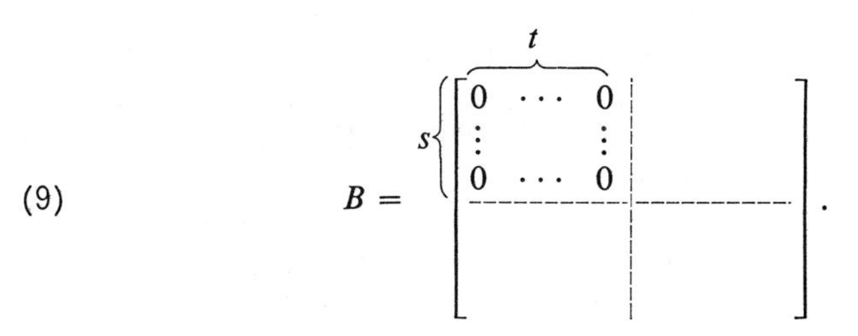

We show that any diagonal of the matrix (9) must contain a 0. For, assume that d is a diagonal of B containing no 0's. Then the elements of d lying in the first s rows of B must lie in the last $n - t$ columns. Thus $s \leq n - t$, and it follows that $s + t \leq n$. But we have assumed that $s + t = n + 1$ and this is a contradiction. Hence every diagonal must contain at least one 0.

We must now show that if every diagonal of B has a 0 in it, then there exists an $s \times t$ submatrix of B whose entries are all 0 and for which $s + t = n + 1$. This proof is by mathematical induction on n (the size of the matrix). In case $n = 1$, B is a 1×1 matrix and the only diagonal it possesses is the single entry. Since we are assuming this entry to be 0, we may take the $s \times t$ zero submatrix to be B itself. For, $s = 1, t = 1$, and $s + t = 1 + 1 = 2 = n + 1$. Thus we assume that the theorem holds for all $k \times k$ matrices, $k \leq n - 1$. That is, if X is any $k \times k$ matrix, $k \leq n - 1$, which has the property that every diagonal contains a 0, then X contains an $s \times t$ submatrix of 0's for which $s + t = k + 1$. Suppose then that B is an $n \times n$ matrix and that every diagonal of B contains a 0. If every entry of B is 0, there is no problem. For, we can simply take the zero submatrix to be the first row of B. Thus assume that B contains at least one 1, and by an interchange of rows and of columns, we may assume that the 1 is the (n, n) entry of the matrix, so that the matrix can be assumed to have the following appearance:

(10)

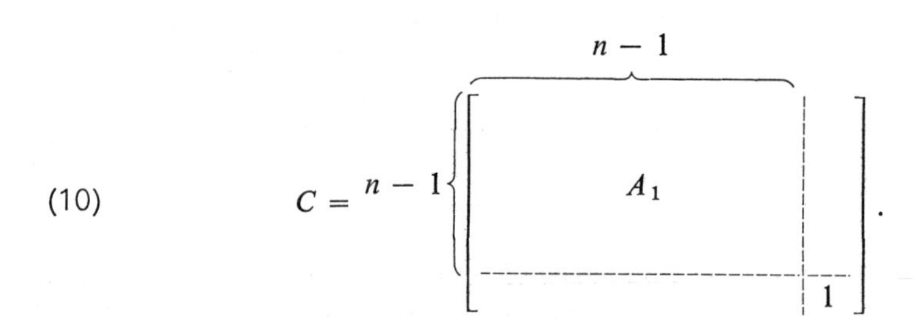

The matrix A_1 in (10) lies in the first $n - 1$ rows and columns of C. Any diagonal of A_1 can obviously be completed to a diagonal of C by adjoining the 1 in the (n, n) position. Since every diagonal of C contains a 0, it now follows that every diagonal of A_1 contains a 0. Thus the induction hypothesis tells us that A_1 must contain an $s_1 \times t_1$ submatrix of 0's for which $s_1 + t_1 = (n - 1) + 1 = n$. By an interchange of rows and columns in the matrix (10) we may put this zero submatrix of A_1 in the upper left corner so that the matrix can now be assumed to have the following form:

$$(11) \qquad D = \begin{array}{r} \\ s_1 \left\{ \vphantom{\begin{array}{c}0\\ \vdots\\ 0\end{array}} \right. \\ \\ n - s_1 = t_1 \left\{ \vphantom{\begin{array}{c}X\\ \\ \end{array}} \right. \end{array} \begin{array}{c} \overbrace{\hphantom{0 \cdots 0}}^{t_1} \quad \overbrace{\hphantom{\quad Y \quad}}^{n - t_1 = s_1} \\ \left[\begin{array}{ccc|c} 0 & \cdots & 0 & \\ \vdots & & \vdots & Y \\ 0 & \cdots & 0 & \\ \hline & X & & \\ & & & \end{array}\right] \end{array}.$$

It is clear from (11) that X is a $t_1 \times t_1$ matrix and Y is an $s_1 \times s_1$ matrix. Suppose first that X contains a diagonal of 1's. If we adjoin to this diagonal any diagonal of the matrix Y, clearly we will obtain a diagonal of D. Since every diagonal of D must contain a 0, it follows that every diagonal of Y must contain a 0. Again using the induction hypothesis, we can conclude that Y contains an $s_2 \times t_2$ submatrix of 0's for which $s_2 + t_2 = s_1 + 1$. By interchanges among the first s_1 rows of D and the last s_1 columns of D, we may bring this zero submatrix of Y into the upper left corner of Y. Moreover, such interchanges will not affect the fact that an $s_1 \times t_1$ submatrix of 0's appears in the upper left corner of D. Thus we can assume that the matrix has the following form:

$$(12) \qquad E = s_1 \left\{ \vphantom{\begin{array}{c}0\\ \vdots\\ 0\\ \vdots\\ 0\end{array}} \right. \begin{array}{c} \overbrace{\hphantom{0 \cdots 0}}^{t_1} \quad \overbrace{\hphantom{0 \cdots 0}}^{t_2} \\ \left[\begin{array}{ccc|ccc|c} 0 & \cdots & 0 & 0 & \cdots & 0 & \\ \vdots & & \vdots & \vdots & & \vdots & \\ 0 & \cdots & 0 & 0 & \cdots & 0 & \\ \cline{4-7} \vdots & & \vdots & & & & \\ 0 & \cdots & 0 & & & & \\ \hline & X & & & & & \\ & & & & & & \end{array}\right] \end{array} \begin{array}{l} \left.\vphantom{\begin{array}{c}0\\ \vdots\\ 0\end{array}}\right\} s_2 \\ \\ \\ \\ \end{array}.$$

Consider the submatrix of 0's in E which lies in the first s_2 rows of E and in the first $t_1 + t_2$ columns of E. If we compute the sum of the number of rows and columns in this zero submatrix, we have

$$\begin{aligned} s_2 + (t_1 + t_2) &= (s_2 + t_2) + t_1 \\ &= (s_1 + 1) + t_1 \\ &= (s_1 + t_1) + 1 \\ &= n + 1. \end{aligned}$$

In other words, the indicated submatrix of 0's has the property that the sum of the number of rows and number of columns is $n + 1$. This completes the induction step in the event the submatrix X in (11) contains a diagonal of 1's. If this is not the case, then every diagonal in X must contain a 0, and we can make a similar argument using the induction hypothesis to obtain an $s_3 \times t_3$ submatrix of 0's in X for which $s_3 + t_3 = t_1 + 1$. By appropriate interchanges of the last t_1 rows and the first t_1 columns in the matrix in (11), we may bring this $s_3 \times t_3$ submatrix into the upper left hand corner of X, so that the matrix can be assumed to have the following form:

(13)
$$E' = \begin{array}{r} s_1\left\{\begin{array}{c} \\ \\ \\ \end{array}\right. \\ s_3\left\{\begin{array}{c} \\ \\ \\ \end{array}\right. \\ \\ \\ \end{array} \left[\begin{array}{ccccc:c} \multicolumn{5}{c}{\overbrace{\hphantom{0 \cdots 0 \cdots 0}}^{t_1}} & \\ 0 & \cdots & 0 & \cdots & 0 & \\ \vdots & & \vdots & & \vdots & Y \\ 0 & \cdots & 0 & \cdots & 0 & \\ \hdashline 0 & \cdots & 0 & & & \\ \vdots & & \vdots & & & \\ 0 & \cdots & 0 & & & \\ \multicolumn{3}{c}{\underbrace{\hphantom{0 \cdots 0}}_{t_3}} & & & \end{array}\right].$$

Consider the submatrix of 0's in (13) which lies in the first $s_1 + s_3$ rows and the first t_3 columns. Then we compute that

$$\begin{aligned} (s_1 + s_3) + t_3 &= s_1 + (s_3 + t_3) \\ &= s_1 + (t_1 + 1) \\ &= (s_1 + t_1) + 1 \\ &= n + 1. \end{aligned}$$

This completes the proof. ▌

Example 4.5 Five hostile Middle Eastern countries have long-range weapons arrayed with respect to one another as follows: each country has its weapons aimed at two different countries and is the target for the weapons of two different countries. Is there a pairing of the five countries, say c_k with c_{i_k}, such that c_k is a target for weapons of c_{i_k}, $k = 1, 2, 3, 4, 5$, where the c_{i_k} are all different. Another way of asking this is: if all the countries fire all their weapons at once, is it certain that they all will be destroyed? We define a 5×5 matrix A such that the (i, j) entry of A is 1 or 0 according as country c_j has weapons aimed at country c_i or not. The fact that each country has its weapons aimed at two other countries means that there are two 1's in each column of the matrix A. The fact that each country is a target for the weapons of two other countries means that there are two 1's in each row of A. We must now interpret the question in terms of the matrix A. The question is whether there exists a pairing of the countries (c_1, c_{i_1}), (c_2, c_{i_2}), (c_3, c_{i_3}), (c_4, c_{i_4}), and (c_5, c_{i_5}) such that c_{i_k} has weapons aimed at c_k, $k = 1, \ldots, 5$. It is clear that this can happen if and only if 1's appear in positions $(1, i_1)$, $(2, i_2)$, $(3, i_3)$, $(4, i_4)$, and $(5, i_5)$, that is, if and only if A possesses a diagonal of 1's. Now suppose that A does not contain a diagonal of 1's. Then every diagonal of A must contain a 0. By Theorem 4.2, we conclude that A contains an $s \times t$ submatrix of 0's for which

$$\begin{aligned} s + t &= n + 1 \\ &= 5 + 1 \\ &= 6. \end{aligned}$$

Now, to interchange rows or to interchange columns merely amounts to considering the countries in a different order. We can therefore bring the $s \times t$ zero submatrix of A into the upper left hand corner of A:

$$A = \begin{array}{r} s\left\{ \vphantom{\begin{matrix} 0 \\ \vdots \\ 0 \end{matrix}} \right. \\ 5 - s\left\{ \vphantom{\begin{matrix} X \\ \\ \end{matrix}} \right. \end{array} \overset{\begin{matrix} \overbrace{\qquad\qquad}^{t} & \overbrace{\qquad\qquad}^{5 - t} \end{matrix}}{\left[\begin{array}{ccc|c} 0 & \cdots & 0 & \\ \vdots & & \vdots & Y \\ 0 & \cdots & 0 & \\ \hline & & & \\ & X & & \\ & & & \end{array} \right]}. \tag{14}$$

Now, each row of A has two 1's in it and thus the 1's in each of the first s rows of A must appear in Y. In other words, each row of Y has two 1's in it and since Y has s rows, it follows that the sum of all the entries in Y is $2s$. Similarly, each column in A has two 1's in it and hence each of the t columns in X has two 1's. It follows that the sum of the entries in X is $2t$. Thus the sum of the entries in X and Y is

$$\begin{aligned} 2s + 2t &= 2(s + t) \\ &= 2(6) \\ &= 12. \end{aligned}$$

Now A has 5 rows with two 1's appearing in each row and hence the sum of the entries in A is 10. This contradicts our conclusion that the sum of the entries in X and Y is 12. We must therefore abandon our assumption that every diagonal of A contains a 0. Hence it is certain that if the countries start firing they will destroy one another in some order.

Example 4.6 Recall the dance problem in Example 4.2 where it is assumed that in a group of n boys and n girls, each boy has been previously introduced to precisely k girls and each girl has been previously introduced to precisely k boys. The problem is to compute the total number of ways in which the boys and girls can pair off into dance partners with no further introductions. Theorem 4.2 answers part of this question. Namely, we show that there exists at least one such pairing. According to Example 4.2, if S_i denotes the set of boys who have been introduced to girl i, $i = 1, \ldots, n$, then the problem is to show that there is an SDR for the sets $S_1, \ldots, S_n$. But according to Theorem 4.1, this question is equivalent to the existence of a diagonal of 1's in the matrix A whose (i, j) entry is 1 if the jth boy has been introduced to the ith girl and 0 otherwise. The conditions of the problem state that each boy has been introduced to precisely k girls and each girl has been introduced to precisely k boys. This means that each row and each column of A contains precisely k one's. Stripped of inessential verbiage, the problem can be stated as follows. Given an $n \times n$ matrix A of 0's and 1's with k one's in each row and each column, does there exist a diagonal consisting solely of 1's? Suppose not. Then every diagonal of A must contain a 0 and hence by Theorem 4.2, A must contain an $s \times t$ submatrix of 0's with $s + t = n + 1$. By an interchange of rows and columns, the submatrix may be brought into the upper left-hand corner of A. As we have discussed before, these interchanges amount to a relabelling of the rows and columns of A, i.e., a reordering of the boys and girls. This clearly leaves the problem unaffected. Thus, unless there exists a diagonal of 1's, we may assume that A has the following form:

$$A = \begin{array}{r} \\ s\left\{\begin{array}{l} \\ \\ \\ \end{array}\right. \\ n - s = t - 1\left\{\begin{array}{l} \\ \\ \end{array}\right. \end{array} \overset{\quad \overbrace{\qquad\qquad}^{t} \quad \overbrace{\qquad\qquad\qquad}^{n - t = s - 1}}{\left[\begin{array}{ccc|c} 0 & \cdots & 0 & \\ \vdots & & \vdots & Y \\ 0 & \cdots & 0 & \\ \hline & X & & \\ & & & \end{array}\right]}. \tag{15}$$

Now each row of the matrix in (15) contains precisely k ones, as does each column. It follows that each row of the $s \times (s - 1)$ submatrix Y contains k ones and each column of the $(t - 1) \times t$ submatrix X contains k ones. Thus if we add the entries in Y, we obtain the sum ks and similarly, if we add the entries in X, we obtain kt. (Conceivably, if $s = 1$, then Y does not appear, or if $t = 1$, then X does not appear in (15); but the argument remains unchanged.) Thus the sum of the entries in X and Y is

$$\begin{aligned} ks + kt &= k(s + t) \\ &= k(n + 1). \end{aligned} \tag{16}$$

But, since every one of the n rows in A contains precisely k ones, the sum of all the entries in A is kn, and this is in conflict with (16).

It follows that A must possess a diagonal of 1's and hence a pairing into dance partners does indeed exist.

Example 4.7 Show that if, in the statement of the dance problem, we assume that each of the n girls has met *at least* k boys and each of the n boys has met *at least* k girls, then a pairing into dance partners does not necessarily exist. For, suppose $n = 4$, i.e., there are four boys and four girls. Assume that the following introductions have been made. Girl 1 has been introduced to all four boys; girls 2, 3, and 4 have been introduced to boy 1; boy 1 has been introduced to all four girls; and boys 2, 3, and 4 have been introduced to girl 1. In this case, each boy has met at least one girl, and each girl has met at least one boy. The incidence matrix A has the following form:

$$A = \begin{bmatrix} 1 & 1 & 1 & 1 \\ 1 & 0 & 0 & 0 \\ 1 & 0 & 0 & 0 \\ 1 & 0 & 0 & 0 \end{bmatrix}.$$

The matrix A contains a 3×3 submatrix of zeros and $3 + 3 = 6 > 4 + 1$. (Of course, the matrix contains a 2×3 submatrix of zeros if it contains a 3×3 submatrix of zeros.) Thus, according to Theorem 4.2, A does not possess a diagonal of 1's and no pairing into dance partners is possible.

Quiz

Answer true or false.

1 Any two different non-empty subsets of a 2-element set possess an SDR.

2 A 2×3 matrix has six diagonals.

3 If A is an $n \times m$ matrix, then A and A^T have the same number of diagonals.

4 If all corresponding diagonals (i.e., corresponding to the same permutations) of two $n \times n$ matrices are the same, then the matrices are equal.

5 The matrix [1 2] contains no submatrices.

6 If A is an $n \times m$ 0,1 matrix such that every diagonal has a 0 in it, then A contains an $s \times t$ submatrix whose entries are all 0 and $s + t$ can never exceed $n + 1$.

7 Every diagonal of the matrix

$$\begin{bmatrix} 0 & 1 & 0 \\ 1 & 1 & 1 \\ 0 & 1 & 0 \end{bmatrix}$$

contains a 0.

8 The matrix $\begin{bmatrix} 1 & 1 \\ 1 & 1 \end{bmatrix}$ is a submatrix of the matrix in Question 7.

9 The matrix in Question 7 has nine different 2×2 submatrices.

†10 If a collection of sets $S_1, \ldots, S_m$ possesses an SDR, then the union of the sets S_i, $i = 1, \ldots, m$, must contain at least m elements.

Exercises

1 List all the submatrices of each of the following matrices:

(a) $[1 \quad 2 \quad 3]$;

(b) $[1 \quad 2 \quad 3]^T$;

(c) $\begin{bmatrix} 1 & 2 & 3 \\ 4 & 5 & 6 \end{bmatrix}$;

(d) $\begin{bmatrix} 0 & 0 \\ 0 & 0 \end{bmatrix}$;

(e) $\begin{bmatrix} 1 & 2 & 3 \\ 4 & 5 & 6 \\ 7 & 8 & 9 \end{bmatrix}$.

2 Prove that there are $\binom{m}{p}\binom{n}{q}$ possible $p \times q$ submatrices of an $m \times n$ matrix.

3 Let $S = \{1, 2, 3, 4\}$. Find all SDR's for the subsets $S_1 = \{1, 4\}$, $S_2 = \{2, 3\}$, $S_3 = \{1, 3\}$.

4 List all diagonals of each of the following $m \times n$ matrices A. In case $m \leq n$ write the diagonals as

$$(a_{1\sigma(1)}, \ldots, a_{m\sigma(m)}),$$

where σ is an m-permutation of $N_n = \{1, \ldots, n\}$. In case $m \geq n$, write the diagonals as

$$(a_{\varphi(1)1}, \ldots, a_{\varphi(n)n}),$$

where φ is an n-permutation of $N_m = \{1, \ldots, m\}$.

(a) $A = \begin{bmatrix} 1 & 2 & 3 & 4 \\ 0 & 2 & 4 & 5 \\ 1 & -1 & -1 & 2 \end{bmatrix}$;

(b) $A = \begin{bmatrix} 1 & 2 \\ 3 & 4 \\ -1 & 1 \end{bmatrix}$;

(c) $A = \operatorname{diag}(1, 2, 3, 4)$;

(d) $A = \begin{bmatrix} 1 \\ 2 \\ 3 \\ 4 \end{bmatrix}$.

5 Show that if A is an $m \times n$ matrix, $m \leq n$, then A has $\dfrac{n!}{(n-m)!}$ diagonals.

6 Five personality characteristics are being observed by a psychologist in a group of six children. The children individually exhibit the characteristics as follows: characteristic 1 is exhibited by the 3rd, 4th, and 5th children; characteristic 2 is exhibited by the 1st child; characteristic 3 is exhibited by the 3rd and 5th children; characteristic 4 is exhibited by the 1st and 2nd children; characteristic 5 is exhibited by the 1st, 4th, and 6th children. Is it possible to choose a group of 5 of the children such that every child exhibits a different characteristic? If so, how many groups of five children are there having this property?

7 An office telephone switchboard permits the following connections among 12 telephones: each of the first three phones can be connected to any one of the 10th, 11th, and 12th phones and each of the 4th, 5th, and 6th phones can be connected to any one of the 7th, 8th, and 9th phones. In how many ways is it possible for each of the first six phones to be connected to precisely one of the phones 7 through 12 in some order?

8 Each of the following matrices is an incidence matrix in which the elements are listed by the columns and the subsets are listed by the rows. Explicitly write out the subsets in each case.

(a) $\begin{bmatrix} 1 & 1 & 0 & 1 & 0 \\ 0 & 1 & 1 & 0 & 1 \\ 0 & 0 & 0 & 0 & 0 \\ 1 & 0 & 0 & 1 & 1 \end{bmatrix}$;

(b) $\begin{bmatrix} 1 & 0 & 1 \\ 0 & 1 & 0 \end{bmatrix}$;

(c) $\begin{bmatrix} 1 & 1 & 0 & 0 & 1 & 1 \\ 1 & 0 & 0 & 1 & 1 & 0 \\ 0 & 0 & 1 & 1 & 0 & 0 \\ 0 & 1 & 1 & 0 & 0 & 1 \end{bmatrix}$;

(d) $\begin{bmatrix} 1 & 0 & 0 & 0 & 1 & 0 & 0 & 0 & 0 & 1 \\ 0 & 0 & 0 & 0 & 0 & 1 & 0 & 0 & 0 & 0 \\ 0 & 0 & 1 & 0 & 0 & 0 & 0 & 0 & 0 & 0 \\ 1 & 0 & 0 & 0 & 0 & 0 & 0 & 0 & 1 & 0 \\ 0 & 0 & 0 & 0 & 0 & 0 & 0 & 1 & 0 & 0 \end{bmatrix}$;

(e) $\begin{bmatrix} 1 & 0 & 0 & 0 \\ 0 & 0 & 1 & 0 \\ 0 & 1 & 0 & 0 \\ 0 & 0 & 0 & 1 \end{bmatrix}$.

9 For each collection of subsets obtained in Exercise 8, count the number of SDR's by using Theorem 4.1.

†10 Prove in detail that if the sets $S_1, \ldots, S_m$ possess an SDR, then the union of any k of the sets must contain at least k elements, $k = 1, \ldots, m$, i.e., show that the inequalities in (8) must hold.

11 Let A be a 4×4 matrix in which each entry is either 1 or 2 and suppose that every diagonal of A contains a 2. Show that the sum of the entries in A is at least 20. (Hint: Imagine the 2's in A replaced by 0's. Then use Theorem 4.2 to conclude that A contains an $s \times t$ submatrix of 2's in which $s + t = 5$. Thus the sum of the entries in A is at least $2st + (16 - st) = 16 + st$. What is the least value that st can be?)

12 Suppose that the five hostile Middle Eastern countries discussed in Example 4.5 have their guns arrayed as follows: two of the countries are the targets for the weapons of at least two different countries, and each country has its weapons aimed at at least two different countries. Is it necessarily true that if all countries fire their weapons at once, they will all be destroyed? If not, exhibit an example.

13 In the dance problem in Example 4.6, suppose the assumptions concerning introductions are modified as follows: each boy has been previously introduced to precisely k girls, and each girl has been previously introduced to at least one boy, $k < n$. Is it necessarily true that a pairing into dance partners is possible with no further introductions?

***14** Five cities are connected to one another by a network of one-way roads as follows: there is a road from each city to precisely three other cities, and each city is the terminal point of roads from precisely three other cities. Is it possible to tour all five cities in such a way that each city is visited precisely once? Can such a tour be made in more than one way?

15 Three countries with similar economies each produce k dollars worth of goods which are sold among the three countries. Moreover, each country buys precisely k dollars worth of goods (there is no balance of payments problem here). Does there exist a pairing of the countries, say country j with country i_j, $j = 1, 2, 3$ (i_1, i_2, i_3 distinct), in which country j sells to country i_j?

2.5

Systems of Distinct Representatives

Our goal in this section is the statement, proof, and application of a very remarkable theorem on systems of distinct representatives. This result (Theorem 5.1) is due to the English mathematician Philip Hall and was originally published in 1935.

In the discussion immediately preceding Theorem 4.2 we noted that if $S_1, \ldots, S_m$ is a collection of sets which possesses SDR, then the union of any k of the sets must contain at least k elements. This is obviously necessary because the k distinct elements from the SDR are themselves in the union. Thus if $S_{i_1}, \ldots, S_{i_k}$ are k of the sets from among $S_1, \ldots, S_m$, then

$$\nu(S_{i_1} \cup S_{i_2} \cup \cdots \cup S_{i_k}) \geq k, \tag{1}$$

$k = 1, 2, \ldots, m$. Hall's theorem tells us that the inequalities in (1), which are obviously necessary for the existence of an SDR, are also sufficient.

Theorem 5.1 *Let $S_1, \ldots, S_m$ be subsets of an n-element set $S = \{x_1, \ldots, x_n\}$, $m \leq n$. Assume, moreover, that each S_i contains at least t elements, $t \geq 1$. Then a necessary and sufficient condition for the subsets $S_1, \ldots, S_m$ to possess an SDR is that the union of any k of the subsets contains at least k elements, i.e., that the inequalities in* (1) *hold. If these inequalities do hold, then there are at least*

(*a*) *$t!$ SDR's if $t \leq m$;*

and

(b) $\dfrac{t!}{(t-m)!}$ *SDR' s if $t > m$.*

Proof We have already proved that if $S_1, \ldots, S_m$ possess an SDR then the inequalities in (1) must hold. We will prove that one of the two alternatives (*a*) and (*b*) must hold, and since $t \geq 1$, it will follow that whether $t \leq m$ or $t > m$, the existence of an SDR will have been demonstrated. The proof that one of the alternatives (*a*) and (*b*) must be the case if the inequalities (1) hold will be made by an induction on m, the number of subsets. If $m = 1$, then there is, of course, no problem. We first write down the incidence matrix

$$A = \begin{matrix} \\ S_1 \\ S_2 \\ \vdots \\ S_m \end{matrix} \begin{matrix} x_1 \quad x_2 \quad \cdots \quad x_n \\ \left[\begin{matrix} & & \\ & & \\ & a_{ij} & \\ & & \\ & & \end{matrix} \right] \end{matrix},$$

in which $a_{ij} = 1$ if $x_j \in S_i$ and $a_{ij} = 0$ otherwise. We can restate the conditions in (1) in terms of the matrix A. The inequalities in (1) imply for the matrix A that given any k rows of A, there exist at least k columns in which 1's appear somewhere among the k given rows, $k = 1, \ldots, m$. In other words, all the 1's that appear in any k rows of A are not "crammed" into fewer than k columns. For example, $\nu(S_1 \cup S_2) \geq 2$ means that $S_1 \cup S_2$ contains at least two elements, say x_{j_1} and x_{j_2}. Thus columns j_1 and j_2 of A each have a 1 occurring in the first or second row. These 1's may appear in the same row. We are also assuming that each row of A contains at least t ones, $t \geq 1$. Our problem is to prove that A has at least $t!$ diagonals of 1's if $t \leq m$ and $\dfrac{t!}{(t-m)!}$ diagonals of 1's if $t > m$. The induction hypothesis states that any matrix with $m - 1$ or fewer rows has the required number of diagonals of 1's. There are two possibilities:

Case I: for some k, $1 \leq k \leq m - 1$, there are k rows of A for which there exist exactly k columns (and no more) in which 1's appear among these k rows;

Case II: given *any* k rows of A, $1 \leq k \leq m - 1$, there are at least $k + 1$ columns in which 1's appear among the k given rows.

In case I we may as well assume that the distinguished k rows are rows $1, 2, \ldots, k$ and that it is the first k columns in which the 1's appear among these rows. This just amounts to a relabelling of the elements $x_1, \ldots, x_n$ and the sets $S_1, \ldots, S_m$, $m \leq n$. Thus A can be assumed to have the following form in case I.

$$A = \begin{array}{c} \\ k\left\{ \right. \\ \\ m-k\left\{ \right. \end{array} \overset{\qquad\overbrace{\qquad k \qquad}\quad\overbrace{\qquad n-k \qquad}}{\left[\begin{array}{c|ccc} & 0 & \cdots & 0 \\ A_1 & \vdots & & \vdots \\ & 0 & \cdots & 0 \\ \hline & & A_2 & \end{array}\right]} \tag{2}$$

The reason that the upper right $k \times (n - k)$ block in A consists of 0's is that we are assuming in case I that among the first k rows, 1's appear only in columns $1, \ldots, k$. Now every row of A and hence every row of A_1 has at least t ones in it and thus $t \leq k$ (there are only k places for 1's in any row of A_1). Hence $t \leq m - 1$. Also it is obvious that given any number of rows of A_1 there exists an equal number of columns in which 1's appear among these given rows. For, after all, this property holds for A and the last $n - k$ entries in each of the first k rows of A are all 0's. Since A_1 has $m - 1$ or fewer rows ($k \leq m - 1$), it follows from the induction hypothesis, since $t \leq m - 1$, that A_1 has at least $t!$ diagonals of 1's. Now let A_2 be the $(m - k) \times (n - k)$ lower right block in A indicated in (2). We are going to use Theorem 4.2 to show that A_2 contains a diagonal of 1's. If every diagonal of A_2 contains a 0, then the same is true for the following $(n - k) \times (n - k)$ matrix that we can construct from A_2 by adjoining $n - m$ rows of 1's to A_2.

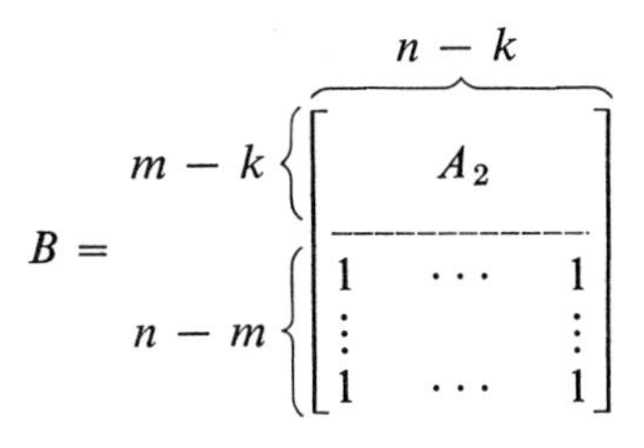

Then the matrix B contains an $s \times r$ submatrix of 0's in which $s + r = n - k + 1$. By an interchange of rows and columns, we may assume that this submatrix is in the upper right-hand corner of A_2. It is clear, then, that by interchanges of the last $m - k$ rows and last $n - k$ columns of (2) (which leaves A_1 alone), the matrix A can be assumed to have the following form.

$$(3) \qquad A = \begin{array}{c} \\ k\left\{\right. \\ \\ m-k\left\{\right. \end{array} \overset{\displaystyle \overbrace{\qquad\quad}^{k} \quad \overbrace{\qquad\qquad\qquad\qquad}^{n-k}}{\left[\begin{array}{c|ccc|ccc} & 0 & \cdots & 0 & 0 & \cdots & 0 \\ A_1 & \vdots & & \vdots & \vdots & & \vdots \\ & 0 & \cdots & 0 & 0 & \cdots & 0 \\ \hline & & & & 0 & \cdots & 0 \\ & & & & \vdots & & \vdots \\ & & & & 0 & \cdots & 0 \\ & & & & \multicolumn{3}{c}{\underbrace{\qquad\qquad}_{r}} \end{array}\right]} \begin{array}{c} \\ \\ \\ \left.\right\} s \\ \\ \end{array}$$

Now consider the first $k + s$ rows of A. It is clear from (3) that there are only $k + (n - k) - r$ columns available in which the 1's in these $k + s$ rows can appear. But

$$\begin{aligned} k + (n - k) - r &= n - r \\ &= k + s - 1. \end{aligned}$$

(Recall that $s + r = n - k + 1$.) In other words, the 1's that appear in the first $k + s$ rows of A must be in the first $k + s - 1$ columns. But this is in contradiction with the assumption we are making that given any p rows of A there exist at least p columns in which 1's appear among the p given rows, $p = 1, \ldots, m$. Thus we must abandon the assumption that every diagonal of A_2 contains a zero, i.e., A_2 must contain a diagonal of 1's. This diagonal of 1's can be adjoined to any of the (at least) $t!$ diagonals of A_1 to produce $t!$ diagonals of A. We remarked that in case I, $t \leq m - 1$ and thus we have proved that in case I, the conclusion that A has at least $t!$ diagonals of 1's is indeed correct. In terms of the original statement of the theorem, this means that in case I, the sets $S_1, \ldots, S_m$

possess at least $t!$ SDR's, proving (a) under these circumstances.

For case II, we are assuming that given any k rows, there are at least $k + 1$ columns in which the 1's appear among these k rows. Observe that in this case, $t > 1$. For, taking $k = 1$, we see that any row must contain at least 2 ones. It is just a matter of relabelling the elements to bring any of the 1's in row 1 of A to the (1, 1) position. There are at least t such 1's and thus this can be done in at least t ways.

$$A = \left[\begin{array}{c|c} 1 & \overbrace{\qquad\qquad}^{n-1} \\ \hline & \\ & A_3 \\ & \end{array}\right]\Bigg\} m - 1 \tag{4}$$

The matrix A_3 is $(m - 1) \times (n - 1)$. If we take any k rows of the matrix A_3, $1 \leq k \leq m - 1$, we know that in case II there must be at least k columns of A_3 in which 1's appear among these k rows. This is because the rows of A_3 are obtained by deleting the first entry in each of the last $m - 1$ rows of A. Case II states that there are at least $k + 1$ columns among which the 1's appear in any k rows of A. Knocking out the first column of A means that there are at least k columns left in which the 1's may appear in any of the k rows of A_3. If $t \leq m$, then $t - 1 \leq m - 1$, and since each row of A_3 contains at least $(t - 1)$ ones, it follows from the induction hypothesis that A_3 has at least $(t - 1)!$ diagonals of 1's. Similarly if $t > m$, then $t - 1 > m - 1$ and again by the induction hypothesis, A_3 has at least $\dfrac{(t-1)!}{((t-1)-(m-1))!}$ diagonals of 1's. We may adjoin the 1 in the (1, 1) position in A to any diagonal of A_3 to obtain a diagonal of A. We have proved now that each of the 1's in the first row of A is contained in

(a) $(t - 1)!$ diagonals of 1's in A if $t \leq m$;

(b) $\dfrac{(t-1)!}{((t-1)-(m-1))!}$ diagonals of 1's in A if $t > m$.

(Recall we moved an arbitrary 1 appearing in the first row of A to the (1, 1) position for convenience.) Since there are at least t ones in the first row of A it means that we have found

(a) $t \cdot (t - 1)! = t!$ diagonals of 1's in A if $t \leq m$ and

(b) $t\left(\dfrac{(t-1)!}{((t-1)-(m-1))!}\right) = \dfrac{t!}{(t-m)!}$ diagonals of 1's in A if $t > m$. This completes the proof. ▌

Example 5.1 The result of Theorem 5.1 can be used to improve our answer to the dance problem. In Example 4.6 we showed that if in a group of n boys and n girls, each boy has been introduced to precisely t girls and each girl has been introduced to precisely t boys, then at least one pairing into dance partners is possible. As we saw, the number of such pairings is precisely equal to the number of diagonals consisting of 1's in the incidence matrix A. The matrix A is $n \times n$ with t ones in each row and in each column (in Example 4.6 we used the letter k instead of t; we use t now to conform to the notation of Theorem 5.1, which we intend to apply shortly). We showed in Example 4.6 that this matrix A contains at least one diagonal of 1's. This means that there exists at least one SDR for the sets S_i, $i = 1, \ldots, n$, where S_i denotes the set of boys who have been introduced to girl i.

According to Theorem 5.1, it follows from the existence of at least one SDR that the union of any k of the sets $S_1, \ldots, S_n$ must contain at least k elements, i.e., that the inequalities in (1) must hold. Now each S_i contains t elements, $t \leq n$, and hence we have satisfied the hypothesis of Theorem 5.1. That is, the necessary inequalities in (1) hold, and each S_i contains t elements. We can therefore conclude that there are at least $t!$ SDR's. (Note that $m = n$ here.) In terms of the incidence matrix A, this means that A contains at least $t!$ diagonals of 1's. We can summarize now by saying that if each boy has been introduced to precisely t girls and each girl to precisely t boys, then there are at least $t!$ different pairings into dance partners.

The result of Theorem 5.1 can be used to analyze probability problems which heretofore have been inaccessible.

Example 5.2 A supermarket carries 6 brands of bread. Six housewives equipped with certain prejudices enter the market to purchase bread. It is equi-probable that a given housewife will purchase any one of precisely 3 brands of bread but definitely will not purchase the other 3. Moreover, any particular brand is among the acceptable choices of precisely 3 of the housewives. Question: What is the probability that all 6 brands will be purchased by the housewives? Also estimate the least value for this number. We label the housewives $1, \ldots, 6$ and the brands of bread $1, \ldots, 6$. Moreover, let p_{ij} be the probability that housewife i will purchase brand j, $i, j = 1, \ldots, 6$. Let P be the 6×6 matrix whose (i, j) entry is p_{ij}. The appropriate probability space U is the set of all 6-tuples

$$(j_1, j_2, \ldots, j_6) \tag{5}$$

in which j_s is the brand that housewife s purchases, $s = 1, \ldots, 6$. The purchase that any housewife makes is clearly independent of the purchase of any other housewife. Hence it is sensible to use the product probability measure: to the elementary event consisting of the 6-tuple in (5) we assign the probability

$$p(\{(j_1, \ldots, j_6)\}) = p_{1j_1}p_{2j_2}\cdots p_{6j_6}. \tag{6}$$

Before verifying that p is indeed a probability measure, we interpret the conditions of the problem in terms of the matrix P. The assumption that housewife i will purchase any one of precisely 3 brands with equal probability and definitely will not purchase any of the remaining 3 brands means that the ith row of the matrix P has precisely 3 entries equal to $\frac{1}{3}$ and the remaining 3 entries equal to 0. The assumption that brand j is among the acceptable choices of precisely 3 of the housewives means that precisely 3 of the entries in column j are equal to $\frac{1}{3}$ and the remaining 3 entries are equal to 0. In other words, P is a matrix each of whose entries are 0 or $\frac{1}{3}$, with precisely 3 entries equal to $\frac{1}{3}$ in each row and in each column. To check that the function p be given by (6) is a probability measure, we need only observe that the value on the right in (6) is a product of 6 entries of P, precisely one from each of the 6 rows of P. Any such product appears precisely once in the expression for the product of the row sums of P:

$$\left(\sum_{j=1}^{6} p_{1j}\right)\left(\sum_{j=1}^{6} p_{2j}\right)\cdots\left(\sum_{j=1}^{6} p_{6j}\right), \tag{7}$$

i.e., we can choose term p_{1j_1} from the first factor in (7), p_{2j_2} from the second factor, $\ldots$, p_{6j_6} from the sixth factor. But after all, each row sum of P is 1 so that the product in (7) has value 1. In other words the sum of the values of p over all elementary events is 1 and thus p is a probability measure. The event in U which interests us is the set X of 6-tuples (5) in which $j_1, \ldots, j_6$ are distinct. For then the right-hand side of (6) is just the probability that housewife s purchases brand j_s, $s = 1, \ldots, 6$, and brands $j_1, \ldots, j_6$ constitute all 6 brands. In other words, $p(X)$ is a sum of terms, each of which is a product of the entries in a diagonal of P. Let P_1 be the matrix obtained from P by replacing each of the entries equal to $\frac{1}{3}$ by 1 and leaving the zero entries alone. Then P_1 is a matrix of 0's and 1's with three 1's in each row and each column. Moreover, a diagonal of 1's in P_1 corresponds to a diagonal of $\frac{1}{3}$'s in P and hence to a term in $p(X)$ equal to $(\frac{1}{3})^6$. We saw in the preceding example that a matrix of 0's and 1's with t ones in each row and column contains at least $t!$ diagonals of 1's. Hence there are at least 3! terms in $p(X)$

with value $(\frac{1}{3})^6$. It follows that

$$\begin{aligned} p(X) &\geq \frac{3!}{3^6} \\ &= \frac{2}{243}. \end{aligned}$$

The probability that the housewives purchase all 6 brands of bread is positive and, in fact, at least $\frac{2}{243}$.

Example 5.3 A nuclear reaction involves n different types of subatomic particles. Assume that it is known that the probability that a particle of type j changes into a particle of type i is p_{ij}, $i, j = 1, \ldots, n$. Moreover, it is known that the reaction is conservative in the following sense: each particle changes into one of the n original types of particles (including its own type, possibly) and every one of the n types is present at the end of the reaction. Show that there is an n-permutation $(j_1, \ldots, j_n)$ of $\{1, \ldots, n\}$ such that the probability that a particle of type j_s changes into a particle of type s, $s = 1, \ldots, n$, is positive. Find a formula in terms of the probabilities p_{ij} for all possible ways in which this can happen.

To solve this problem, let P be the $n \times n$ matrix whose (i, j) entry is p_{ij}, $i, j = 1, \ldots, n$. The conditions of the problem can be restated as follows.

(i) Each particle changes into one of the n original types of particles. Thus,

$$p_{1j} + p_{2j} + \cdots + p_{nj} = 1, \tag{8}$$

$j = 1, \ldots, n$, for, (8) is the statement that the probability that particles of type j change into at least one of the n types of particles is 1.

(ii) The fact that every one of the original types of particles must be present at the end of the reaction is just the statement that

$$p_{i1} + p_{i2} + \cdots + p_{in} = 1, \tag{9}$$

$i = 1, \ldots, n$. For, (9) asserts that particles of type i are present at the end of the reaction and we know that they came from particles of types $1, \ldots, n$.

The conditions of the problem imply that P is an $n \times n$ matrix whose entries are nonnegative numbers. Moreover, each row sum and each column sum of P is 1. Precisely as in Example 5.2, the appropriate probability space U is the set of all n-tuples

$$(j_1, j_2, \ldots, j_n) \tag{10}$$

in which particles of type s result from particles of type j_s in the reaction. We assume that the behavior of any individual particle in the reaction is independent of the behavior of any of the remaining particles and hence it is reasonable to assign the probability

$$p(\{(j_1, \ldots, j_n)\}) = p_{1j_1}p_{2j_2}\cdots p_{nj_n} \tag{11}$$

to the elementary event consisting of the n-tuple in (10). Again as in Example 5.2, we can verify that the function p in (11) yields a probability measure on the space U. Let X be the event in U consisting of all n-tuples of the form (10) such that $(j_1, \ldots, j_n)$ is an n-permutation of $\{1, \ldots, n\}$. The set X corresponds to the totality of ways in which particles of types $1, \ldots, n$ can come from particles of types $1, \ldots, n$ in some order, i.e., particles of type s come from particles of type j_s, $s = 1, \ldots, n$, and all n integers appear among $j_1, \ldots, j_n$. Thus the probability of the event X is given by

$$p(X) = \sum p_{1j_1} \cdots p_{nj_n}, \tag{12}$$

in which the summation in (12) is over all n-permutations $(j_1, \ldots, j_n)$ of the integers $1, \ldots, n$. We can rewrite (12) somewhat more succinctly as follows:

$$p(X) = \sum_{\sigma \in S_n} \prod_{i=1}^{n} p_{i\sigma(i)}. \tag{13}$$

The summation in (13) is over the set of all n-permutations of $\{1, \ldots, n\}$, denoted here by S_n. Equivalently, the right-hand side of (13) is the sum of all $n!$ products formed by multiplying together the elements in a given diagonal of P. The formula (13) is the required expression for the probability that the n different types of particles remaining after the reaction came from the n original types of particles in some order. We show that there exists at least one diagonal product in P which is positive. For, suppose that the product of the elements of every diagonal in P is 0. It follows that every diagonal of P must contain at least one 0. Now let P_1 be the matrix obtained from P by replacing each of the strictly positive entries by 1 and leaving the 0 entries unaltered. Then every diagonal of P_1 contains a 0 and we may apply Theorem 4.2 to conclude that P_1 contains an $s \times t$ submatrix whose entries are all 0 and for which $s + t = n + 1$. It is obvious that P must have this property as well, so that by a rearrangement of the rows and columns we may assume that the $s \times t$ submatrix of 0's appears in P as follows.

$$P = \begin{array}{c} \\ s\left\{\begin{array}{c} \\ \\ \\ \end{array}\right. \\ \\ \end{array}\left[\begin{array}{ccc|c} \multicolumn{3}{c|}{\overbrace{\qquad\qquad}^{t}} & \\ 0 & \cdots & 0 & \\ \vdots & & \vdots & Y \\ 0 & \cdots & 0 & \\ \hline & X & & \\ \end{array}\right] \tag{14}$$

The submatrix X has t columns and each column sum of P is 1 It follows from (14) that each column sum of X is 1 and hence the sum of the entries in X is t. Similarly, each row sum of Y is 1 and hence the sum of the entries in Y is s. Thus the sum of the entries in X and Y together is $s + t = n + 1$. But each row sum of P is 1 and hence the sum of all the entries in P is n. This is a contradiction and it follows that P possesses at least one diagonal consisting of strictly positive entries.

In the preceding examples, we studied $n \times n$ matrices with nonnegative entries in which all the row sums and all the column sums are 1. Such matrices and their immediate generalizations are extremely important in many applications that we consider here and in later sections.

Definition 5.1 ***Stochastic matrices*** Let A be an $m \times n$ matrix with nonnegative entries. If the sum of the entries in each row of A is 1, then A is called a *row stochastic* matrix. If the sum of the entries in each column of A is 1, then A is called a *column stochastic* matrix. If A is both row stochastic and column stochastic, then A is called a *doubly stochastic* matrix.

Observe that if A is doubly stochastic, then A must be a square matrix. For, summing all the entries in A by adding all row sums, we obtain m, and summing all the entries by adding all column sums, we obtain n. Thus $m = n$.

In an $m \times n$ matrix A containing only 0's and 1's as entries, it is frequently of importance to find the minimal number of rows and columns which contain all the 1's.

Example 5.4 In a model of a segment of an economy, 5 manufacturing companies, $M_1, \ldots, M_5$, are in partial competition with 7 other companies, $C_1, \ldots, C_7$. However, it is known that certain modifications of distribution policies are possible which will result in a greater net profit for the entire economy. As a particular example, we can summarize the information in a 5×7 matrix A in which the (i, j) entry is 1 if a change in the distribution policy in either M_i or C_j results in an increase in the net profit for the entire economy; if no such change is possible, the (i, j) entry is 0. Moreover, it is assumed that if $a_{ij} = 1$ then the increase in net profit is the same whether M_i or C_j affects a change in its distribution policy.

$$A = \begin{array}{c} \\ M_1 \\ M_2 \\ M_3 \\ M_4 \\ M_5 \end{array} \begin{array}{c} \begin{array}{ccccccc} C_1 & C_2 & C_3 & C_4 & C_5 & C_6 & C_7 \end{array} \\ \begin{bmatrix} 0 & 0 & 1 & 0 & 0 & 1 & 0 \\ 1 & 0 & 1 & 0 & 1 & 0 & 1 \\ 0 & 0 & 0 & 0 & 0 & 1 & 0 \\ 1 & 1 & 1 & 1 & 0 & 0 & 1 \\ 0 & 0 & 1 & 0 & 0 & 1 & 0 \end{bmatrix} \end{array} \tag{15}$$

The problem is to find the minimum number of companies among the M_i and C_j which must modify their distribution policies so that the profit for this entire segment of the economy will be maximum. In terms of the matrix A, this question amounts to finding the minimum number of rows and columns in A so that all the 1's in the matrix appear among these rows and columns. For then it would only be necessary for the companies M_i corresponding to the given rows and the companies C_j corresponding to the given columns to change their distribution practices. Observe that if four companies, M_2, M_4, C_3, and C_6, change their distribution policies, then the maximum profit for the economy will be achieved, since all the 1's in the matrix appear in rows 2 and 4 and columns 3 and 6. Thus the minimum number of companies necessary is no greater than 4.

Before we can completely answer the question posed in Example 5.4, we prove a remarkable theorem which follows directly from Theorem 5.1. To facilitate the statement we introduce two convenient definitions due to H. J. Ryser. A *line* in a matrix refers to either a row or a column. The *term rank* of a 0, 1 matrix A (i.e., a matrix whose entries are either 0 or 1) is the largest integer k for which it is possible to find k ones lying in k different rows and in k different columns of A. In other words, the term rank of a matrix A is just the largest integer k for which there exists a $k \times k$ submatrix of A containing a diagonal of 1's. For example, the matrix in (15) contains four 1's in the following positions: (1, 3), (2, 1), (3, 6), (4, 4). However, it is not possible to find five 1's lying in different rows and columns of A. For, the 1st, 3rd, and 5th rows of A have 1's only in columns 3 and 6, so that the choice of 1's from rows 1, 3, and 5 must be from columns 3 and 6. Hence only 2 of these 3 rows can be used for the choice. In other words, the term rank of the matrix A in (15) is 4. Ryser's theorem tells us that the term rank of a matrix is precisely the minimum number of lines in the matrix containing all the 1's. A set of lines is said to *cover* the 1's in a matrix if all the 1's lie among the rows or columns of the set of lines.

Theorem 5.2 *Let A be an $m \times n$ matrix with entries 0 or 1. The term rank of A is equal to the minimal number of lines in A that contain all the 1's in A.*

Proof Let r denote the term rank of A. Then we know that there are r ones lying in r different rows and r different columns of A. Thus any set of lines which cover all the 1's in A must cover these r 1's. In other words, if ℓ denotes the minimal number of lines in A that contain all the 1's in A, then

$$\ell \geq r. \tag{16}$$

Now, to prove that $r \geq \ell$, we will find an $\ell \times \ell$ submatrix which has a diagonal of 1's. Suppose that in a minimal covering of the 1's in A by ℓ lines, there are s rows and t columns, so that

$$s + t = \ell.$$

(Notice that since $\ell \leq n$ it follows that $s \leq n - t$ and $t \leq n - s$.) Before proceeding, we remark that an interchange of rows or an interchange of columns does not affect the number of lines in a minimal covering, nor does it affect the term rank of the matrix A. For, such an interchange simply amounts to a renumbering of the lines that contain all the 1's. Moreover, a set of 1's in A in different rows and different columns will obviously still lie in different rows and different columns after such interchanges are performed. Thus we may assume that the s rows and t columns in the minimal covering of 1's are the first s rows and t columns in A, so that A can be assumed to have the following form.

$$A = \left[\begin{array}{c:c} \overbrace{\quad A_1 \quad}^{t} & A_2 \\ \hdashline A_3 & A_4 \end{array}\right] \quad \text{(A_1, A_2 span the first s rows)} \tag{17}$$

We first prove that A_2 has term rank s. To see this, define s subsets, $S_1, \ldots, S_s$, of the set of integers $t + 1, \ldots, n$, as follows: an integer $j \in S_k$ if and only if the (k, j) entry of A is 1, $k = 1, \ldots, s$,

$j = t + 1, \ldots, n$. Then clearly A_2 is the incidence matrix for the sets $S_1, \ldots, S_s$. It is also clear that no S_k can be empty, for if this were the case, all the 1's in A could be covered by the first t columns in A and the $s - 1$ rows in A numbered $1, 2, \ldots, k - 1, k + 1, \ldots, s$. In other words, A could be covered by $s - 1 + t = \ell - 1$ lines, and this contradicts the fact that ℓ is the minimal number of lines necessary to contain all the 1's. Now suppose that no SDR for the sets $S_1, \ldots, S_s$ exists. According to Theorem 5.1, this means that there exist k of the sets, say $S_{i_1}, \ldots, S_{i_k}$, such that the union of these k sets contains fewer than k elements, i.e., one of the inequalities in (1) must fail. In terms of the matrix A_2, this means that in the k rows numbered $i_1, \ldots, i_k$, the 1's appear in fewer than k columns, say in columns $j_1, \ldots, j_q$, where q is less than k. Now consider the following set of lines in A: columns $1, \ldots, t$, columns $j_1, \ldots, j_q$, and the $s - k$ rows obtained by deleting rows $i_1, \ldots, i_k$ from among the first s rows. This set of lines contains all the 1's in A. But since $k > q$, there are only

$$\begin{aligned} t + q + s - k &= s + t - (k - q) \\ &= \ell - (k - q) \\ &< \ell \end{aligned}$$

lines in this set, once again contradicting the minimality of ℓ. Hence we have proved that the sets $S_1, \ldots, S_s$ must possess an SDR, which means that the matrix A_2 contains a diagonal of s ones. Similarly, we can prove (see Exercise 26) that the matrix A_3 must contain a diagonal of t ones. If we put these two diagonals together, we will have a set of 1's in the matrix A lying in $s + t = \ell$ different rows and ℓ different columns. But then the term rank of A must be at least ℓ, that is,

$$r \geq \ell. \tag{18}$$

Combining (16) and (18) we see that $r = \ell$, and the theorem is proved. ∎

We can now complete the analysis of Example 5.4. We saw that the matrix in (15) contains four 1's in positions (1, 3), (2, 1), (3, 6), and (4, 4). We also proved that the matrix in (15) does not contain five 1's lying in different rows and in different columns. Hence the term rank of the matrix in (15) is 4, and it follows that the four lines, rows 2 and 4, and columns 3 and 6, are a minimal covering of the 1's in the sense that no fewer than four lines in the matrix can contain all the 1's. Thus, no fewer than four companies must modify their distribution policies in order that the profit for the entire segment of the economy is maximum.

Quiz

Answer true or false.

1 Let $S_1 = \{1, 2\}$, $S_2 = \{2, 3\}$, $S_3 = \{1, 2, 3\}$ be three subsets of $\{1, 2, 3\}$. Then the sets S_1, S_2, S_3 possess three SDR's.

2 The matrix $\begin{bmatrix} \frac{1}{2} & \frac{1}{2} & 0 \\ \frac{1}{2} & \frac{1}{2} & 0 \end{bmatrix}$ is doubly stochastic.

3 If S_1, S_2, and S_3 are subsets of a 5-element set and each S_i contains at least 4 elements, then there are at least 24 SDR's for the subsets.

4 If S_1, S_2, S_3, and S_4 are subsets of a 10-element set, each containing at least 7 elements, then there are at least 840 SDR's for the subsets.

5 The matrix

$$\begin{bmatrix} \frac{1}{2} & \frac{1}{2} & 0 \\ \frac{1}{2} & \frac{1}{2} & 0 \\ 0 & 0 & 1 \end{bmatrix}$$

is doubly stochastic.

6 If A is $n \times n$ row stochastic and B is $n \times n$ column stochastic, then $\frac{1}{2}A + \frac{1}{2}B$ is doubly stochastic.

7 If a 4×4 matrix A of 0's and 1's has precisely two 1's in each row and each column, then the term rank of A is 4.

8 An $m \times n$ matrix contains $m + n$ lines.

9 If A and B are two 3×3 matrices of 0's and 1's and the term rank of A exceeds the term rank of B, then A must contain at least as many 1's as B.

10 Let A be a 2×2 matrix of 0's and 1's. In the matrix AA^T, replace each positive entry by 1 and leave the 0's alone. Then the term rank of the resulting matrix is at least equal to the term rank of A.

Exercises

1 Each of the following matrices are incidence matrices in which the elements are listed across the columns and the sets are listed down the rows. The (i, j) entry is 1 or 0 according as the jth element belongs to the ith set. Compute the minimum number of SDR's guaranteed by Theorem 5.1.

(a) $\begin{bmatrix} 1 & 1 & 0 & 0 & 0 \\ 0 & 1 & 0 & 0 & 1 \\ 0 & 0 & 1 & 1 & 1 \end{bmatrix}$;

(b) $\begin{bmatrix} 0 & 1 & 1 & 0 \\ 0 & 0 & 1 & 1 \\ 1 & 0 & 0 & 1 \\ 1 & 1 & 0 & 0 \end{bmatrix}$;

(c) $\begin{bmatrix} 1 & 1 & 1 & 1 & 1 & 1 & 0 & 0 \\ 0 & 1 & 1 & 1 & 1 & 1 & 1 & 1 \\ 1 & 1 & 1 & 0 & 1 & 1 & 0 & 1 \\ 1 & 1 & 1 & 1 & 1 & 0 & 1 & 1 \end{bmatrix}$;

(d) $\begin{bmatrix} 0 & 0 & 0 & 1 & 1 & 0 \\ 1 & 1 & 0 & 0 & 1 & 0 \\ 0 & 1 & 1 & 0 & 0 & 1 \\ 1 & 0 & 0 & 1 & 0 & 0 \\ 1 & 1 & 1 & 0 & 0 & 0 \end{bmatrix}$.

2 Construct an $n \times n$ 0,1 matrix A of term rank $n - 1$ such that A^2 has term rank $n - 2$, A^3 has term rank $n - 3$, etc. (Hint: Consider the $n \times n$ matrix

$$A = \begin{bmatrix} 0 & 1 & 0 & \cdots & 0 \\ 0 & 0 & 1 & \cdots & 0 \\ \vdots & \vdots & \vdots & \cdots & \vdots \\ 0 & & 0 & \cdots & 0 \\ 0 & & 0 & \cdots & 1 \\ 0 & & 0 & \cdots & 0 \end{bmatrix}.)$$

3 Let A be a 2×2 doubly stochastic matrix. Show that A can be written in the form

$$A = s\begin{bmatrix} 1 & 0 \\ 0 & 1 \end{bmatrix} + (1 - s)\begin{bmatrix} 0 & 1 \\ 1 & 0 \end{bmatrix}$$

where $0 \leq s \leq 1$.

4 Show that for any $n \times n$ doubly stochastic matrix A, the inequality

$$\sum_{i=1}^{n} a_{ii} \leq n$$

must hold. What can be said about A if equality holds?

5 Let A be a 2×2 doubly stochastic matrix. Show that

$$a_{11}a_{22} + a_{12}a_{21} \geq \tfrac{1}{2}$$

and equality can hold if and only if

$$A = \begin{bmatrix} \frac{1}{2} & \frac{1}{2} \\ \frac{1}{2} & \frac{1}{2} \end{bmatrix}.$$

(Hint: Use the result of Exercise 3.)

†6 Write out in detail the proof that any doubly stochastic matrix must be square.

7 Show that if A is an $n \times n$ 0,1 matrix and the term rank of A is 1, then the sum of the entries in A is at most n.

8 Three mutually hostile countries maintain extensive intelligence networks that allow each of the three countries to estimate the probability that they will be attacked by any one of the others. Country 1 estimates that it will be attacked by Country 2 with probability $\frac{1}{2}$ and by Country 3 with the probability $\frac{1}{3}$. Country 2 estimates that it will be attacked by Country 1 with probability $\frac{1}{4}$ and by Country 3 with probability $\frac{1}{8}$. Country 3 estimates that it will be attacked by Country 1 with probability $\frac{2}{3}$ and by Country 2 with probability $\frac{3}{8}$. Find the probability that each of the countries will attack another country in such a way that all three countries are under attack in some order. (Hint: The matrix for this situation is

$$A = \begin{bmatrix} 0 & \frac{1}{2} & \frac{1}{3} \\ \frac{1}{4} & 0 & \frac{1}{8} \\ \frac{2}{3} & \frac{3}{8} & 0 \end{bmatrix}.$$

The required probability is the sum of the products of the entries in all diagonals in A.)

9 Three archers shoot arrows at three targets. On the basis of previous performance, it is estimated that the first archer will shoot his arrows in the bullseyes in targets 1, 2, and 3 with probabilities $\frac{1}{2}$, $\frac{1}{2}$, and $\frac{2}{3}$, respectively; the second archer will shoot his arrows in the bullseyes of the three targets with probabilities $\frac{2}{3}$, $\frac{3}{8}$, and $\frac{1}{4}$, respectively; and the third archer will shoot his arrows in the bullseyes of the three targets with probabilities $\frac{1}{5}$, $\frac{1}{8}$, and $\frac{1}{2}$, respectively. If all three archers shoot their arrows simultaneously, what is the probability that the bullseyes of all three targets will be hit?

***10** Let A be a 5×5 matrix in which each row and each column contains precisely three 1's and two 0's. Prove that

$$\sum_{\sigma \in S_5} \prod_{i=1}^{5} a_{i\sigma(i)} \geq 6.$$

(Hint: The matrix A can be thought of as an incidence matrix. Theorem 4.2 will imply that A contains at least one diagonal of 1's. Then Theorem 5.1 can be used.)

11 In a dance class consisting of 5 boys and 5 girls, each of the boys has been introduced to precisely 3 girls, and each girl has been introduced to precisely 3 boys. Show that there are at least 6 pairings into 5 dance partners without any further introductions.

†12 Let J_n be the $n \times n$ matrix every one of whose entries is 1. Let A be an $m \times n$ matrix with nonnegative entries. Prove that A is row stochastic if and only if every entry of AJ_n is 1. Show that A is column stochastic if and only if every entry of J_mA is 1. If A is an arbitrary $m \times n$ matrix, express the entries of AJ_n in terms of the row sums of A. Similarly, express the entries of J_mA in terms of the column sums of A. (Hint: See Exercise 10, Section 2.3.)

†13 Let A be an $m \times n$ row stochastic matrix and B be an $n \times q$ row stochastic matrix. Show that AB is an $m \times q$ row stochastic matrix. Prove a similar result for column stochastic matrices. Prove that the product of any two doubly stochastic matrices is a doubly stochastic matrix.

†14 Let A and B be $m \times n$ row stochastic matrices and let c be a number satisfying $0 \leq c \leq 1$. Show that $cA + (1 - c)B$ is a row stochastic matrix. Prove the same result for column stochastic and doubly stochastic matrices.

15 In each of the following doubly stochastic matrices exhibit a diagonal consisting of positive entries:

(a) $\begin{bmatrix} \frac{1}{2} & \frac{1}{2} & 0 \\ 0 & \frac{1}{2} & \frac{1}{2} \\ \frac{1}{2} & 0 & \frac{1}{2} \end{bmatrix}$;

(b) $\begin{bmatrix} 0 & 0 & \frac{1}{3} & \frac{2}{3} \\ 0 & 0 & \frac{2}{3} & \frac{1}{3} \\ \frac{1}{2} & \frac{1}{2} & 0 & 0 \\ \frac{1}{2} & \frac{1}{2} & 0 & 0 \end{bmatrix}$;

(c) $\begin{bmatrix} 0 & \frac{1}{4} & \frac{1}{4} & 0 & \frac{1}{2} \\ \frac{1}{2} & 0 & \frac{1}{4} & \frac{1}{4} & 0 \\ 0 & \frac{1}{2} & 0 & \frac{1}{4} & \frac{1}{4} \\ \frac{1}{4} & 0 & \frac{1}{2} & 0 & \frac{1}{4} \\ \frac{1}{4} & \frac{1}{4} & 0 & \frac{1}{2} & 0 \end{bmatrix}$.

†16 An $n \times n$ *permutation matrix* is a matrix with precisely one 1 and $n - 1$ zeros in each row and each column. Show that if $f\colon N_n \to N_n$ is an n-permutation of the set $N_n = \{1, \ldots, n\}$, then the incidence matrix for f, $A(f)$, is a permutation matrix. Conversely, show that if A is an $n \times n$ permutation matrix, then A is the incidence matrix for precisely one permutation $f\colon N_n \to N_n$.

17 Write down all 3×3 permutation matrices.

†18 Let $c_1, \ldots, c_p$ be nonnegative numbers with sum 1, i.e., $\sum_{i=1}^{p} c_i = 1$. Let $A_1, \ldots, A_p$ be a set of $n \times n$ permutation matrices. Show that $c_1A_1 + \cdots + c_pA_p$ is a doubly stochastic matrix. (Hint: $A_iJ_n = J_nA_i = J_n$. Hence,

$$\begin{aligned} J_n\left(\sum_{i=1}^{p} c_iA_i\right) &= \sum_{i=1}^{p} c_iJ_nA_i \\ &= \left(\sum_{i=1}^{p} c_i\right)J_n \\ &= J_n. \end{aligned}$$

Similarly, $\left(\sum_{i=1}^{p} c_iA_i\right)J_n = J_n$. Apply the result of Exercise 12.)

19 Write each of the doubly stochastic matrices in Exercise 15 in the form $\sum_{i=1}^{p} c_iA_i$, where $\sum_{i=1}^{p} c_i = 1$, $c_1, \ldots, c_p$ are positive numbers, and $A_1, \ldots, A_p$ are permutation matrices. (Hint: e.g.,

$$\begin{bmatrix} \frac{1}{2} & \frac{1}{2} & 0 \\ 0 & \frac{1}{2} & \frac{1}{2} \\ \frac{1}{2} & 0 & \frac{1}{2} \end{bmatrix} = \frac{1}{2}\begin{bmatrix} 1 & 0 & 0 \\ 0 & 1 & 0 \\ 0 & 0 & 1 \end{bmatrix} + \frac{1}{2}\begin{bmatrix} 0 & 1 & 0 \\ 0 & 0 & 1 \\ 1 & 0 & 0 \end{bmatrix}.$$

In this case $p = 2$, $c_1 = \frac{1}{2}$, $c_2 = \frac{1}{2}$,

$$A_1 = \begin{bmatrix} 1 & 0 & 0 \\ 0 & 1 & 0 \\ 0 & 0 & 1 \end{bmatrix}, \quad \text{and}$$

$$A_2 = \begin{bmatrix} 0 & 1 & 0 \\ 0 & 0 & 1 \\ 1 & 0 & 0 \end{bmatrix}.)$$

†**20** Show that if A is an $n \times n$ doubly stochastic matrix, then A contains at least one diagonal consisting of positive entries. (Hint: The argument is the same as that used in Example 5.3. Suppose the assertion is not true. Then every diagonal of A must contain a zero. Then by Theorem 4.2, we conclude that A contains an $s \times t$ submatrix whose entries are all zero and $s + t = n + 1$. We bring this $s \times t$ submatrix into the upper left-hand corner of A by row interchanges and column interchanges. Clearly the doubly stochastic property of A is unaltered by interchanges, i.e., the resulting matrix is still doubly stochastic.

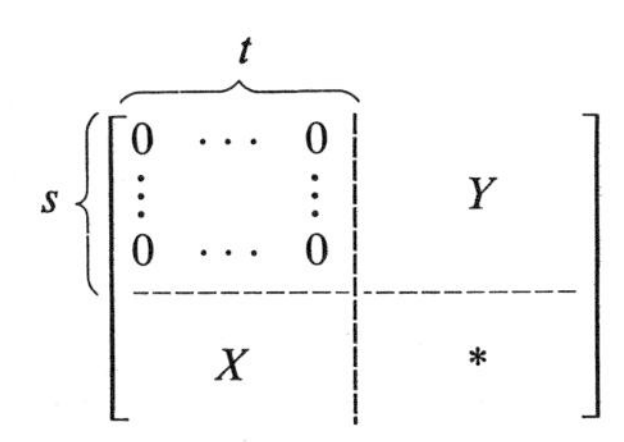

Each row sum in Y is 1, and hence the sum of all the entries in Y is s. (Why?) Similarly, the sum of all entries in X is t. (Why?) Hence, the sum of all the entries in X and Y is $s + t = n + 1$. But the sum of the entries in a doubly stochastic $n \times n$ matrix is exactly n, and therefore we have a contradiction.)

21 Find the term rank of each of the following matrices:

(a) $\begin{bmatrix} 1 & 0 & 0 \\ 0 & 0 & 1 \\ 0 & 0 & 1 \end{bmatrix}$;

(b) $\begin{bmatrix} 0 & 1 & 0 & 0 \\ 0 & 1 & 0 & 0 \\ 1 & 1 & 0 & 0 \end{bmatrix}$;

(c) $\begin{bmatrix} 0 & 0 & 1 \\ 1 & 0 & 0 \\ 0 & 0 & 1 \\ 0 & 1 & 0 \end{bmatrix}$.

22 Prove that if A is an $n \times n$ 0,1 matrix then A and A^T have the same term rank.

23 Show that if A is an $n \times n$ 0,1 matrix, and every row sum and every column sum of A is 2, then A has term rank n.

24 Let A and B be two $n \times n$ 0,1 matrices. Replace each positive entry in $A + B$ by 1 and leave the zero entries unaltered. Show that the term rank of the resulting matrix is at least as great as the term rank of either A or B.

25 Find the minimal number of lines containing all of the 1's in each of the matrices in Exercise 21.

†26 Prove that the matrix A_3 appearing in (17) in the proof of Theorem 5.2 must contain a diagonal of t ones.

27 It is discovered by personality testing involving 7 young men and 7 young women that the offspring of certain unions will be neurotic. This data is summarized in the following 7×7 matrix in which the (i, j) entry is 1 if the union of man i and woman j will produce a neurotic offspring, and is 0 otherwise.

$$\begin{bmatrix} 1 & 0 & 0 & 0 & 1 & 1 & 0 \\ 1 & 1 & 1 & 1 & 1 & 1 & 1 \\ 1 & 1 & 1 & 1 & 1 & 1 & 1 \\ 1 & 0 & 0 & 0 & 1 & 1 & 0 \\ 1 & 0 & 0 & 0 & 1 & 1 & 0 \\ 1 & 0 & 0 & 0 & 1 & 1 & 0 \\ 1 & 1 & 1 & 1 & 1 & 1 & 1 \end{bmatrix}$$

Assume that psychiatric treatment of either member of a couple which will produce neurotic offspring will result in a change in personality so that the offspring will be well-balanced. Find the minimum number of people to whom it is necessary to give psychiatric counselling so that every member of the next generation will be well-balanced, whatever unions may take place.

2.6

Introduction to Linear Equations

The simplest kinds of equations that arise in mathematics are linear. A single linear equation in a single unknown has the form

$$Ax = b. \tag{1}$$

In (1), A and b are fixed numbers, and it is required to find a number x that makes Ax the same as b. For example, if $A = 2$, $b = 3$, then $x = A^{-1}b = \frac{1}{2} \cdot 3 = \frac{3}{2}$. It is quite clear that if $A \neq 0$, then a unique x satisfying equation (1) can always be found. On the other hand, if $A = 0$ and $b \neq 0$, then no number x exists which satisfies (1).

The form in which (1) appears suggests the formulation of a similar problem involving vectors and matrices. Thus, if A is a given $m \times n$ matrix and b is a given m-vector, then the fundamental problem in the theory of linear equations is to find all n-vectors x such that (1) is satisfied. As we saw above, a solution need not exist, so that a major problem implicit in finding all solutions to (1) involves determining whether any such solutions exist.

Example 6.1 Let $u = (2, 3)$ and $v = (-1, 1)$. Find all numbers x_1 and x_2 such that the linear combination $x_1u + x_2v$ is the vector $(1, 0)$. We compute that

$$\begin{aligned} x_1u + x_2v &= (2x_1 - x_2, 3x_1 + x_2) \\ &= (1, 0). \end{aligned}$$

Equating components, we have the two equations

$$\begin{aligned} 2x_1 - x_2 &= 1, \\ 3x_1 + x_2 &= 0. \end{aligned} \tag{2}$$

From the first equation in (2), we see that $x_2 = 2x_1 - 1$, and from the second equation, $x_2 = -3x_1$. Hence, if (2) is to be satisfied, it follows that

$$\begin{aligned} 2x_1 - 1 &= -3x_1, \\ 5x_1 &= 1, \\ x_1 &= \tfrac{1}{5}, \end{aligned}$$

and hence,

$$x_2 = 2 \cdot \tfrac{1}{5} - 1 = -\tfrac{3}{5}.$$

Example 6.2 Write the following system of equations in the form $Ax = b$, where A is a matrix and x and b are vectors:

$$\begin{aligned} x_1 + x_2 + x_3 - x_4 &= 1, \\ x_2 + x_4 &= -2, \\ 2x_1 - x_3 &= 5. \end{aligned} \tag{3}$$

Let A be the 3×4 matrix of coefficients which appear in the system (3):

$$A = \begin{bmatrix} 1 & 1 & 1 & -1 \\ 0 & 1 & 0 & 1 \\ 2 & 0 & -1 & 0 \end{bmatrix}.$$

Let $x = (x_1, x_2, x_3, x_4)$ and let $b = (1, -2, 5)$. Then the three equations in (3) can be written as a single matrix-vector equation:

$$Ax = b.$$

Example 6.3 A set of n cities, $c_1, \ldots, c_n$, form a closed economic unit in the sense that they trade only among themselves according to the following scheme: c_i sells what it produces at an average price of x_i dollars per unit produced, $i = 1, \ldots, n$. Also, city c_i purchases a certain number of units from city c_j, call it b_{ij}, $i, j = 1, \ldots, n$. Then

$$b_{ij}x_j$$

is the amount of money that c_i pays to c_j for its purchases from c_j, $i, j = 1, \ldots, n$. Thus, the total amount spent by c_i is

$$\sum_{j=1}^{n} b_{ij}x_j,$$

for total purchases of $\sum_{k=1}^{n} b_{ik}$ units.

Thus the average amount spent by c_i per unit purchased is

$$\frac{\sum_{j=1}^{n} b_{ij}x_j}{\sum_{k=1}^{n} b_{ik}} = \sum_{j=1}^{n} \left(\frac{b_{ij}}{\sum_{k=1}^{n} b_{ik}} \right) x_j.$$

Set

$$a_{ij} = \frac{b_{ij}}{\sum_{k=1}^{n} b_{ik}}.$$

Observe that a_{ij} represents the fraction of the total purchases that c_i makes from c_j. Then the average amount spent by c_i per unit purchased is

$$\sum_{j=1}^{n} a_{ij}x_j. \tag{4}$$

If the economy is to be stable, no city should spend on the average more or less per purchased unit than the price per produced unit. In other words, the total average expenditure (4) per item purchased by c_i must be equal to x_i, $i = 1, \ldots, n$. Hence,

$$\sum_{j=1}^{n} a_{ij}x_j = x_i, \qquad i = 1, \ldots, n. \tag{5}$$

If we let A be the $n \times n$ matrix whose (i, j) entry is a_{ij} and set $x = (x_1, x_2, \ldots, x_n)$, then the conditions (5) for a stable economy can be summarized in the following single matrix-vector equation:

$$Ax = x. \tag{6}$$

The problem is to find the "price vector" x that satisfies (6). The equation (6) may be put in the form

$$Ax = b \tag{7}$$

in which x appears only on the left by writing $x = I_n x$, so that (6) becomes

$$\begin{aligned} Ax &= I_n x, \\ I_n x - Ax &= 0, \\ (I_n - A)x &= 0. \end{aligned} \tag{8}$$

Observe that (8) is in the form (7), in which the role of A is now played by $I_n - A$, and the role of b by the zero vector.

Example 6.4 Consider an elastic beam supported at each end.

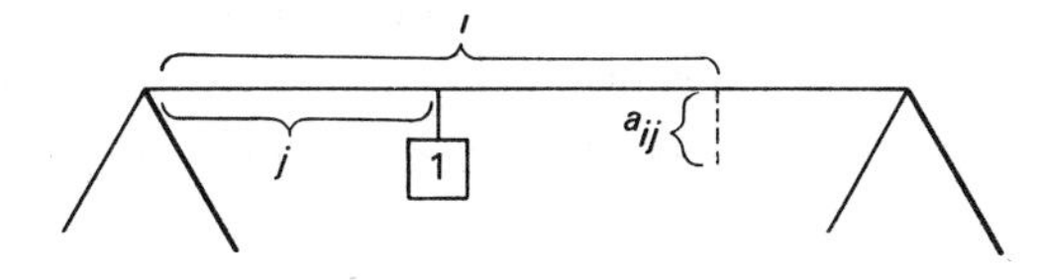

Suppose it is known that if a one pound weight is placed j units from the left end of the beam then at i units from the left end, the beam will deflect through a distance of a_{ij} units. Suppose it is observed that simultaneously loading the beam at distances of 1, 2, 3, and 4 units from the left end produces a total deflection of b_1, b_2, b_3, and b_4 units at each of these respective points. The problem is to determine what weights were attached at each of these points. Let x_j denote the weight in pounds attached at the point which is j units from the left end, $j = 1, 2, 3, 4$. Then Hooke's Law states that if the elastic limits of the beam are not exceeded, then the deflection at point i due to a weight of x_j at point j is $a_{ij}x_j$. Thus the total deflection at point i due to the four weights x_1, x_2, x_3, x_4 attached at each of the points 1, 2, 3, and 4 units from the left end is

$$a_{i1}x_1 + a_{i2}x_2 + a_{i3}x_3 + a_{i4}x_4 = b_i, \qquad i = 1, 2, 3, 4. \tag{9}$$

Hence, if we let A be the 4×4 matrix whose (i, j) entry is a_{ij}, x the 4-vector $x = (x_1, x_2, x_3, x_4)$, and $b = (b_1, b_2, b_3, b_4)$, then the system of four equations (9) can be written as the single matrix-vector equation

$$Ax = b.$$

Example 6.5 When the digits in a 2-digit whole number are reversed, the number is increased by 36 and the sum of the digits in the number is 10.

Find the number. Let x_1 denote the first digit in the number and x_2 the second digit. Then the number is $10x_1 + x_2$. When the digits are reversed, the number is $10x_2 + x_1$. The conditions of the problem state that

$$x_1 + x_2 = 10$$

and

$$10x_2 + x_1 = 10x_1 + x_2 + 36,$$

or

$$\begin{aligned} x_1 + x_2 &= 10, \\ x_1 - x_2 &= -4. \end{aligned}$$

In matrix-vector language, these last two equations become

$$Ax = b,$$

where $A = \begin{bmatrix} 1 & 1 \\ 1 & -1 \end{bmatrix}$, $x = (x_1, x_2)$, and $b = (10, -4)$. It is, of course, extremely easy to compute that the solution to this system of equations is $x = (3, 7)$. Thus the required number is 37.

Example 6.6 An amount of \$10,000 is put into three investments at rates of 3%, 4%, and 6% per year, respectively. The income from the first two investments is \$80 less than the income from the third investment. The total annual income from the three investments is \$460. Find the amount of each investment. Let x_1, x_2, and x_3 denote the amounts invested at 3%, 4%, and 6%, respectively. Then

$$.03x_1 + .04x_2 + .06x_3 = 460$$

and

$$.03x_1 + .04x_2 + 80 = .06x_3.$$

Moreover,

$$x_1 + x_2 + x_3 = 10{,}000.$$

These three equations may be written in the matrix-vector form

$$Ax = b,$$

where

$$A = \begin{bmatrix} 3 & 4 & 6 \\ 3 & 4 & -6 \\ 1 & 1 & 1 \end{bmatrix},$$

$x = (x_1, x_2, x_3)$, and $b = (46 \times 10^3, -8 \times 10^3, 10^4)$. By an elementary computation, we see that the solution is

$$x = (3 \times 10^3, 2.5 \times 10^3, 4.5 \times 10^3).$$

Example 6.7 A large food market chain has n retail outlets, $o_1, \ldots, o_n$, located around the city. The chain also maintains m warehouses, $h_1, \ldots, h_m$, which have capacities of $s_1, \ldots, s_m$ tons of food, respectively. The outlets place orders with the warehouses which will empty all of them. Let x_{ij} denote the amount (in tons) shipped from h_i to o_j. The conditions that the warehouses will be emptied by these orders take the form

$$\sum_{j=1}^{n} x_{ij} = s_i, \qquad i = 1, \ldots, m. \tag{10}$$

For, the ith equation in (10) simply states that the total amount shipped from h_i to the outlets $o_1, \ldots, o_n$ is precisely the amount s_i. Suppose in particular that o_j orders a total of b_j tons to be delivered from the warehouses. Suppose further that it is known that the cost of shipping one ton of food from h_i to o_j is a_{ij} dollars. The condition that o_j orders a total of b_j tons to be delivered from the warehouses can be expressed by the equations

$$\sum_{i=1}^{m} x_{ij} = b_j, \qquad j = 1, \ldots, n. \tag{11}$$

The delivery of x_{ij} tons from h_i to o_j costs $a_{ij}x_{ij}$ dollars, and hence the total cost for operating the chain is

$$c = \sum_{i=1}^{m} \sum_{j=1}^{n} a_{ij}x_{ij}. \tag{12}$$

Thus the problem is to find mn numbers x_{ij} such that the cost c in (12) is minimum, subject to the $m + n$ linear conditions expressed in (10) and (11). The procedure for solving this problem amounts to finding all solutions to the equations (10) and (11) which have nonnegative values, and then finding which of these solutions cause the function c in (12) to have the least value. This type of problem is called a Hitchcock Transportation Problem.

Example 6.8 In the case of two warehouses and three retail outlets in Example 6.7, express the problem in matrix-vector notation. In this case x_{11}, $x_{12}, x_{13}, x_{21}, x_{22}$, and x_{23} are the unknown quantities. Denote these six numbers by $y_1, \ldots, y_6$, respectively. The equations in (10) then become

$$\begin{aligned} y_1 + y_2 + y_3 &= s_1, \\ y_4 + y_5 + y_6 &= s_2. \end{aligned} \tag{13}$$

The equations in (11) become

$$\begin{aligned} y_1 + y_4 &= b_1, \\ y_2 + y_5 &= b_2, \\ y_3 + y_6 &= b_3. \end{aligned} \tag{14}$$

If we set $d_1 = s_1, d_2 = s_2, d_3 = b_1, d_4 = b_2, d_5 = b_3$,

$$M = \begin{bmatrix} 1 & 1 & 1 & 0 & 0 & 0 \\ 0 & 0 & 0 & 1 & 1 & 1 \\ 1 & 0 & 0 & 1 & 0 & 0 \\ 0 & 1 & 0 & 0 & 1 & 0 \\ 0 & 0 & 1 & 0 & 0 & 1 \end{bmatrix},$$

$y = (y_1, \ldots, y_6)$, and $d = (d_1, \ldots, d_5)$, then the equations in (13) and (14) take the form

(15) $$My = d.$$

Let $a_{11} = \alpha_1, a_{12} = \alpha_2, a_{13} = \alpha_3, a_{21} = \alpha_4, a_{22} = \alpha_5, a_{23} = \alpha_6$. Then the cost function (12) becomes

(16) $$c = \sum_{k=1}^{6} \alpha_k y_k.$$

Thus the transportation problem reduces to finding vectors y which satisfy (15), have nonnegative components, and minimize (16). Observe that in this case, five linear conditions are imposed on the six numbers $y_1, \ldots, y_6$ and hence it is reasonable to expect that there will be more than one possible solution to the equation in (15).

The preceding examples show that the problem of solving systems of linear equations can arise in a number of realistic situations. The rest of this section and the next section will be devoted to developing the techniques necessary to solve an arbitrary system of linear equations.

The theory behind the analysis of systems of linear equations is based on three extremely elementary facts. Let α, β, γ, and δ be in R^n, i.e., they are fixed n-vectors.

(i) The pair of equalities

$$\begin{cases} \alpha = \beta \\ \delta = \gamma \end{cases}$$

holds if and only if the pair of equalities

$$\begin{cases} \delta = \gamma \\ \alpha = \beta \end{cases}$$

holds. In other words, the validity of a pair of equalities does not depend upon the order in which they appear.

(ii) The pair of equalities

$$\begin{cases} \alpha = \beta \\ \delta = \gamma \end{cases}$$

holds if and only if the pair of equalities

$$\begin{cases} \alpha = \beta \\ \delta + c\alpha = \gamma + c\beta \end{cases}$$

holds for any number c.

(iii) The equality $\alpha = \beta$ holds if and only if the equality $c\alpha = c\beta$ holds for any non-zero number c.

These facts are virtually self-evident for $n = 1$ and are trivial consequences of the definitions of vector addition and scalar multiplication when $n > 1$. Nevertheless, facts (i), (ii), and (iii) provide us with the essential tools for solving any system of linear equations.

Definition 6.1 **Equivalent systems** Let A be an $m \times n$ matrix and M a $k \times n$ matrix. Consider the two systems of equations in matrix-vector notation:

$$Ax = b \tag{17}$$

and

$$Mx = c, \tag{18}$$

where $x = (x_1, \ldots, x_n)$ is an n-vector, b is an m-vector, and c is a k-vector. A *solution* to the system (17) is an n-vector $x = (x_1, \ldots, x_n)$ which makes (17) true. The components of x are sometimes called the *unknowns*. The two systems (17) and (18) are said to be *equivalent* if and only if they have precisely the same set of solution vectors x. In other words, for the two systems (17) and (18) to be equivalent, we require that for any vector x, $Ax = b$ if and only if $Mx = c$.

The crucial idea involved in solving the system of equations $Ax = b$ is quite simple. Perform a series of elementary operations corresponding to (i), (ii), and (iii) above to produce an equivalent system $Mx = c$ in which the solutions are obvious. We exhibit this idea in the following example.

Example 6.9 Find all solutions to the system of equations

$$\begin{aligned} 2x_1 + 3x_2 - x_3 + x_4 &= 3, \\ 3x_1 + 2x_2 - 2x_3 + 2x_4 &= 4, \\ 5x_1 - 4x_3 + 4x_4 &= 6. \end{aligned} \tag{19}$$

Subtract the first equation in (19) from the second, i.e., add to the second equation -1 times the first equation. This is an application of (ii) in which c is -1. Next, subtract 2 times the first equation from the third equation. This produces the equivalent system of equations

$$\begin{aligned} 2x_1 + 3x_2 - x_3 + x_4 &= 3, \\ x_1 - x_2 - x_3 + x_4 &= 1, \\ x_1 - 6x_2 - 2x_3 + 2x_4 &= 0. \end{aligned} \tag{20}$$

Interchange the first and second equations in (20), an application of (i). The resulting equivalent system is

$$\begin{aligned} x_1 - x_2 - x_3 + x_4 &= 1, \\ 2x_1 + 3x_2 - x_3 + x_4 &= 3, \\ x_1 - 6x_2 - 2x_3 + 2x_4 &= 0. \end{aligned} \tag{21}$$

In the system (21), subtract twice the first equation from the second and then subtract the first equation from the third (applications of (ii)). The resulting equivalent system is

$$\begin{aligned} x_1 - x_2 - x_3 + x_4 &= 1, \\ 5x_2 + x_3 - x_4 &= 1, \\ -5x_2 - x_3 + x_4 &= -1. \end{aligned} \tag{22}$$

Now add the second equation in (22) to the third to produce the equivalent system

$$\begin{aligned} x_1 - x_2 - x_3 + x_4 &= 1, \\ 5x_2 + x_3 - x_4 &= 1, \\ 0 &= 0. \end{aligned} \tag{23}$$

Of course, we can drop the last equation in (23). We can then multiply the second equation in (23) by $\frac{1}{5}$ (a type (iii) operation) to produce the equivalent system

$$\begin{aligned} x_1 - x_2 - x_3 + x_4 &= 1, \\ x_2 + \tfrac{1}{5}x_3 - \tfrac{1}{5}x_4 &= \tfrac{1}{5}. \end{aligned} \tag{24}$$

Finally, add the second equation to the first to produce the equivalent system

$$\begin{aligned} x_1 \qquad - \tfrac{4}{5}x_3 + \tfrac{4}{5}x_4 &= \tfrac{6}{5}, \\ x_2 + \tfrac{1}{5}x_3 - \tfrac{1}{5}x_4 &= \tfrac{1}{5}. \end{aligned} \tag{25}$$

Thus a 4-vector $x = (x_1, x_2, x_3, x_4)$ is a solution to (25) and hence to (19) if and only if

$$\begin{aligned} x &= (x_1, x_2, x_3, x_4) \\ &= (\tfrac{6}{5} + \tfrac{4}{5}x_3 - \tfrac{4}{5}x_4, \tfrac{1}{5} - \tfrac{1}{5}x_3 + \tfrac{1}{5}x_4, x_3, x_4) \\ &= (\tfrac{6}{5}, \tfrac{1}{5}, 0, 0) + x_3(\tfrac{4}{5}, -\tfrac{1}{5}, 1, 0) + x_4(-\tfrac{4}{5}, \tfrac{1}{5}, 0, 1). \end{aligned} \tag{26}$$

Let $u = (\frac{6}{5}, \frac{1}{5}, 0, 0)$, $v = (\frac{4}{5}, -\frac{1}{5}, 1, 0)$, $w = (-\frac{4}{5}, \frac{1}{5}, 0, 1)$. We conclude then that the solution set of the system (19) is precisely the set of linear combinations of the form

$$u + x_3 v + x_4 w$$

for any values of the scalars x_3 and x_4. Hence it is possible for a system of linear equations to have an infinite number of solutions.

If we examine the method of solution in Example 6.9, we see that the unknowns x_i are really quite irrelevant. To clarify this remark, consider the matrix of coefficients in (19) augmented by a fifth column formed from the right-hand sides of the equations (19):

$$\begin{bmatrix} 2 & 3 & -1 & 1 & 3 \\ 3 & 2 & -2 & 2 & 4 \\ 5 & 0 & -4 & 4 & 6 \end{bmatrix}. \tag{27}$$

The process of solving the system (19) can be thought of as a sequence of operations on the matrix in (27). Thus, if we perform a certain sequence of "row operations" on (27), we will obtain the matrix

$$B = \begin{bmatrix} 1 & 0 & -\frac{4}{5} & \frac{4}{5} & \frac{6}{5} \\ 0 & 1 & \frac{1}{5} & -\frac{1}{5} & \frac{1}{5} \\ 0 & 0 & 0 & 0 & 0 \end{bmatrix}. \tag{28}$$

The matrix B in (28) corresponds to the system (25) in precisely the same way as the matrix in (27) corresponds to the system (19). The point is that the system corresponding to the matrix B in (28) is easy to solve.

To summarize, the matrix in (28) results from the matrix in (27) by the following sequence of elementary operations on the rows of (27).

(a) Subtract the first row from the second.
(b) Subtract twice the first row from the third.
(c) Interchange the first and second rows.
(d) Subtract twice the first row from the second.
(e) Subtract the first row from the third.
(f) Add the second row to the third.
(g) Multiply the second row by $\frac{1}{5}$.
(h) Add the second row to the first row.

The student will verify that this sequence of operations on the matrix in (27) does indeed yield the matrix in (28).

Definition 6.2 ***Elementary row operations, row equivalence*** Let A be a matrix. The three types of *elementary row operations* on A are:

I. interchange two rows,
II. add to a row a constant multiple of a different row,
III. multiply a row by a non-zero constant.

If B is a matrix obtained from A by a sequence of elementary row operations, then B is said to be *row equivalent* to A.

If we refer to the matrix B in (28), we see that it has the following properties: the first two rows of B are non-zero and the remaining row is the zero vector. There are two columns of B, columns 1 and 2, such that

$$B^{(1)} = e_1$$

and

$$B^{(2)} = e_2,$$

where $e_i \in R^3$ is the 3-vector with 1 as the ith component and 0 for the remaining components, $i = 1, 2, 3$.

Definition 6.3 ***Hermite normal form, row rank*** Let H be an $m \times n$ matrix. Then H is in *Hermite normal form* if there exists an integer r such that:

(i) the first r rows of H are non-zero and the remaining $m - r$ rows are zero;
(ii) there are r columns of H numbered $n_1, \ldots, n_r$ $(1 \leq n_1 < n_2 < \cdots < n_r \leq n)$ such that $H^{(n_i)} = e_i$ and $h_{ij} = 0$, $j = 1, \ldots, n_i - 1$, $i = 1, \ldots, r$. The vector e_i is the m-vector with ith component 1 and remaining components 0. The number r is called the *row rank* of H.

Example 6.10 **(a)** Consider the matrix

$$H = \begin{bmatrix} 0 & 1 & 3 & 1 & 0 & 0 & 4 \\ 0 & 0 & 0 & 0 & 1 & 0 & 3 \\ 0 & 0 & 0 & 0 & 0 & 1 & 2 \\ 0 & 0 & 0 & 0 & 0 & 0 & 0 \end{bmatrix}.$$

This matrix is in Hermite normal form. In the notation of Definition 6.3, $n_1 = 2$, $n_2 = 5$, $n_3 = 6$, and the row rank r is 3.

(b) Consider the matrix

$$H = \begin{bmatrix} 0 & 0 & 1 & 2 & 0 & -3 & 8 & 0 & 5 & 4 & 3 & 0 & 7 & 9 \\ 0 & 0 & 0 & 0 & 1 & 2 & -4 & 0 & -6 & 7 & 5 & 0 & 8 & 5 \\ 0 & 0 & 0 & 0 & 0 & 0 & 0 & 1 & 2 & 9 & 3 & 0 & -3 & -6 \\ 0 & 0 & 0 & 0 & 0 & 0 & 0 & 0 & 0 & 0 & 0 & 1 & -2 & -4 \\ 0 & 0 & 0 & 0 & 0 & 0 & 0 & 0 & 0 & 0 & 0 & 0 & 0 & 0 \\ 0 & 0 & 0 & 0 & 0 & 0 & 0 & 0 & 0 & 0 & 0 & 0 & 0 & 0 \end{bmatrix}.$$

In this case, $n_1 = 3, n_2 = 5, n_3 = 8, n_4 = 12$, and the row rank is 4.

(c) Consider the matrix

$$H = \begin{bmatrix} 0 & 1 & 2 & 0 & 0 \\ 0 & 0 & 0 & 1 & 0 \\ 0 & 0 & 0 & 0 & 1 \\ 0 & 0 & 0 & 0 & 0 \\ 0 & 0 & 0 & 0 & 0 \\ 0 & 0 & 0 & 0 & 0 \\ 0 & 0 & 0 & 0 & 0 \\ 0 & 0 & 0 & 0 & 0 \\ 0 & 0 & 0 & 0 & 0 \\ 0 & 0 & 0 & 0 & 0 \end{bmatrix}.$$

Here $n_1 = 2, n_2 = 4, n_3 = 5$, and the row rank of H is 3.

Theorem 6.1 *Any matrix A is row equivalent to a matrix in Hermite normal form.*

Proof The argument that we use to establish this result is constructive in that it outlines exactly how to reduce any $m \times n$ matrix A to Hermite normal form by a sequence of elementary row operations. We assume that A is not the zero matrix (otherwise it is already in Hermite normal form).

(i) Find the first column in A which contains a non-zero element. Suppose it is column n_1. By a type I operation, bring a non-zero entry into the first row position and by a type III operation, reduce it to 1. Then, by adding appropriate multiples of row 1 to the remaining rows which contain non-zero entries in column n_1, reduce column n_1 to the m-vector e_1.

(ii) Scan rows 2 through m of the matrix resulting from (i) and find the first column, say column n_2, in which there appears a non-zero entry in any position below the first row. Observe that the manner of choosing n_2 implies that in every row numbered 2 through m, there must be zero entries in the columns preceding column n_2.

$$\begin{bmatrix} 0 & 0 & \cdots & 0 & \overset{n_1}{1} & \times & \cdots & \times & \overset{n_2}{\times} & \cdots & \times \\ 0 & 0 & & 0 & 0 & 0 & & 0 & \times & & \times \\ 0 & 0 & & 0 & 0 & 0 & & 0 & \times & & \times \\ 0 & 0 & & 0 & 0 & 0 & & 0 & \times & & \times \\ \vdots & \vdots & & \vdots & \vdots & \vdots & & \vdots & \vdots & & \vdots \\ 0 & 0 & \cdots & 0 & 0 & 0 & \cdots & 0 & \times & \cdots & \times \end{bmatrix}$$

Bring a non-zero entry in column n_2 from somewhere in rows 2 through m to the $(2, n_2)$ position by a type I operation, make it a 1 by a type III operation, and then, by successive type II operations, reduce every entry in column n_2 to 0 except the 1 in the $(2, n_2)$ position. Then column n_2 of the resulting matrix is the m-vector e_2.

(iii) Scan the matrix resulting from (ii) for the first column beyond column n_2, say column n_3, which contains a non-zero entry in rows 3 through m. Bring a non-zero entry in column n_3 from somewhere in rows 3 through m to the $(3, n_3)$ position by a type I operation. Reduce this entry to 1 by a type III operation, and then reduce to 0 the remaining non-zero entries in column n_3 by type II operations. Column n_3 in the resulting matrix is the m-vector e_3.

A continuation of this process will ultimately terminate in a matrix in Hermite normal form. ▮

Example 6.11 Reduce the following matrix to Hermite normal form:

$$A = \begin{bmatrix} 0 & 2 & 0 & -1 & 3 & 0 & 0 & 1 & -2 \\ 0 & -1 & 0 & 2 & 0 & 3 & 0 & -1 & 2 \\ 0 & 1 & 0 & -1 & 1 & -1 & 0 & 2 & -4 \\ 0 & 1 & 0 & -1 & 1 & -1 & 0 & 1 & -2 \end{bmatrix}.$$

The first non-zero column is column 2 and thus $n_1 = 2$. Interchanging rows 1 and 3 places a 1 in the (1, 2) entry. Then add row 1 to row 2, -2 times row 1 to row 3, and -1 times row 1 to row 4, which results in the matrix

$$A_1 = \begin{bmatrix} 0 & 1 & 0 & -1 & 1 & -1 & 0 & 2 & -4 \\ 0 & 0 & 0 & 1 & 1 & 2 & 0 & 1 & -2 \\ 0 & 0 & 0 & 1 & 1 & 2 & 0 & -3 & 6 \\ 0 & 0 & 0 & 0 & 0 & 0 & 0 & -1 & 2 \end{bmatrix}.$$

The first column in A_1 containing non-zero entries below row 1 is column 4, and thus $n_2 = 4$. Now add row 2 to row 1 and add -1

times row 2 to row 3 to produce the matrix

$$A_2 = \begin{bmatrix} 0 & 1 & 0 & 0 & 2 & 1 & 0 & 3 & -6 \\ 0 & 0 & 0 & 1 & 1 & 2 & 0 & 1 & -2 \\ 0 & 0 & 0 & 0 & 0 & 0 & 0 & -4 & 8 \\ 0 & 0 & 0 & 0 & 0 & 0 & 0 & -1 & 2 \end{bmatrix}.$$

We scan the matrix A_2 for the first column which contains a nonzero entry below row 2. Clearly this is column $n_3 = 8$. Multiply row 3 of A_2 by $-\frac{1}{4}$ to make the (3, 8) entry 1. Then add -3 times row 3 to row 1, add -1 times row 3 to row 2, and add row 3 to row 4. This results in the matrix

$$H = \begin{bmatrix} 0 & 1 & 0 & 0 & 2 & 1 & 0 & 0 & 0 \\ 0 & 0 & 0 & 1 & 1 & 2 & 0 & 0 & 0 \\ 0 & 0 & 0 & 0 & 0 & 0 & 0 & 1 & -2 \\ 0 & 0 & 0 & 0 & 0 & 0 & 0 & 0 & 0 \end{bmatrix}.$$

Clearly H is in Hermite normal form: $n_1 = 2$, $n_2 = 4$, $n_3 = 8$, and the row rank of H is $r = 3$.

As a final remark, observe that if H is an m-square matrix in Hermite normal form and H has row rank m, then H must, in fact, be the identity matrix I_m. For, there must be m columns of H which are, in succession, the m-vectors $e_1, \ldots, e_m$. But since H has only m columns, it follows that $H = I_m$.

In the next section, we shall see how the reduction to Hermite normal form of the matrix of coefficients of a system of linear equations easily provides us with a complete set of solution vectors for the system.

Quiz

Answer true or false.

1 A system of linear equations with more unknowns than equations always has a solution.

2 Any vector in R^2 can be expressed as a linear combination of the vectors $u = (2, 3)$ and $v = (-1, 1)$.

3 Any solution vector $x = (x_1, x_2)$ to the system of equations $x_1 + x_2 = 1$ is a linear combination of the vectors $(1, -1)$ and $(0, 1)$.

4 The systems

$$\begin{aligned} x_1 + x_2 &= 1, \\ x_1 - x_2 &= 2 \end{aligned}$$

and

$$\begin{aligned} x_1 + x_2 &= 0, \\ x_1 - x_2 &= 2 \end{aligned}$$

are equivalent.

5 For any non-zero $m \times n$ matrix A, the only solution to the system of equations $Ax = 0_m$ is the zero vector 0_n. (Assume $m + n > 2$.)

6 The matrix

$$\begin{bmatrix} 1 & 0 \\ 0 & 1 \end{bmatrix}$$

is in Hermite normal form.

7 The matrix

$$\begin{bmatrix} 1 & 0 \\ 0 & -1 \end{bmatrix}$$

is in Hermite normal form.

8 The matrices in 6 and 7 are row equivalent.

9 A non-zero matrix can be reduced to an Hermite normal form which contains only zero entries.

10 Two row equivalent 2×2 matrices in Hermite normal form are equal.

Exercises

(Exercises 7–18 *are essential to the theory of linear equations and are used in subsequent sections.)*

1 (a) Let $u = (1, 0)$ and $v = (0, -1)$. Express the vector $(6, -2)$ as a linear combination of u and v.
(b) Let $u = (1, 0, 1)$, $v = (0, -1, 1)$, and $w = (-1, -1, 1)$. Express $(2, 3, 4)$ as a linear combination of u, v, and w.
(c) Let $u = (1, -1, 1)$, $v = (0, 3, 5)$, and $w = (1, 2, 6)$. Exhibit a vector in R^3 which is not a linear combination of u, v, and w.

2 Write the following systems of equations in the form $Ax = b$, in which A is a matrix and x and b are vectors:

(a) $x_1 + x_2 + x_3 = 5;$

(b)
$$\begin{aligned} x_1 + x_2 &= -1, \\ x_1 - x_3 &= 1, \\ x_2 + x_3 &= 4; \end{aligned}$$

(c)
$$\begin{aligned} x_1 + 2x_2 + 3x_3 - x_4 &= 0, \\ 2x_1 + x_3 - x_4 &= 6, \\ x_4 &= -1, \\ x_2 + x_4 &= 3; \end{aligned}$$

(d)
$$\begin{aligned} x_1 &= 1, \\ x_2 &= 2, \\ x_3 &= 3; \end{aligned}$$

(e)
$$\begin{aligned} 2x_1 + 3x_2 &= 4x_3 - x_4, \\ x_1 + x_4 &= 2x_3 + x_5, \\ 3x_1 + x_5 &= 7x_2 + x_1; \end{aligned}$$

(f)
$$\begin{aligned} 3x_1 + x_4 &= 2x_1 - x_3, \\ x_1 + x_2 &= x_3, \\ x_1 + x_2 + x_3 + x_4 &= 1, \\ 3x_2 + 5x_4 &= x_3; \end{aligned}$$

(g) $$\begin{aligned} 2x_1 - x_2 - 3x_3 + x_5 &= 2, \\ 4x_1 - 7x_3 + 5x_4 &= 6, \\ -x_1 - x_2 - x_5 &= x_4, \\ 3x_1 + x_2 &= x_5, \\ x_4 &= 0, \\ x_5 &= 0; \end{aligned}$$

(h) $$\begin{aligned} x_1 - x_2 + 2x_3 - x_4 &= 0, \\ 3x_1 - x_2 + x_3 + x_4 &= 0, \\ -2x_1 + 2x_2 - 3x_3 &= 0, \\ x_4 &= 0, \\ x_1 + x_2 + x_3 + x_4 &= 0. \end{aligned}$$

3 Reduce each of the following matrices to Hermite normal form and record the elementary row operations used:

(a) $\begin{bmatrix} 2 & 1 & 3 & 4 \\ 0 & -1 & -1 & 2 \\ 2 & 0 & 2 & 6 \\ 4 & 0 & 4 & 12 \\ 0 & -1 & -1 & 2 \end{bmatrix}$; (b) $\begin{bmatrix} 2 & 1 & -1 & 3 & 0 \\ 2 & 1 & -1 & 3 & 0 \end{bmatrix}$;

(c) $\begin{bmatrix} 1 & 1 \\ 1 & 1 \end{bmatrix}$; (d) $\begin{bmatrix} 1 & -1 \\ 1 & 1 \end{bmatrix}$;

(e) $\begin{bmatrix} 1 & 0 & 0 \\ 0 & 0 & 1 \end{bmatrix}$; (f) $\begin{bmatrix} 1 & 0 & 1 \\ 0 & 1 & 1 \end{bmatrix}$;

(g) $\begin{bmatrix} 1 \\ 1 \\ 1 \end{bmatrix}$; (h) $\begin{bmatrix} 1 & 1 & 1 & 1 \\ 1 & 1 & 1 & 1 \\ 1 & 1 & 1 & 1 \\ 1 & 1 & 1 & -1 \end{bmatrix}$;

(i) $\begin{bmatrix} 1 & 4 & 5 & 3 \\ 2 & 3 & 5 & 1 \\ 3 & 2 & 5 & -1 \\ 4 & 1 & 5 & -3 \end{bmatrix}$; (j) $\begin{bmatrix} 1 & 2 & 3 & 4 \\ 4 & 1 & 2 & 3 \\ 3 & 4 & 1 & 2 \\ 2 & 3 & 4 & 1 \end{bmatrix}$.

4 Three cities, c_1, c_2, and c_3, form a closed economic unit, trading only among themselves. City c_i sells what it produces at an average price of x_i dollars per unit, $i = 1, 2, 3$. Moreover, c_1 purchases $\frac{1}{2}$ of its total purchases from itself, $\frac{1}{4}$ from c_2, and $\frac{1}{4}$ from c_3. Similarly, c_2 purchases $\frac{3}{4}$ from itself and $\frac{1}{8}$ each from c_1 and c_3. Finally, c_3 purchases $\frac{1}{3}$ of its total purchases from itself and each of the other two cities.

(a) Let $x = (x_1, x_2, x_3)$ be the price vector. What system of linear equations must be satisfied in order that the economy be stable? (See Example 6.3.)

(b) Express the system in (a) in the form $Ax = b$.

(c) What additional conditions must be satisfied by any solution vector x?

†5 A plane in R^3 is defined as the totality of points $x = (x_1, x_2, x_3)$ which satisfy an equation of the form $\alpha x_1 + \beta x_2 + \gamma x_3 = c$, in which α, β, γ, and c are constants.

(a) Find the equation of the plane which contains the points $(0, 0, 1)$, $(-1, 3, 5)$, and $(1, -1, 0)$.
(b) Intuition tells us that, in general, it is not possible to find a plane in R^3 which contains any four points. Find an example of four points in R^3 which do not lie in a plane.

6 A retail grocery company has two outlets, o_1 and o_2, and two warehouses, h_1 and h_2. The cost of shipping one ton of food from h_1 to o_1 is \$5, from h_1 to o_2 is \$7, from h_2 to o_1 is \$6, and from h_2 to o_2 is \$4. Moreover, h_1 has a capacity of 10 tons and h_2 has a capacity of 15 tons. The outlets place orders with the warehouses which will empty both of them: o_1 orders 12 tons and o_2 orders 13 tons. Find the amount that should be shipped from each of the warehouses to each of the outlets in order that the shipping costs be minimized. Set up the problem without finding the solution. (Hint: See Examples 6.7 and 6.8.)

†7 Let $U = \{u_1, \ldots, u_m\}$ be a set of m vectors in R^n. The set U is said to be *linearly dependent* if there exist constants $c_1, \ldots, c_m$, not all of which are 0, such that

$$\sum_{i=1}^{m} c_i u_i = 0.$$

Otherwise the set U is said to be *linearly independent*. For each of the following sets U, determine whether it is linearly dependent or linearly independent:

(a) $\{(1, 0), (-1, -1)\}$;
(b) $\{(1, 0), (-1, -1), (0, 0)\}$;
(c) $\{(1, 0, 0), (0, 1, 0), (0, 0, 1)\}$;
(d) $\{(1, 0, 0, 0), (0, -1, 0, 0), (0, 0, 1, 0), (0, 0, 0, -1)\}$;
(e) $\{(1, 1), (2, 2), (3, 3)\}$.

†8 Let $U = \{u_1, \ldots, u_m\}$ be a set of m vectors in R^n. Show that any of the following three operations on vectors in U will not alter the linear independence or dependence of U:

(i) an interchange of two of the vectors;
(ii) multiplying a vector in U by a non-zero constant;
(iii) adding to a vector in U a constant multiple of another vector in U.

In other words:

(i) if $j \neq k$, then the set

$$U = \{u_1, \ldots, u_j, \ldots, u_k, \ldots, u_m\}$$

is linearly independent (resp. dependent) if and only if the set

$$U_1 = \{u_1, \ldots, u_k, \ldots, u_j, \ldots, u_m\}$$

is linearly independent (resp. dependent);
(ii) if c is a non-zero real number then U is linearly independent (resp. dependent) if and only if

$$U_1 = \{u_1, \ldots, u_{j-1}, cu_j, u_{j+1}, \ldots, u_m\}$$

is linearly independent (resp. dependent);

(iii) if $j \neq k$ and c is a real number, then U is linearly independent (resp. dependent) if and only if

$$U_1 = \{u_1, \ldots, u_j, \ldots, u_{k-1}, u_k + cu_j, u_{k+1}, \ldots, u_m\}$$

is linearly independent (resp. dependent).

†9 Let U be the set considered in the preceding two exercises. Define the *rank* of U to be the largest integer k such that there exist k of the vectors in U which form a linearly independent set. Denote the rank of U by $\rho(U)$. Find $\rho(U)$ for each of the following sets:

(a) $\{(1, 1), (1, -1)\}$;
(b) $\{(1, -1), (0, 0), (2, 2)\}$;
(c) $\{(1, 0, 0), (0, 1, 0), (1, 1, 0)\}$;
(d) $\{(1, 0, 0), (0, 1, 0), (0, 0, 1), (1, 0, 2), (0, -1, 1)\}$;
(e) $\{(1, 1), (1, 0), (0, 1), (2, 2), (3, 3), (-1, -1)\}$.

†10 In the notation of Exercise 9, show that if the zero vector is in U, then $\rho(U) \leq m - 1$.

11 If A is an $m \times n$ matrix, then the *row rank* of A, also denoted by $\rho(A)$, is the rank of the set U whose elements are the m rows of A regarded as vectors in R^n, i.e., if $A_{(1)}, A_{(2)}, \ldots, A_{(m)}$ are the m rows of A, then

$$U = \{A_{(1)}, A_{(2)}, \ldots, A_{(m)}\}.$$

Find the row rank of each of the following matrices:

(a) $\begin{bmatrix} 1 & 0 \\ 0 & 1 \end{bmatrix}$;

(b) $\begin{bmatrix} 1 & 0 & 0 \\ 0 & 0 & 1 \end{bmatrix}$;

(c) $\begin{bmatrix} 0 & 1 & 0 & 0 \\ 0 & 0 & 1 & -1 \\ 0 & 0 & 0 & 0 \end{bmatrix}$;

(d) $\begin{bmatrix} 0 & 1 & 3 & 1 & 0 & 0 & 4 \\ 0 & 0 & 0 & 0 & 1 & 0 & 3 \\ 0 & 0 & 0 & 0 & 0 & 1 & 2 \\ 0 & 0 & 0 & 0 & 0 & 0 & 0 \end{bmatrix}$;

(e) $\begin{bmatrix} 0 & 0 & 1 & 2 & 0 & -3 & 8 & 0 & 5 & 4 & 3 & 0 & 7 & 9 \\ 0 & 0 & 0 & 0 & 1 & 2 & -4 & 0 & -6 & 7 & 5 & 0 & 8 & 5 \\ 0 & 0 & 0 & 0 & 0 & 0 & 0 & 1 & 2 & 9 & 3 & 0 & -3 & -6 \\ 0 & 0 & 0 & 0 & 0 & 0 & 0 & 0 & 0 & 0 & 0 & 1 & 2 & -4 \\ 0 & 0 & 0 & 0 & 0 & 0 & 0 & 0 & 0 & 0 & 0 & 0 & 0 & 0 \\ 0 & 0 & 0 & 0 & 0 & 0 & 0 & 0 & 0 & 0 & 0 & 0 & 0 & 0 \end{bmatrix}$;

(f) $\begin{bmatrix} 0 & 1 & 2 & 0 & 0 \\ 0 & 0 & 0 & 1 & 0 \\ 0 & 0 & 0 & 0 & 1 \\ 0 & 0 & 0 & 0 & 0 \\ 0 & 0 & 0 & 0 & 0 \\ 0 & 0 & 0 & 0 & 0 \end{bmatrix}$;

(g) $\begin{bmatrix} 1 & 1 & 1 \\ 1 & 1 & 2 \\ 0 & 0 & 1 \end{bmatrix}$.

†12 Let A be an $m \times n$ matrix and let B be a matrix obtained from A by any one of the three types of elementary row operations. Show that $\rho(B) = \rho(A)$. (Hint: Use the result of Exercise 8. In particular, we can begin by assuming that if $\rho(A) = k$, then $A_{(1)}, \ldots, A_{(k)}$ are linearly independent and any set of $k + t$ rows, $t > 0$, which contains these k rows is linearly dependent. Then, for example, a type II elementary row operation on A replaces the set of rows

$$U = \{A_{(1)}, \ldots, A_{(k)}, A_{(k+1)}, \ldots, A_{(m)}\}$$

by one of the following sets:

$$\{A_{(1)}, \ldots, A_{(i-1)}, A_{(i)} + cA_{(j)}, A_{(i+1)}, \ldots, A_{(k)}, A_{(k+1)}, \ldots, A_{(m)}\},$$

where $1 \leq i \leq k$ and either $1 \leq j \leq k$ or $j > k$; or

$$\{A_{(1)}, \ldots, A_{(k)}, A_{(k+1)}, \ldots, A_{(p-1)}, A_{(p)} + cA_{(j)}, A_{(p+1)}, \ldots, A_{(m)}\},$$

where $k + 1 \leq p \leq m$ and either $1 \leq j \leq k$ or $j > k$. In each case, exhibit a set of k linearly independent rows. This will show that $\rho(B)$ is always at least as large as $\rho(A)$, i.e., $\rho(B) \geq \rho(A)$. Then, since A can also be obtained from B by an elementary row operation, it follows that $\rho(B) \leq \rho(A)$. Hence $\rho(B) = \rho(A)$. The arguments for type I and type III elementary row operations are very much simpler.)

†13 Let H be a matrix in Hermite normal form as described in Definition 6.3. Show that $\rho(H)$ is the number of non-zero rows in H. Hence $\rho(H)$ is what we called the row rank of H in Definition 6.3.

†14 Show that if A is reduced to Hermite normal form H, then $\rho(A) = \rho(H)$. (Hint: Apply the result of Exercise 12 repeatedly.)

†15 Find the row rank of each of the following matrices by reducing it to Hermite normal form and using Exercise 14:

(a) $\begin{bmatrix} 1 & 1 & 1 & 1 \\ 1 & 0 & 2 & 0 \\ 2 & 1 & 3 & 0 \\ 0 & -1 & 1 & -1 \end{bmatrix}$;

(b) $\begin{bmatrix} 3 & 1 & 1 & 1 \\ 1 & 3 & 1 & 1 \\ 1 & 1 & 3 & 1 \\ 1 & 1 & 1 & 3 \end{bmatrix}$;

(c) $\begin{bmatrix} 1 & 0 & 1 & 0 & -1 \\ 2 & 0 & 3 & 4 & 1 \\ 3 & 0 & 4 & 4 & 0 \\ 4 & 0 & 5 & 4 & -1 \end{bmatrix}$;

(d) $\begin{bmatrix} 2 & 3 & 1 & 1 & 1 \\ 5 & 6 & 2 & 3 & 4 \\ 1 & 0 & -1 & 0 & 2 \\ 2 & 1 & 3 & -1 & 5 \\ 3 & 1 & 2 & -1 & 7 \end{bmatrix}$.

†16 Let A be an $n \times n$ matrix and suppose $\rho(A) = n$. Prove that A is row equivalent to the identity matrix I_n. (Hint: Let H be the Hermite normal form of A. Then by Exercise 14, $\rho(H) = n$. Observe the remark immediately following Example 6.11.)

†17 Show that if A is an $n \times n$ matrix and $\rho(A) = n$, then the only solution to the system $Ax = 0$ is the n-vector $x = 0$. (Hint: According to Exercise 16, the system $Ax = 0$ is equivalent to the system $I_n x = 0$.)

†18 Let $U = \{u_1, \ldots, u_p\}$ be a set of vectors in R^m. Show that if $p > m$ then the set of vectors U is linearly dependent. In other words, if the number of vectors in a set $U \subset R^m$ exceeds m, then the vectors must be linearly dependent. (Hint: Let A be the $p \times m$ matrix whose rows are the vectors $u_1, \ldots, u_p$, i.e.,

$$A = \begin{bmatrix} u_1 \rightarrow \\ u_2 \rightarrow \\ \vdots \\ u_p \rightarrow \end{bmatrix}.$$

Since $p > m$ the number of rows of A strictly exceeds the number of columns. Suppose that the set U is linearly independent. This means (see Exercises 9 and 11) that $\rho(A) = p$. Now reduce A to Hermite normal form H. According to Exercise 14, $\rho(H) = p$. But then (see Definition 6.3) the p-tuples $e_1, \ldots, e_p$ must occur among the columns of H. However, this requires that H have at least p columns and it does not.)

2.7

Applications of the Hermite Normal Form

In this section we discuss a systematic way of solving linear equations using the Hermite normal form of the coefficient matrix. Thus, let A be an $m \times n$ matrix, and let b be an m-vector. We want to find the set of all n-vectors x which are solutions to the system of linear equations

$$Ax = b. \tag{1}$$

The corresponding system of equations

$$Ax = 0 \tag{2}$$

with b replaced by the zero vector is called the *homogeneous* system associated with (1). We formulate a very elementary result concerning the relationship between (1) and (2).

Theorem 7.1 *Any two solutions of* (1) *differ by a solution to the homogeneous system* (2). *Moreover, if u_0 is a solution to* (1), *then the set of all solutions to* (1) *is precisely the set of all n-vectors of the form $u_0 + v$, where v runs over the set of solutions to* (2).

Proof Let z be a solution to (1). Then since u_0 is a solution to (1),

$$\begin{aligned} A(z - u_0) &= Az - Au_0 \\ &= b - b \\ &= 0. \end{aligned}$$

Hence, $z - u_0$ satisfies (2), i.e., $z - u_0 = v$ where v is in the set of solutions to (2). Moreover, if $Av = 0$, then $A(u_0 + v) = Au_0 + Av = b + 0 = b$, and hence $u_0 + v$ satisfies (1). This completes the proof. ▌

If we adjoin the m-vector b on the right as another column of the matrix A, then the resulting $m \times (n + 1)$ matrix is called the *augmented coefficient matrix* for (1), and is denoted by

(3) $$[A:b].$$

Theorem 7.2 *Let H be an Hermite normal form for the $m \times n$ matrix A. Perform a series of elementary row operations on the augmented matrix $[A:b]$ that reduces A to H, and let $[H:c]$ be the resulting matrix. If H has row rank r, then the system* (1) *has a solution u_0 if and only if $c_{r+1} = \cdots = c_m = 0$, where $c = (c_1, \ldots, c_m)$.*

Proof We first observe that an elementary row operation of type I on the matrix $[A:b]$ corresponds to an interchange of the order in which a pair of the equations in $Ax = b$ occurs. Similarly, a type II elementary row operation on $[A:b]$ corresponds to adding to an equation a multiple of another equation in $Ax = b$. Finally, a type III elementary row operation on $[A:b]$ corresponds to multiplying both sides of one of the equations in $Ax = b$ by a non-zero constant. It is obvious that each of these operations results in an equivalent system of equations. Hence $[H:c]$ is the augmented matrix for a system of equations equivalent to the original system $Ax = b$, that is, for the system

(4) $$Hx = c.$$

The system (4) has exactly the same set of solution vectors as does $Ax = b$. The solutions to $Hx = c$, however, are easy to obtain, because H is in a greatly simplified form. Recall from Definition 6.3 that the first r rows of H are non-zero and the remaining $m - r$ rows are all zero; there are r columns of H, numbered $n_1, \ldots, n_r$ $(1 \leq n_1 < n_2 < \cdots < n_r \leq n)$, such that column n_i of H is the m-vector e_i, $i = 1, \ldots, r$; the first $n_i - 1$ entries in row i of H are zero, $i = 1, \ldots, r$. Suppose that one of $c_{r+1}, \ldots, c_m$ is not 0, say, $c_{r+1} \neq 0$. Then, since the $(r + 1)$st row of H consists entirely of 0's, it follows that the $(r + 1)$st equation in $Hx = c$ is

$$0x_1 + 0x_2 + \cdots + 0x_n = c_{r+1},$$

which obviously cannot be satisfied for any n-vector x. Therefore $Hx = c$ (and hence the equivalent system $Ax = b$) has no solutions. On the other hand, suppose that $c_{r+1} = \cdots = c_m = 0$. Now construct an n-vector

$$u_0 = (0, \ldots, 0, \underset{\substack{\uparrow \\ n_1}}{c_1}, 0, \ldots, 0, \underset{\substack{\uparrow \\ n_2}}{c_2}, 0, \ldots, 0, \underset{\substack{\uparrow \\ n_r}}{c_r}, 0, \ldots, 0) \tag{5}$$

in which $c_1, \ldots, c_r$ appear in positions $n_1, \ldots, n_r$, respectively, and 0 appears in every other position. Then it follows immediately from the definition of matrix-vector multiplication that

$$\begin{aligned} Hu_0 &= \sum_{i=1}^{r} c_i H^{(n_i)} \\ &= \sum_{i=1}^{r} c_i e_i \\ &= c. \end{aligned}$$

Thus u_0 is a solution to $Hx = c$ and hence to the equivalent system $Ax = b$. This completes the proof. ∎

Theorem 7.2 gives us a method for deciding when a system of linear equations has a solution and, moreover, the result provides us with a constructive technique for finding a solution, if one exists. Theorem 7.1 tells us that if we can find all the solutions v to the homogeneous system $Ax = 0$, then any solution of the original system $Ax = b$ is of the form $u_0 + v$ where $u_0 \in R^n$ is defined in (5). Thus the problem reduces to finding all solutions v to $Ax = 0$. Fortunately, the reduction of A to Hermite normal form H virtually provides us with a complete set of solutions to $Ax = 0$. For, H has the following form.

$$
\begin{array}{r c}
 & \begin{array}{ccccccccccccc} & n_1 & & n_2 & & n_3 & \ldots & n_r & \end{array} \\
(6) \quad \text{row } r \rightarrow &
\left[\begin{array}{ccccccccccccccccccc}
0 & \ldots & 0 & 1 & * & \ldots & * & 0 & * & \ldots & * & 0 & * & \ldots & * & 0 & * & \ldots & * \\
0 & \ldots & 0 & 0 & 0 & \ldots & 0 & 1 & * & \ldots & * & 0 & * & \ldots & * & 0 & * & \ldots & * \\
0 & \ldots & & 0 & & \ldots & 0 & 0 & 0 & \ldots & 0 & 1 & * & \ldots & * & 0 & * & \ldots & * \\
\vdots & & & \vdots & & & & \vdots & & & & \vdots & & & & \vdots & & & \\
0 & \ldots & & 0 & & \ldots & & 0 & & \ldots & & 0 & & \ldots & 0 & 1 & * & \ldots & * \\
\hline
0 & & & & & \cdot & & & & \cdot & & & & \cdot & & & & & 0 \\
0 & & & & & & & & & & & & & & & & & & 0 \\
\vdots & & & & & & & & & 0 & & & & & & & & & \vdots \\
0 & & & & & & & & & & & & & & & & & & 0 \\
0 & \ldots & & 0 & & \ldots & & 0 & & \ldots & & 0 & & \ldots & & 0 & & \ldots & 0
\end{array}\right]
\end{array}
$$

(The asterisks indicate unspecified entries.) If we write out the system $Hx = 0$ explicitly, we have

$$
\begin{aligned}
x_{n_1} + \; * \quad\; * \quad\; * &= 0, \\
x_{n_2} + \; * \;\; * \;\; * &= 0, \\
&\;\vdots \\
x_{n_r} + *** &= 0.
\end{aligned} \tag{7}
$$

In the first equation in (7), the only one of $x_{n_1}, x_{n_2}, \ldots, x_{n_r}$ which actually appears is x_{n_1}, and in general, x_{n_k} is the only one of $x_{n_1}, x_{n_2}, \ldots, x_{n_r}$ which occurs in the kth equation in (7), $k = 1, \ldots, r$. Thus we can solve (7) for each of $x_{n_1}, \ldots, x_{n_r}$ in terms of the remaining x_j, where j is none of $n_1, \ldots, n_r$:

$$
x_{n_k} = \sum_j{}' \, d_{kj} x_j, \; k = 1, \ldots, r. \tag{8}
$$

(The d_{kj} arise from the coefficients in (7).) The prime on the sigma in (8) means simply that we are to sum over only those values of j different from $n_1, \ldots, n_r$. Now let

$$
x = (x_1, x_2, \ldots, \underset{\substack{\uparrow \\ n_1}}{x_{n_1}}, \ldots, \underset{\substack{\uparrow \\ n_2}}{x_{n_2}}, \ldots, \underset{\substack{\uparrow \\ n_r}}{x_{n_r}}, \ldots, x_n) \tag{9}
$$

(in which we have distinguished components numbered $n_1, \ldots, n_r$ by indicating their position) be a solution vector to (7). Then, if we replace each x_{n_k} in (9) by the right-hand side of (8), the resulting vector will have the following form:

$$
\left(x_1, x_2, \ldots, \overset{\substack{n_1 \\ \downarrow}}{\sum_j{}'} d_{1j}x_j, \ldots, \overset{\substack{n_2 \\ \downarrow}}{\sum_j{}'} d_{2j}x_j, \ldots, \overset{\substack{n_r \\ \downarrow}}{\sum_j{}'} d_{rj}x_j, \ldots, x_n\right). \tag{10}
$$

It is clear from the way we formed the vector in (10) that none of $x_{n_1}, \ldots, x_{n_r}$ appears in (10). Moreover, we know that any vector of the form (10) must satisfy the system (7), which is, of course, equivalent to the system $Ax = 0$. In other words, a vector can satisfy the homogeneous system $Ax = 0$ if and only if it satisfies the equivalent system (7). But as we have seen, x satisfies (7) if and only if it is of the form (10). We see, then, that in general there will be a solution to $Ax = 0$ for any assignment of values to those x_j which appear in (10). We summarize this discussion as follows. To find all solutions to the system $Ax = b$, perform the following steps.

(i) Reduce the matrix A to Hermite normal form H by performing the required sequence of elementary row operations on the augmented matrix $[A:b]$, and suppose that the matrix $[H:c]$ results. If the row rank of H is r, then $c_{r+1} = \cdots = c_n = 0$ is a necessary and sufficient condition for the system $Ax = b$ to have a solution. If this condition holds, a particular solution to $Ax = b$ is given by the vector u_0 in (5).

(ii) To the vector u_0 obtained in (i) add any vector v obtained by assigning values to the x_j appearing in (10). As we have seen, these vectors v constitute the totality of solutions to the homogeneous system $Hx = 0$, and hence to the equivalent system $Ax = 0$.

Example 7.1 Find all solutions to the following system of linear equations:

$$\begin{aligned} 3x_1 + 3x_2 - x_3 + 4x_4 - 2x_5 &= 14, \\ x_1 - x_2 + 7x_3 - x_4 &= -2, \\ 5x_1 + x_2 + 13x_3 + 2x_4 - 2x_5 &= 10, \\ 2x_1 + 4x_2 - 8x_3 + 5x_4 - 2x_5 &= 16. \end{aligned} \tag{11}$$

The augmented matrix of coefficients for the system (11) is

$$\left[\begin{array}{rrrrr|r} 3 & 3 & -1 & 4 & -2 & 14 \\ 1 & -1 & 7 & -1 & 0 & -2 \\ 5 & 1 & 13 & 2 & -2 & 10 \\ 2 & 4 & -8 & 5 & -2 & 16 \end{array}\right]. \tag{12}$$

We perform the following sequence of elementary row operations on the matrix in (12).

(a) Interchange the first two rows.

(b) Add -3 times the first row to the second row, then add -5 times the first row to the third row, and finally add -2 times the first row to the fourth row.

(c) Add -1 times the second row to the third and fourth rows.

(d) Multiply the second row by $\frac{1}{6}$.

(e) Add the second row to the first row.

At this stage, the resulting matrix is

$$\begin{bmatrix} 1 & 0 & \frac{10}{3} & \frac{1}{6} & -\frac{1}{3} & \frac{4}{3} \\ 0 & 1 & -\frac{11}{3} & \frac{7}{6} & -\frac{1}{3} & \frac{10}{3} \\ 0 & 0 & 0 & 0 & 0 & 0 \\ 0 & 0 & 0 & 0 & 0 & 0 \end{bmatrix}. \tag{13}$$

The submatrix consisting of all the rows and all but the last column in (13) is in Hermite normal form, and thus we have completed step (i) above. The row rank of H is 2, $n_1 = 1$, $n_2 = 2$, and $c = (\frac{4}{3}, \frac{10}{3}, 0, 0)$. Thus the particular solution u_0 given by (5) is

$$u_0 = (\tfrac{4}{3}, \tfrac{10}{3}, 0, 0, 0). \tag{14}$$

To proceed to step (ii), we write down the system (7) in this case:

$$\begin{aligned} x_1 + \tfrac{10}{3}x_3 + \tfrac{1}{6}x_4 - \tfrac{1}{3}x_5 &= 0, \\ x_2 - \tfrac{11}{3}x_3 + \tfrac{7}{6}x_4 - \tfrac{1}{3}x_5 &= 0, \end{aligned} \tag{15}$$

or, solving for x_1 and x_2 in (15), we obtain the equalities corresponding to (8). These are:

$$\begin{aligned} x_1 &= -\tfrac{10}{3}x_3 - \tfrac{1}{6}x_4 + \tfrac{1}{3}x_5, \\ x_2 &= \tfrac{11}{3}x_3 - \tfrac{7}{6}x_4 + \tfrac{1}{3}x_5. \end{aligned} \tag{16}$$

The vector in (10) in this case becomes

$$(-\tfrac{10}{3}x_3 - \tfrac{1}{6}x_4 + \tfrac{1}{3}x_5, \tfrac{11}{3}x_3 - \tfrac{7}{6}x_4 + \tfrac{1}{3}x_5, x_3, x_4, x_5). \tag{17}$$

Observe that the vector in (17) involves only those x_j different from x_1 and x_2. We rewrite the vector in (17) in the following form:

$$x_3(-\tfrac{10}{3}, \tfrac{11}{3}, 1, 0, 0) + x_4(-\tfrac{1}{6}, -\tfrac{7}{6}, 0, 1, 0) + x_5(\tfrac{1}{3}, \tfrac{1}{3}, 0, 0, 1). \tag{18}$$

Let v^1, v^2, and v^3 be the vectors appearing in (18) next to x_3, x_4, and x_5, respectively. Then we have proved that any linear combination of these vectors is a solution to the homogeneous system of equations corresponding to (11). Thus the set of solutions to the system (11) is precisely the totality of vectors

$$\begin{aligned} &u_0 + \alpha_1 v^1 + \alpha_2 v^2 + \alpha_3 v^3 = \\ &\quad (\tfrac{4}{3}, \tfrac{10}{3}, 0, 0, 0) + \alpha_1(-\tfrac{10}{3}, \tfrac{11}{3}, 1, 0, 0) + \alpha_2(-\tfrac{1}{6}, -\tfrac{7}{6}, 0, 1, 0) + \alpha_3(\tfrac{1}{3}, \tfrac{1}{3}, 0, 0, 1) \end{aligned} \tag{19}$$

obtained by assigning arbitrary values to α_1, α_2, and α_3. We have used α's as coefficients in (19) rather than the original "unknowns" x_3, x_4, and x_5 that appear in (18) to emphasize that an arbitrary linear combination of the vectors v^1, v^2, and v^3 satisfies the homogeneous system.

Example 7.2 Find all solutions (if any exist) of the system of equations

$$
\begin{aligned}
2x_1 + 2x_2 + x_3 + x_4 &= 20,\\
x_1 + x_2 \qquad + x_4 &= 12,\\
x_2 + x_3 + x_4 &= 14,\\
2x_1 + x_2 \qquad\qquad &= 7.
\end{aligned} \tag{20}
$$

The augmented matrix for the system (20) is

$$
\begin{bmatrix} 2 & 2 & 1 & 1 & 20 \\ 1 & 1 & 0 & 1 & 12 \\ 0 & 1 & 1 & 1 & 14 \\ 2 & 1 & 0 & 0 & 7 \end{bmatrix}. \tag{21}
$$

Perform the following sequence of elementary row operations on the matrix in (21).

(a) Interchange rows 1 and 2.
(b) Add -2 times row 1 to row 2.
(c) Add -2 times row 1 to row 4.
(d) Interchange rows 2 and 3.
(e) Add row 2 to row 4.
(f) Add -1 times row 2 to row 1.
(g) Add -1 times row 3 to row 4.
(h) Add -1 times row 3 to row 2.
(i) Add row 3 to row 1.

The resulting matrix is

$$
\begin{bmatrix} 1 & 0 & 0 & -1 & -6 \\ 0 & 1 & 0 & 2 & 18 \\ 0 & 0 & 1 & -1 & -4 \\ 0 & 0 & 0 & 0 & 1 \end{bmatrix}. \tag{22}
$$

In this case $r = 3$ but $c_4 = c_{r+1} = 1 \neq 0$, and hence the system of equations in (20) has no solution.

Example 7.3 We refer to Example 6.3. Suppose that three cities, c_1, c_2, and c_3, form a closed economic unit, trading only among themselves. Assume that city c_i sells what it produces at an average price of x_i dollars per unit produced, $i = 1, 2, 3$. Also, c_i buys a fractional amount of its total purchases from c_j at an average price of x_j dollars per unit. Let this fractional amount be a_{ij}, $i, j = 1, 2, 3$. In Example 6.3 we discovered that given the matrix A whose (i, j)

entry is a_{ij}, the conditions on a price vector $x = (x_1, x_2, x_3)$ that insure that the economy is stable can be expressed by a system of linear equations

$$(I_3 - A)x = 0. \tag{23}$$

Suppose that c_1 obtains $\frac{2}{3}$ of its purchases from itself and $\frac{1}{6}$ each from c_2 and c_3; c_2 obtains $\frac{1}{4}$ from itself, $\frac{1}{2}$ from c_1, and $\frac{1}{4}$ from c_3; and c_3 obtains $\frac{1}{2}$ from itself and $\frac{1}{2}$ from c_2. The system of equations (23) then becomes

$$\begin{bmatrix} \frac{1}{3} & -\frac{1}{6} & -\frac{1}{6} \\ -\frac{1}{2} & \frac{3}{4} & -\frac{1}{4} \\ 0 & -\frac{1}{2} & \frac{1}{2} \end{bmatrix} x = 0. \tag{24}$$

We reduce the coefficient matrix in (24) to Hermite normal form.

(a) In order to simplify the computations, we first multiply row 1 by 6, rows 2 and 3 by 4 and 2, respectively, to obtain

$$\begin{bmatrix} 2 & -1 & -1 \\ -2 & 3 & -1 \\ 0 & -1 & 1 \end{bmatrix}.$$

(b) Add row 1 to row 2 to obtain

$$\begin{bmatrix} 2 & -1 & -1 \\ 0 & 2 & -2 \\ 0 & -1 & 1 \end{bmatrix}.$$

(c) Now multiply row 2 by $\frac{1}{2}$, add row 2 to rows 1 and 3 and multiply row 1 by $\frac{1}{2}$ to produce the matrix

$$\begin{bmatrix} 1 & 0 & -1 \\ 0 & 1 & -1 \\ 0 & 0 & 0 \end{bmatrix},$$

which is in Hermite normal form.

For this matrix, $n_1 = 1, n_2 = 2, r = 2$, and the system corresponding to (7) becomes

$$x_1 - x_3 = 0,$$
$$x_2 - x_3 = 0.$$

Thus the solutions to (23) consist of all vectors

$$(x_3, x_3, x_3) = x_3(1, 1, 1),$$

for any value assigned to x_3. In other words, whatever average price c_3 charges for the goods it produces, c_2 and c_1 must charge the same.

Example 7.4 The psychology department orders four monkeys to be delivered from the university animal vivarium. The differences in responses of the monkeys are to be studied as a function of different diets made from ingredients A, B, and C. The first monkey will be fed a daily diet consisting of 20 units of A, 30 units of B, and 10 units of C. The amounts for the second, third, and fourth monkeys are 40 units of A and no units of B and C; 30 units of A, 10 units of B, and 30 units of C; and 10 units of A, 10 units of B, and 50 units of C, respectively. The department has 150 units of A, 200 units of B, and 220 units of C on hand. How far can the experiment proceed before the food supply must be replenished? Let x_1, x_2, x_3, and x_4 be the number of days that each of the monkeys remain in the psychology department. Then

$$\begin{aligned} &20x_1 + 40x_2 + 30x_3 + 10x_4, \\ &30x_1 + \ 0x_2 + 10x_3 + 10x_4, \\ &10x_1 + \ 0x_2 + 30x_3 + 50x_4 \end{aligned} \tag{25}$$

are the amounts of ingredients A, B, and C, respectively, that will be consumed during the monkeys' stay. The x_i, $i = 1, 2, 3, 4$, are determined by the system of linear equations obtained by setting the expressions in (25) equal to 150, 200, and 220, respectively. The resulting system of linear equations has the following augmented coefficient matrix:

$$\begin{bmatrix} 20 & 40 & 30 & 10 & 150 \\ 30 & 0 & 10 & 10 & 200 \\ 10 & 0 & 30 & 50 & 220 \end{bmatrix}. \tag{26}$$

Perform the following sequence of elementary row operations on the matrix in (26).

(a) Multiply each of the rows by $\frac{1}{10}$.

(b) Interchange rows 1 and 3.

(c) Add -3 times row 1 to row 2 and -2 times row 1 to row 3.

(d) Interchange rows 2 and 3.

(e) Multiply row 3 by $-\frac{1}{2}$.

(f) Add $\frac{3}{4}$ times row 3 to row 2.

(g) Add $-\frac{3}{4}$ times row 3 to row 1.

(h) Multiply rows 2 and 3 by $\frac{1}{4}$.

The resulting matrix is

$$\begin{bmatrix} 1 & 0 & 0 & -\frac{1}{4} & \frac{19}{4} \\ 0 & 1 & 0 & -\frac{15}{16} & -\frac{47}{16} \\ 0 & 0 & 1 & \frac{7}{4} & \frac{23}{4} \end{bmatrix}.$$

Hence the row rank is $r = 3$, and $n_1 = 1$, $n_2 = 2$, and $n_3 = 3$; the system corresponding to (7) now becomes

$$x_1 - \tfrac{1}{4}x_4 = 0,$$
$$x_2 - \tfrac{15}{16}x_4 = 0,$$
$$x_3 + \tfrac{7}{4}x_4 = 0.$$

The vector corresponding to (10) is

$$(\tfrac{1}{4}x_4, \tfrac{15}{16}x_4, -\tfrac{7}{4}x_4, x_4).$$

The particular solution u_0 given by (5) is

$$u_0 = (\tfrac{19}{4}, -\tfrac{47}{16}, \tfrac{23}{4}, 0).$$

Thus, the set of solutions to the system consists of the totality of vectors

$$(\tfrac{19}{4}, -\tfrac{47}{16}, \tfrac{23}{4}, 0) + \alpha(\tfrac{1}{4}, \tfrac{15}{16}, -\tfrac{7}{4}, 1) \tag{27}$$

obtained by assigning arbitrary values to α. However, the required solutions must be nonnegative, i.e., each monkey is involved in the experiment for a nonnegative number of days. Thus we must restrict α so that each component of (27) is nonnegative:

$$\frac{19}{4} + \frac{\alpha}{4} \geq 0,$$
$$-\frac{47}{16} + \frac{15\alpha}{16} \geq 0,$$
$$\frac{23}{4} - \frac{7\alpha}{4} \geq 0,$$
$$\alpha \geq 0.$$

The second inequality implies the first and fourth ones, and the second and third inequalities impose the following restrictions on α:

$$\tfrac{47}{15} \leq \alpha \leq \tfrac{23}{7}. \tag{28}$$

Thus any choice of α satisfying (28) yields an acceptable solution vector (27).

If A is a non-zero number, then we can write the solution to the linear equation

$$Ax = b$$

as

$$x = A^{-1}b.$$

It is an attractive possibility that we may be able to do the same thing with the solution to systems of linear equations. The important property that the number A^{-1} possesses is that it satisfies the equation

(29)
$$\begin{aligned} A^{-1}A &= AA^{-1} \\ &= 1. \end{aligned}$$

Thus if A and A^{-1} are matrices, the equations corresponding to (29) become

(30)
$$\begin{aligned} A^{-1}A &= AA^{-1} \\ &= I, \end{aligned}$$

where I is the identity matrix of the appropriate dimension. If A is $m \times n$, then in order that the product $A^{-1}A$ be defined, A^{-1} must have m columns, and thus I must be I_n. It follows that the number of rows of A^{-1} must be n since $A^{-1}A = I_n$. On the other hand, in order that the product AA^{-1} be defined, A^{-1} must have n rows. Now, AA^{-1} has m rows, and thus I must also be I_m. It follows that $m = n$ and hence A must be n-square if an equation of the form (30) is to hold at all. Suppose, then, that A is n-square, A^{-1} is n-square, and that

(31)
$$\begin{aligned} A^{-1}A &= AA^{-1} \\ &= I_n. \end{aligned}$$

Now suppose that B is an n-square matrix satisfying $BA = I_n$. Then

(32)
$$\begin{aligned} B &= BI_n \\ &= B(AA^{-1}) \\ &= (BA)A^{-1} \\ &= I_nA^{-1} \\ &= A^{-1}. \end{aligned}$$

This little calculation shows that if a matrix A^{-1} exists satisfying (31), then it is unique.

Definition 7.1 ***Inverse of a matrix*** Let A be an $n \times n$ matrix. If there exists an $n \times n$ matrix X such that

(33)
$$\begin{aligned} AX &= XA \\ &= I_n, \end{aligned}$$

then X is called the *inverse* of A.

In view of the computation (32), we see that the equations in (33) imply that if B is any n-square matrix satisfying

$$BA = I_n, \tag{34}$$

then $B = X$. In other words, if a matrix X exists which satisfies (33), then it is uniquely determined. We denote the inverse of A when it exists by the symbol

$$A^{-1}.$$

If A is an $n \times n$ matrix which has an inverse, then A is said to be *nonsingular*, or *regular*, or a *unit matrix*. There are a number of elementary results concerning nonsingular matrices.

Theorem 7.3

(a) *The product AB of two nonsingular matrices A and B is nonsingular and*

$$(AB)^{-1} = B^{-1}A^{-1}. \tag{35}$$

(b) *If A is nonsingular, then so are A^T and A^{-1}, and*

$$(A^T)^{-1} = (A^{-1})^T, \tag{36}$$

$$(A^{-1})^{-1} = A. \tag{37}$$

Proof

(a) Since we are assuming that both A and B are nonsingular and that the product AB is defined, it follows that both A and B are n-square. (Why?) Then

$$\begin{aligned}(AB)(B^{-1}A^{-1}) &= A(BB^{-1})A^{-1}\\ &= AI_nA^{-1}\\ &= AA^{-1}\\ &= I_n,\end{aligned}$$

and similarly,

$$(B^{-1}A^{-1})(AB) = I_n.$$

Hence, $B^{-1}A^{-1}$ satisfies the defining equation (33) for the existence of the unique inverse of AB.

(b) We compute (see Exercise 11, Section 2.1) that

$$\begin{aligned}A^T(A^{-1})^T &= (A^{-1}A)^T\\ &= I_n^T\\ &= I_n,\end{aligned}$$

and similarly $(A^{-1})^T A^T = I_n$. Thus, $(A^{-1})^T$ satisfies the defining equation (33) for the existence of the unique inverse of A^T.

Since A is nonsingular, A satisfies equation (33):

$$AA^{-1} = A^{-1}A = I_n.$$

But this is also a defining relation for the existence of an inverse for A^{-1}. Hence A^{-1} is nonsingular and, by the uniqueness of the inverse,

$$(A^{-1})^{-1} = A. \blacksquare$$

Example 7.5 **(a)** The matrix

$$A = \begin{bmatrix} 1 & -1 \\ 1 & -1 \end{bmatrix}$$

does not have an inverse, i.e., it is *singular*. For suppose

$$X = \begin{bmatrix} x_{11} & x_{12} \\ x_{21} & x_{22} \end{bmatrix}$$

satisfies (33). Then

$$\begin{aligned} AX &= \begin{bmatrix} 1 & -1 \\ 1 & -1 \end{bmatrix} \begin{bmatrix} x_{11} & x_{12} \\ x_{21} & x_{22} \end{bmatrix} \\ &= \begin{bmatrix} x_{11} - x_{21} & x_{12} - x_{22} \\ x_{11} - x_{21} & x_{12} - x_{22} \end{bmatrix} \\ &= \begin{bmatrix} 1 & 0 \\ 0 & 1 \end{bmatrix}. \end{aligned}$$

Equating the (1, 1) entries and the (2, 1) entries, we obtain two obviously incompatible conditions:

$$\begin{aligned} x_{11} - x_{21} &= 1, \\ x_{11} - x_{21} &= 0. \end{aligned}$$

Hence no such X can exist.

(b) Find the inverse of the matrix

$$A = \begin{bmatrix} 1 & 1 \\ 1 & -1 \end{bmatrix}.$$

If X is to satisfy $AX = I_2$, and x denotes the first column of X, then

$$Ax = e_1,$$

where $e_1 = (1, 0)$. Similarly, if y denotes the second column of X, then

$$Ay = e_2,$$

where $e_2 = (0, 1)$. Thus, finding X for which $AX = I_2$ amounts to solving the preceding two systems of linear equations. Writing out $Ax = e_1$ explicitly, we have

$$x_1 + x_2 = 1,$$
$$x_1 - x_2 = 0,$$

which has the solution

$$x_1 = \tfrac{1}{2},$$
$$x_2 = \tfrac{1}{2}.$$

Similarly, the equation $Ay = e_2$ has the solution $y_1 = \frac{1}{2}, y_2 = -\frac{1}{2}$. Hence

$$X = \begin{bmatrix} \frac{1}{2} & \frac{1}{2} \\ \frac{1}{2} & -\frac{1}{2} \end{bmatrix}.$$

Observe that

$$AX = \begin{bmatrix} 1 & 1 \\ 1 & -1 \end{bmatrix} \begin{bmatrix} \frac{1}{2} & \frac{1}{2} \\ \frac{1}{2} & -\frac{1}{2} \end{bmatrix} = \begin{bmatrix} 1 & 0 \\ 0 & 1 \end{bmatrix}$$

and

$$XA = \begin{bmatrix} \frac{1}{2} & \frac{1}{2} \\ \frac{1}{2} & -\frac{1}{2} \end{bmatrix} \begin{bmatrix} 1 & 1 \\ 1 & -1 \end{bmatrix} = \begin{bmatrix} 1 & 0 \\ 0 & 1 \end{bmatrix}.$$

Thus $XA = AX = I_2$ and X is the unique inverse of A.

(c) Let

$$E = \begin{bmatrix} 0 & 1 & 0 \\ 1 & 0 & 0 \\ 0 & 0 & 1 \end{bmatrix}.$$

We compute that if A is any $3 \times n$ matrix, then

$$EA = \begin{bmatrix} 0 & 1 & 0 \\ 1 & 0 & 0 \\ 0 & 0 & 1 \end{bmatrix} \begin{bmatrix} a_{11} & a_{12} & a_{13} & a_{14} \\ a_{21} & a_{22} & a_{23} & a_{24} \\ a_{31} & a_{32} & a_{33} & a_{34} \end{bmatrix} = \begin{bmatrix} a_{21} & a_{22} & a_{23} & a_{24} \\ a_{11} & a_{12} & a_{13} & a_{14} \\ a_{31} & a_{32} & a_{33} & a_{34} \end{bmatrix}.$$

(We have taken $n = 4$.) Putting this in a different way, for *any* $3 \times n$ matrix A, EA is obtained from A by interchanging rows 1 and 2 of A. But E itself is obtained from I_3 by interchanging rows 1 and 2 of I_3. If we interchange rows 1 and 2 of EA we of course recover A:

$$E(EA) = A. \tag{38}$$

By setting $A = I_3$ in (38) we obtain

$$EE = I_3 \tag{39}$$

and hence $E = E^{-1}$.

(d) Let

$$E = \begin{bmatrix} 1 & 0 & 0 \\ 0 & 1 & 0 \\ c & 0 & 1 \end{bmatrix}.$$

We compute that for any $3 \times n$ matrix (take $n = 4$ again),

$$\begin{aligned} EA &= \begin{bmatrix} 1 & 0 & 0 \\ 0 & 1 & 0 \\ c & 0 & 1 \end{bmatrix} \begin{bmatrix} a_{11} & a_{12} & a_{13} & a_{14} \\ a_{21} & a_{22} & a_{23} & a_{24} \\ a_{31} & a_{32} & a_{33} & a_{34} \end{bmatrix} \\ &= \begin{bmatrix} a_{11} & a_{12} & a_{13} & a_{14} \\ a_{21} & a_{22} & a_{23} & a_{24} \\ a_{31} + ca_{11} & a_{32} + ca_{12} & a_{33} + ca_{13} & a_{34} + ca_{14} \end{bmatrix}. \end{aligned}$$

Thus for any $3 \times n$ matrix A, EA is obtained from A by adding c times row 1 to row 3. But E itself results from I_3 by performing this operation. Also, if we add $-c$ times row 1 of EA to row 3 of EA we will recover A. Thus if

$$E_1 = \begin{bmatrix} 1 & 0 & 0 \\ 0 & 1 & 0 \\ -c & 0 & 1 \end{bmatrix},$$

then

$$E_1(EA) = A.$$

Setting $A = I_3$ we obtain $E_1E = I_3$. We similarly conclude that $EE_1 = I_3$ and hence that E is nonsingular, $E^{-1} = E_1$.

(e) Let $c \neq 0$ and set

$$E = \begin{bmatrix} 1 & 0 & 0 \\ 0 & c & 0 \\ 0 & 0 & 1 \end{bmatrix}.$$

Then if A is any $3 \times n$ matrix, we can compute easily that the matrix EA is obtained from A by multiplying row 2 of A by c. If we set

$$E_1 = \begin{bmatrix} 1 & 0 & 0 \\ 0 & c^{-1} & 0 \\ 0 & 0 & 1 \end{bmatrix},$$

then it is easy to check that $E_1E = EE_1 = I_3$, i.e., $E^{-1} = E_1$.

In Examples 7.5 (c), (d), and (e) we learned that a type I, II, or III elementary row operation on a matrix A can be accomplished by multiplying A on the left by an appropriate nonsingular matrix E. Moreover, the same matrix E works for any A, as long as multiplication is compatible, and E is the matrix that results by performing the required elementary row operation on the identity matrix. The computations in these examples are perfectly general and clearly do not depend on the fact that we are dealing with $3 \times n$ matrices. These remarks lead us to the following definition and theorem.

Definition 7.2 **Elementary matrices** Let E be the $m \times m$ matrix that results by performing an elementary row operation on I_m. Then E is called an *elementary matrix*. An elementary matrix is said to be of the same type as the elementary row operation used to produce it.

Theorem 7.4 *Let A be an $m \times n$ matrix and let E be an m-square elementary matrix. Then EA is the matrix that results by performing the elementary row operation on A that produced E from I_m. Moreover, the elementary matrices are nonsingular:*

(a) *if E is a type I elementary matrix, then*

$$E^{-1} = E;$$

(b) *if E is a type* II *elementary matrix that results from I_m by adding c times row s to row t, $s \neq t$, then E^{-1} is also of type* II *and results from I_m by adding $-c$ times row s to row t;*

(c) *if E is of type* III *and results from I_m by multiplying row s by c, $c \neq 0$, then E^{-1} is of type* III *and results from I_m by multiplying row s of I_m by c^{-1}.*

We know that any $m \times n$ matrix A can be reduced to Hermite normal form by a sequence of elementary row operations. Moreover, each one of these elementary operations can be achieved by multiplying A on the left by the corresponding m-square elementary matrix. Thus we can state the following result.

Theorem 7.5 *If A is an $m \times n$ matrix, then there exists a sequence of m-square elementary matrices, $E_1, \ldots, E_k$, such that*

$$E_k E_{k-1} \cdots E_1 A = H, \tag{40}$$

where H is an $m \times n$ matrix in Hermite normal form.

If A is an $m \times m$ matrix, Theorem 7.5 can be used to prove a very practical result which can be used constructively to decide whether A is nonsingular and, if so, to produce A^{-1}.

Theorem 7.6 *Let A be an m-square matrix and suppose A is reduced to Hermite normal form H. Then A is nonsingular if and only if*

$$H = I_m.$$

Proof Let $E_1, \ldots, E_k$ be a sequence of $m \times m$ elementary matrices such that (40) holds:

$$E_k E_{k-1} \cdots E_1 A = H. \tag{41}$$

Since each E_i is nonsingular, $i = 1, \ldots, k$, we know by applying finite induction to Theorem 7.3(a) that

$$P = E_k E_{k-1} \cdots E_1 \tag{42}$$

is nonsingular and

$$P^{-1} = E_1^{-1} E_2^{-1} \cdots E_k^{-1}. \tag{43}$$

If we multiply both sides of (42) on the left by P^{-1}, we obtain

$$A = P^{-1} H. \tag{44}$$

Suppose, now, that A is nonsingular. We recall an observation that was made immediately following Example 6.11, namely that if the Hermite normal form H of an m-square matrix is not the identity

I_m, then the last row of H must consist entirely of 0's. Since

$$PA = H \tag{45}$$

and we are assuming that A is nonsingular, we can multiply (45) on the right by A^{-1} to obtain

$$P = HA^{-1}. \tag{46}$$

Thus if the last row of H consists of 0's, the same is true of HA^{-1}, and hence the last row of the matrix P consists entirely of 0's. (Why?) But this cannot be the case for a nonsingular matrix P, for if X is any matrix, then PX also has a zero last row and hence there does not exist an X such that $PX = I_m$, i.e., P cannot have an inverse. This conflict tells us that if A is nonsingular, then $H = I_m$, and the equation (46) becomes

$$\begin{aligned} E_kE_{k-1}\cdots E_1 &= P \\ &= HA^{-1} \\ &= I_mA^{-1} \\ &= A^{-1}. \end{aligned} \tag{47}$$

In other words, the inverse of A is precisely the product $E_kE_{k-1}\cdots E_1$.

On the other hand, if $H = I_m$, then (45) becomes

$$PA = I_m.$$

But from (44), $A = P^{-1}$ when $H = I_m$, and hence

$$\begin{aligned} AP &= P^{-1}P \\ &= I_m. \end{aligned}$$

It follows that A is nonsingular; the proof is complete. ▌

We take a closer look at the implications of equation (47),

$$A^{-1} = E_kE_{k-1}\cdots E_1I_m, \tag{48}$$

where we have added the factor I_m on the right in (48). (Of course, this does not change the value of the product.) However, according to Theorem 7.4, the right side of (48) is the matrix that results from I_m by performing the elementary row operations that correspond to the elementary matrices $E_1, E_2, \ldots, E_k$, in that order. In other words, if A has an inverse, then the same sequence of elementary row operations that reduce A to its Hermite normal form (which happens to be I_m) will reduce I_m to A^{-1}. This provides us with a very neat computational method for obtaining A^{-1}.

Example 7.6 Determine whether A has an inverse, and if so, find it:

$$A = \begin{bmatrix} 3 & 1 & 1 & 1 \\ 1 & 3 & 1 & 1 \\ 1 & 1 & 3 & 1 \\ 1 & 1 & 1 & 3 \end{bmatrix}.$$

We first adjoin I_4 to A to obtain a 4×8 matrix:

$$B = \left[\begin{array}{cccc|cccc} 3 & 1 & 1 & 1 & 1 & 0 & 0 & 0 \\ 1 & 3 & 1 & 1 & 0 & 1 & 0 & 0 \\ 1 & 1 & 3 & 1 & 0 & 0 & 1 & 0 \\ 1 & 1 & 1 & 3 & 0 & 0 & 0 & 1 \end{array}\right].$$

Then, according to our previous discussion, when we perform a sequence of elementary row operations on B that will reduce A to its Hermite normal form H, the submatrix of B consisting of the last four columns must be A^{-1} (if it exists). If A^{-1} does not exist, H will have a zero last row. Perform the following sequence of elementary row operations on B.

(a) Interchange rows 1 and 4.

(b) Add -1 times row 1 to rows 2 and 3 and then add -3 times row 1 to row 4.

(c) Multiply rows 2, 3, and 4 by $\frac{1}{2}$.

(d) Add row 2 to row 4 and add -1 times row 2 to row 1.

(e) Add row 3 to row 4 and add -1 times row 3 to row 1.

(f) Multiply row 4 by $-\frac{1}{6}$.

(g) Add row 4 to rows 2 and 3 and add -5 times row 4 to row 1.

The resulting matrix is

$$\left[\begin{array}{cccc|cccc} 1 & 0 & 0 & 0 & \frac{5}{12} & -\frac{1}{12} & -\frac{1}{12} & -\frac{1}{12} \\ 0 & 1 & 0 & 0 & -\frac{1}{12} & \frac{5}{12} & -\frac{1}{12} & -\frac{1}{12} \\ 0 & 0 & 1 & 0 & -\frac{1}{12} & -\frac{1}{12} & \frac{5}{12} & -\frac{1}{12} \\ 0 & 0 & 0 & 1 & -\frac{1}{12} & -\frac{1}{12} & -\frac{1}{12} & \frac{5}{12} \end{array}\right].$$

Hence the Hermite normal form of A is I_4 and

$$A^{-1} = \begin{bmatrix} \frac{5}{12} & -\frac{1}{12} & -\frac{1}{12} & -\frac{1}{12} \\ -\frac{1}{12} & \frac{5}{12} & -\frac{1}{12} & -\frac{1}{12} \\ -\frac{1}{12} & -\frac{1}{12} & \frac{5}{12} & -\frac{1}{12} \\ -\frac{1}{12} & -\frac{1}{12} & -\frac{1}{12} & \frac{5}{12} \end{bmatrix} \tag{49}$$

$$= \frac{1}{12}\begin{bmatrix} 5 & -1 & -1 & -1 \\ -1 & 5 & -1 & -1 \\ -1 & -1 & 5 & -1 \\ -1 & -1 & -1 & 5 \end{bmatrix}$$

$$= \frac{1}{12}(6I_4 - J_4).$$

Of course, A is similar in form to (49):

$$A = 2I_4 + J_4. \tag{50}$$

We check that

$$\begin{aligned}(2I_4 + J_4)[\tfrac{1}{12}(6I_4 - J_4)] &= \tfrac{1}{12}(12I_4 + 4J_4 - J_4^2)\\ &= \tfrac{1}{12}(12I_4 + 4J_4 - 4J_4)\\ &= I_4.\end{aligned}$$

Similarly, we confirm that

$$[\tfrac{1}{12}(6I_4 - J_4)](2I_4 + J_4) = I_4.$$

Quiz

Answer true or false.

1 The homogeneous system associated with

$$\begin{aligned}2x_1 + x_3 &= 1,\\ x_1 + x_2 &= 5\end{aligned}$$

is $Ax = 0$, where

$$A = \begin{bmatrix} 2 & 1 \\ 1 & 2 \end{bmatrix}.$$

2 Any homogeneous system of linear equations has a solution.

3 The only solution to the system of equations

$$\begin{bmatrix} 1 & 1 & 1 \\ 1 & 1 & 1 \\ 1 & 1 & 1 \end{bmatrix} x = (1, 1, 1)$$

is $x = (\frac{1}{3}, \frac{1}{3}, \frac{1}{3})$.

4 The Hermite normal form for the matrix

$$\begin{bmatrix} 1 & 1 & 1 \\ 1 & 1 & 0 \\ 0 & 0 & 1 \end{bmatrix}$$

is I_3.

5 The Hermite normal form for the n-square matrix J_n, each of whose entries is 1, is the matrix with (1, 1) entry 1 and every other entry equal to 0.

6 The matrices

$$\begin{bmatrix} 1 & 1 \\ 1 & -1 \end{bmatrix}$$

and

$$\begin{bmatrix} 1 & 0 \\ 0 & -1 \end{bmatrix}$$

are row equivalent.

7 Any matrix is row equivalent to itself.

8 If A is row equivalent to B and B is row equivalent to C, then A is row equivalent to C.

9 If A and B are 2×2 nonsingular matrices, then $A^2 + B^2$ is a nonsingular matrix.

†10 The inverse of

$$\begin{bmatrix} -1 & 1 \\ 1 & 1 \end{bmatrix}$$

is

$$-\tfrac{1}{2}\begin{bmatrix} 1 & -1 \\ -1 & -1 \end{bmatrix}.$$

Exercises

1 Solve (if possible) each of the following systems of linear equations by reducing the coefficient matrix to Hermite normal form:

(a)
$$\begin{aligned} x_1 + 7x_2 - 3x_3 &= 2, \\ x_1 - 2x_2 - x_3 &= 1, \\ x_1 - x_2 + x_3 &= 0, \\ 2x_1 + 2x_2 - 2x_3 &= 3; \end{aligned}$$

(b)
$$\begin{aligned} 2x_1 + x_2 + x_3 + x_4 &= 0, \\ 4x_1 - x_2 + 5x_3 + 3x_4 &= 1, \\ x_1 - x_2 + 2x_3 + x_4 &= 0, \\ 5x_1 - 2x_2 + 6x_3 + 2x_4 &= 3; \end{aligned}$$

(c)
$$\begin{aligned} 3x_1 + 2x_2 - x_3 &= 0, \\ x_2 + x_3 &= 0, \\ x_1 + 4x_2 + 7x_3 + 2x_4 + 3x_5 &= 0, \\ 5x_1 + 3x_2 - x_3 + x_4 + 2x_5 &= 0, \\ 6x_1 + x_2 - 2x_3 + 3x_4 + 6x_5 &= 0; \end{aligned}$$

(d)
$$\begin{aligned} 3x_1 - 2x_2 + x_3 - 6x_4 &= -4, \\ 3x_1 + 4x_2 + 3x_3 + x_4 &= 1, \\ x_1 - 6x_2 + 3x_3 - x_4 &= 1; \end{aligned}$$

(e)
$$\begin{aligned} 3x_1 + 2x_2 - 4x_3 - 3x_4 - 9x_5 &= 7, \\ 3x_1 + 3x_4 + 9x_5 &= 10; \end{aligned}$$

(f)
$$\begin{aligned} x_1 + 2x_2 - 2x_3 + x_4 - x_5 &= -1, \\ 2x_1 + 3x_2 + x_3 - 4x_4 + 2x_5 &= -1, \\ 3x_1 + 5x_2 - x_3 + x_5 &= -1, \\ x_1 + x_2 + 3x_3 - 2x_4 + 3x_5 &= -1. \end{aligned}$$

2 Let $d = (1, 1, 1, 1, 1)$. Find all vectors y which satisfy $My = \mathrm{d}$, where

$$M = \begin{bmatrix} 1 & 1 & 1 & 0 & 0 & 0 \\ 0 & 0 & 0 & 1 & 1 & 1 \\ 1 & 0 & 0 & 1 & 0 & 0 \\ 0 & 1 & 0 & 0 & 1 & 0 \\ 0 & 0 & 1 & 0 & 0 & 1 \end{bmatrix}.$$

Compare with Example 6.8.

3 Let P be the n-square matrix whose $(1, 2), (2, 3), (3, 4), \ldots, (n - 1, n)$ entries are all 1 and whose remaining entries are 0. Find $(I_n + P)^{-1}$.

4 Find the inverses of each of the following matrices:

(a) $\begin{bmatrix} 1 & -2 & 0 & 3 \\ -1 & 3 & 1 & 2 \\ 2 & -4 & -1 & -1 \\ 3 & -3 & 0 & 4 \end{bmatrix}$;

(b) $\begin{bmatrix} -2 & 1 & 0 & 1 \\ 1 & 0 & 2 & -1 \\ -4 & 1 & -3 & 1 \\ -1 & 0 & -2 & 2 \end{bmatrix}$;

(c) $\begin{bmatrix} -1 & -1 & -1 & 1 \\ -1 & -5 & -5 & 2 \\ 2 & 21 & 20 & -7 \\ 0 & -3 & -3 & 1 \end{bmatrix}$;

(d) $\begin{bmatrix} 2 & 1 & 1 & 1 & 1 \\ 1 & 2 & 1 & 1 & 1 \\ 1 & 1 & 2 & 1 & 1 \\ 1 & 1 & 1 & 2 & 1 \\ 1 & 1 & 1 & 1 & 2 \end{bmatrix}$.

5 Find all 3-vectors x such that $Ax = x$, where $A = I_3 + \frac{1}{2}P + \frac{1}{2}P^2$ and

$$P = \begin{bmatrix} 0 & 1 & 0 \\ 0 & 0 & 1 \\ 1 & 0 & 0 \end{bmatrix}.$$

6 Let A be an $m \times n$ row stochastic matrix and let e be the m-vector $e = (1, 1, \ldots, 1)$. (Recall that if A is a row stochastic matrix, then each row sum is 1, i.e., $\sum_{j=1}^{n} a_{ij} = 1$, $i = 1, \ldots, m$.) Show that $Ax = e$ always has a solution.

7 Solve each of the following systems of equations, using the results of Exercise 4:

(a)
$$\begin{aligned} x_1 - 2x_2 + 3x_4 &= 1, \\ -x_1 + 3x_2 + x_3 + 2x_4 &= 0, \\ 2x_1 - 4x_2 - x_3 - x_4 &= -1, \\ 3x_1 - 3x_2 + 4x_4 &= 0; \end{aligned}$$

(b)
$$\begin{aligned} -2x_1 + x_2 + x_4 &= 1, \\ x_1 + 2x_3 - x_4 &= 1, \\ -4x_1 + x_2 - 3x_3 + x_4 &= 1, \\ -x_1 - 2x_3 + 2x_4 &= 1; \end{aligned}$$

(c)
$$\begin{aligned} -x_1 - x_2 - x_3 + x_4 &= 2, \\ -x_1 - 5x_2 - 5x_3 + 2x_4 &= 0, \\ 2x_1 + 21x_2 + 20x_3 - 7x_4 &= -2, \\ -3x_2 - 3x_3 + x_4 &= -1; \end{aligned}$$

(d) $$\begin{aligned} 2x_1 + x_2 + x_3 + x_4 + x_5 &= 0,\\ x_1 + 2x_2 + x_3 + x_4 + x_5 &= 0,\\ x_1 + x_2 + 2x_3 + x_4 + x_5 &= 0,\\ x_1 + x_2 + x_3 + 2x_4 + x_5 &= 0,\\ x_1 + x_2 + x_3 + x_4 + 2x_5 &= 1. \end{aligned}$$

8 Find the inverses of each of the following elementary matrices:

(a) $\begin{bmatrix} 0 & 1 & 0 \\ 0 & 0 & 1 \\ 1 & 0 & 0 \end{bmatrix}$; (b) $\begin{bmatrix} 2 & 0 & 0 \\ 0 & 1 & 0 \\ 0 & 0 & 1 \end{bmatrix}$;

(c) $\begin{bmatrix} 1 & 0 & 0 & 0 \\ 0 & 1 & 0 & 0 \\ 0 & 0 & 1 & -4 \\ 0 & 0 & 0 & 1 \end{bmatrix}$; (d) $\begin{bmatrix} 1 & 0 & 0 & 0 \\ 0 & 0 & 0 & 1 \\ 0 & 0 & 1 & 0 \\ 0 & 1 & 0 & 0 \end{bmatrix}$.

9 For each of the matrices in Exercise 8, describe the corresponding elementary row operation.

†10 Prove that if A is an $n \times n$ nonsingular matrix, then A can be written as a product of elementary matrices. (Hint: Look carefully at the proof of Theorem 7.6.)

11 Express each of the following nonsingular matrices as a product of elementary matrices:

(a) $\begin{bmatrix} 1 & 2 \\ 0 & -1 \end{bmatrix}$;

(b) $\begin{bmatrix} 1 & 1 \\ -1 & 1 \end{bmatrix}$;

(c) $\begin{bmatrix} 2 & 0 & 1 \\ 0 & 3 & 1 \\ 1 & 0 & 0 \end{bmatrix}$;

(d) $-I_n$;

(e) $\operatorname{diag}(c_1, \ldots, c_n)$, where $c_i \neq 0$, $i = 1, \ldots, n$.

12 Prove that the transpose of an elementary matrix is an elementary matrix.

13 Let A be a nonsingular matrix. Show by induction on p that

$$(A^{-1})^p = (A^p)^{-1},$$

for any positive integer p.

†14 Let A and B be $n \times n$ matrices. Show that AB is nonsingular if and only if both A and B are nonsingular.

†15 A *lower triangular* n-square matrix has the property that every entry above the main diagonal is zero, i.e., $a_{ij} = 0$ whenever $i < j$. Prove that if $\prod_{i=1}^{n} a_{ii} \neq 0$ and A is lower triangular, then A is nonsingular. (Hint: Show that the Hermite normal form of A is I_n by indicating the appropriate sequence of elementary row operations. What must the main diagonal entries of A^{-1} be? Note that the sum and product of two lower triangular matrices are again lower triangular.)

†**16** An *upper triangular* n-square matrix has the property that every entry below the main diagonal is zero, i.e., $a_{ij} = 0$ whenever $i > j$. Show that if $\prod_{i=1}^{n} a_{ii} \neq 0$ and A is upper triangular, then A is nonsingular. (Hint: Use Theorem 7.3 (b) and Exercise 15.)

***17** Show that if A is an $n \times n$ nonsingular lower (upper) triangular matrix, then A is a product of elementary matrices, each of which is lower (upper) triangular.

18 Find the inverses of each of the following matrices using the results of Exercises 15 and 16:

(a) $\begin{bmatrix} 1 & 3 \\ 0 & 5 \end{bmatrix}$; (b) $\begin{bmatrix} 6 & 0 \\ 1 & -1 \end{bmatrix}$;

(c) $\begin{bmatrix} 1 & 3 & 1 \\ 0 & -1 & 1 \\ 0 & 0 & 4 \end{bmatrix}$; (d) $\begin{bmatrix} -5 & 0 & 0 \\ 2 & 2 & 0 \\ 3 & -4 & 4 \end{bmatrix}$;

(e) $\begin{bmatrix} 2 & 3 & 3 & 2 \\ 0 & 1 & -3 & 4 \\ 0 & 0 & 6 & 5 \\ 0 & 0 & 0 & 1 \end{bmatrix}$; (f) $\begin{bmatrix} 1 & 1 & 0 & 0 & 0 \\ 0 & 1 & 1 & 0 & 0 \\ 0 & 0 & 1 & 1 & 0 \\ 0 & 0 & 0 & 1 & 1 \\ 0 & 0 & 0 & 0 & 1 \end{bmatrix}$.

19 Let X be an n-element set and let f be an n-permutation of X. Show that the incidence matrix for f, $A(f)$, is nonsingular and prove that

$$(A(f))^{-1} = A(f^{-1}).$$

(Hint: See Theorem 3.2, Section 2.3.)

20 Let A be an $m \times m$ nonsingular matrix. Show that if B is an $m \times n$ matrix and the jth column of AB is the zero m-vector, then the jth column of B must also be the zero m-vector. (Hint: Recall that the jth column of AB is $AB^{(j)}$. Thus if $AB^{(j)} = 0$, then $B^{(j)} = 0$. Why?)

***21** Let A and B be arbitrary matrices for which the product AB is defined. Show that the ith row of AB is the product $A_{(i)}B$ where, as usual, $A_{(i)}$ is the ith row of A, thought of as a matrix with one row. Using this result, show that if B is nonsingular, then $\rho(AB) = \rho(A)$, where $\rho(A)$ denotes the rank of A. (Hint: Recall the definition of ρ given in Exercise 11, Section 2.6. Suppose that some set of k rows of A are linearly independent, say rows $A_{(1)}, \ldots, A_{(k)}$. Then the corresponding rows of AB are $A_{(1)}B, \ldots, A_{(k)}B$. Suppose that $c_1, \ldots, c_k$ are numbers such that

$$\sum_{i=1}^{k} c_i A_{(i)} B = 0.$$

Then

$$\left(\sum_{i=1}^{k} c_i A_{(i)}\right) B = 0,$$

and hence (Why?)

$$\sum_{i=1}^{k} c_i A_{(i)} = 0.$$

But then the linear independence of $A_{(1)}, \ldots, A_{(k)}$ implies that $c_1 = c_2 = \cdots = c_k = 0$. (See Exercise 7, Section 2.6.) Thus for any set of k rows of A that are linearly independent, the corresponding set of k rows of AB are linearly independent. It follows that $\rho(AB) \geq \rho(A)$. Now apply this inequality with the role of A being taken by AB and the role of B being taken by B^{-1}.)

†22 Define an *elementary column operation* precisely as in Definition 6.2, replacing each occurrence of the word "row" by the word "column." Define *column equivalence* of two matrices in precisely the same way.

(a) Show that performing an elementary column operation on the identity matrix results in an elementary matrix as defined in Definition 7.2. (Hint: Observe that an elementary column operation on A is the same as an elementary row operation on A^T.)

(b) Show that if A is an $m \times n$ matrix and E is an n-square elementary matrix, then AE is the matrix obtained from A by performing the elementary column operation on A which corresponds to E. (Column operations must be achieved by multiplication on the right.)

†23 Show that if A is an $m \times n$ matrix and $\rho(A) = r$, then A can be reduced by a sequence of elementary row and column operations to an $m \times n$ matrix D satisfying $d_{ii} = 1$, $i = 1, \ldots, r$, and $d_{ij} = 0$ for all other values of i and j.

†24 Show that in the notation of Exercise 23 there exist nonsingular m-square and n-square matrices P and Q such that

$$PAQ = D.$$

(Hint: Each row and column operation in Exercise 23 can be achieved by multiplication on the left and right by appropriate elementary matrices.)

25 In the notation of Exercise 23, reduce each of the following matrices A to the form D and find $\rho(A)$:

(a) $\begin{bmatrix} 1 & 0 & 2 & 3 & 4 \\ 0 & 1 & -1 & 2 & 3 \\ 1 & 1 & 1 & 5 & 7 \end{bmatrix}$;

(b) $\begin{bmatrix} 2 & 1 & 1 & 1 & 1 \\ 1 & 2 & 1 & 1 & 1 \\ 1 & 1 & 2 & 1 & 1 \\ 1 & 1 & 1 & 2 & 1 \\ 1 & 1 & 1 & 1 & 2 \end{bmatrix}$;

(c) $\begin{bmatrix} 1 & 1 & 1 & 1 \\ 1 & 1 & 1 & 1 \\ 1 & 1 & 1 & 1 \\ 1 & 1 & 1 & 1 \end{bmatrix}$;

(d) $\begin{bmatrix} 1 & 3 & 6 & 2 & 5 & 9 \\ 0 & 1 & -7 & 4 & -6 & 5 \\ 0 & 0 & 1 & 1 & -1 & 2 \\ 0 & 0 & 0 & 1 & 10 & 5 \\ 0 & 0 & 0 & 0 & 1 & -8 \\ 0 & 0 & 0 & 0 & 0 & 1 \end{bmatrix}$.

convexity

chapter 3

3.1

Introduction to Linear Programming

Linear programming has to do with the problem of optimization. For example, in running a business in which the operations can be performed in several alternative ways, each at different costs, it is reasonable and desirable to find a way of operating at a minimum cost. Problems of this kind are easy to construct.

Example 1.1 Suppose that the Viber Automobile Company owns two manufacturing plants, I and II, both of which produce three types of cars: the Viber 4, the Viber 6, and the Viber 8. The following table displays the daily production figures of the two plants.

	V4	V6	V8
I	10	30	50
II	20	20	20

In a six-month period, the company requires at least 800 units of the $V4$, 1600 units of the $V6$, and 2000 units of the $V8$ in order to meet the demands of its distributors. Moreover, the daily cost of operation of each of the plants is \$20,000. The question is this:

how many days should each of the plants operate so as to minimize costs? Let x_1 and x_2 denote the number of days that plants I and II operate, respectively. Then, of course, $x_1 \geq 0$, $x_2 \geq 0$, and the production requirements become

$$\begin{aligned} 10x_1 + 20x_2 &\geq 800, \\ 30x_1 + 20x_2 &\geq 1600, \\ 50x_1 + 20x_2 &\geq 2000. \end{aligned} \tag{1}$$

The cost of operating the two plants is

$$f(x) = 20{,}000x_1 + 20{,}000x_2, \tag{2}$$

where x is the 2-vector $x = (x_1, x_2)$. In other words, the problem is to minimize the function f, subject to the restrictions in (1). The inequalities in (1) define a region in the first quadrant (i.e., $x_1 \geq 0$, $x_2 \geq 0$) of the x_1, x_2-plane, namely, that region consisting of all 2-vectors x which satisfy (1). We can simplify the inequalities in (1) by dividing each of them by 10 to obtain an "equivalent" system of inequalities:

$$\begin{aligned} x_1 + 2x_2 &\geq 80, \\ 3x_1 + 2x_2 &\geq 160, \\ 5x_1 + 2x_2 &\geq 200. \end{aligned} \tag{3}$$

The function f can also be modified by dividing by 20,000, i.e., it is clear that the minimum of f will be known as soon as we know the minimum of the function

$$g(x) = x_1 + x_2. \tag{4}$$

We can graph the region defined by the inequalities in (3). Consider the line $x_1 + 2x_2 = 80$. This can be rewritten as

$$x_2 = 40 - \tfrac{1}{2}x_1. \tag{5}$$

Obviously $x_1 + 2x_2 \geq 80$ is equivalent to

$$x_2 \geq 40 - \tfrac{1}{2}x_1. \tag{6}$$

The line (5) has the following graph.

(7)

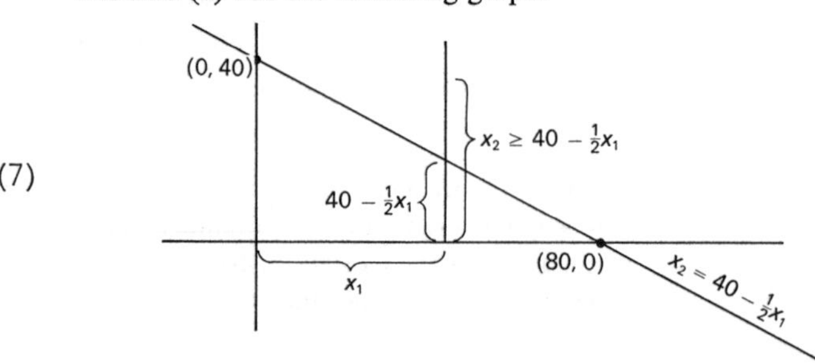

Thus the points defined by (6) are precisely those 2-vectors $x = (x_1, x_2)$ which lie above the line in (7). Since both x_1 and x_2 are restricted to be nonnegative, we are only interested in those points above the line with nonnegative components, i.e., the intersection of the area above the line with the first quadrant. If we similarly graph the second and third lines in (3), we will again obtain certain regions in the first quadrant of the plane. Since all three inequalities must be satisfied, the required region will be the intersection of the three regions bounded by each of the lines separately. The picture

(8)

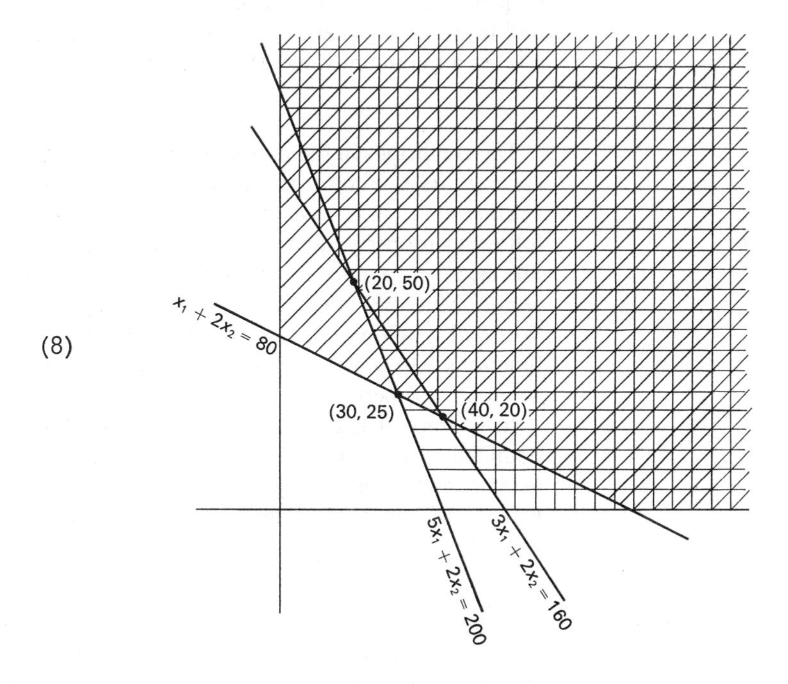

in (8) tells us that this crosshatched region will consist of all vectors in the unbounded region above or on all three lines and the positive x_1 and x_2 axes. In the present and subsequent sections we will develop the necessary methods for minimizing the function g in (4) for those 2-vectors which lie in the indicated region. However, an ad hoc argument can be given for this problem.

Consider an arbitrary vector $a = (a_1, a_2)$ in the region. The first two inequalities in (3) must be satisfied by a:

$$a_1 + 2a_2 \geq 80,$$
$$3a_1 + 2a_2 \geq 160.$$

Adding these two inequalities, we have

$$4(a_1 + a_2) \geq 240,$$

which is equivalent to

$$a_1 + a_2 \geq 60.$$

Now, $g(a) = a_1 + a_2$ and $g((40, 20)) = 40 + 20 = 60$. Therefore if a is any point in the region, the value of g at a exceeds (or equals) the value of g at the point (40, 20). We can conclude that the function g (and hence the original function f) assumes its minimum value at the point (40, 20). This means that plant I should operate 40 days and plant II should operate 20 days in order that the cost of producing the required units be minimum. The student might find it interesting that (40, 20) is the only point in the region where g takes a minimum value. (See Section 3.3.)

The conditions in the preceding example imply that the function f achieves its minimum value on a "corner" of the region in (8). Furthermore, it is clear that the region in diagram (8) has the following property: given any two points in the region, the entire line segment joining the two points is completely contained in the region. Moreover, the region is bounded by straight line segments. We shall develop general methods for minimizing and maximizing functions like g when they are defined over regions having such geometric properties.

In Example 1.1, the inequalities (3) defined a region in the plane bounded by straight lines. It is important for us to have systematic techniques for describing straight lines, not only in R^2 but in R^n, $n > 2$. We motivate the formal definition with the following example.

Example 1.2 As we saw in Section 2.1, if u is a 2-vector and t is a number, then tu is a vector which points in the same direction as u if t is positive, and in the opposite direction if t is negative. Moreover, the length of tu can be obtained from the length of u by multiplying the length of u by the numerical value of t.

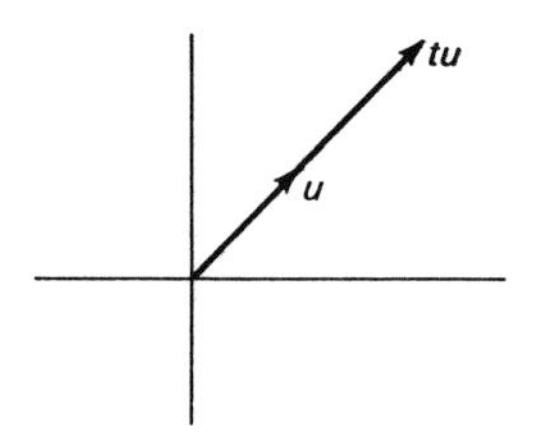

Before proceeding, the student should review the material on vectors in Section 2.1. Let a be a vector in R^2. Then for any value of the number t, the sum of the two vectors a and tu can be obtained by forming the diagonal of the parallelogram constructed by translating tu parallel to itself until its initial point coincides with the endpoint of a. As t varies, it is clear that the "head" of the vector

(9)

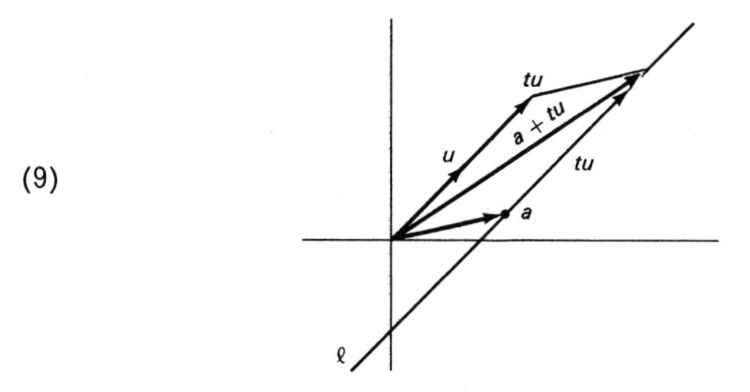

$a + tu$ hits every point on the line ℓ through the point a in the direction of u. This elementary geometric example in R^2 suggests the following general definition.

Definition 1.1 ***Straight line*** Let u and a be fixed vectors in R^n, $u \neq 0$. Then the *line through a in the direction of u* is the totality of points x of the form

$$x = a + tu \tag{10}$$

as t takes on all real values. The formula (10) is called the *parametric equation* of the line. (The variable t is called the *parameter*.)

Example 1.3 Find the parametric equation of a line in R^2 which goes through the points (1, 1) and (3, 4). Consider this diagram.

(11)

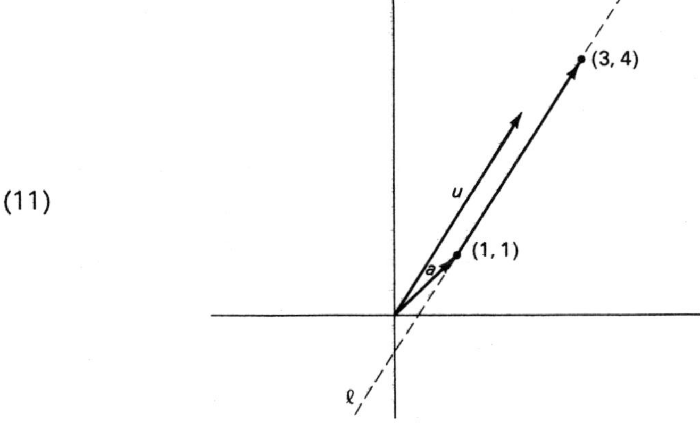

Let a be the vector (1, 1). It is clear from the diagram (11) that the required vector $u = (u_1, u_2)$ must have the property that $u + a = (3, 4)$, or equivalently that its coordinates u_1 and u_2 must satisfy the equations

$$\begin{aligned} u_1 + 1 &= 3 \\ u_2 + 1 &= 4, \end{aligned}$$

that is,

$$\begin{aligned} u_1 &= 3 - 1 = 2 \\ u_2 &= 4 - 1 = 3. \end{aligned}$$

Thus $u = (2, 3)$ and therefore any point x on the line ℓ can be denoted

$$\begin{aligned} x &= (1, 1) + t(2, 3) \\ &= a + tu. \end{aligned}$$

Observe that if $t = 0$ then $x = a$, and if $t = 1$ then $x = a + u = a + [(3, 4) - a] = (3, 4)$. Hence the line ℓ does indeed go through the points (1, 1) and (3, 4).

Definition 1.2 ***Standard inner product*** Let u and v be vectors in R^n, $u = (u_1, \ldots, u_n)$ and $v = (v_1, \ldots, v_n)$. Then the (*standard*) *inner product* of u and v is the number

$$\sum_{i=1}^{n} u_i v_i. \tag{12}$$

The inner product is denoted by (u, v).

A very important property of the inner product is due to the fact that it is simply a sum of numbers. Suppose $w = (w_1, \ldots, w_n) \in R^n$. Then

$$\begin{aligned} (u + w, v) &= \sum_{i=1}^{n} (u_i + w_i)v_i \\ &= \sum_{i=1}^{n} (u_i v_i + w_i v_i) \\ &= \sum_{i=1}^{n} u_i v_i + \sum_{i=1}^{n} w_i v_i \\ &= (u, v) + (w, v). \end{aligned}$$

A similar computation yields

$$(u, v + w) = (u, v) + (u, w).$$

Also, if α is a number, we can check that

$$\begin{aligned}(\alpha u, v) &= \sum_{i=1}^{n} (\alpha u_i) \cdot v_i \\ &= \sum_{i=1}^{n} \alpha(u_i v_i) \\ &= \alpha \sum_{i=1}^{n} u_i v_i \\ &= \alpha(u, v),\end{aligned}$$

and that

$$(u, \alpha v) = \alpha(u, v).$$

If $u = (u_1, u_2)$ is a vector in R^2, then $(u, u)^{1/2} = (u_1 \cdot u_1 + u_2 \cdot u_2)^{1/2} = (u_1^2 + u_2^2)^{1/2}$.

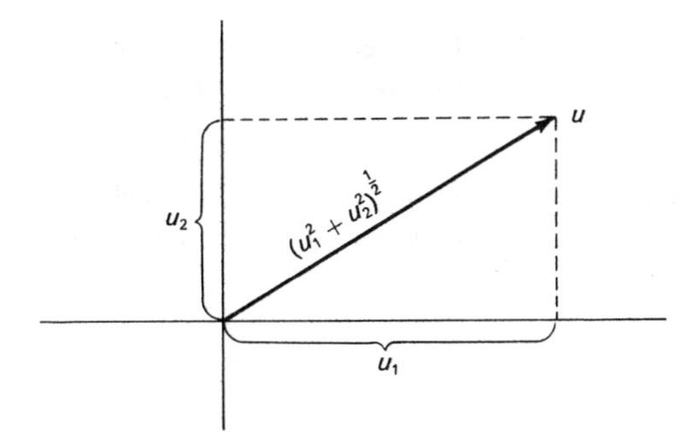

Thus $(u, u)^{1/2}$ is just the length of the vector u.

Suppose that u and v are two vectors in R^2. We know from elementary plane geometry that u and v will be perpendicular if and only if

$$(u + v, u + v) = (u, u) + (v, v). \tag{13}$$

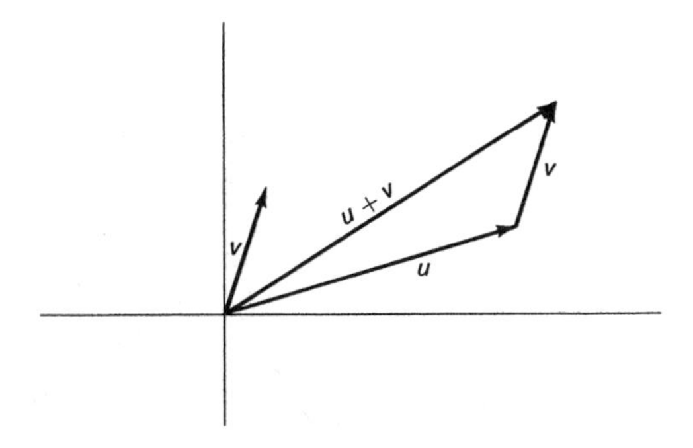

For, it is clear from the preceding diagram that equation (13) simply states that the triangle whose sides are denoted by u, v, and $u + v$ is a right triangle if and only if the sum of the squares of the lengths of the sides, $(u, u) + (v, v)$, is equal to the square of the length of the hypotenuse, $(u + v, u + v)$. If we write out (13) explicitly in terms of the components of $u = (u_1, u_2)$ and $v = (v_1, v_2)$, we have as necessary and sufficient conditions for the perpendicularity of u and v that

$$(u_1 + v_1)^2 + (u_2 + v_2)^2 = u_1^2 + u_2^2 + v_1^2 + v_2^2. \tag{14}$$

Expanding (14) and cancelling common terms from both sides, we have the equivalent formula

$$u_1v_1 + u_2v_2 = 0,$$

or

$$(u, v) = 0.$$

In other words, we see that a necessary and sufficient condition that two vectors be perpendicular is that their inner product be zero.

Definition 1.3 ***Orthogonality*** Two vectors u and v in R^n are said to be *orthogonal* or *perpendicular* if their inner product is zero:

$$(u, v) = 0. \tag{15}$$

Example 1.4

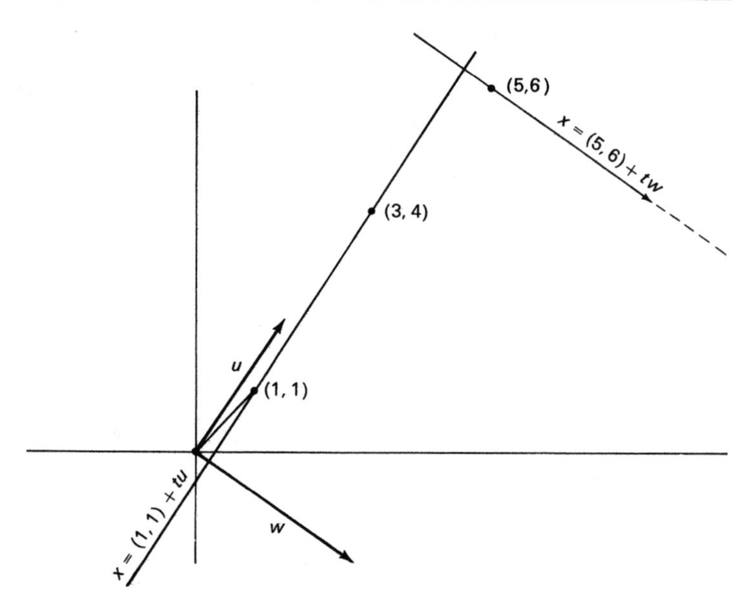

Find the equation of the line in R^2 which is orthogonal to the line through (1, 1) and (3, 4) and goes through the point (5, 6). If we refer to diagram (11), we see that the required line must go through

(5, 6) in the direction of a vector w, where w is perpendicular to $u = (3, 4) - (1, 1) = (2, 3)$. Thus any w which satisfies $(w, u) = 0$ will do, e.g., $w = (3, -2)$. Then the equation of the line is

$$x = (5, 6) + t(3, -2).$$

Example 1.5 Find the point of intersection of the two perpendicular lines in the preceding example. In order to solve this problem we must find numbers t_1 and t_2 such that

$$(1, 1) + t_1 u = (5, 6) + t_2 w. \tag{16}$$

Since $u = (2, 3)$ and $w = (3, -2)$, equation (16) becomes

$$(1 + 2t_1, 1 + 3t_1) = (5 + 3t_2, 6 - 2t_2),$$

or, equating components,

$$1 + 2t_1 = 5 + 3t_2$$

and

$$1 + 3t_1 = 6 - 2t_2.$$

We find that $t_1 = \frac{23}{13}$ and $t_2 = -\frac{2}{13}$. Hence the point of intersection is

$$\begin{aligned} (1, 1) + t_1 u &= (1, 1) + \tfrac{23}{13}(2, 3) \\ &= (\tfrac{59}{13}, \tfrac{82}{13}). \end{aligned}$$

Let u be a fixed vector in R^2 and let ℓ be a line through a point a, perpendicular to u.

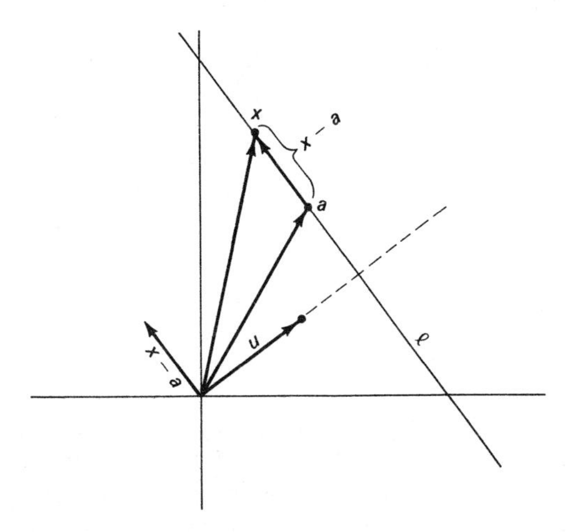

It is clear from the diagram that x lies on ℓ if and only if $x - a$ is perpendicular to u, i.e., $(x - a, u) = 0$. Using the properties of the inner product, we see that

$$\begin{aligned} 0 &= (x - a, u) \\ &= (x + (-a), u) \\ &= (x, u) + (-a, u) \\ &= (x, u) - (a, u), \end{aligned}$$

or equivalently,

$$(x, u) = (a, u).$$

If we let $(a, u) = c$, then x lies on ℓ if and only if

$$(x, u) = c. \tag{17}$$

Thus a line in R^2 can also be written in the form (17). Similarly, if it is required to find the equation of a plane Π in R^3 perpendicular to a fixed vector u and passing through a fixed point a, then Π is the totality of $x \in R^3$ that satisfy

$$(x - a, u) = 0$$

or

$$(x, u) = c, \tag{18}$$

where we have set

$$(a, u) = c.$$

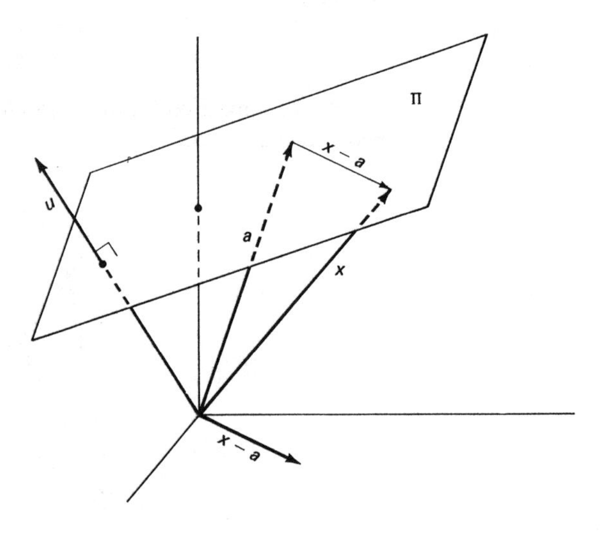

Example 1.6 Find an equation of the form (18) for the plane in R^3 which goes through the three points $(1, 0, 0)$, $(0, 1, 0)$, and $(0, 0, 1)$. Let $u = (u_1, u_2, u_3)$. Then if we successively replace the vector x in (18) by each of the three preceding points, we obtain $u_1 = u_2 = u_3 = c$, e.g., $((1, 0, 0), u) = 1 \cdot u_1 + 0 \cdot u_2 + 0 \cdot u_3 = c$. In other words,

$u = c(1, 1, 1)$. Thus (18) becomes

$$c(x_1 + x_2 + x_3) = c,$$

and, cancelling c from both sides, we see that the equation of the required plane is

$$x_1 + x_2 + x_3 = 1.$$

It is clear from the geometry that $u = (1, 1, 1)$ is indeed perpendicular to the plane.

Definition 1.4 ***Hyperplane*** Let u be a fixed vector in R^n and let c be a number. Then the totality of vectors x in R^n that satisfy

$$(u, x) = c$$

is called a *hyperplane*. The vector u is called the *normal* to the hyperplane.

Example 1.7 Find the distance from the point $a = (a_1, a_2) \in R^2$ to the line ℓ whose equation is $(x, u) = c$. The vector u is perpendicular to the line ℓ, and thus the equation of the line ℓ' through a perpendicular to ℓ is given by

$$x = a + tu.$$

Now, the point of intersection of ℓ and ℓ' is determined by that value of t for which $a + tu$ satisfies the equation of the line ℓ:

$$(a + tu, u) = c. \tag{19}$$

We check that

$$\begin{aligned}(a + tu, u) &= (a, u) + (tu, u)\\ &= (a, u) + t(u, u),\end{aligned}$$

and hence (19) becomes

$$(a, u) + t(u, u) = c$$

or

$$t = \frac{c - (a, u)}{(u, u)}.$$

Thus the point of intersection of ℓ and ℓ' is

$$z = a + \left(\frac{c - (a, u)}{(u, u)}\right) u. \tag{20}$$

The distance d from a to the point in (20) is the length of the vector

$z - a$, that is,

$$
\begin{aligned}
d^2 &= (z - a, z - a) \\
&= \left(\frac{c - (a, u)}{(u, u)} u, \frac{c - (a, u)}{(u, u)} u\right) \\
&= \frac{c - (a, u)}{(u, u)} \left(u, \frac{c - (a, u)}{(u, u)} u\right) \\
&= \left(\frac{c - (a, u)}{(u, u)}\right)^2 (u, u) \\
&= \frac{(c - (a, u))^2}{(u, u)}. \qquad (21)
\end{aligned}
$$

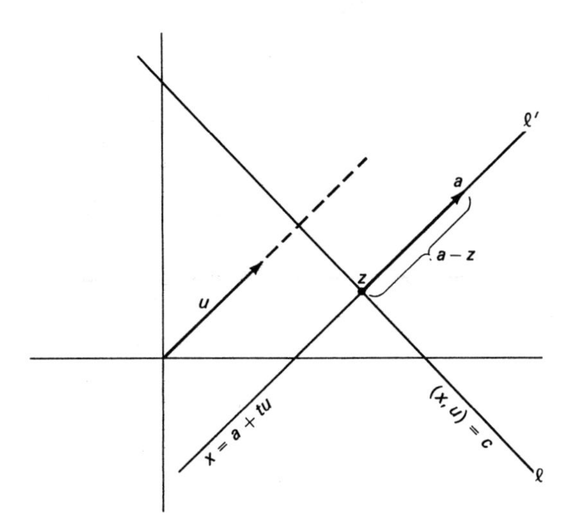

A hyperplane separates R^n into three disjoint sets whose union is all of R^n, i.e., the points on the hyperplane and those on either side of the hyperplane.

Definition 1.5 ***Half-spaces*** Let $(x, u) = c$ be the equation of a hyperplane Π in R^n. Then the set of all $x \in R^n$ such that $(x, u) < c$ is called the *open negative half-space* determined by Π. Similarly, the set of all $x \in R^n$ such that $(x, u) > c$ is called the *open positive half-space* determined by Π. If the above strict inequalities are changed to non-strict inequalities, i.e., $(x, u) \leq c$ and $(x, u) \geq c$, then the half-spaces are called *closed negative* and *closed positive*, respectively.

We want to show that geometrically, a half-space for a hyperplane Π corresponds to the set of all points in R^n which lie on "one side" of Π. In order to confirm this for R^2 and R^3, we require the general notion of a *line segment* joining two points a and b in R^n. The line segment joining a and b is defined to be the set all points x of the form

$$x = (1 - t)a + tb, \tag{22}$$

where t takes on all values $0 \leq t \leq 1$. Observe that if $t = 0$ then $x = a$, and if $t = 1$ then $x = b$. The diagram (23) of a line segment joining a and b in R^2 is indicated with a solid line.

(23)

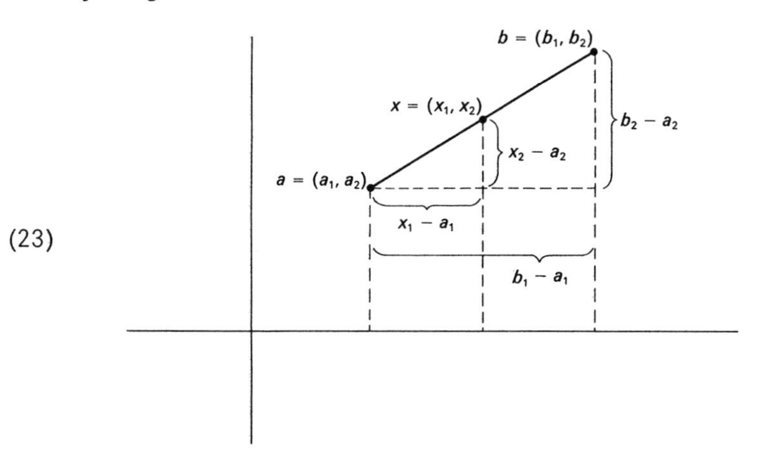

We see by similar triangles that

$$\frac{x_2 - a_2}{b_2 - a_2} = \frac{x_1 - a_1}{b_1 - a_1}. \tag{24}$$

It is clear from diagram (23) that each of the fractions in (24) is nonnegative and less than or equal to 1. If we set both of the fractions in (24) equal to t, we obtain

$$x_1 - a_1 = t(b_1 - a_1),$$
$$x_2 - a_2 = t(b_2 - a_2),$$

or

$$x_1 = tb_1 + (1 - t)a_1,$$
$$x_2 = tb_2 + (1 - t)a_2.$$

In vector notation the preceding equations become

$$x = (1 - t)a + tb,$$

which is the formula in (22). A similar argument can be made in 3 dimensions, thereby motivating our definition of a line segment in R^n.

Let Π be the hyperplane $(u, x) = c$. Suppose that $x_0 \notin \Pi$, that is,

x_0 does not satisfy the defining equation for Π. Then $(u, x_0) \neq c$, say, for example, $(u, x_0) = \alpha > c$. Now let x' be another point on the "same side" of Π as x_0. We want to show that $(u, x') > c$.

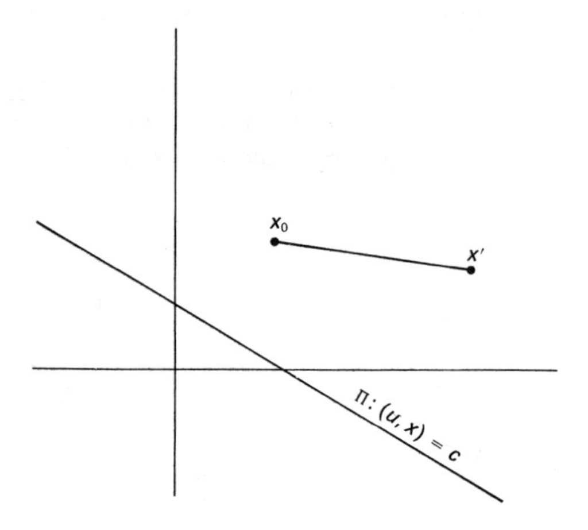

Suppose that $(u, x') = \beta < c$. It is obvious from the diagram that the entire line segment joining x_0 and x' lies on the same side of Π. First note that any point on the line segment can be written $(1 - t)x_0 + tx'$, and so

$$\begin{aligned}(u, (1 - t)x_0 + tx') &= (1 - t)(u, x_0) + t(u, x') \\ &= (1 - t)\alpha + t\beta.\end{aligned}$$

Then it is easy to see that there exists a number t satisfying $0 \leq t \leq 1$ such that $(1 - t)\alpha + t\beta = c$. For, if we solve this equation for t, we obtain

$$t = \frac{\alpha - c}{\alpha - \beta},$$

and since $\beta < c < \alpha$, it follow that $0 \leq \dfrac{\alpha - c}{\alpha - \beta} \leq 1$. In other words, we can find a point on the line segment joining x_0 and x' which lies on the hyperplane Π. But this is in conflict with our observation that the entire line segment joining x_0 and x' lies on one side of Π. Thus, in R^2 the positive open half-space is precisely the set of all points which lie above the hyperplane.

Consider a hyperplane Π orthogonal to the vector u in R^n: $(u, x) = c$. Let x_0 be the intersection point of the hyperplane and the line ℓ through the origin in the direction of u.

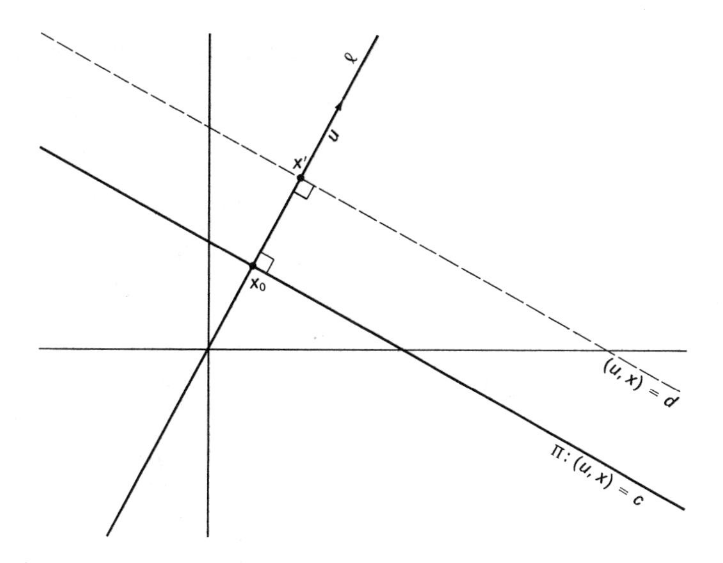

Let x' be a point on the line ℓ which satisfies $x' = \lambda x_0$, $\lambda > 1$, i.e., x' lies "farther out" than x_0 on ℓ in the direction of u. Then $(u, x') = (u, \lambda x_0) = \lambda(u, x_0) = \lambda c > c$. Thus the equation of the hyperplane through x' orthogonal to u is of the form

$$(u, x) = d,$$

where $d > c$. We can summarize these observations in a somewhat different way. Think of the hyperplane $(u, x) = c$ as being moved parallel to itself. If it is moved in the direction of u, the right side of the defining equation increases. If it is moved in the direction of $-u$, then, of course, the right side decreases. These remarks provide us with a geometric method for finding the maximum (and/or minimum) values of the function

$$f(x) = (u, x),$$

where x varies over some region in R^n bounded by hyperplanes. We illustrate this in the next example.

Example 1.8 Find the maximum and minimum values of the function

$$f(x) = 2x_1 + 7x_2,$$

as x varies over the region defined by the inequalities

$$\begin{aligned} x_1 + x_2 &\le 3, \\ x_2 - 4x_1 &\ge -7, \\ x_1 &\ge 1, \\ x_2 &\ge -1. \end{aligned} \tag{25}$$

We first graph the region defined by the preceding inequalities.

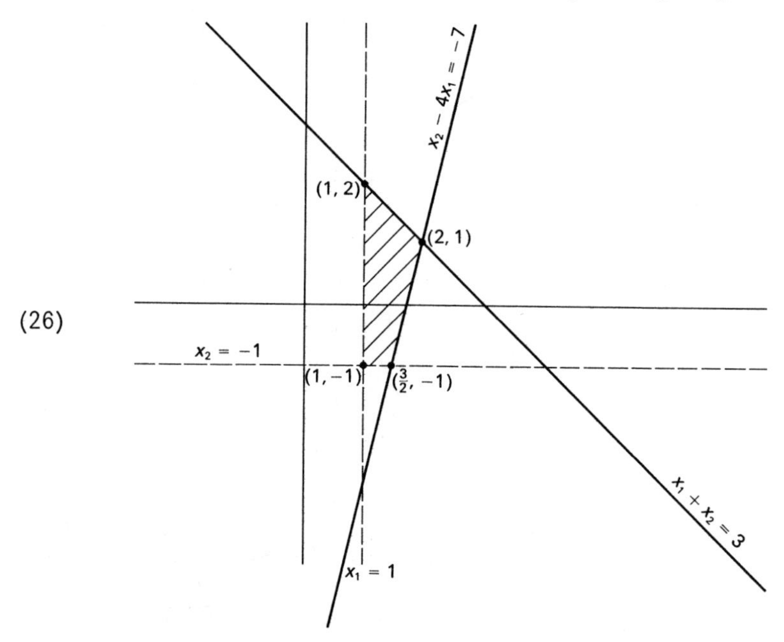

(26)

It is clear from the diagram that the inequalities in (25) define the shaded region whose vertices (i.e., corner points) are indicated. Next let $u = (2, 7)$, so that $f(x)$ can be written in the form $f(x) = (u, x)$. We reproduce the shaded region in (26) together with a typical line $(u, x) = c$.

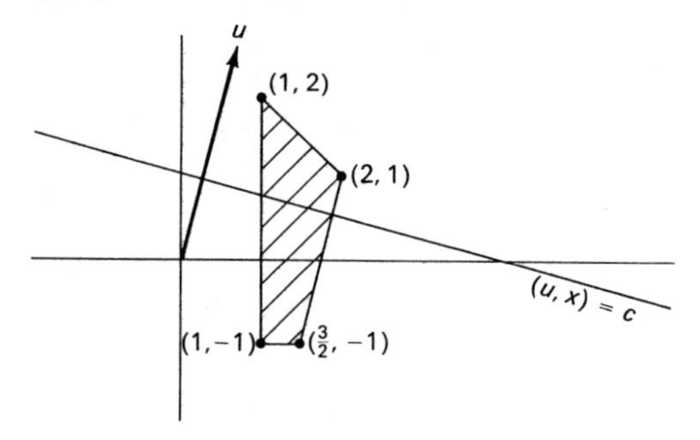

If we imagine the line $(u, x) = c$ as moving across the shaded region in the direction of u and constantly parallel to itself, we know from the discussion preceding this example that when c is largest (or smallest) then $(u, x) = c$ intersects the boundary of the region. It is obvious that this happens when $(u, x) = c$ goes through one of the four vertices. For, as $(u, x) = c$ moves across the shaded region in the direction of u, it first enters the region at a vertex and finally leaves the region at a vertex, and the value of c steadily increases as the line moves across the region. It follows that we can determine the maximum and minimum values of $f(x)$ for x in the shaded region simply by computing the values of $f(x)$ at each of the four vertices and observing the largest and smallest of these values:

$$\begin{aligned} f((1, 2)) &= 2 \cdot 1 + 7 \cdot 2 = 16, \\ f((2, 1)) &= 2 \cdot 2 + 7 \cdot 1 = 11, \\ f((1, -1)) &= 2 \cdot 1 + 7 \cdot (-1) = -5, \\ f((\tfrac{3}{2}, -1)) &= 2 \cdot \tfrac{3}{2} + 7 \cdot (-1) = -4. \end{aligned}$$

Thus the largest and smallest values of $f(x)$ are 16 and -5, respectively.

Quiz

Answer true or false.

1 If u and a are points in R^n, then the line through a in the direction of u has the equation $x = u + ta$.

2 The inner product of two non-zero vectors with nonnegative components is always positive.

3 The vectors $(2, 3)$ and $(-3, 2)$ are orthogonal in R^2.

4 If u and v are vectors in R^2, then a necessary and sufficient condition that u and v be orthogonal is that

$$(u + v, u + v) = (u, u) + (v, v).$$

5 The intersection of the two open half-spaces determined by the hyperplane Π is empty.

6 The intersection of the positive and negative closed half-spaces of Π is Π.

7 The intersection of the positive open half-space of Π and the positive closed half-space of Π is the positive open half-space of Π.

8 The intersection of the positive open half-space of Π and the negative closed half-space of Π is empty.

9 Any two non-parallel line segments in R^2 intersect.

10 The point $(1, 1)$ is on the line segment joining $(2, 2)$ and $(3, 3)$.

Exercises

1 Find the parametric equation of the line ℓ which:

(a) goes through $(-1, 1)$ and $(3, 5)$;
(b) goes through $(2, 1, 3)$ and $(-1, 1, 0)$;
(c) goes through $(1, 1, 1, 1)$ in the direction of $(-1, 2, 1, 1)$;
(d) goes through $(1, 1)$ and is perpendicular to the line in (a);

(e) goes through (1, 1, 1) and is perpendicular to the plane $x_1 + x_2 + x_3 = 1$;
(f) goes through (1, 1) and the point of intersection of the two lines $x_1 + x_2 = 0$ and $2x_1 + x_2 = -1$;
(g) goes through the origin in the direction of the normal to $2x_1 + 3x_2 + 4x_3 = 2$;
(h) goes through the point (1, 1) and lies entirely in the open positive half-space of $x_1 + x_2 = -1$;
(i) goes through the point (−5, −5) and lies entirely in the open negative half-space of $x_1 + x_2 = -1$;
(j) goes through (1, 2, 3) and is parallel to the two planes $x_1 + x_2 + x_3 = 1$ and $x_1 + x_2 - x_3 = 0$;
(k) is the intersection of the two planes in (j) (Hint: the required line ℓ is in the direction of a vector perpendicular to the normals of both planes);
(l) is parallel to the line $x_1 + x_2 = -1$ and goes through the point (10, 10).

2 Graph the regions determined by the following systems of inequalities:

(a) $x_1 + x_2 \geq -1$, $x_1 - x_2 \geq 2$, $x_2 \leq 0$;

(b) $x_2 - x_1 \leq 1$, $x_1 + x_2 \leq 1$, $x_2 \geq 0$;

(c) $x_1 + x_2 \leq 1$, $x_2 - x_1 \leq \frac{1}{2}$, $x_1 \geq 0$, $x_2 \geq 0$;

(d) $x_1 + x_2 \leq 1$, $x_1 - x_2 \geq 1$, $x_1 + x_2 \geq -1$, $x_2 - x_1 \leq 1$;

(e) $x_1 \geq 0$, $x_2 \geq 0$, $x_1 + x_2 \geq 1$;

(f) $x_2 - x_1 \geq 0$, $x_2 - 2x_1 \leq 0$, $x_1 \geq 0$, $x_2 \geq 0$;

(g) $x_2 - x_1 \geq 0$, $x_2 - 2x_1 \leq 0$, $x_1 + x_2 \leq 1$, $x_1 \geq 0$, $x_2 \geq 0$;

(h) $x_2 - x_1 \geq 0$, $x_2 - 2x_1 \leq 0$, $x_1 + x_2 \leq 1$, $x_1 + x_2 \geq \frac{1}{2}$, $x_1 \geq 0$, $x_2 \geq 0$;

(i) $x_2 - x_1 \geq 0$, $x_2 - 2x_1 \leq 0$, $x_1 + x_2 \leq 1$, $x_1 + x_2 \geq \frac{1}{2}$, $x_2 - \frac{3}{2}x_1 \geq \frac{1}{12}$, $x_1 \geq 0$, $x_2 \geq 0$.

3 (a) Find the maximum and minimum values of $f(x) = 5x_1 + 3x_2$ for $x = (x_1, x_2)$ in the region in Exercise 2(a), using the methods demonstrated in Example 1.8.
(b) Work the preceding exercise for the region in Exercise 2(b) and the function $f(x) = x_1 - \frac{1}{2}x_2$.
(c) Work the preceding exercise for the region in Exercise 2(d) and the function $f(x) = x_1$.

(d) Find the minimum value of the function $f(x) = 2x_1 + 3x_2$ on the region in Exercise 2(e).

(e) Does the function in (d) have a maximum in the region? Why?

(f) Find the minimum value of the function $f(x) = x_1 + 5x_2$ in the region in Exercise 2(f).

(g) Find the minimum and maximum values of the function $f(x) = x_1 + x_2$ in the region in Exercise 2(h).

†4 Let (u, v) denote the standard inner product of u and v in R^2.

(a) Show that $(u, u) \geq 0$, and that equality holds if and only if u is the zero vector.

(b) Show that $(u, v) = (v, u)$.

(c) Show that if w is a vector in R^2, then $(u, v + w) = (u, v) + (u, w)$.

(d) Show that if α is a number, then $(u, \alpha v) = \alpha(u, v)$.

(e) Show that $|(u, v)| \leq (u, u)^{1/2}(v, v)^{1/2}$, with equality holding if and only u and v are linearly dependent vectors. (Hint: Let $u = (u_1, u_2)$ and $v = (v_1, v_2)$; then the required inequality is equivalent to

$$(u_1v_1 + u_2v_2)^2 \leq (u_1^2 + u_2^2)(v_1^2 + v_2^2).$$

Multiplying out both sides and cancelling similar terms, we obtain

$$2(u_1v_2)(u_2v_1) \leq (u_1v_2)^2 + (u_2v_1)^2. \tag{27}$$

Now let $a = u_1v_2$ and $b = u_2v_1$. Then the inequality (27) is equivalent to

$$2ab \leq a^2 + b^2$$

or

$$a^2 + b^2 - 2ab \geq 0,$$

that is,

$$(a - b)^2 \geq 0.$$

Since a and b are real numbers, this last inequality does indeed hold, with equality if and only if $a = b$. This inequality is called the *Cauchy-Schwarz Inequality* and will be proved in general for R^n in the next section.)

5 Find the distance from each of the points to each of the lines in the following problems:

(a) $(0, 0)$, $x_1 + x_2 = 1$;

(b) $(1, 1)$, $(u, x) = 0$, where $u = (3, -2)$;

(c) the point of intersection of the lines given by $x = (2, 3) + t(1, -1)$ and $x = (1, 4) + t(2, 2)$, and the line $(u, x) = 3$, where $u = (2, \frac{1}{2})$.

6 (a) Let a and v be fixed vectors in R^2. Find the point on the line given by the parametric equation $x = a + tv$ such that the point is closest to the origin. Express it in terms of the components of a and v. (Hint: Use the result of Example 1.7 by rewriting the equation for the line in the form $(x, u) = c$.)

(b) Find the point on the line $x = (2, 3) + t(1, 1)$ which is closest to the origin.
(c) Find the point on the line $(u, x) = 3$ closest to the point $(3, -2)$, where $u = (1, \frac{1}{2})$.

7 (a) Find three lines such that the intersection of appropriate closed half-spaces of these lines is the triangle in R^2 whose vertices are $(1, -1)$, $(2, 2)$, and $(3, 1)$.
(b) Find the maximum and minimum values of the function $f(x) = 3x_1 + 5x_2 + 2$ over the triangle in part (a). (Hint: Find the maximum and minimum values of the function $g(x) = 3x_1 + 5x_2$ and add 2 to each of them.)

8 (a) Find four lines such that the intersection of appropriate closed half-spaces of these lines is the quadrilateral (four-sided polygon) in R^2 whose vertices are $(1, 1)$, $(2, -1)$, $(-1, -2)$, and $(-3, -1)$.
(b) Find the maximum and minimum values of the function $f(x) = 2x_1 - x_2 + 3$ over the quadrilateral in part (a).

9 (a) Graph the region in the plane defined by the inequalities

$$\begin{aligned} 4x_1 - x_2 &\leq 5, \\ 5x_1 - 4x_2 &\geq -2, \\ x_1 - 3x_2 &\leq 4. \end{aligned}$$

(b) Find the minimum and maximum values of the function $f(x) = 4x_1 - x_2$ over this region.
(c) Find the set of all points $x = (x_1, x_2)$ in the region that satisfy $f(x) \geq f(y)$ for all y in the region.

10 A 30-minute television program consists of a solo sitar performance, a folk singer, and commercials sponsored by a well-known sandal company. The sitar player insists on performing at least half as long as the folk singer. The sponsor requires at least 3 minutes of commercials, and the station requires the folk singing performance to last at least 4 times as long as the commercial. If the sitar player, the folk singer, and the commercial cost \$250, \$180, and \$110 per minute, respectively, what is the minimal cost of the program? (Hint: Let x_1 and x_2 be the number of minutes in the sitar performance and the folk music performance, respectively. The commercials then take $30 - (x_1 + x_2)$ minutes. The conditions can then be translated into the following inequalities:

$$\begin{aligned} x_1 &\geq \tfrac{1}{2}x_2, \\ 30 - (x_1 + x_2) &\geq 3, \\ x_2 &\geq 4(30 - (x_1 + x_2)). \end{aligned}$$

Let $f((x_1, x_2)) = 250x_1 + 180x_2 + 110(30 - (x_1 + x_2))$ and minimize $f(x)$ over the region determined by the above inequalities.)

11 A typical laboratory mouse requires 7, 10, and 4 units of vitamins x, y, and z, respectively. Two commercial food mixtures A and B contain these vitamins in the following amounts per sack: mixture A contains 5, 2, and 1 units of x, y, and z, respectively, and mixture B contains 1, 2, and 4 units, respectively, of the vitamins. Mixture A sells for \$4/sack and mixture B sells for \$3/sack. In ordering 100 sacks, how many sacks of each mixture should be purchased in order to minimize feeding costs?

12 A 100-pound batch of a mixture is to be made of ingredients x, y, and z which cost \$6, \$3, and \$4 per pound, respectively. What are the least and largest costs of the mixture if the quantity of x exceeds the quantity of y by at least 10 pounds and must not exceed the quantity of z by more than 40 pounds, and if the quantity of z cannot exceed the quantity of y?

13 (a) Show that the line segment joining (2, 3) and (5, 7) lies entirely in the closed positive half-space of the line $(u, x) = -1$, where $u = (1, 1)$.

(b) Show that the line segment joining (1, 5) and (−5, −7) intersects the line in part (a).

14 Let $(u, x) = c$ be a hyperplane in R^n. Let a and b be any two points lying in the same (a) closed positive; (b) open positive; (c) open negative; (d) closed negative half-space. Show that the line segment joining a and b is entirely contained in each of the appropriate half-spaces.

15 Let $(u, x) = c$ be a hyperplane in R^n. Let a and b be any two points lying in the (a) open positive and closed negative; (b) open negative and closed positive; (c) open negative and open positive; (d) closed negative and closed positive half-spaces, respectively. Show that the line segment joining a and b must intersect the hyperplane.

16 Find all vectors $u = (u_1, u_2, u_3)$ in R^3 which are orthogonal to the vectors (1, 0, 1) and (1, 1, −1). (Hint: If $u = (u_1, u_2, u_3)$ then the following equations must be satisfied:

$$u_1 + u_3 = 0,$$
$$u_1 + u_2 - u_3 = 0.)$$

17 Two hyperplanes are said to be *parallel* if they are orthogonal to the same vector u. Let $u = (2, 1, 3, -4)$ and find the equation of the hyperplane in R^4 parallel to the hyperplane $(u, x) = 5$ and containing the point (1, 1, 0, 1).

18 Let $u = (1, 1, 1)$, $v = (1, -1, 1)$, and $w = (-1, 1, 1)$. Show that the only vector $x = (x_1, x_2, x_3)$ in R^3 that is orthogonal to each of u, v, and w is $x = 0$.

19 (a) Let $u = (1, 1, 1)$, $v = (1, -1, 1)$, and $w = (-1, 1, 1)$. Show that the three hyperplanes $(u, x) = c_1$, $(v, x) = c_2$, and $(w, x) = c_3$ intersect in precisely one point, for any values of c_1, c_2, c_3.

(b) Find the point of intersection if $c_1 = 1$, $c_2 = 2$, and $c_3 = 0$.

20 (a) Let u and b be two vectors in R^2. Let θ be a number satisfying $0 \leq \theta \leq 1$. Find the coordinates of a point which is θ of the distance from a to b along the line segment joining a and b.

(b) Let $a = (1, 1)$ and $b = (2, 3)$. Find the equation of the line which is perpendicular to the line segment joining a and b and intersects the line segment at a point $\frac{2}{3}$ of the distance from a to b along the line segment.

3.2

Geometry in R^n

In the preceding section, we defined the inner product of two vectors $u = (u_1, \ldots, u_n)$ and $v = (v_1, \ldots, v_n)$ in R^n to be

(1)
$$(u, v) = \sum_{i=1}^{n} u_i v_i.$$

If we set $u = v$ in (1), we obtain

(2)
$$(u, u) = \sum_{i=1}^{n} u_i^2.$$

In R^2 and R^3, we know that (2) is just the square of the length of the line segment from the origin to the point (u_1, u_2).

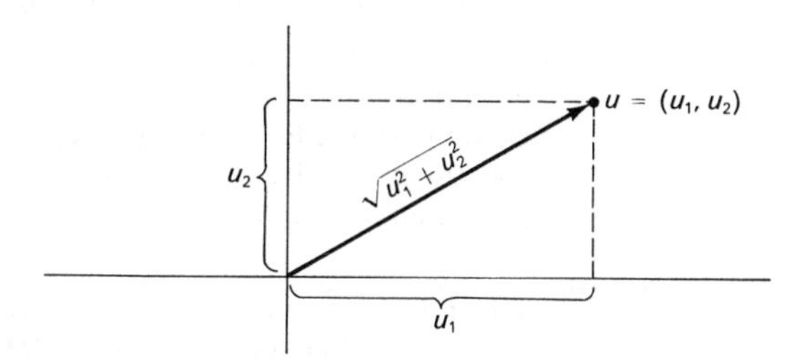

In general, we call the number $(u, u)^{1/2}$ the *length* or the *norm* of the vector u, denoted by

(3)
$$(u, u)^{1/2} = \|u\|.$$

In Exercise 4, Section 3.1, we developed some of the elementary properties of the inner product in R^2. It turns out that these are true in R^n.

Theorem 2.1 *The inner product in R^n satisfies the following properties.*

(a) *(Symmetry) For any $u, v \in R^n$, $(u, v) = (v, u)$.*

(b) *(Linearity) For any three vectors u, v, and w in R^n and numbers c and d,*

(4)
$$(cu + dv, w) = c(u, w) + d(v, w)$$

and

$$(w, cu + dv) = c(w, u) + d(w, v).$$

(c) *For any $u \in R^n$,*

$$\|u\| \geq 0,$$

and equality holds if and only if u is the zero vector.

Proof

(a) Let $u = (u_1, \ldots, u_n)$, $v = (v_1, \ldots, v_n)$, and compute their inner product:

$$\begin{aligned}(u, v) &= \sum_{i=1}^{n} u_i v_i \\ &= \sum_{i=1}^{n} v_i u_i \\ &= (v, u).\end{aligned}$$

(b) Let $w = (w_1, \ldots, w_n)$. Let u and v be as in (a). Observe that

$$\begin{aligned}(cu + dv, w) &= \sum_{i=1}^{n} (cu_i + dv_i)\, w_i \\ &= \sum_{i=1}^{n} (cu_i w_i + dv_i w_i) \\ &= \sum_{i=1}^{n} cu_i w_i + \sum_{i=1}^{n} dv_i w_i \\ &= c \sum_{i=1}^{n} u_i w_i + d \sum_{i=1}^{n} v_i w_i \\ &= c(u, w) + d(v, w).\end{aligned}$$

By symmetry,

$$c(u, w) + d(v, w) = c(w, u) + d\,(w, v).$$

(c) From (2) and (3) we know that $\|u\| \geq 0$. If

$$\begin{aligned}\|u\| &= \left(\sum_{i=1}^{n} u_i^2\right)^{1/2} \\ &= 0,\end{aligned}$$

then since the u_i^2's are nonnegative numbers, $i = 1, \ldots, n$, it follows that their sum is zero only if $u_1 = u_2 = \cdots = u_n = 0$, i.e., only if u is the zero vector. On the other hand, it is clear that if u is the zero vector, $\|u\| = 0$. ▌

The set of all vectors u in R^n which satisfy $\|u - a\| \leq r$ is called the *closed sphere of radius r with center at a*. This nomenclature is a sensible extension of our familiar circle in R^2 and sphere in R^3, for consider the number

$$\|u - a\| = \left(\sum_{i=1}^{n} (u_i - a_i)^2\right)^{1/2}. \tag{5}$$

In R^2 and R^3, the right side of (5) is just the formula for the distance from $u = (u_1, \ldots, u_n)$ to $a = (a_1, \ldots, a_n)$, for $n = 2$ or 3. Thus $\|u - a\| \leq r$ is just the assertion that the distance from u to a is no larger than r.

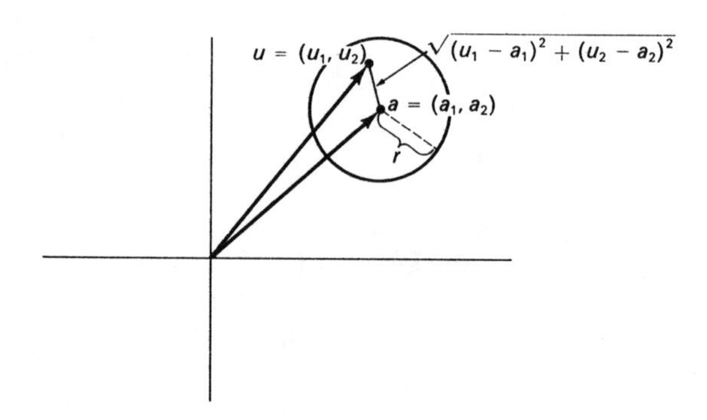

In general, the *distance* between two vectors u and v in R^n is the number

$$\|u - v\| = \left(\sum_{i=1}^{n} (u_i - v_i)^2\right)^{1/2}. \tag{6}$$

A set S in R^n is said to be *bounded* if there exists a sphere such that S is entirely contained within the sphere. Otherwise the set is said to be *unbounded.*

Example 2.1 The set of points in R^3 lying on the line ℓ whose parametric equation is $x = v + tu = (1, 0, 1) + t(1, 1, 1)$ is not bounded. For, let $a \in R^3$, $a = (a_1, a_2, a_3)$. If x is a point on ℓ, then

$$\begin{aligned} \|x - a\|^2 &= \|v + tu - a\|^2 \\ &= \|(v - a) + tu\|^2 \\ &= ((v - a) + tu, (v - a) + tu) \\ &= (v - a, v - a) + 2t(v - a, u) + t^2(u, u) \\ &= \|v - a\|^2 + 2t(v - a, u) + t^2\|u\|^2. \end{aligned} \tag{7}$$

Let $\alpha = \|v - a\|^2$ and $\beta = (v - a, u)$. Also note that $\|u\|^2 = 3$. Thus (7) becomes

$$\begin{aligned} \|x - a\|^2 &= \alpha + 2\beta t + 3t^2, \\ &= t^2\left(\frac{\alpha}{t^2} + \frac{2\beta}{t} + 3\right). \end{aligned} \tag{8}$$

As t increases positively, it is clear that the second factor in (8) approaches the value 3. Since we are multiplying by t^2 to obtain $\|x - a\|^2$, it follows that $\|x - a\|^2$ can be made to exceed any given number r^2 by choosing t large enough. Thus surely there is no number r such that $\|x - a\| \leq r$, i.e., the line ℓ is unbounded. This coincides with our intuitive notion that a line is "infinite" in extent.

Example 2.2 Let u be a *unit vector* in R^n, i.e., a vector of length 1, $\|u\| = 1$. If v is any vector in R^n, interpret geometrically the numerical or *absolute value* of the inner product (u, v). We remind the reader that the absolute value of a number c is c if $c \geq 0$ and $-c$ if $c < 0$, e.g., the absolute value of 3 is 3 and the absolute value of -3 is also 3. The absolute value of a number c is denoted by

$$|c|.$$

Let w be a vector in R^n perpendicular to u and consider the parametric equation of the line ℓ through v in the direction of w,

$$x = v + tw.$$

The line ℓ_1 through the origin in the direction of u has the parametric equation

$$x = su$$

(here s is a variable). The lines ℓ and ℓ_1 intersect when there are values of s and t that satisfy

$$su = v + tw. \tag{9}$$

Now compute the inner product of both sides of equation (9) with the vector u:

$$\begin{aligned}(u, su) &= (u, v + tw)\\ &= (u, v) + t(u, w).\end{aligned}$$

Since w is perpendicular to u it follows that $(u, w) = 0$, and hence the preceding equation becomes

$$(u, su) = (u, v)$$

or

$$s\|u\|^2 = (u, v).$$

But u is a unit vector, so that $\|u\|^2 = 1$, and hence $s = (u, v)$. Thus the point on the line ℓ_1 at which ℓ and ℓ_1 intersect is $su = (u, v)u$. The distance from the origin to this point of intersection is

$$\begin{aligned}((u, v)u, (u, v)u)^{1/2} &= |(u, v)|\ \|u\|\\ &= |(u, v)|.\end{aligned}$$

In other words, $|(u, v)|$ is the *length of the projection of the vector v in the direction of u.*

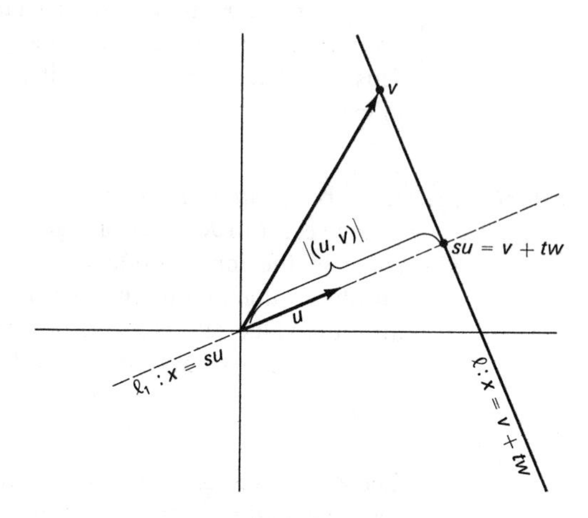

Let u and v be arbitrary non-zero vectors in R^n. Then $\frac{1}{\|u\|} u$ and $\frac{1}{\|v\|} v$ are unit vectors in R^n, and thus $\left(\frac{1}{\|u\|} u, \frac{1}{\|v\|} v\right)$ is a measure (to within sign) of the length of the projection of $\frac{u}{\|u\|}$ in the direction of $\frac{v}{\|v\|}$ or the length of the projection of $\frac{v}{\|v\|}$ in the direction of $\frac{u}{\|u\|}$. Since both vectors are unit vectors, it seems plausible that

(10) $$\left|\left(\frac{1}{\|u\|} u, \frac{1}{\|v\|} v\right)\right| \leq 1,$$

and that equality can hold only when one of the vectors is a multiple of the other, i.e., when they point in the same or in "opposite" directions.

(11)

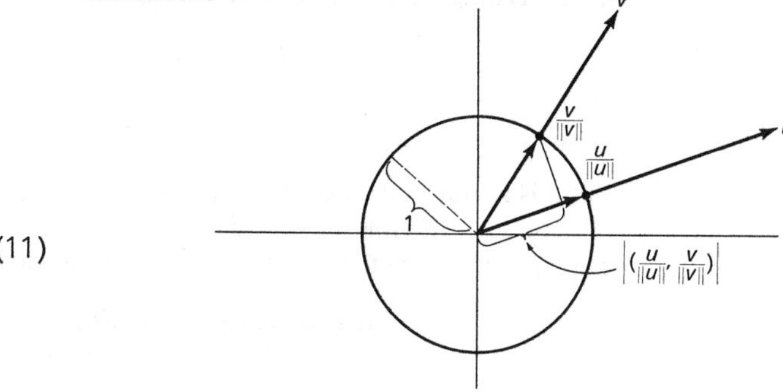

Notice from diagram (11) that the inner product on the left side of (10), i.e., the length of the projection of $\frac{v}{\|v\|}$ in the direction of $\frac{u}{\|u\|}$,

appears to vary from 1 to 0 for u remaining fixed and v rotating from an original position coincident with u to a position perpendicular to u. The inequality (10) can be rewritten

$$|(u, v)| \leq \|u\| \, \|v\|.$$

These geometric remarks can be formalized in the following important and far-reaching result.

Theorem 2.2 **Cauchy-Schwarz Inequality** *Let u and v be non-zero vectors in R^n. Then*

$$|(u, v)| \leq \|u\| \, \|v\|. \tag{12}$$

Equality can hold in (12) *if and only if u and v are linearly dependent.*

Proof Let $\epsilon = \pm 1$, so chosen that

$$(u, v) = \epsilon |(u, v)|,$$

i.e., if $(u, v) \geq 0$ let $\epsilon = 1$, and if $(u, v) < 0$ let $\epsilon = -1$. Note that $\epsilon^2 = 1$ and $|(u, v)| = \epsilon(u, v)$. Consider the number $\|\epsilon u + tv\|^2$ which is nonnegative for any real value of t. Then we compute that

$$\begin{aligned} \|\epsilon u + tv\|^2 &= (\epsilon u + tv, \epsilon u + tv) \\ &= \epsilon^2(u, u) + \epsilon t(u, v) + t\epsilon(v, u) + t^2(v, v) \\ &= \|u\|^2 + 2\epsilon t(u, v) + t^2\|v\|^2 \\ &= \|u\|^2 + 2t|(u, v)| + t^2\|v\|^2. \end{aligned} \tag{13}$$

The quadratic polynomial on the right side of (13) must also be nonnegative for any value of t. To simplify the notation, let $a = \|v\|^2$, $b = 2|(u, v)|$, and $c = \|u\|^2$. Then this nonnegative quadratic polynomial is given by

$$at^2 + bt + c. \tag{14}$$

This means that (14) can have at most one root (by the quadratic formula), i.e., either (14) is always positive for any value of t and

$$b^2 - 4ac < 0, \tag{15}$$

or (14) is zero for an appropriate value of t, $t = t_0$, and

$$b^2 - 4ac = 0. \tag{16}$$

If we write out (15) we obtain

$$4|(u, v)|^2 - 4\|u\|^2\|v\|^2 < 0$$

or

$$|(u, v)| < \|u\|\ \|v\|.$$

Equality can hold if and only if

$$at_0^2 + bt_0 + c = 0,$$

or, in view of (13),

$$\|\epsilon u + t_0 v\|^2 = 0.$$

But, according to Theorem 2.1(c), this latter equality can hold if and only if

$$\epsilon u + t_0 v = 0,$$

i.e., if and only if u and v are linearly dependent vectors. (See Exercise 7, Section 2.6.) ▮

Example 2.3 Show that if u_1, u_2, and u_3 are positive numbers, then

$$3 \leq (u_1 + u_2 + u_3)^{1/2}\left(\frac{1}{u_1} + \frac{1}{u_2} + \frac{1}{u_3}\right)^{1/2}.$$

To see this, let u be the vector $u = (u_1^{1/2}, u_2^{1/2}, u_3^{1/2})$ in R^3, and v the vector $v = \left(\left(\frac{1}{u_1}\right)^{1/2}, \left(\frac{1}{u_2}\right)^{1/2}, \left(\frac{1}{u_3}\right)^{1/2}\right)$. Then

$$\begin{aligned}(u, v) &= u_1^{1/2} \cdot \left(\frac{1}{u_1}\right)^{1/2} + u_2^{1/2} \cdot \left(\frac{1}{u_2}\right)^{1/2} + u_3^{1/2} \cdot \left(\frac{1}{u_3}\right)^{1/2} \\ &= 3.\end{aligned}$$

On the other hand,

$$\begin{aligned}\|u\|^2 &= (u_1^{1/2})^2 + (u_2^{1/2})^2 + (u_3^{1/2})^2 \\ &= u_1 + u_2 + u_3\end{aligned}$$

and similarly,

$$\|v\|^2 = \frac{1}{u_1} + \frac{1}{u_2} + \frac{1}{u_3}.$$

Then (12) becomes

$$\begin{aligned}3 &= |(u, v)| \\ &\leq \|u\|\ \|v\| \\ &= (u_1 + u_2 + u_3)^{1/2}\left(\frac{1}{u_1} + \frac{1}{u_2} + \frac{1}{u_3}\right)^{1/2}.\end{aligned}$$

For example, if $u_1 = 1$, $u_2 = 2$, and $u_3 = 1$, then

$$
\begin{aligned}
(u_1 + u_2 + u_3)^{1/2}\left(\frac{1}{u_1} + \frac{1}{u_2} + \frac{1}{u_3}\right)^{1/2} &= (4)^{1/2}(2\tfrac{1}{2})^{1/2} \\
&= 2 \cdot \sqrt{\tfrac{5}{2}} \\
&= \sqrt{10} \\
&> 3.
\end{aligned}
$$

Example 2.4 In Example 1.7, we found a formula for the distance from the point $a = (a_1, a_2)$ to the line ℓ whose equation is $(x, u) = c$ in R^2. In deriving this formula, we used the obvious geometric fact in R^2 that the distance (i.e., shortest distance) from a to ℓ is along the line perpendicular to ℓ passing through a. This fact is obvious in R^2 and R^3, but requires justification in R^n. Thus, suppose that $(x, u) = c$ is the equation of a hyperplane in R^n, and let $a \in R^n$. Let v be a vector in R^n and consider the parametric equation of the line ℓ through a in the direction of v:

$$x = a + tv.$$

Then ℓ intersects the hyperplane for the value of t so chosen that

$$(a + tv, u) = c,$$

that is

$$(a, u) + t(v, u) = c.$$

In order that there be a point of intersection, assume that v is chosen so that $(v, u) \neq 0$ (i.e., v and u are not orthogonal). We then obtain

$$t = \frac{c - (a, u)}{(v, u)},$$

and thus the point of intersection is

$$x = a + \frac{c - (a, u)}{(v, u)} v.$$

The distance between this point x and a is

$$\left\| \left(a + \frac{c - (a, u)}{(v, u)} v\right) - a \right\| = |c - (a, u)| \cdot \frac{\|v\|}{|(v, u)|}. \tag{17}$$

Now, by the Cauchy-Schwarz Inequality, we know that

$$|(v, u)| \leq \|v\| \, \|u\|$$

and hence

$$\frac{\|v\|}{|(v, u)|} \geq \frac{1}{\|u\|}. \tag{18}$$

It follows from (17) that the distance from a to the point of intersection of the line ℓ and the hyperplane is always at least

$$\frac{|c - (a, u)|}{\|u\|}. \tag{19}$$

On the other hand, we know from the case of equality in the Cauchy-Schwarz Inequality that if u and v are linearly dependent (i.e., if v is a multiple of u) then (18) becomes equality. The number (19) is now exactly the distance from the point a to the intersection of the line ℓ and the hyperplane, where ℓ is in the direction of the normal to the hyperplane.

As an example, we find the distance from $(1, 0, -1, 2)$ to the hyperplane defined by

$$2x_1 + 3x_2 - x_3 + 2x_4 = -1.$$

In this case, $a = (1, 0, -1, 2)$, $u = (2, 3, -1, 2)$, and $c = -1$. Then

$$\begin{aligned}(a, u) &= 1 \cdot 2 + 0 \cdot 3 + (-1) \cdot (-1) + 2 \cdot 2 \\ &= 2 + 1 + 4 \\ &= 7,\end{aligned}$$

and

$$\begin{aligned}\|u\| &= (2^2 + 3^2 + (-1)^2 + 2^2)^{1/2} \\ &= 3\sqrt{2}.\end{aligned}$$

Thus the distance from the point a to the hyperplane is given by (19) as

$$\frac{|(-1) - 7|}{3\sqrt{2}} = \frac{8}{3\sqrt{2}}. \tag{20}$$

We know from elementary geometry that the sum of the lengths of two sides of a triangle in R^2 is always at least as great as the length of the third side. This interesting geometric result has a precise counterpart in R^n.

Theorem 2.3 ***Triangle Inequality*** *Let $u = (u_1, \ldots, u_n)$ and $v = (v_1, \ldots, v_n)$ be two vectors in R^n. Then*

$$\|u + v\| \leq \|u\| + \|v\|. \tag{21}$$

Equality holds if and only if one of the vectors is the zero vector, or neither is the zero vector and $u = cv$ for some number $c > 0$.

Proof Using the properties of the inner product, we compute that

$$\begin{aligned}\|u + v\|^2 &= (u + v, u + v)\\ &= (u, u) + (v, u) + (u, v) + (v, v)\\ &= \|u\|^2 + (u, v) + (u, v) + \|v\|^2\\ &= \|u\|^2 + 2(u, v) + \|v\|^2.\\ &\leq \|u\|^2 + 2|(u, v)| + \|v\|^2.\end{aligned}$$

Now by (12), $|(u, v)| \leq \|u\| \, \|v\|$, so that

$$\begin{aligned}\|u + v\|^2 &\leq \|u\|^2 + 2|(u, v)| + \|v\|^2 && (22)\\ &\leq \|u\|^2 + 2\|u\| \, \|v\| + \|v\|^2 && (23)\\ &= (\|u\| + \|v\|)^2,\end{aligned}$$

and (21) follows.

In order for equality to hold in (21), equality must hold in both (22) and (23). Equality in (22) implies $(u, v) = |(u, v)|$, i.e., $(u, v) \geq 0$. Equality in (23) can hold if and only if u and v are linearly dependent. This is the case when one of them is the zero vector, or if neither u nor v is the zero vector, when $u = cv$, $c \neq 0$. Also, since $0 \leq (u, v) = (u, cu) = c(u, u) = c\|u\|$ and $\|u\| > 0$, it follows that $c > 0$. ▮

In order that we be able to make a more systematic analysis of problems in linear programming, we introduce an important classification of sets.

Definition 2.1 ***Convex set*** A set $S \subset R^n$ is said to be *convex* if the line segment joining any two vectors a and b in S is entirely contained in S.

Recall that the line segment joining a and b is the totality of points of the form

$$(1 - t)a + tb, \quad 0 \leq t \leq 1. \tag{24}$$

Thus the set S is convex if and only if each of the points in (24) is in S whenever a and b are in S.

Example 2.5 **(a)** The following are pictures of sets in R^2, all of which are convex.

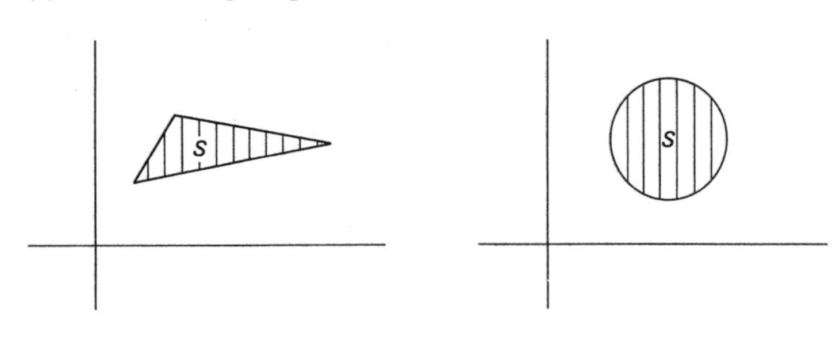

(b) The following are pictures of sets in R^2, *none* of which is convex.

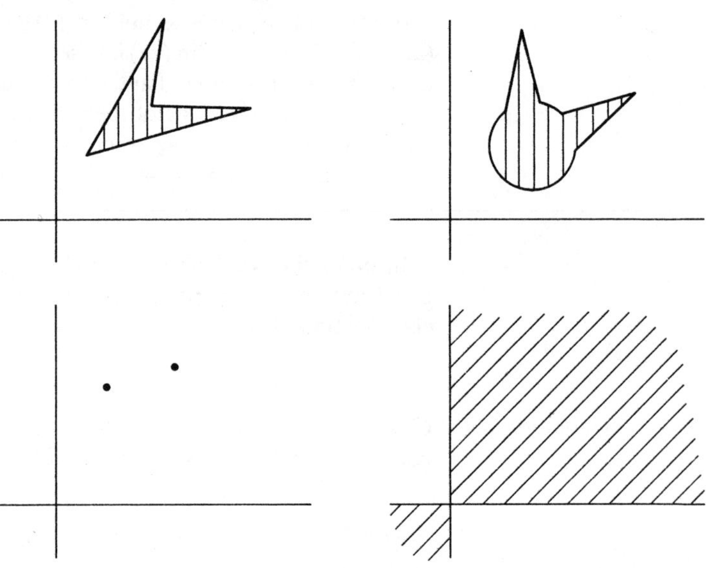

(c) It is geometrically plausible to conclude that the closed sphere S of radius r with center at u is convex. The proof of this fact provides us with an interesting application of the triangle inequality in R^n. Recall that S consists of the totality of $x \in R^n$ which satisfy

$$\|x - u\| \leq r. \tag{25}$$

Suppose that (25) is satisfied for $x = a$ and $x = b$, that is,

$$\|a - u\| \leq r$$

and

$$\|b - u\| \leq r.$$

We must show that for any t, $0 \leq t \leq 1$, $((1 - t)a + tb) \in S$.

Since $u = (1 - t)u + tu$, we compute that

$$\begin{aligned}\|((1 - t)a + tb) - u\| &= \|((1 - t)a + tb) - ((1 - t)u + tu)\|\\ &= \|(1 - t)(a - u) + t(b - u)\|\\ &\leq \|(1 - t)(a - u)\| + \|t(b - u)\|\\ &= (1 - t)\|a - u\| + t\|b - u\|\\ &\leq (1 - t)r + tr\\ &= r.\end{aligned}$$

Thus $((1 - t)a + tb) \in S$ and S is convex.

(d) A hyperplane and a half-space are always convex sets in R^n. For, suppose that $(x, u) = c$ is the equation of a hyperplane Π and that a and b are two members of Π. This means that

$$(a, u) = c$$

and

$$(b, u) = c.$$

If $0 \leq t \leq 1$, then

$$\begin{aligned}((1 - t)a + tb, u) &= (1 - t)(a, u) + t(b, u)\\ &= (1 - t)c + tc\\ &= c;\end{aligned}$$

hence $((1 - t)a + tb) \in \Pi$. We verify that the negative open half-space of Π is convex and leave the verification of the convexity of the remaining types of half-spaces as exercises for the student (see Exercise 9(g)). Thus, suppose that

$$(a, u) < c$$

and

$$(b, u) < c.$$

Then

$$\begin{aligned}((1 - t)a + tb, u) &= (1 - t)(a, u) + t(b, u)\\ &< (1 - t)c + tc\\ &= c,\end{aligned}$$

so that $(1 - t)a + tb$ is again in the open negative half-space belonging to Π.

We require some notation in order to state the next theorem. Let A be an $m \times n$ matrix and $b \in R^m$. If $x \in R^n$ and the ith component of Ax is less than or equal to the ith component of b, $i = 1, \ldots, m$, we write

$$Ax \leq b. \tag{26}$$

If the ith component of Ax is strictly less than the ith component of b, $i = 1, \ldots, m$, we write

$$Ax < b. \tag{27}$$

We refer to (26) as a *system of linear inequalities*, and to (27) as a system of *strict* linear inequalities. If $x \in R^n$ and each component of x is nonnegative we write $x \geq 0$. If $x \geq 0$ and $x \neq 0$ then we sometimes write $x > 0$.

Example 2.6 Write the system of inequalities

$$\begin{aligned} 2x_1 + x_2 - \ x_3 &\leq 1, \\ x_1 + 2x_2 &\geq 3, \\ x_1 - x_2 + \ x_3 &\leq 0, \\ x_2 + 3x_3 &\leq -1 \end{aligned}$$

in the form (26). The second inequality can be rewritten in the form

$$-x_1 - 2x_2 \leq -3.$$

If we define the 4×3 matrix

$$A = \begin{bmatrix} 2 & 1 & -1 \\ -1 & -2 & 0 \\ 1 & -1 & 1 \\ 0 & 1 & 3 \end{bmatrix},$$

the 4-vector $b = (1, -3, 0, -1)$, and the 3-vector $x = (x_1, x_2, x_3)$, then the system of inequalities has the form $Ax \leq b$.

Theorem 2.4 (a) *If $\mathfrak{A}$ is a family of subsets of R^n and each $S \in \mathfrak{A}$ is convex, then*

$$\bigcap_{S \in \mathfrak{A}} S \tag{28}$$

is convex.

(b) *If A is an $m \times n$ matrix, $x \in R^n$, and $b \in R^m$, then the totality of solutions of the equation*

$$Ax = b \tag{29}$$

is a convex set in R^n.

(c) *The set of solutions to either of the systems of inequalities* (26) *or* (27) *is a convex set.*

Proof

(a) Let a and b be points in the intersection (28). Then a and b are in each of the sets $S \in \mathfrak{A}$. Since any $S \in \mathfrak{A}$ is convex by assumption, it follows that $(1 - t)a + tb \in S$ for $0 \leq t \leq 1$. Since this is true for any $S \in \mathfrak{A}$, it follows that $(1 - t)a + tb \in \bigcap_{S \in \mathfrak{A}} S$. In other words, the line segment joining any two points in the intersection is entirely contained in the intersection.

(b) We use the result proved in (a) to prove (b). Let $A_{(i)}$ denote the ith row of A. Then the equation in (29) is equivalent to the system of equalities

(30) $$(A_{(i)}, x) = b_i, \qquad i = 1, \ldots, m.$$

According to Example 2.5(d), any hyperplane in R^n is a convex set, and (30) states that x is a solution to (29) if and only if x is in each of the convex hyperplanes $(A_{(i)}, x) = b_i$, $i = 1, \ldots, m$. Hence x satisfies (29) if and only if x is in the intersection of the m hyperplanes in (30), and since each of these hyperplanes is convex, the result follows from part (a).

(c) If we use the notation in part (b), we are able to rewrite the system (26):

(31) $$(A_{(i)}, x) \leq b_i, \qquad i = 1, \ldots, m.$$

Putting this in a different way, x is a solution to (26) if and only if x is contained in each of the m closed half-spaces defined in (31). By Example 2.5(d), each of these closed half-spaces is convex, and therefore we can apply part (a) exactly as before. The same argument applies to the m open half-spaces $(A_{(i)}, x) < b_i$, $i = 1, \ldots, m$, formed from (27). ▮

In general, the set of solutions to a system of linear inequalities

$$Ax \leq b$$

is not necessarily bounded. For example, the set of solutions to the inequality

$$x_1 - x_2 \leq 0$$

is not contained in any circle. On the other hand, in many linear programming problems, some of which we considered in Section 3.1, we saw that the set of solutions to a system of linear inequalities can be bounded. We argued in this case (at least in the case of R^2) that the maximum and minimum values of a function of the form

$$f(x) = (u, x)$$

were assumed on the vertices of the polygonal set of solutions to the inequalities $Ax \leq b$. In order to handle somewhat more complicated linear programming problems, we want to generalize these ideas to R^n. As a first step, we show how the solution set to a system of linear inequalities can be thought of as a solution set to a system of linear equations.

Example 2.7 Consider the system of inequalities

$$\begin{aligned} x_1 + 2x_2 &\leq 3, \\ x_1 &\geq 0, \\ x_2 &\geq 0. \end{aligned}$$

We introduce a *dummy* or *slack variable* x_3 so that

$$\begin{aligned} x_1 + 2x_2 + x_3 &= 3, \\ x_1 &\geq 0, \\ x_2 &\geq 0, \\ x_3 &\geq 0. \end{aligned}$$

Thus, rather than consider the set of nonnegative solutions to the inequality

$$x_1 + 2x_2 \leq 3, \tag{32}$$

we can consider the set of nonnegative solutions to the equation

$$x_1 + 2x_2 + x_3 = 3. \tag{33}$$

Observe that (x_1, x_2, x_3) is in the solution set to (33) if and only if (x_1, x_2) is in the solution set to (32), assuming that x_1, x_2, and x_3 are nonnegative.

More generally, suppose we want to convert the system of inequalities

$$Ax \leq b, \quad x \geq 0, \tag{34}$$

to a system of equations. Assume A is an $m \times n$ matrix, $x \in R^n$, $b \in R^m$. Define m new dummy variables, $x_{n+1}, \ldots, x_{n+m}$. Consider the system of linear equations

$$(A_{(i)}, x) + x_{n+i} = b_i, \qquad i = 1, \ldots, m. \tag{35}$$

Then if an n-tuple $x = (x_1, \ldots, x_n)$ with nonnegative components is a solution to (34), we can find an $(n + m)$-tuple of the form $(x_1, \ldots, x_n, x_{n+1}, \ldots, x_{n+m})$ with nonnegative components which is a solution to the system (35) by simply defining

$$x_{n+i} = b_i - (A_{(i)}, x) \geq 0, \qquad i = 1, \ldots, m.$$

Conversely, if $x = (x_1, \ldots, x_n, x_{n+1}, \ldots, x_{n+m})$ is a solution to the system (35), $x_i \geq 0$, $i = 1, \ldots, n + m$, then $(x_1, \ldots, x_n)$ is a solution to (34), $x_i \geq 0$, $i = 1, \ldots, n$. We say that the system of inequalities (34) has been put in *equality form* in (35).

Example 2.8 Write the following system of linear inequalities in equality form:

$$\begin{aligned} 2x_1 + 3x_2 &\leq 1, \\ x_1 - x_2 &\leq 3, \\ 2x_1 - 3x_2 &\leq 5, \\ x_1 &\geq 0, \\ x_2 &\geq 0. \end{aligned}$$

To do this, we define three nonnegative slack variables, x_3, x_4, and x_5, and the preceding system can be written in equality form as follows:

$$\begin{aligned} 2x_1 + 3x_2 + x_3 &= 1, \\ x_1 - x_2 + x_4 &= 3, \\ 2x_1 - 3x_2 + x_5 &= 5, \\ x_1 &\geq 0, \\ x_2 &\geq 0, \\ x_3 &\geq 0, \\ x_4 &\geq 0, \\ x_5 &\geq 0. \end{aligned} \tag{36}$$

We can write the preceding system in the following abbreviated form. Let A be the matrix of coefficients for the system of inequalities and let $\tilde{A}$ be the matrix obtained by adjoining a 3×3 identity matrix immediately to the right of A, that is,

$$A = \begin{bmatrix} 2 & 3 \\ 1 & -1 \\ 2 & -3 \end{bmatrix},$$

and

$$\tilde{A} = \begin{bmatrix} 2 & 3 & 1 & 0 & 0 \\ 1 & -1 & 0 & 1 & 0 \\ 2 & -3 & 0 & 0 & 1 \end{bmatrix}.$$

Also, let $\tilde{x} = (x_1, x_2, x_3, x_4, x_5)$ and $b = (1, 3, 5)$. Then (36) can be written in the form

$$\begin{aligned} \tilde{A}\tilde{x} &= b, \\ \tilde{x} &\geq 0, \end{aligned}$$

where, as usual, 0 is the zero 5-vector and the second inequality means that each component of $\tilde{x}$ is greater than zero.

We want to generalize the notion of a vertex of a convex polygon in R^2. Observe that a vertex of a polygon is characterized by the property that it does not lie on any line segment joining any two distinct points in the polygon, except possibly at one of the ends (i.e., if it lies on a line segment, it must be one of the endpoints of that segment; see diagram below).

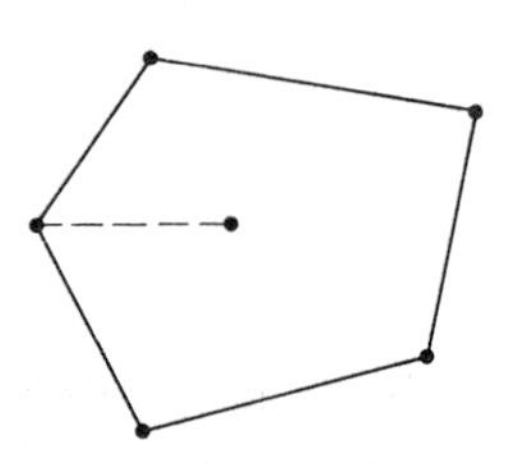

Definition 2.2 ***Extreme point of a convex set*** If S is a convex set in R^n and $a \in S$, then a is an *extreme point* or a *vertex* of S if and only if there do not exist distinct points u and v in S and a number t, $0 < t < 1$, such that $a = (1 - t)u + tv$.

The problem we want to solve is that of determining the extreme points of the convex set of solutions to the equation

$$Ax = b, \quad x \geq 0.$$

Theorem 2.5 *Let A be an $m \times n$ matrix and $b \in R^m$. Let S be the set of solutions to the system*

$$Ax = b, \quad x \geq 0. \tag{37}$$

(a) *If $b = 0$ then $x = 0$ is the only extreme point of S.*
(b) *If $b \neq 0$ and $x \in S$ then $x > 0$, say $x_{i_j} > 0$, $j = 1, \ldots, p$, and the remaining $n - p$ components of x are zero. A necessary and sufficient condition that this x be an extreme point of S is that the columns $A^{(i_1)}, \ldots, A^{(i_p)}$ of A be linearly independent.*
(c) *No extreme point of S can have more than m positive components.*
(d) *The set S has only a finite number of extreme points.*

Proof

(a) Clearly $x = 0 \in S$ when $b = 0$. Moreover, if $u \geq 0$ and $v \geq 0$ are two points in S and

$$0 = \theta u + (1 - \theta)v$$

for some $0 < \theta < 1$, then it is clear from the nonnegativity of the components of u and v that $u = v = 0$, i.e., $x = 0$ is an extreme point of S. Moreover, if $y > 0$ is a solution to (37) then $u = \frac{1}{2}y$ and $v = \frac{3}{4}y$ are also in S (remember that $b = 0$). But

$$y = \tfrac{1}{2}u + \tfrac{1}{2}v$$

and hence y is not an extreme point; in other words, $x = 0$ is the only extreme point of S.

(b) Assume that $x \in S$ and that x has p positive components, and the other $n - p$-components are all zero. We lose no generality in assuming that it is the first p components that are positive, that is, $i_j = j, j = 1, \ldots, p$. First, assume that the corresponding columns $A^{(1)}, \ldots, A^{(p)}$ of A are linearly independent. We prove that x is an extreme point of S by contradiction. If x is not extreme, then

$$x = (1 - t)u + tv, \tag{38}$$

where u and v are distinct vectors in S and $0 < t < 1$, i.e., x lies on a line segment joining two vectors u and v in S and is distinct from u and v. Now, u and v both have nonnegative components (they are in S and hence satisfy (37)) and, considering (38),

$$\begin{aligned} 0 &= x_i \\ &= (1 - t)u_i + tv_i, \qquad i = p + 1, \ldots, n, \end{aligned}$$

and hence $u_i = v_i = 0$ for $i = p + 1, \ldots, n$. In other words, the last $n - p$ components of u and v must also be zero. Now, $Au = b$ and $Av = b$, and thus

$$\begin{aligned} b &= u_1A^{(1)} + u_2A^{(2)} + \cdots + u_pA^{(p)} \\ &= v_1A^{(1)} + v_2A^{(2)} + \cdots + v_pA^{(p)}. \end{aligned}$$

Therefore

$$(u_1 - v_1)A^{(1)} + \cdots + (u_p - v_p)A^{(p)} = 0.$$

The linear independence of the columns $A^{(1)}, \ldots, A^{(p)}$ now implies that $(u_1 - v_1) = \cdots = (u_p - v_p) = 0$. Putting this in other words, the first p components of u are the same as the first p components of v, and the remaining $n - p$ components of both vectors are zero. But u and v were supposed to be distinct vectors, therefore this conflict implies that the assumption that x is not an extreme point is false.

To prove necessity, we must show that if x is an extreme point of S and precisely p of the components of x are positive, say, $x_1, \ldots, x_p$ (the rest of the x_i being zero), then the corresponding columns of A, i.e., $A^{(1)}, \ldots, A^{(p)}$, must be linearly independent. Suppose, on the contrary, that $A^{(1)}, \ldots, A^{(p)}$ are linearly dependent. This means that there exist constants $c_1, \ldots, c_p$, not all of which are zero, such that

$$c_1A^{(1)} + c_2A^{(2)} + \cdots + c_pA^{(p)} = 0. \tag{39}$$

Now since x satisfies (37), we have

$$x_1A^{(1)} + x_2A^{(2)} + \cdots + x_pA^{(p)} = b. \tag{40}$$

Let s be any non-zero number and observe that (39) and (40) imply that

$$\begin{aligned}(41)\qquad b &= x_1A^{(1)} + \cdots + x_pA^{(p)} \pm s\cdot 0\\ &= x_1A^{(1)} + \cdots + x_pA^{(p)} \pm s[c_1A^{(1)} + \cdots + c_pA^{(p)}]\\ &= (x_1 \pm sc_1)A^{(1)} + \cdots + (x_p \pm sc_p)A^{(p)}.\end{aligned}$$

Since $x_1, \ldots, x_p$ are positive, it follows that we can find a positive number s so small that for both choices of sign, $x_1 \pm sc_1 > 0$, $x_2 \pm sc_2 > 0, \ldots, x_p \pm sc_p > 0$. With such a choice of s, we can conclude from (41) that the two n-vectors

$$u = (x_1 + sc_1, \ldots, x_p + sc_p, 0, \ldots, 0)$$

and

$$v = (x_1 - sc_1, \ldots, x_p - sc_p, 0, \ldots, 0)$$

are in S. Moreover, since at least one of the c_i is not zero, we know that u and v are distinct, i.e., $x_i + sc_i \neq x_i - sc_i$ if $c_i \neq 0$. However,

$$\begin{aligned}\tfrac{1}{2}u + \tfrac{1}{2}v &= \left(\frac{x_1 + sc_1 + x_1 - sc_1}{2}, \ldots, \frac{x_p + sc_p + x_p - sc_p}{2}, 0, \ldots, 0\right)\\ &= (x_1, \ldots, x_p, 0, \ldots, 0),\end{aligned}$$

which satisfies the equation in (38) (with $t = \frac{1}{2}$), contradicting the fact that x is an extreme point of S. Hence $A^{(1)}, \ldots, A^{(p)}$ are linearly independent.

(c) Observe that if an extreme point of S has p positive components, then the corresponding columns of A must be linearly independent. But these columns are m-tuples, so that if p were greater than m, we would have $p > m$ vectors in R^m which are linearly independent. As we saw in Exercise 18, Section 2.6, this cannot happen; hence $p \leq m$.

(d) The fact that S has only a finite number of extreme points can be seen by an argument very much like the one used in (b). Given p linearly independent columns of A, say $A^{(i_1)}, \ldots, A^{(i_p)}$, there may or may not be an extreme point $x \in S$ with precisely the positive components $x_{i_1}, \ldots, x_{i_p}$. But if there is, there is at most one. For if $y \in S$ and the positive components of y are precisely $y_{i_1}, \ldots, y_{i_p}$, then

$$\begin{aligned}b &= Ax\\ &= x_{i_1}A^{(i_1)} + \cdots + x_{i_p}A^{(i_p)},\end{aligned}$$

and also

$$b = y_{i_1}A^{(i_1)} + \cdots + y_{i_p}A^{(i_p)}.$$

Hence, equating the two,

$$(x_{i_1} - y_{i_1})A^{(i_1)} + \cdots + (x_{i_p} - y_{i_p})A^{(i_p)} = 0,$$

and the linear independence of $A^{(i_1)}, \ldots, A^{(i_p)}$ implies that $x_{i_1} = y_{i_1}, \ldots, x_{i_p} = y_{i_p}$. Since the remaining components of x and y are zero, it follows that $x = y$. We have proved that associated with each set of p columns of the matrix A there is at most one extreme point of S. But, after all, there are only a finite number of ways of choosing a set of columns from an $m \times n$ matrix. ▌

Theorem 2.5 provides us with a constructive method for finding all of the extreme points of the set S of solutions to (37). Suppose that $m \leq n$. (In putting a system of linear inequalities into equality form, it is always the case that the number of rows is at most the number of columns, and it is for these systems that we want to find the extreme points for the set of solutions.) Choose any m columns of A (this can be done in $\binom{n}{m}$ ways, of course). Suppose $j_1 < j_2 < \cdots < j_m$ are the numbers of the columns chosen. Now, consider the system of m linear equations for the determination of $x_{j_1}, \ldots, x_{j_m}$:

$$x_{j_1}A^{(j_1)} + \cdots + x_{j_m}A^{(j_m)} = b. \tag{42}$$

If we can find nonnegative $x_{j_1}, \ldots, x_{j_m}$ satisfying (42), we can form the n-vector x whose components numbered $j_1, \ldots, j_m$ are $x_{j_1}, \ldots, x_{j_m}$ and whose remaining components are zero. Clearly such an x satisfies

$$Ax = b, \quad x \geq 0.$$

According to Theorem 2.5, x will be an extreme point if and only if those columns of A corresponding to the positive components of x are linearly independent, and the question of the linear independence of any set of columns can be settled by the routine methods established at the end of Chapter 2. Thus we can find all possible extreme points by this protracted but nevertheless finite method.

Example 2.9 Reduce the following system of linear inequalities to equality form and find the extreme points:

$$\begin{aligned} 2x_2 - x_1 &\leq 1, \\ x_2 + x_1 &\leq 2, \\ x_1 &\geq 0, \\ x_2 &\geq 0. \end{aligned} \tag{43}$$

We introduce two nonnegative dummy variables x_3 and x_4 and note that the equality form of the preceding system in matrix-vector notation becomes

$$\begin{bmatrix} -1 & 2 & 1 & 0 \\ 1 & 1 & 0 & 1 \end{bmatrix} x = b, \quad x \geq 0, \tag{44}$$

where $b = (1, 2)$ and $x = (x_1, x_2, x_3, x_4)$. The set of solutions to (44) is the same as the set of solutions to the equivalent system in Hermite normal form:

$$x_1 - \tfrac{1}{3}x_3 + \tfrac{2}{3}x_4 = 1, \quad x_2 + \tfrac{1}{3}x_3 + \tfrac{1}{3}x_4 = 1. \tag{45}$$

According to the discussion preceding this example, we can find all the extreme points of the set of solutions to (45) by considering the $\binom{4}{2} = 6$ subsystems of two equations in two unknowns obtained by setting any two of the x_i equal to zero. We do this systematically as follows.

(i) $x_1 = x_2 = 0$:

$$-\tfrac{1}{3}x_3 + \tfrac{2}{3}x_4 = 1, \quad \tfrac{1}{3}x_3 + \tfrac{1}{3}x_4 = 1.$$

The solution to this sytem is $x_3 = 1$, $x_4 = 2$. This is a positive solution, and, moreover, the two columns $(-\tfrac{1}{3}, \tfrac{1}{3})$ and $(\tfrac{2}{3}, \tfrac{1}{3})$ are clearly linearly independent. Hence, according to Theorem 2.5, $(0, 0, 1, 2)$ is an extreme point of the set of solutions to (44).

(ii) $x_1 = x_3 = 0$:

$$\tfrac{2}{3}x_4 = 1, \quad x_2 + \tfrac{1}{3}x_4 = 1.$$

Thus $x_4 = \tfrac{3}{2}$ and $x_2 = \tfrac{1}{2}$. Also, the two columns $(0, 1)$ and $(\tfrac{2}{3}, \tfrac{1}{3})$ are linearly independent, and hence $(0, \tfrac{1}{2}, 0, \tfrac{3}{2})$ is another extreme point.

(iii) $x_1 = x_4 = 0$: by a similar calculation, we find that $x_2 = 2$ and $x_3 = -3$ and hence we do not have an extreme point (since x_3 is negative).

(iv) $x_2 = x_3 = 0$: in this case $x_1 = -1$, $x_4 = 3$ in the resulting system, and again we do not have an extreme point.

(v) $x_2 = x_4 = 0$: here we get $x_1 = 2$ and $x_3 = 3$, the columns $(1, 0)$ and $(-\tfrac{1}{3}, \tfrac{1}{3})$ are obviously linearly independent, and so we obtain the extreme point $(2, 0, 3, 0)$.

(vi) $x_3 = x_4 = 0$: in this case $x_1 = 1$ and $x_2 = 1$, the columns $(1, 0)$ and $(0, 1)$ are linearly independent, and this produces the extreme point $(1, 1, 0, 0)$.

Summarizing, we see that the set of extreme points for the convex set of solutions to (44) is

$$\{(0, 0, 1, 2), (0, \tfrac{1}{2}, 0, \tfrac{3}{2}), (2, 0, 3, 0), (1, 1, 0, 0)\}.$$

The extreme points for the convex set of solutions to (43) are obtained by simply suppressing the last two components corresponding to the slack variables in the preceding set of four vectors:

$$\{(0, 0), (0, \tfrac{1}{2}), (2, 0), (1, 1)\}.$$

We will prove this assertion in general in Theorem 3.2 and Exercise 20 in the next section. However we can verify it in the present problem by diagramming the set defined by the inequalities in (43).

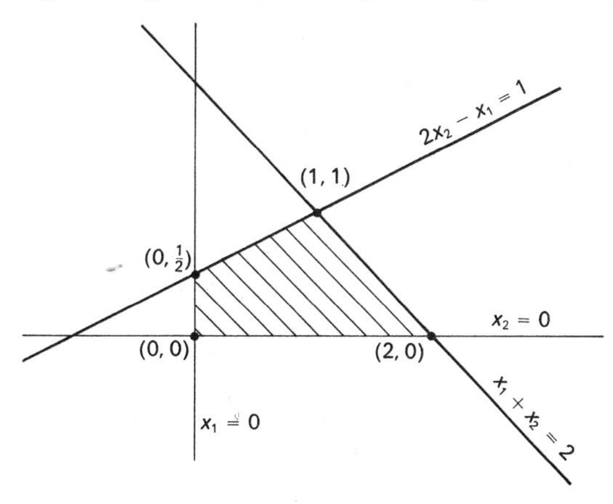

Quiz

Answer true or false.

1 If $\|u\| = 0$ then u is the zero vector.

2 The closed sphere of radius r with center at a in R^n is the totality of vectors $u = (u_1, \ldots, u_n)$ which satisfy $\sum_{i=1}^{n} (u_i - a_i)^2 \leq r^2$.

3 The intersection of two bounded sets of vectors in R^n is a bounded set.

4 The union of two bounded sets in R^n is not necessarily a bounded set.

5 If $|c| = -c$ then $c = 0$.

6 If $|c| + c = 0$ then $c \leq 0$.

7 If u is a unit vector then $|(u, v)|$ can be interpreted as the length of the projection of the vector v in the direction of u.

8 The Cauchy-Schwarz Inequality states that $|(u, v)| \geq \|u\| \, \|v\|$.

9 The Triangle Inequality states that $\|u + v\| \leq \|u\| + \|v\|$.

10 If $\mathfrak{A}$ is a family of convex sets, then $\bigcap_{A \in \mathfrak{A}} A$ is a convex set.

Exercises

1 Compute the length of each of the following vectors, using the definition in formula (3):

(a) $(1, 1, 0)$;
(b) $(2, 3, 4, 5)$;
(c) $(0, 0, 0)$;
(d) $(0, 1, 1)$;
(e) $(9, 8, 7, 6, 5)$;
(f) $(7, 0, 0, 0, 1)$;
(g) $(7, 0, 0, 0, 0)$;
(h) (3);
(i) $(0, 0, 0, 0, 0, 0, 0, 0, 0)$;
(j) $(3, 3, 3)$.

2 Determine whether each of the following points u lies in the sphere of radius r with center at a:

(a) $u = (1, 1, 1),\ a = (1, 1, 2),\ r = 1$;
(b) $u = (1, 1, 1),\ a = (1, 1, 2),\ r = \frac{1}{2}$;
(c) $u = (4, 3, 2, 1),\ a = (1, 2, 3, 4),\ r = 5$;
(d) $u = (7, 10),\ a = (20, 20),\ r = 16$;
(e) $u = (0, 0, 0, 0, 0),\ a = (1, 1, 1, 1, 1),\ r = 2$;
(f) $u = (1, 0, 1),\ a = (0, 1, 0),\ r = 2$;
(g) $u = (2, 4, 6, 8),\ a = (2, 1, 0, 7),\ r = 7$;
(h) $u = (0, 0, 0, 1),\ a = (0, 0, 0, 3),\ r = 2$;
(i) $u = (-2, 0, 3, -1),\ a = (1, 3, -5, 2),\ r = 9$;
(j) $u = (-1, 0, -1, -1),\ a = (1, 0, 1, 1),\ r = 0$.

3 Determine which of the following sets is bounded in R^n:

(a) $\{u \mid u \in R^2 \wedge |u_1| \leq 1\}$;
(b) $\{u \mid u \in R^3 \wedge u_1^2 + u_2^2 + u_3^2 \leq 1\}$;
(c) $\{u \mid u \in R^2 \wedge u_1 u_2 = 1\}$;
(d) $\{u \mid u \in R^3 \wedge |u_1 u_2 u_3| \leq 1\}$;
(e) $\{u \mid u \in R^2 \wedge u_1 + u_2 \leq 1 \wedge u_1 \geq 0 \wedge u_2 \geq 0\}$;
(f) $\{u \mid u \in R^2 \wedge u_1 + 2u_2 \leq 1 \wedge u_1 - u_2 \geq 3\}$;
(g) $\{u \mid u \in R^2 \wedge u_1 \geq 0 \wedge u_2 \geq 0 \wedge u_1^2 \leq u_2\}$;
(h) the line in R^3 whose parametric equation is $x = (0, 1, 1) + t(-1, 1, 0)$;
(i) the line segment in R^4 joining the points $(1, 1, 1, 1)$ and $(-1, 1, -1, 1)$.

4 Find the length of the projection of v in the direction of u in each of the following:

(a) $u = (1, 0, 0),\ v = (2, 7, -3)$;
(b) $u = \dfrac{1}{\sqrt{6}}(2, 1, 0, 1),\ v = (0, 2, 1, 1)$;

(c) $u = \left(\frac{1}{\sqrt{2}}, \frac{1}{\sqrt{2}}\right)$, $v = (\sqrt{2}, \sqrt{2})$;
(d) $u = \left(\frac{1}{\sqrt{3}}, 0, \frac{1}{\sqrt{3}}, 0, \frac{1}{\sqrt{3}}\right)$, $v = (3, -1, 0, 4, -7)$;
(e) $u = (0, 0, 1)$, $v = (-1, 0, 0)$.

5 For each of the following pairs of vectors u and v, verify the Cauchy-Schwarz Inequality. In the cases of equality, confirm the linear dependence of the vectors.

(a) $u = (1, 0)$, $v = (0, 1)$;
(b) $u = (1, 2, 3, 4)$, $v = (4, 2, 3, 1)$;
(c) $u = (2, 3, -1)$, $v = (-4, -6, 2)$;
(d) $u = (0, 0, 0)$, $v = (5, 2, 1)$;
(e) $u = (7, -2, 1, 1)$, $v = (3, -1, \frac{1}{2}, \frac{1}{2})$;
(f) $u = (-1, 6, -3, -1)$, $v = (2, 12, 6, 2)$;
(g) $u = (-1, 6, -3, -1)$, $v = (2, -12, 6, 2)$;
(h) $u = (1, 0, 1)$, $v = (0, 1, 1)$;
(i) $u = (1)$, $v = (2)$;
(j) $u = (1000, 2000)$, $v = (1, 2)$;
(k) $u = (1, 2, 3)$, $v = (4, 5, 6)$.

6 Using formula (19), find the distance from the point a to the hyperplane $(x, u) = c$ in each of the following:

(a) $a = (1, 1, 1)$, $u = (3, 0, 3)$, $c = 5$;
(b) $a = (1, -2)$, $u = (5, 7)$, $c = 20$;
(c) $a = (3, 4, 4, 3)$, $u = (-2, 1, 0, 3)$, $c = 2$;
(d) $a = (1, 0, 1, 0)$, $u = (0, 1, 0, 1)$, $c = 1$;
(e) $a = (2, 2, 4, 1)$, $u = (1, 1, 2, \frac{1}{2})$, $c = 3$;
(f) $a = (3, 0, 2, 1)$, $u = (1, 1, 3, 2)$, $c = 11$;
(g) $a = (1, 0, 0)$, $u = (2, 4, 6)$, $c = 10$;
(h) $a = (3)$, $u = (4)$, $c = 6$;
(i) $a = (0, 0, 0, 0, 0)$, $u = (1, 2, 3, 4, 5)$, $c = 5$;
(j) $a = (3, 1, 0, 7)$, $u = (3, 4, 3, 9)$, $c = 2$.

7 Verify the Triangle Inequality for each of the following pairs of vectors u and v. Confirm the conditions of Theorem 2.3 when equality holds.

(a) $u = (1, 1, 0)$, $v = (0, 1, 1)$;
(b) $u = (3, 2, 1, 4)$, $v = (6, 3, 2, 8)$;
(c) $u = (1, 4, -2, 7)$, $v = (-2, -8, 4, -14)$;
(d) $u = (0, 0, 0, 0, 0, 0, 0)$, $v = (3, 9, -5, 1, 1, -2, 6)$;
(e) $u = (3)$, $v = (27)$.

8 Which of the following figures in R^2 is convex?

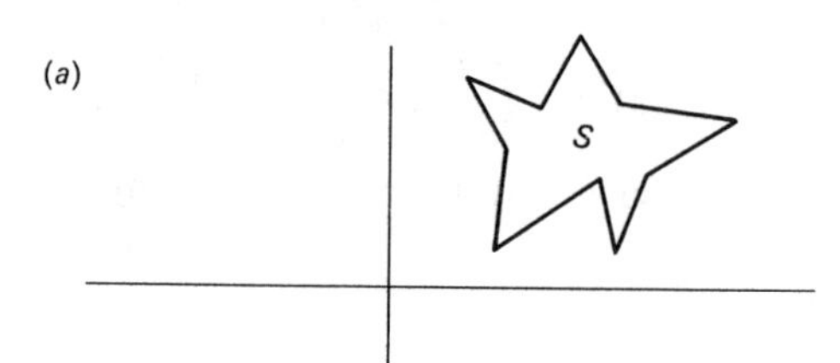

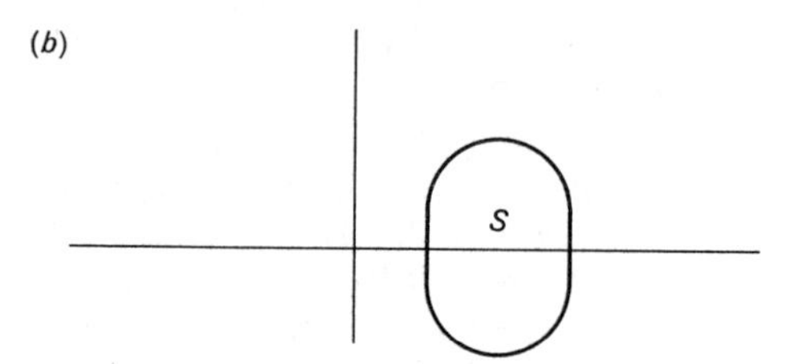

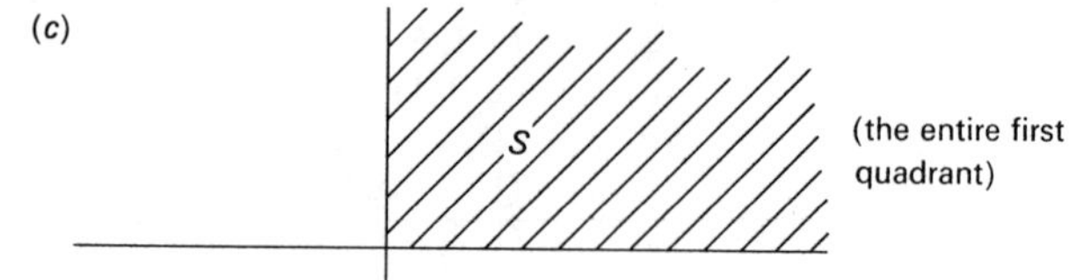

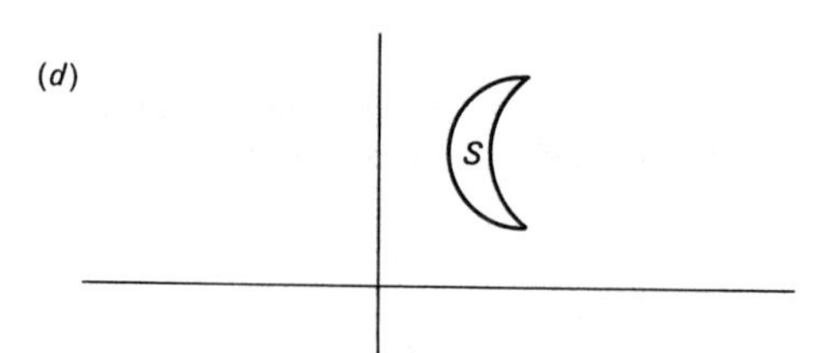

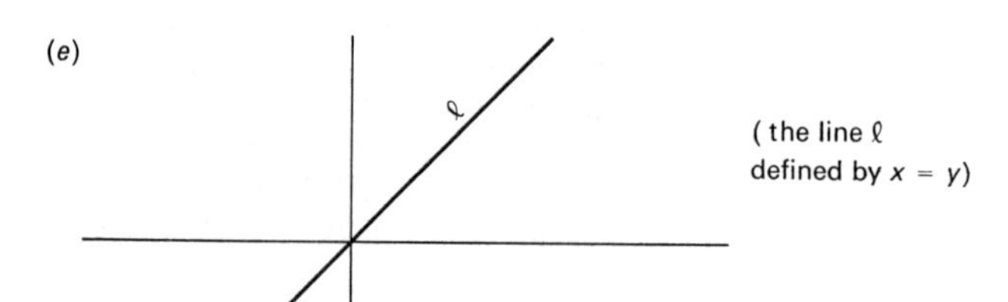

9 Which of the following sets is convex?

(a) The line in R^n whose parametric equation is $x = a + tu$;
(b) $\{x \mid x \in R^n \wedge 1 \leq \|x - a\| \leq 2\}$;
(c) $\{u \mid u \in R^2 \wedge u_1 + u_2 \leq 1 \wedge u_1 - u_2 \geq 1 \wedge u_1 \geq 0 \wedge u_2 \geq 0\}$;
(d) $\{u \mid u \in R^3 \wedge |u_1| \leq 1 \wedge |u_2| \leq 2 \wedge u_3 = 0\}$;
(e) $\{u \mid u \in R^3 \wedge u_i \geq 0, i = 1, 2, 3\}$;
(f) $\{u \mid u \in R^n \wedge |(u, a)| \leq 1\}$;
†(g) The remaining types of half-spaces in Example 2.5(d).

10 Write each of the following systems of inequalities in the form $Ax \le b$:

(a)
$$\begin{aligned} x_1 - 3x_2 + x_4 &\le -2, \\ x_2 + x_3 - 5x_4 &\ge 1, \\ x_1 + 7x_3 - x_4 &= 2; \end{aligned}$$

(b)
$$\begin{aligned} 3x_1 + x_2 &\le 7, \\ x_1 - x_2 &\ge -4; \end{aligned}$$

(c)
$$\begin{aligned} x_1 + 2x_2 + 4x_4 &\le 8, \\ x_2 - x_3 &= 0, \\ 3x_1 - 4x_2 + 5x_3 &\le 6; \end{aligned}$$

(d)
$$\begin{aligned} x_1 + x_2 &\le 4, \\ x_2 + x_3 &\ge 4, \\ x_1 + x_3 &\le 5; \end{aligned}$$

(e)
$$\begin{aligned} -2x_1 + 7x_2 + 3x_3 - 4x_4 - x_5 &\le 10, \\ x_1 + x_2 + x_3 + x_4 + 18x_5 &\le -3, \\ x_1 - x_2 + 5x_3 + 7x_4 &\ge 9, \\ x_2 - 4x_3 + 3x_4 - 9x_5 + x_1 &= 2. \end{aligned}$$

11 By introducing nonnegative slack variables, put each of the systems in the preceding problem into equality form. Write out the coefficient matrices.

12 Describe the extreme points of each of the following sets:

(a) $\{x \mid x \in R^3 \wedge \|x - a\| \le 1\}$, where $a = (1, 1, 0)$;
(b) $\{x \mid x \in R^n \wedge x = (1 - t)a + tb,\ 0 \le t \le 1\}$;
(c) $\{x \mid x \in R^2 \wedge x_1 + x_2 \le 1 \wedge x_1 \ge 0 \wedge x_2 \ge 0\}$;
(d) $\{x \mid x \in R^2 \wedge x_1 + x_2 < 1 \wedge x_1 \ge 0 \wedge x_2 \ge 0\}$.

13 By introducing slack variables and using the method of Example 2.9, find all extreme points of the equality form of each of the following systems of linear inequalities:

(a)
$$\begin{aligned} 2x_1 + x_2 &\ge 3, \\ x_1 - 2x_2 &\le -1, \\ x_2 &\le 3; \end{aligned}$$

(b)
$$\begin{aligned} x_2 &\le x_1, \\ x_1 &\ge 0, \\ x_1 &\le 1, \\ x_2 &\ge 0; \end{aligned}$$

(c)
$$\begin{aligned} x_2 &\le 2x_1, \\ x_2 &\ge \tfrac{1}{2}x_1, \\ x_1 + 2x_2 &\le 3; \end{aligned}$$

(d)
$$\begin{aligned} x_2 &\le 2x_1, \\ x_2 &\ge \tfrac{1}{2}x_1, \\ x_1 + 2x_2 &\le 3, \\ x_1 + 2x_2 &\ge 1. \end{aligned}$$

14 If S and T are sets of vectors in R^n and α is a number, define $S + T = \{x + y \mid x \in S \wedge y \in T\}$ and $\alpha S = \{\alpha x \mid x \in S\}$. Show that if S is a convex set and α and β are positive numbers, then $\alpha S + \beta S = (\alpha + \beta)S$. (Hint: Note that $\dfrac{\alpha}{\alpha + \beta} + \dfrac{\beta}{\alpha + \beta} = 1$.)

3.3

Linear Functions on Convex Polyhedra

In this section, we shall discuss the general problem of finding the maximum and minimum values of functions which are defined on the set of solutions to a system of linear inequalities. In order to do this, we introduce an important notion which generalizes the idea of a line segment joining two points.

Example 3.1 Express the points contained in the triangle in R^2 with vertices (2, 2), (0, 1), and (1, 0) in terms of these vertices. Of course, we must clarify precisely what is being asked here. We know that the points on the line segment joining (0, 1) and (1, 0) are all of the form

$$u = (u_1, u_2) = (1 - t)(0, 1) + t(1, 0), \tag{1}$$

where $0 \leq t \leq 1$. We want a formula analogous to (1) which will represent a typical point x in the triangle in terms of the vertices.

(2)

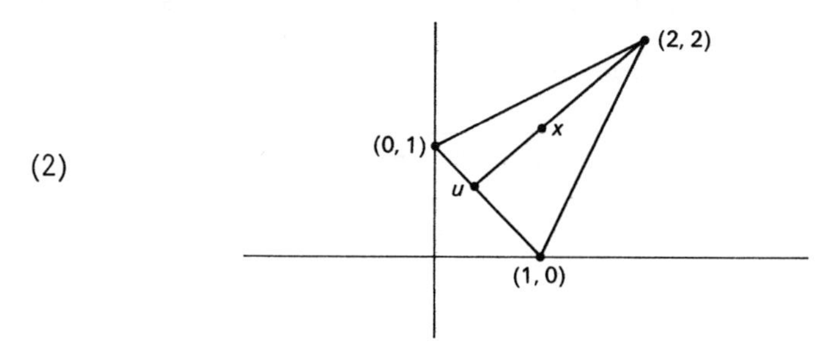

It is clear from the diagram in (2) that any point x in the triangle lies on the line segment joining (2, 2) to an appropriate point u lying on the line segment joining (0, 1) and (1, 0):

$$x = (1 - s)(2, 2) + su, \qquad 0 \leq s \leq 1. \tag{3}$$

However, if we substitute into (3) the equation for u in (1), we obtain

$$\begin{aligned} x &= (1 - s)(2, 2) + s[(1 - t)(0, 1) + t(1, 0)] \\ &= (1 - s)(2, 2) + s(1 - t)(0, 1) + st(1, 0). \end{aligned} \tag{4}$$

Consider the three numbers which appear as coefficients in (4), namely $(1 - s)$, $s(1 - t)$, and st. Since both s and t lie in the interval between 0 and 1, it is clear that $1 - s$, $s(1 - t)$, and st are nonnegative. But more is true:

$$\begin{aligned} (1 - s) + s(1 - t) + st &= (1 - s) + s(1 - t + t) \\ &= 1 - s + s \\ &= 1. \end{aligned}$$

If we set $\sigma_1 = 1 - s$, $\sigma_2 = s(1 - t)$, and $\sigma_3 = st$, then (4) becomes

$$x = \sigma_1(2, 2) + \sigma_2(0, 1) + \sigma_3(1, 0), \tag{5}$$

where $\sigma_i \geq 0$, $i = 1, 2, 3$, and

$$\sigma_1 + \sigma_2 + \sigma_3 = 1. \tag{6}$$

In other words, we can express an arbitrary point in the triangle as a linear combination of the three vertices. Furthermore, the coefficients will be nonnegative and add to 1. But the converse of this statement also holds: if any point x is of the form (5) where the coefficients σ_i are nonnegative and sum to 1, then this point must lie in the triangle. This can be seen as follows. If $\sigma_1 = 1$, then clearly $x = (2, 2)$, a point in the triangle. Thus assume $\sigma_1 < 1$, and write the right-hand side of (5) as

$$\sigma_1(2, 2) + (1 - \sigma_1)\left[\frac{\sigma_2}{1 - \sigma_1}(0, 1) + \frac{\sigma_3}{1 - \sigma_1}(1, 0)\right]. \tag{7}$$

Consider the point

$$v = \frac{\sigma_2}{1 - \sigma_1}(0, 1) + \frac{\sigma_3}{1 - \sigma_1}(1, 0).$$

Clearly

$$\begin{aligned} \frac{\sigma_2}{1 - \sigma_1} + \frac{\sigma_3}{1 - \sigma_1} &= \frac{\sigma_2 + \sigma_3}{1 - \sigma_1} \\ &= \frac{1 - \sigma_1}{1 - \sigma_1} \\ &= 1. \end{aligned}$$

Thus the point v satisfies equation (1) and hence is on the line segment joining (0, 1) and (1, 0). But then (7) becomes

$$\sigma_1(2, 2) + (1 - \sigma_1)v,$$

a point on the line segment joining (2, 2) to v. We have proved that the right-hand side of (5) is always a point on the line segment between the point (2, 2) and a point on the line segment joining (0, 1) and (1, 0). Hence, it must lie in the triangle.

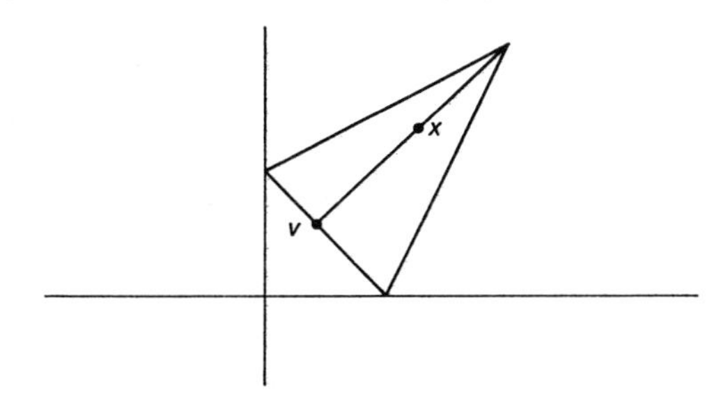

In conclusion, any point in the triangle can be expressed in the form (5), and any point of the form (5) must lie in the triangle.

Definition 3.1 ***Convex combination, convex polyhedron*** Let $a^1, a^2, \ldots, a^k$ be k vectors in R^n (a^t should not be confused with the tth component of a vector). If $\sigma_1, \ldots, \sigma_k$ are nonnegative numbers such that

$$\sigma_1 + \sigma_2 + \cdots + \sigma_k = 1,$$

then the linear combination

$$\sigma_1 a^1 + \sigma_2 a^2 + \cdots + \sigma_k a^k$$

is called a *convex combination* of $a^1, \ldots, a^k$. The totality of convex combinations of the set of vectors $a^1, \ldots, a^k$ is called the *convex polyhedron spanned by* $a^1, \ldots, a^k$. We denote this latter set by $H(a^1, \ldots, a^k)$. (In Exercise 6 the student is asked to prove that $H(a^1, \ldots, a^k)$ is indeed a convex set.)

Observe that $H(a^1, a^2)$ is what we have previously referred to as the line segment joining a^1 and a^2.

The main result of this section is contained in the following theorem.

Theorem 3.1 *Let A be an $m \times n$ matrix, b an m-vector and x an n-vector. If the set S of solutions to the system*

$$Ax = b, \qquad x \geq 0, \tag{8}$$

is bounded, then S is the convex polyhedron of its set of extreme points.

Proof According to Theorem 2.5, we know that the set S has at most a finite number of extreme points; call these $a^1, \ldots, a^k$. Since $a^i \in S$, it follows that $Aa^i = b$, $i = 1, \ldots, k$. Then a typical point $x \in H(a^1, \ldots, a^k)$ is of the form

$$x = \sigma_1 a^1 + \cdots + \sigma_k a^k,$$

where $\sigma_1 + \cdots + \sigma_k = 1$. It follows that

$$\begin{aligned} Ax &= A(\sigma_1 a^1 + \cdots + \sigma_k a^k) \\ &= \sigma_1 A a^1 + \cdots + \sigma_k A a^k \\ &= \sigma_1 b + \cdots + \sigma_k b \\ &= (\sigma_1 + \cdots + \sigma_k) b \\ &= b. \end{aligned}$$

We conclude that any point in $H(a^1, \ldots, a^k)$ is in S, i.e.,

(9) $$H(a^1, \ldots, a^k) \subset S.$$

The fact that (9) is indeed equality can be argued geometrically. The argument that follows in fact mimics the actual proof of this result. The details would carry us too far afield to be included here. Suppose that $v \in S$ but $v \notin H(a^1, \ldots, a^k)$.

(10)

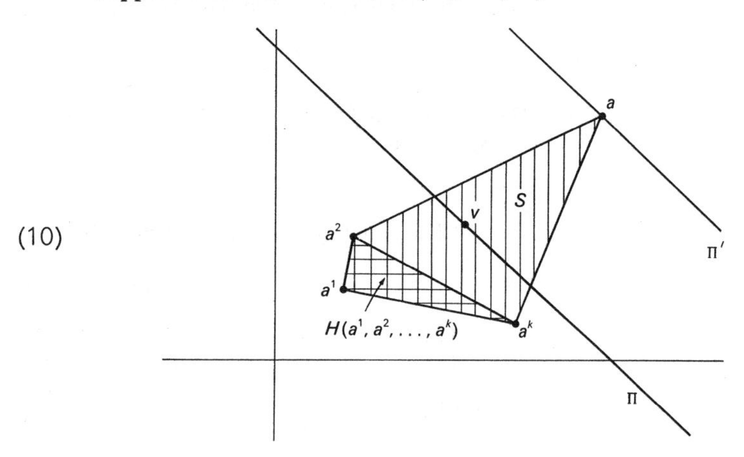

Pass a hyperplane Π through v such that $H(a^1, \ldots, a^k)$ is entirely contained in the open negative half-space of Π. Move the hyperplane Π parallel to itself and away from $H(a^1, \ldots, a^k)$ until it comes into contact with S for the last time. Call this "last" plane Π′. Now, Π′ will either be in contact with a vertex of S or an edge of S. In either case, the plane Π′ contains an extreme point of S. Therefore this extreme point is not contained in the open negative half-space of Π′. But the set $\{a^1, \ldots, a^k\}$ is supposed to be the complete set of extreme points of S and by assumption these all lie in the open negative half-space of Π and therefore in the open negative half-space of Π′. This is a contradiction. ▌

In most of the preceding examples of linear programming problems, we were required to find the maximum or minimum value of a function of the form $f(x) = (u, x)$, where x is in the set of solutions to a system of linear inequalities of the form

(11) $$Ax \leq b, \qquad x \geq 0.$$

Here A is an $m \times n$ matrix, x is an n-vector, and b is an m-vector. We already know how to find the extreme points of the set S of solutions to a system of *linear equations*. As we shall see shortly, it is easy to show that the maximum and minimum values of most of the functions which we study are taken on at extreme points. The question that remains is this: how do we find the maximum and minimum values of the function $f(x) = (u, x)$ on the set of solutions to the system of *linear inequalities* (11)? We answer this question in the next two theorems.

Theorem 3.2 *Let $u = (u_1, \ldots, u_n)$ be an extreme point of the set S of solutions of* (11). *Then there exists an m-tuple $(u_{n+1}, \ldots, u_{n+m})$ such that*

$$u' = (u_1, \ldots, u_n, u_{n+1}, \ldots, u_{n+m}) \tag{12}$$

is an extreme point of the set S' of solutions of the equality form of the system (11).

Proof The equality form of the system (11) is

$$m\left\{\left[\overbrace{\begin{matrix} a_{11} & \cdots & a_{1n} \\ & & \\ & & \\ \vdots & & \vdots \\ & & \\ a_{m1} & \cdots & a_{mn} \end{matrix}}^{n}\ \overbrace{\begin{matrix} 1 & 0 & 0 & \cdots & 0 & 0 \\ 0 & 1 & 0 & \cdots & 0 & 0 \\ 0 & & 1 & . & & \\ \vdots & & & \ddots & \vdots & \vdots \\ 0 & & & & 1 & 0 \\ 0 & 0 & 0 & \cdots & 0 & 1 \end{matrix}}^{m}\right]\right. \begin{pmatrix} x_1 \\ \vdots \\ x_n \\ x_{n+1} \\ \vdots \\ x_{n+m} \end{pmatrix} = b, \tag{13}$$

where $b = (b_1, \ldots, b_m)$. Since u satisfies (11), we know that

$$\sum_{j=1}^{n} a_{ij}u_j \leq b_i, \qquad i = 1, \ldots, m.$$

Define the nonnegative numbers $u_{n+1}, \ldots, u_{n+m}$ by

$$\begin{aligned} u_{n+1} &= b_1 - \sum_{j=1}^{n} a_{1j}u_j \geq 0, \\ &\vdots \\ u_{n+m} &= b_m - \sum_{j=1}^{n} a_{mj}u_j \geq 0. \end{aligned}$$

We claim that u' given in (12) is an extreme point of S'. For, suppose that

$$u' = (1 - t)w' + tv', \tag{14}$$

where

$$w' = (w_1, \ldots, w_n, w_{n+1}, \ldots, w_{n+m})$$

and

$$v' = (v_1, \ldots, v_n, v_{n+1}, \ldots, v_{n+m})$$

are in S', $0 < t < 1$. Let $w = (w_1, \ldots, w_n)$ and $v = (v_1, \ldots, v_n)$. Then

$$\sum_{j=1}^{n} a_{ij}w_j + w_{n+i} = b_i, \qquad i = 1, \ldots, m, \tag{15}$$

and since $w_{n+i} \geq 0$, we know that

$$\sum_{j=1}^{n} a_{ij}w_j \leq b_i, \qquad i = 1, \ldots, m,$$

i.e., $w \in S$. Similarly $v \in S$, and from (14), it is clear that

$$u = (1 - t)w + tv. \tag{16}$$

But u is an extreme point of S and since $0 < t < 1$, we conclude that $w = v$, i.e., u does not lie in the interior of any line segment in S. Since w and v are the same, it follows from (15) that $w_{n+i} = v_{n+i}$, $i = 1, \ldots, m$; for,

$$\begin{aligned} w_{n+i} &= b_i - \sum_{j=1}^{n} a_{ij}w_j \\ &= b_i - \sum_{j=1}^{n} a_{ij}v_j \\ &= v_{n+i}, \qquad i = 1, \ldots, m. \end{aligned}$$

Hence $w' = v'$. In other words, we have proved that if u' can be written in the form (14) in which $0 < t < 1$, then $w' = v'$ and hence u' cannot be in the interior of any line segment in S', i.e., u' is an extreme point of S'. This completes the proof. ∎

Theorem 3.2 together with Theorem 2.5 in the previous section shows us how to find all the extreme points of the set of solutions to a system of linear inequalities

$$Ax \leq b, \qquad x \geq 0. \tag{17}$$

These extreme points can all be systematically obtained as follows.

I. Put the system (17) into equality form and then find all the extreme points of the solution set. These will be vectors of the form

$$u' = (u_1, \ldots, u_n, u_{n+1}, \ldots, u_{n+m}) \in R^{n+m}. \tag{18}$$

II. For each of these extreme points u', consider the vector in R^n consisting of the first n components, i.e., "chop off" the last m components of u'. Then by Theorem 3.2, the extreme points of the system (17) must occur among these "truncated" vectors of the form

$$u = (u_1, \ldots, u_n) \in R^n.$$

Example 3.2 Find all extreme points of the set of solutions to the inequalities

$$\begin{aligned} 2x_2 - x_1 &\leq 1, \\ x_2 + x_1 &\leq 2, \\ x_1 &\geq 0, \\ x_2 &\geq 0. \end{aligned} \tag{19}$$

Let x_3 and x_4 denote nonnegative slack variables, so the equality form of the system (19) is

$$\begin{bmatrix} -1 & 2 & 1 & 0 \\ 1 & 1 & 0 & 1 \end{bmatrix} x = b, \quad x \geq 0.$$

This is precisely the system (44) considered in Example 2.9 in the preceding section. We found there that the complete set of extreme points is

$$\{(0, 0, 1, 2), (0, \tfrac{1}{2}, 0, \tfrac{3}{2}), (2, 0, 3, 0), (1, 1, 0, 0)\}.$$

According to the preceding theorem, the extreme points of the set of solutions to (19) must occur among the set

$$\{(0, 0), (0, \tfrac{1}{2}), (2, 0), (1, 1)\}. \tag{20}$$

Actually, if we graph the set of solutions to the system (19), we find that the extreme points are in fact all of those given in (20).

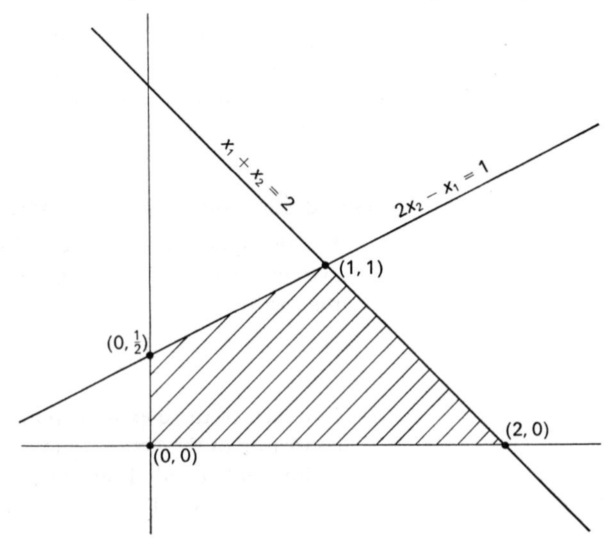

This is generally the case, namely, if we compute the set of all extreme points of S' in Theorem 3.2, then the "truncated" vectors, those obtained by "chopping off" the last m components of each of the vectors in S', are precisely the extreme points of S. This argument is very much like the one in Theorem 3.2 and will be relegated to an exercise.

Any function defined on R^n and of the form

$$f(x) = (u, x),$$

where u is some fixed n-vector, is called a *linear* function. Observe that the properties of the inner product (see Theorem 2.1, previous section) immediately imply that for any two vectors x and y in R^n and any numbers α and β,

$$f(\alpha x + \beta y) = \alpha f(x) + \beta f(y).$$

In fact, if $a^1, \ldots, a^k$ are vectors and $\sigma_1, \ldots, \sigma_k$ are numbers, then

$$f\left(\sum_{i=1}^{k} \sigma_i a^i\right) = \sum_{i=1}^{k} \sigma_i f(a^i). \tag{21}$$

We are now in a position to state and prove an important result in the applications to linear programming.

Theorem 3.3 *Let $a^1, \ldots, a^k$ be the set of extreme points of the set S of solutions to the system of inequalities*

$$Ax \leq b, \qquad x \geq 0. \tag{22}$$

Let

$$f(x) = (u, x) \tag{23}$$

be any linear function. If S is bounded, then the maximum (minimum) value of $f(x)$ for $x \in S$ is the largest (smallest) of the numbers $f(a^j)$, $j = 1, \ldots, k$.

Proof We know from Theorem 3.1 that

$$S = H(a^1, \ldots, a^k).$$

Thus any $x \in S$ has the form

$$x = \sum_{j=1}^{k} \sigma_j a^j,$$

in which $\sigma_j \geq 0, j = 1, \ldots, k$, and $\sum_{j=1}^{k} \sigma_j = 1$. Then by (21),

$$\begin{aligned} f(x) &= f\left(\sum_{j=1}^{k} \sigma_j a^j\right) \\ &= \sum_{j=1}^{k} \sigma_j f(a^j). \end{aligned}$$

Let $r_j = f(a^j)$, $j = 1, \ldots, k$. Then without loss of generality we can assume that r_1 is the largest of the r_j and r_k is the smallest. We compute that

$$\begin{aligned} r_1 &= r_1 \cdot 1 \\ &= r_1 \cdot \sum_{j=1}^{k} \sigma_j \\ &= \sum_{j=1}^{k} r_1 \sigma_j \\ &\geq \sum_{j=1}^{k} r_j \sigma_j \\ &= f(x). \end{aligned}$$

In other words, $f(x) \leq r_1 = f(a^1)$, so that the function f assumes its maximum value at the extreme point a^1. Similarly, we see that

$$\begin{aligned} r_k &= r_k \cdot 1 \\ &= r_k \cdot \sum_{j=1}^{k} \sigma_j \\ &= \sum_{j=1}^{k} r_k \sigma_j \\ &\leq \sum_{j=1}^{k} r_j \sigma_j \\ &= f(x), \end{aligned}$$

and thus the minimum value of f is $f(a^k)$. ▮

Theorem 3.2 tells us how to compute all the extreme points of the set S of solutions of the system (22) and Theorem 3.3 shows us that if S is bounded, then any linear function assumes its maximum and minimum values at the extreme points. In linear programming problems, the linear function f is usually called the *objective* function.

Example 3.3 A candy company has on hand 50, 80, and 90 pounds of cornstarch, cocoa, and sugar, respectively, from which two kinds of fudge are concocted. A six pound box of the first kind of fudge requires 2, 1, and 3 pounds each of cornstarch, cocoa, and sugar, respectively, while a five pound box of the second kind requires 1, 2, and 2 pounds of the respective ingredients. The first kind of fudge sells for \$3/box and the second kind sells for \$2.50/box. How many boxes of each kind of fudge should be made so that the company maximizes its profits? Let x_1 and x_2 be the number of boxes of the first and second types of fudge, respectively. Then the amount of cornstarch used in preparing the two kinds of fudge is $2x_1 + x_2$; the amount of cornstarch available is 50 pounds. Hence

$$2x_1 + x_2 \leq 50,$$

i.e., the amount of cornstarch used must not exceed the amount available. Similarly, the amount of cocoa used is $x_1 + 2x_2$; the amount available is 80 pounds, and hence

$$x_1 + 2x_2 \leq 80.$$

Finally, the amount of sugar used is $3x_1 + 2x_2$; the amount available is 90 pounds, so that

$$3x_1 + 2x_2 \leq 90.$$

These three inequalities form a system of linear inequalities together with the obvious requirements $x_1 \geq 0$, $x_2 \geq 0$. This system defines a convex region in the plane with easily computed extreme points. We solve the problem by two methods: first graphically, then using Theorem 3.2. If we graph the three lines

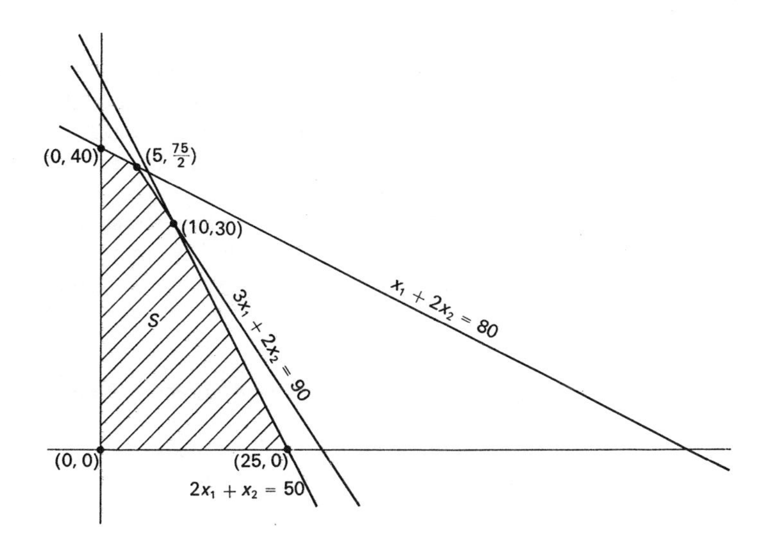

$$(24)\qquad \begin{aligned} 2x_1 + x_2 &= 50, \\ x_1 + 2x_2 &= 80, \\ 3x_1 + 2x_2 &= 90, \end{aligned}$$

and solve for the points of intersection in pairs, we find that the region S defined by the inequalities and the restrictions $x_1 \geq 0$, $x_2 \geq 0$ has as vertices the points $(0, 40)$, $(5, \frac{75}{2})$, $(10, 30)$, $(25, 0)$, $(0, 0)$.

The candy company receives \$3/box for the first kind of fudge and \$2.50/box for the second and hence its total financial return in dollars is

$$f(x) = 3x_1 + \tfrac{5}{2}x_2.$$

In order to maximize this function, we need only compare the values of f at the five vertices of S. We compute that

$$\begin{aligned} f((0, 40)) &= 3 \cdot 0 + (\tfrac{5}{2}) \cdot (40) \\ &= 100, \\ f((5, \tfrac{75}{2})) &= 3 \cdot 5 + (\tfrac{5}{2}) \cdot (\tfrac{75}{2}) \\ &= 15 + (\tfrac{375}{4}) \\ &= 15 + 93.75 \\ &= 108.75, \\ f((10, 30)) &= 3 \cdot 10 + (\tfrac{5}{2}) \cdot (30) \\ &= 30 + 75 \\ &= 105, \\ f((25, 0)) &= 3 \cdot 25 + (\tfrac{5}{2}) \cdot 0 \\ &= 75, \\ f((0, 0)) &= 0. \end{aligned}$$

Clearly the maximum return will be obtained when 5 boxes of the first kind of fudge and $37\frac{1}{2}$ boxes of the second kind of fudge are produced. If we assume that an integral number must be produced, then, of course, we can manufacture only 37 boxes of the second kind of fudge.

We can also compute the extreme points of the convex region S by using the methods in Theorem 2.5. By putting the inequalities into equality form, the matrix-vector form of the system is

$$(25)\qquad \begin{bmatrix} 2 & 1 & 1 & 0 & 0 \\ 1 & 2 & 0 & 1 & 0 \\ 3 & 2 & 0 & 0 & 1 \end{bmatrix} x = \begin{bmatrix} 50 \\ 80 \\ 90 \end{bmatrix}, \qquad x \geq 0.$$

Suppressing the variables two at a time in (25), we find that the resulting systems of linear equations produce the 5-tuples $(0, 0, 50, 80, 90)$, $(0, 50, 0, -20, -10)$, $(0, 40, 10, 0, 10)$, $(0, 45, 5, -10, 0)$,

(25, 0, 0, 55, 15), (80, 0, −110, 0, −150), (30, 0, −10, 50, 0), $(\frac{20}{3}, \frac{110}{3}, 0, 0, -\frac{10}{3})$, (10, 30, 0, 10, 0), $(5, \frac{75}{2}, \frac{5}{2}, 0, 0)$. According to Theorem 2.5, the extreme points of the set S of solutions to (25) are those in which all the components are nonnegative. (Any set of 3 columns of the coefficient matrix in (25) is linearly independent, as can be checked directly.) Thus by Theorem 3.2 the set of extreme points of S' is

$$\{(0, 0, 50, 80, 90), (0, 40, 10, 0, 10), (25, 0, 0, 55, 15), (10, 30, 0, 10, 0), (5, \tfrac{75}{2}, \tfrac{5}{2}, 0, 0)\}. \tag{26}$$

We can obtain the extreme points of the set S of solutions by simply dropping the last three components of the points in (26). The resulting set of extreme points of S is then

$$\{(0, 0), (0, 40), (25, 0), (10, 30), (5, \tfrac{75}{2})\}.$$

These latter points, of course, coincide exactly with those determined by the graphical method.

It is often the case in linear programming problems that one is required to find the minimum value of a linear function defined on a region which is not necessarily bounded but which nevertheless is the set of solutions to a system of the form

$$Ax \leq b, \qquad x \geq 0.$$

In fact, in Example 1.1, Section 3.1, we were required to minimize the objective function

$$g(x) = x_1 + x_2$$

over the region consisting of the solutions to the system of linear inequalities

$$\begin{aligned} x_1 + 2x_2 &\geq 80, \\ 3x_1 + 2x_2 &\geq 160, \\ 5x_1 + 2x_2 &\geq 200, \\ x_1 &\geq 0, \\ x_2 &\geq 0. \end{aligned} \tag{27}$$

It was seen that the set of solutions to (27) is unbounded (see Figure (8), Section 3.1), and thus we do not have the conditions of the hypotheses in Theorem 3.3, i.e., that the set of solutions to the system (27) be bounded. Thus we cannot find the maximum and/or minimum values of g. But we can bring this problem under the purview of Theorem 3.3 quite easily. For, in the statement of Example 1.1, it was stipulated that the production quotas were to be met in a six-month period. This means that both x_1 and x_2 (the numbers of days that plants I and II operate, respectively) cannot exceed 180, i.e., $x_1 \leq 180$ and $x_2 \leq 180$. Thus we can rewrite the system (27) in the standard form:

$$\begin{aligned} -x_1 - 2x_2 &\leq -80, \\ -3x_1 - 2x_2 &\leq -160, \\ -5x_1 - 2x_2 &\leq -200, \\ x_1 &\leq 180, \\ x_2 &\leq 180, \\ x_1 &\geq 0, \\ x_2 &\geq 0. \end{aligned} \tag{28}$$

If we graph the region defined by this last system, we find that it is a polygon in the plane whose vertices are (0, 180), (0, 100), (20, 50), (40, 20), (80, 0), (180, 0), (180, 180).

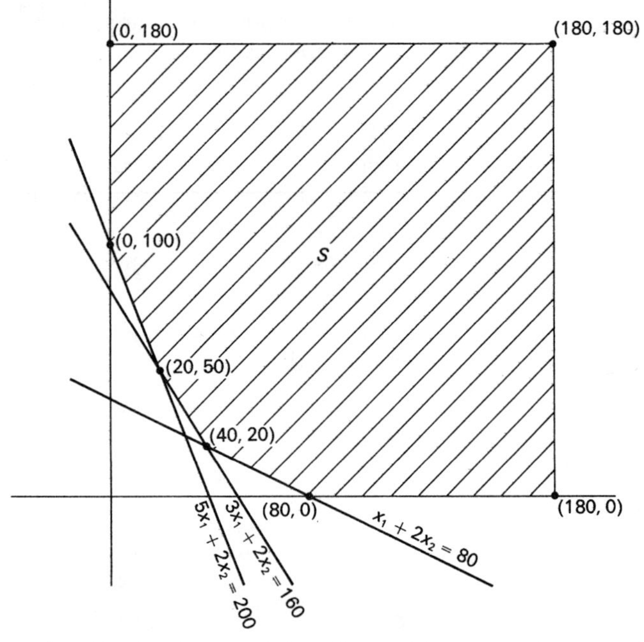

The region S is clearly bounded and hence Theorem 3.3 now applies. We compute that

$$\begin{aligned} g((0, 180)) &= 0 + 180 \\ &= 180, \\ g((0, 100)) &= 100, \\ g((20, 50)) &= 70, \\ g((40, 20)) &= 60, \\ g((80, 0)) &= 80, \\ g((180, 0)) &= 180, \\ g((180, 180)) &= 360. \end{aligned}$$

The minimum of g is clearly 60, which is achieved at the vertex (40, 20).

We conclude this section by showing how the techniques developed so far, in particular the results of Theorem 3.3, can be used to solve a linear programming problem which ostensibly sounds rather different from the ones we have thus far considered. Consider Example 6.7 in Section 2.6, which we briefly restate in our next example.

Example 3.4 (Example 6.7, Chapter 2, revisited) A food chain store has n retail outlets $o_1, \ldots, o_n$ and m warehouses $h_1, \ldots, h_m$ of capacities $s_1, \ldots, s_m$, respectively. The outlets place orders with the warehouses which will empty them all: o_j orders a total of b_j tons to be delivered from the m warehouses. The cost of shipping one ton from h_i to o_j is a_{ij} dollars. Let x_{ij} denote the amount in tons shipped from h_i to o_j. The delivery of x_{ij} tons from h_i to o_j costs $a_{ij}x_{ij}$ dollars and hence the total cost for operating the chain is the sum of all the numbers $a_{ij}x_{ij}$. We shall assume here that the number of outlets and the number of warehouses is the same: $n = m$. Moreover, we will assume that the capacities of all the warehouses are the same, i.e., $s_1 = s_2 = \cdots = s_n = s$, and that each outlet orders the same amount, i.e., $b_1 = b_2 = \cdots = b_n = b$. The fact that the warehouses will be emptied by these orders can be expressed as

$$\sum_{j=1}^{n} x_{ij} = s, \qquad i = 1, \ldots, n, \tag{29}$$

and the fact that each store orders a total of b tons can be expressed

$$\sum_{i=1}^{n} x_{ij} = b, \qquad j = 1, \ldots, n. \tag{30}$$

It follows from (29) and (30) that $b = s$. (Why? See Exercise 21.) Each x_{ij} is nonnegative, and if we let $y_{ij} = \dfrac{x_{ij}}{s}$, $i, j = 1, \ldots, n$, then the conditions in (29) and (30) say that the matrix Y whose (i, j) entry is y_{ij} is a doubly stochastic matrix. (See Definition 5.1, Chapter 2). The objective function that we want to minimize is

$$\sum_{i=1}^{n} \sum_{j=1}^{n} a_{ij}x_{ij}. \tag{31}$$

Clearly, minimizing (31) is the same problem as minimizing the objective function

$$f(Y) = \sum_{i=1}^{n} \sum_{j=1}^{n} a_{ij}y_{ij}, \tag{32}$$

where Y varies over the set of all doubly stochastic matrices. To complete this problem, we draw upon a result contained in Exercise 20 in Section 2.5 in which we proved that every doubly stochastic matrix contains at least one diagonal of positive entries. We shall use this result to prove that any doubly stochastic matrix is a convex combination of permutation matrices (see Exercise 16, Section 2.5), that is,

$$Y = \sum_{k=1}^{N} \gamma_k P_k, \tag{33}$$

where $\gamma_k \geq 0, k = 1, \ldots, N, \sum_{k=1}^{N} \gamma_k = 1$, and the $P_k, k = 1, \ldots, N$, are all the $N = n!$ n-square permutation matrices. Deferring this argument momentarily, we return to the function $f(Y)$ in (32) and immediately verify (see Exercise 22) that for any n-square matrices Y and Z and for any numbers α and β,

$$f(\alpha Y + \beta Z) = \alpha f(Y) + \beta f(Z). \tag{34}$$

It follows immediately from (33) and (34) that

$$\begin{aligned} f(Y) &= f\left(\sum_{k=1}^{N} \gamma_k P_k\right) \\ &= \sum_{k=1}^{N} \gamma_k f(P_k). \end{aligned} \tag{35}$$

Precisely as in the proof of Theorem 3.3, we can prove that the minimum value is the least of the numbers $f(P_k)$, $k = 1, \ldots, N$, call it $f(P_1)$. Now, P_1 is a permutation matrix, so there exists an n-permutation σ of $\{1, \ldots, n\}$ such that the (i, j) entry of P_1 is 1 if $\sigma(j) = i$ and 0 otherwise. Hence

$$f(P_1) = \sum_{j=1}^{n} a_{\sigma(j),j}, \tag{36}$$

because $a_{ij}y_{ij} = 0$ unless $i = \sigma(j)$, in which case $a_{ij}y_{ij} = a_{\sigma(j),j}$. But (36) is just the sum down a diagonal of $A = (a_{ij})$. It follows that the minimum cost for operating this chain is the least sum obtained as follows. Add the elements in each of the $n! = N$ diagonals of A; call these sums $\alpha_1, \ldots, \alpha_N$. The least of the numbers $s\alpha_i$, $i = 1, \ldots, N$, is the minimum cost of supplying the chain of outlets from the warehouses. (Recall that we modified the original x_{ij}'s by dividing by s.) The only thing that remains to be resolved is the proof that any doubly stochastic matrix is of the form (33). We leave this as Exercise 23, which is accompanied by a lengthy hint.

Quiz

Answer true or false.

1 The vector $(0, 0)$ is a convex combination of $(-1, -1)$ and $(1, 1)$ in R^2.

2 If S is a line segment in R^n joining a^1 and a^2, then $S = H(a^1, a^2)$.

3 The convex polyhedron in R^2 spanned by $(1, 0)$, $(0, 1)$, $(-1, 0)$, and $(0, -1)$ consists of the interior and boundary of a square.

4 The function $f(x) = 2x_1 + 3x_2 + 1$ defined on R^2 is linear.

5 If $f(x) = (u, x)$ and $g(x) = (v, x)$ for all $x \in R^n$ and $h = f + g$, i.e., $h(x) = f(x) + g(x)$, then $h(x) = (u + v, x)$ for all $x \in R^n$.

6 The set of solutions $S \subset R^n$ of a system of linear inequalities $Ax \leq b$, $x \geq 0$, is always the convex polyhedron spanned by the extreme points of S.

7 If S^n is a set of vectors of the form $a = (a_1, \ldots, a_n)$ in which $a_j \geq 0$, $j = 1, \ldots, n$, and $\sum_{j=1}^{n} a_j = 1$, then S^n is a convex polyhedron.

8 If $a^1 = (2, 1)$, $a^2 = (1, 2)$, $a^3 = (1, 1)$, and $a^4 = (2, 2)$ then $H(a^1, a^2) = H(a^3, a^4)$.

9 If $a^1 = (1, 0)$, $a^2 = (0, 1)$, and $a^3 = (\frac{1}{2}, \frac{1}{2})$, then $H(a^1, a^2, a^3) = H(a^1, a^2)$.

10 If $H(a^1, \ldots, a^k) = H(b^1, \ldots, b^r)$ in R^n, then each b^i is a convex combination of $a^1, \ldots, a^k$, $i = 1, \ldots, r$.

Exercises

1 Sketch the convex polyhedron in R^2 or R^3 spanned by each of the following sets of points:

(a) $(1, 0), (0, 1)$;
(b) $(0, 0), (1, 0), (0, 1)$;
(c) $(1, 1), (2, 2), (\frac{1}{2}, \frac{1}{2})$;
(d) $(0, 1), (1, 0), (-1, 0), (0, -1)$;
(e) $(1, 1, 1), (0, 0, 1), (1, 0, 0)$;
(f) $(1, 0, 0), (0, 0, 1), (0, 1, 0)$;
(g) $(1, 0, 0), (0, 0, 1), (0, 1, 0), (0, 0, 0)$;
(h) $(-1, -1), (1, 1), (1, -1), (-1, 1)$;
(i) $(1, 1), (2, 2), (3, 3), (4, 4)$;
(j) $(1, 1), (2, 2), (3, 3), (4, 4), (1, -1), (2, -2), (3, -3), (4, -4)$.

2 Express the indicated vector v as a convex combination of the vectors $a^1, a^2, \ldots, a^k$ in each of the following:

(a) $v = (\frac{1}{2}, \frac{1}{2})$, $a^1 = (1, 1)$, $a^2 = (0, 0)$;
(b) $v = (0, 0)$, $a^1 = (1, 0)$, $a^2 = (0, 1)$, $a^3 = (-1, 0)$, $a^4 = (0, -1)$;
(c) $v = (-1, -2)$, $a^1 = (-4, 0)$, $a^2 = (0, 0)$, $a^3 = (0, -3)$;
(d) $v = (\frac{1}{4}, \frac{1}{2}, \frac{1}{4})$, $a^1 = (1, 0, 0)$, $a^2 = (0, 1, 0)$, $a^3 = (0, 0, 1)$;
(e) $v = (\frac{5}{2}, \frac{1}{4})$, $a^1 = (1, 1)$, $a^2 = (3, -2)$, $a^3 = (5, 1)$.

3 Which of the following figures in R^2 is a convex polyhedron?

(*a*)

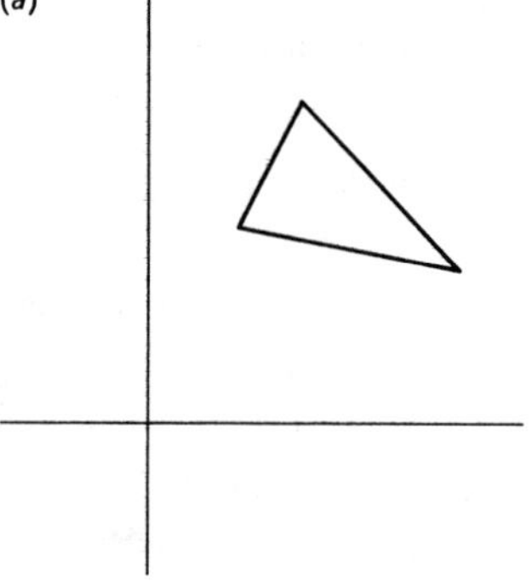

(*b*)

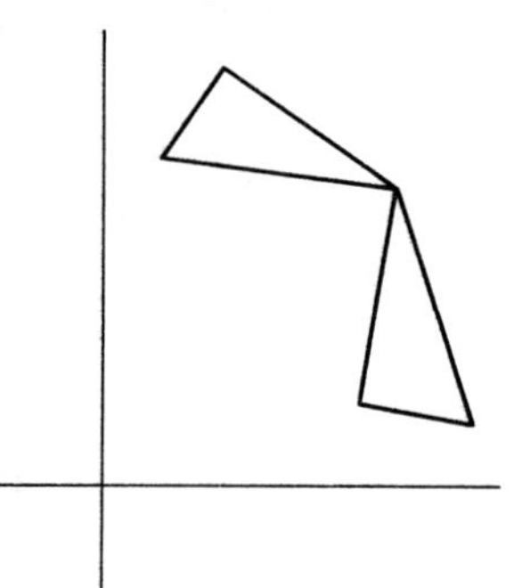

(*c*)

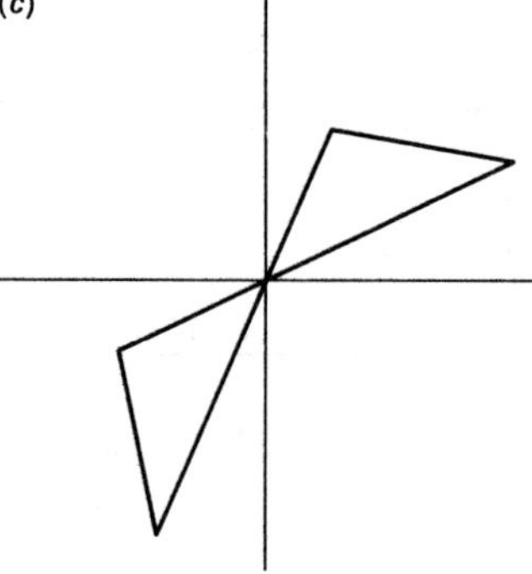

(*d*)

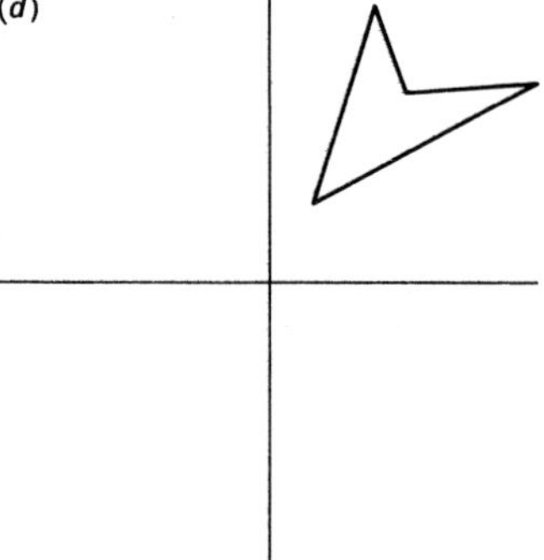

(*e*)

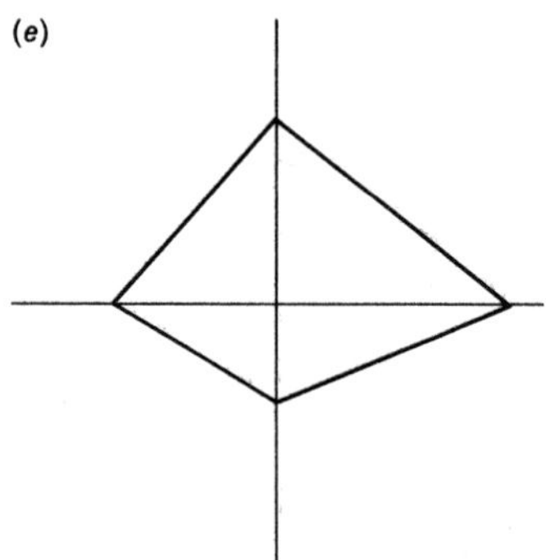

(*f*)

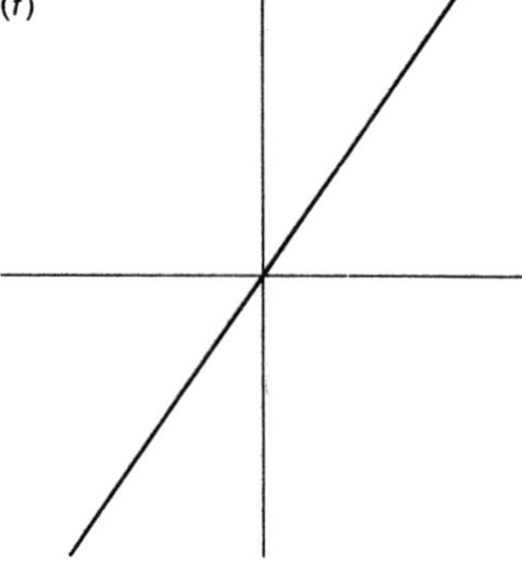

(*g*)

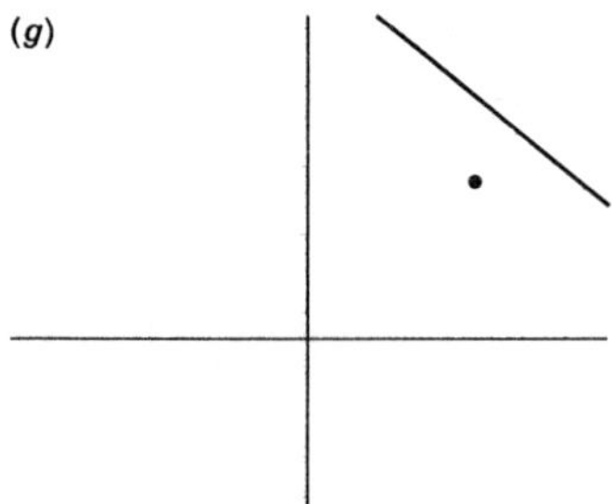

(*h*)

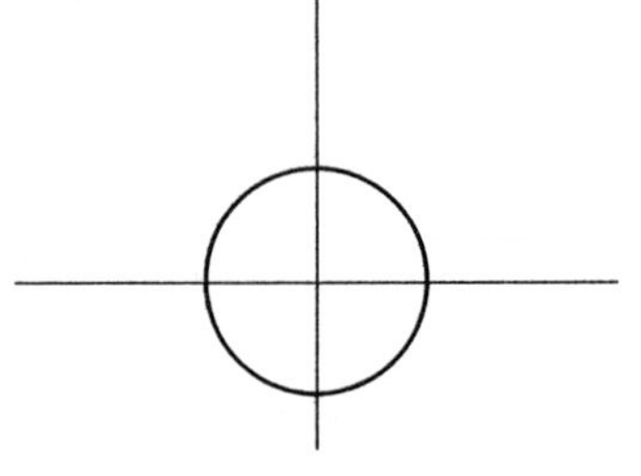

4 Let a^1, a^2, and a^3 be three vectors in R^2 such that $a^2 - a^1$ and $a^3 - a^2$ are linearly independent. Show that any vector $v \in H(a^1, a^2, a^3)$ can be expressed *uniquely* as

$$v = \sigma_1 a^1 + \sigma_2 a^2 + \sigma_3 a^3$$

in which $\sigma_1 + \sigma_2 + \sigma_3 = 1$, $\sigma_i \geq 0$, $i = 1, 2, 3$.

5 Express each of the following convex polyhedra $H(a^1, \ldots, a^k)$ as the intersection of half-spaces in R^2:
(a) $a^1 = (1, -1)$, $a^2 = (2, 2)$, $a^3 = (3, 1)$;
(b) $a^1 = (1, 1)$, $a^2 = (2, -1)$, $a^3 = (-1, -2)$, $a^4 = (-3, -1)$;
(c) $a^1 = (9, 18)$, $a^2 = (\frac{60}{7}, \frac{120}{7})$, $a^3 = (15, 12)$;
(d) $a^1 = (1, 2)$, $a^2 = (-1, 1)$, $a^3 = (0, -1)$, $a^4 = (2, 2)$.

†6 Prove that the convex polyhedron $H(a^1, \ldots, a^k)$ spanned by $a^1, \ldots, a^k$ in R^n is a convex set.

†7 Let X be any subset of R^n. Define the set $H(X) \subset R^n$ to be the totality of convex combinations of finite subsets of X, i.e.,

$$H(X) = \{x \mid x \in H(a^1, \ldots, a^k) \text{ for some vectors } a^1, \ldots, a^k \text{ in } X\}.$$

(a) Show that $H(X)$ is always a convex set.
(b) Show that $H(X)$ is the smallest convex set containing X, i.e., if S is a convex set such that $X \subset S$, then $X \subset H(X) \subset S$.
(c) Show that X is convex if and only if $H(X) = X$.
(d) Show that $H(H(X)) = H(X)$ for any set X.

The set $H(X)$ is called the *convex hull* of X.

8 Find the minimum and maximum values for each of the functions $f(x)$ defined on the set of all $x \in R^n$ satisfying $x \geq 0$ and $(x, e) = 1$, $e = (1, \ldots, 1) \in R^n$:
(a) $f(x) = 2x_1 + 3x_2$ $(n = 2)$;
(b) $f(x) = 2x_1 + 3x_2$ $(n = 3)$;
(c) $f(x) = -x_1 + 2x_2 + x_4$ $(n = 4)$;
(d) $f(x) = x_1 + 2x_2 + 3x_3 + 4x_4$ $(n = 4)$;
(e) $f(x) = (x_1 + 2x_2 + x_3)^2$ $(n = 3)$;
(f) $f(x) = x_1 + 2x_2 - x_3 - x_4 + 5$ $(n = 4)$.

9 Find the maximum and minimum values of $f(x) = (u, x)$ on $H(a^1, \ldots, a^k)$:
(a) $u = (1, 2, -1)$, $a^1 = (0, 0, 0)$, $a^2 = (-3, 0, 1)$, $a^3 = (1, 0, 1)$;
(b) $u = (1, 1)$, $a^1 = (-1, 1)$, $a^2 = (1, -1)$, $a^3 = (1, 1)$, $a^4 = (-1, 0)$, $a^5 = (0, -1)$;
(c) $u = (0, 0, 0)$, $a^1 = (1, 2, 3)$, $a^2 = (5, -9, 23)$, $a^3 = (-10, 5, 18)$;
(d) $u = (1, 1, 1, 1)$, $a^1 = (1, -1, 1, -1)$, $a^2 = (-1, -1, 1, 1)$, $a^3 = (-1, 1, -1, 1)$, $a^4 = (1, 1, -1, -1)$, $a^5 = (-1, 1, 1, -1)$;
(e) $u = (1, 2, 3)$, $a^1 = (4, 5, 6)$, $a^2 = (5, 4, 6)$, $a^3 = (6, 5, 4)$, $a^4 = (4, 6, 5)$, $a^5 = (6, 4, 5)$, $a^6 = (5, 6, 4)$.

10 Draw a sketch depicting the convex hull of each of the following sets X.

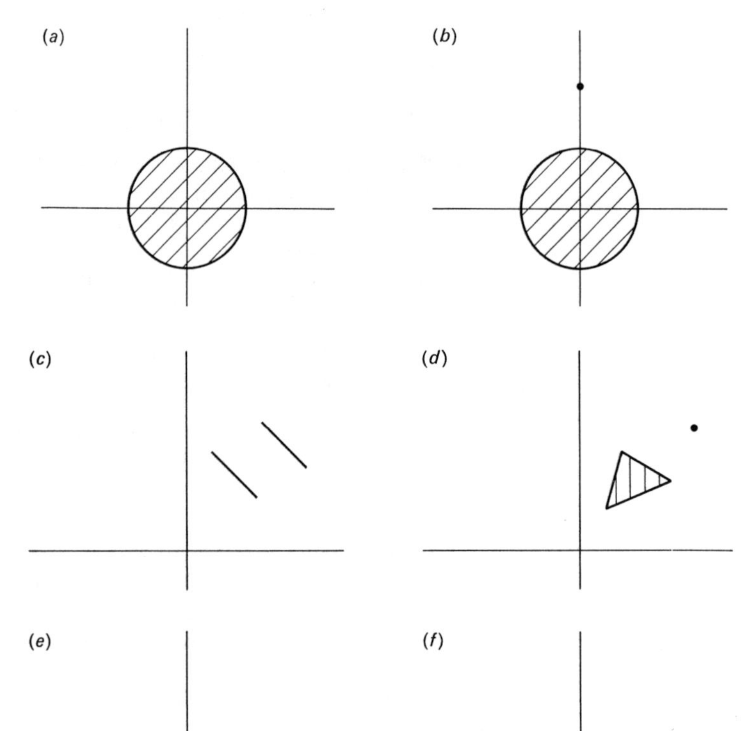

†11 Let $r_1, \ldots, r_n$ be n numbers and set $R = \max_k r_k$, $r = \min_k r_k$ (i.e., R and r are the largest and smallest of the r_k, $k = 1, \ldots, n$, respectively). Let $\alpha = \sum_{k=1}^{n} \sigma_k r_k$ be a convex combination of $r_1, \ldots, r_n$ in which $\sigma_k > 0$, $k = 1, \ldots, n$, $\sum_{k=1}^{n} \sigma_k = 1$, and let $\sigma = \min_k \sigma_k$. Show that $\sigma r + (1 - \sigma)R \geq \alpha \geq \sigma R + (1 - \sigma)r$. (Hint: Suppose that $R = r_t$. Then

$$\begin{aligned}\alpha &= \sum_{k=1}^{n} \sigma_k r_k \\ &= \sigma_t r_t + \sum_{k \neq t} \sigma_k r_k \\ &= \sigma_t R + \sum_{k \neq t} \sigma_k r_k\end{aligned}$$

$$\begin{aligned}
&\geq \sigma_t R + \sum_{k \neq t} \sigma_k r \\
&= \sigma_t R + (1 - \sigma_t)r \\
&= \sigma_t(R - r) + r \\
&\geq \sigma(R - r) + r \\
&= \sigma R + (1 - \sigma)r.
\end{aligned}$$

Similarly, suppose that $r = r_s$. Then

$$\begin{aligned}
\alpha &= \sum_{k=1}^{n} \sigma_k r_k \\
&= \sigma_s r_s + \sum_{k \neq s} \sigma_k r_k \\
&= \sigma_s r + \sum_{k \neq s} \sigma_k r_k \\
&\leq \sigma_s r + \sum_{k \neq s} \sigma_k R \\
&= \sigma_s r + (1 - \sigma_s)R \\
&= \sigma_s(r - R) + R \\
&\leq \sigma(r - R) + R \qquad (\text{since } r - R \leq 0) \\
&= \sigma r + (1 - \sigma)R.)
\end{aligned}$$

†12 Let S be an $n \times n$ column stochastic matrix with positive entries. Let $R_i(p)$ be the largest entry in the ith row of S^p and let $r_i(p)$ be the smallest entry in the ith row of S^p. Let σ be the least entry in S. Prove that

$$R_i(p + 1) - r_i(p + 1) \leq (1 - 2\sigma)(R_i(p) - r_i(p)),$$

$p = 1, 2, \ldots$. (Hint: Let $s_{ij}^{(p)}$ denote the (i, j) entry in S^p. Then

$$s_{ij}^{(p+1)} = \sum_{k=1}^{n} s_{ik}^{(p)} s_{kj}, \qquad j = 1, \ldots, n.$$

Now, $\sum_{k=1}^{n} s_{kj} = 1$. Let s_{kj} play the role of σ_k and $s_{ik}^{(p)}$ play the role of r_k in Exercise 11, $k = 1, \ldots, n$. Then if $\sigma_j' = \min_k s_{kj}$, we have

$$\begin{aligned}
s_{ij}^{(p+1)} &\geq \sigma_j' R_i(p) + (1 - \sigma_j')r_i(p) \\
&\geq \sigma R_i(p) + (1 - \sigma)r_i(p), \qquad j = 1, \ldots, n.
\end{aligned}$$

Similarly,

$$s_{ij}^{(p+1)} \leq \sigma r_i(p) + (1 - \sigma)R_i(p), \qquad j = 1, \ldots, n.$$

Hence, in particular,

$$\begin{aligned}
r_i(p + 1) &\geq \sigma R_i(p) + (1 - \sigma)r_i(p), \\
R_i(p + 1) &\leq \sigma r_i(p) + (1 - \sigma)R_i(p),
\end{aligned}$$

and thus

$$R_i(p + 1) - r_i(p + 1) \leq (1 - 2\sigma)(R_i(p) - r_i(p)).$$

This last formula is valid even if the least element of the matrix S is 0, i.e., even if $\sigma = 0$. (Why?) In this case, the formula will be

$$R_i(p + 1) - r_i(p + 1) \leq R_i(p) - r_i(p).)$$

†13 Let σ be the least entry in an $n \times n$ column stochastic matrix S. Prove that $\sigma \leq \frac{1}{n}$. (Hint: Each column sum is 1. If $\sigma = a_{st}$, then $a_{it} \geq a_{st}$, $i = 1, \ldots, n$, and hence

$$1 = \sum_{i=1}^{n} a_{it} \geq na_{st} = n\sigma.)$$

14 Find the maximum of the function $f(x) = (u, x)$, where $u = (1, 10)$, defined on the set S of all solutions to the system of inequalities

$$\begin{aligned} x_1 + 5x_2 &\leq 15, \\ 3x_1 + 5x_2 &\leq 20, \\ x_1 &\leq 5, \\ x_1 &\geq 0, \\ x_2 &\geq 0. \end{aligned}$$

15 Find the minimum value of the function $f(x) = (u, x)$, $u = (3, 5)$, defined on the set S of all solutions to the system of linear inequalities

$$\begin{aligned} 2x_1 - 3x_2 &\leq 6, \\ -3x_1 + 2x_2 &\leq 6, \\ x_1 &\geq 0, \\ x_2 &\geq 0. \end{aligned}$$

Does f have a maximum on S? Is S bounded? If the additional inequalities $x_1 \leq 10$ and $x_2 \leq 5$ are included, find the maximum and minimum values of f on this modified set.

16 The Ace Tennis Racket Company manufactures two different types of metal rackets, the Fault (I) and the Snowshoe (II). For a given hour's assembly line production, the following amounts of aluminum (A), steel (S), and magnesium (M) are available: 16, 11, and 15 units, respectively. The construction of a single type I racket requires 1, 2, and 3 units of A, S, and M, respectively, while the construction of a single racket of type II requires 2, 1, and 1 units of A, S, and M, respectively. How many of each of the two types of rackets should the company construct each hour in order to maximize the amount of money earned if the Fault sells for \$30 and the Snowshoe for \$50? (Assume the demand is unlimited.)

17 The Olfactory Perfume Factory manufactures two different colognes, Nostrils Forever and Enchanted Sinus, in which ingredients A, B, and C are used in the following amounts per bottle: 2 oz., 3 oz., and 1 oz. of the respective ingredients are used in the first kind of cologne, and 3 oz., 1 oz., and 2 oz. of the respective ingredients are used in the second kind. There are 150 oz., 120 oz., and 90 oz. of ingredients A, B, and C, respectively, in the storeroom. Nostrils Forever sells for \$3.50/6 oz. bottle, and Enchanted Sinus sells for \$5/6 oz. bottle. What proportions of the two colognes should be manufactured in order that the Olfactory Factory's market is most lucrative? (There is unlimited demand for these two scents.)

18 A company owns two manufacturing plants, I and II, having manufacturing capacities of 700 and 500 units daily. The firm must supply three distributors, d_1, d_2, and d_3, with 250, 600, and 350 units daily, respectively, and the cost in dollars of transporting one unit is given in the following table.

	I	II
d_1	3	5
d_2	2	3
d_3	5	7

How many units should be shipped daily from each plant to each distributor so as to minimize transportation costs?

†19 Let S be a convex subset of R^n. A real-valued function f defined on S is said to be *convex* if for any x and y in S and any number t, $0 \leq t \leq 1$,

$$f((1 - t)x + ty) \leq (1 - t)f(x) + tf(y).$$

(a) Show that the linear function $f(x) = (u, x)$ is convex.

(b) Show that for $n = 1$ and $S = R^1$, $f(x_1) = x_1^2$ is a convex function.

(c) Show that if $\varphi_1(s), \ldots, \varphi_n(s)$ are convex functions on R^1, then $f(x) = f((x_1, \ldots, x_n)) = \varphi_1(x_1) + \cdots + \varphi_n(x_n)$ is a convex function on R^n.

(d) Show that if f and g are convex functions on S, then $f + g$ is a convex function on S, where $(f + g)(x) = f(x) + g(x)$.

(e) Show by constructing an example that the product of two convex functions is not necessarily a convex function.

(f) Let f and g be convex functions defined on S, and define $h(x)$ to be the largest of $f(x)$ and $g(x)$ for each $x \in S$, i.e., $h(x) = \max(f(x), g(x))$. Show that h is a convex function defined on S.

(g) Let $a^1, \ldots, a^k$ be vectors in S. Prove that the largest value taken on by a convex function on the polyhedron $H(a^1, \ldots, a^k)$ is $\max_j f(a^j)$.

†20 Let S' be the set of all solutions of the equality form of the system

$$Ax \leq b, \quad x \geq 0.$$

Show that the "truncated" vectors obtained by "chopping off" the last m components of all of the vectors which are extreme points of S' are precisely all the extreme points of the set S of solutions to the system of inequalities

$$Ax \leq b, \quad x \geq 0.$$

(Hint: Carefully examine the proof of Theorem 3.2.)

21 In reference to equations (29) and (30) prove that $b = s$.

†22 Verify that the function $f(Y)$ in formula (32) is *linear in* Y, i.e., if Y and Z are two matrices and α and β are two numbers, then

$$f(\alpha Y + \beta Z) = \alpha f(Y) + \beta f(Z).$$

†23 Let Y be an $n \times n$ doubly stochastic matrix. Show that Y is a convex combination of permutation matrices, as indicated in formula (33). (Hint: According to Problem 20 in Section 2.5, we know Y must contain at least one diagonal of positive entries, that is, there are n entries of the matrix Y, precisely one in each row and each column, all of which are positive. Let σ denote the smallest positive entry in this diagonal. Suppose $\sigma = 1$. Then, since Y is doubly stochastic, it follows that every entry in the diagonal is 1 (Why?) and hence that Y is a permutation matrix. (Why?) Let P be an $n \times n$ permutation matrix with 1's in the positions of this positive diagonal. Assuming now that $\sigma < 1$, consider the matrix $Y' = \dfrac{1}{1 - \sigma}(Y - \sigma P)$. It is very easy to verify that the entries in Y' are all nonnegative and that every row and column sum of Y' is 1. In other words, Y' is a doubly stochastic matrix. But Y' has at least one more zero entry than Y. (Why?) We are now able to write a finite induction argument in which the induction is made on the number of positive entries in Y. The student will provide details of the argument.)

3.4

Game Theory

In this and the next section, we shall see how the techniques from the theory of convex sets, linear algebra, and probability theory can be brought together to analyze whole new classes of problems. In the present section, we consider the elementary aspects of the theory of games, a field invented in 1928 by the American mathematician John von Neumann.

A *game*, in most general terms, is a situation of conflict between people or groups of people (such as countries or bridge partnerships or chess players) which can be rationally analyzed. A class of games which can be studied with some degree of completeness is the class of so-called *matrix games*. Suppose A is an $m \times n$ matrix whose (i, j) entry is a_{ij}. Define a game associated with this matrix as follows. Two players, denoted by R and C, play a game: R chooses a row of the matrix A, and simultaneously, C chooses a column of A; then C pays R the amount which lies at the intersection of the chosen row and column. That is, if R chooses row i and C chooses column j, then C pays R the amount a_{ij}. If a_{ij} is positive, it represents a payment from C to R; if a_{ij} is negative, it represents a payment from R to C. The number a_{ij} is called the *payoff*.

Suppose we want to determine the safest possible way the two players can play the game. Let us say R chooses $A_{(i)}$. With this choice of row, R realizes that there is a possibility that he will receive the least payment in $A_{(i)}$. This amount is

$$\alpha_i = \min_j a_{ij}.$$

On the other hand, R can always guarantee himself a payment which is the largest of the α_i. Thus R can assure himself of winning at least

$$\alpha = \max_i \min_j a_{ij}.$$

We can analyze the situation for C by a similar argument; for, if C chooses column j, then he realizes that there is a possibility of having to pay R the largest amount in $A^{(j)}$, that is, C may have to pay

$$\beta_j = \max_i a_{ij}.$$

Thus the smartest thing for C to do is to minimize this payoff, i.e., C can assure that the payoff is the smallest of the β_j's:

$$\beta = \min_j \max_i a_{ij}.$$

These conservative strategies for R and C are called the *maximin* and *minimax* strategies. The number α is called the *lower value* of the game; the number β is called the *upper value* of the game.

Example 4.1 Find the upper and lower values of the game defined by the matrix

$$\begin{bmatrix} 0 & 1 & -2 & 4 \\ 5 & 0 & 1 & -1 \\ 2 & -1 & 0 & 3 \end{bmatrix}.$$

Observe that $\alpha_1 = \min\{0, 1, -2, 4\} = -2$. Similarly, $\alpha_2 = -1$ and $\alpha_3 = -1$. Thus

$$\begin{aligned} \alpha &= \max\{\alpha_1, \alpha_2, \alpha_3\} \\ &= \max\{-2, -1, -1\} \\ &= -1 \\ &= \alpha_2 \text{ or } \alpha_3. \end{aligned}$$

Similarly, $\beta_1 = \max\{0, 5, 2\} = 5$, and $\beta_2 = 1$, $\beta_3 = 1$, $\beta_4 = 4$. Thus

$$\begin{aligned} \beta &= \min\{\beta_1, \beta_2, \beta_3, \beta_4\} \\ &= \min\{5, 1, 1, 4\} \\ &= 1 \\ &= \beta_2 \text{ or } \beta_3. \end{aligned}$$

Hence in playing this game in the most conservative way, R can guarantee a payoff of -1 from C regardless of what C does (i.e., he has to pay at most 1 to C). Similarly, if C plays the most conservative strategy, he can assure that the amount paid to R is at most 1.

Some games are *strictly determined* in the sense that $\alpha = \beta$. We can prove that a game is strictly determined if and only if there is an element in the game matrix A which is smallest in its row and largest in its column. We first prove that the upper and lower values of a game must satisfy

$$\alpha \leq \beta. \tag{1}$$

The number $\alpha = \max_i \min_j a_{ij}$ is the largest of the smallest elements in all of the rows, and hence α must be some entry of A, say

$$\alpha = a_{st}.$$

Similarly, β must be some entry of A, say

$$\beta = a_{pq}.$$

Consider the entry a_{sq}:

$$\begin{array}{c} \\ s \end{array}\begin{array}{c} q \\ \left[\begin{array}{ccccccc} & & \vdots & & & & \\ & & a_{pq} & & & & \\ & & \vdots & & & & \\ \cdots & & a_{sq} & \cdots & a_{st} & \cdots & \\ & & \vdots & & & & \end{array}\right]. \end{array}$$

Now, $\beta = a_{pq}$ is the largest entry in column q, and hence

$$\beta = a_{pq} \geq a_{sq}; \tag{2}$$

$\alpha = a_{st}$ is the smallest entry in row s, and thus

$$a_{sq} \geq a_{st} = \alpha. \tag{3}$$

It follows immediately from (2) and (3) that $\alpha \leq \beta$. Now, let $a_{k\ell}$ be an entry of A which is simultaneously smallest in its row and largest in its column (if such an entry exists). Then

$$\begin{aligned} a_{k\ell} &= \min_j a_{kj} \\ &= \alpha_k \\ &\leq \max_i \alpha_i \\ &= \alpha \end{aligned} \tag{4}$$

$$\begin{aligned}
&\leq \beta \\
&= \min_j \beta_j \\
&\leq \beta_\ell \\
&= \max_i a_{i\ell} \\
&= a_{k\ell}.
\end{aligned}$$

It follows from (4) that

$$a_{k\ell} = \alpha = \beta.$$

We have proved that if there is an entry $a_{k\ell}$ of A which is smallest in its row and largest in its column, then $\alpha = \beta = a_{k\ell}$ and the game is strictly determined. Conversely, suppose $\alpha = \beta$. Now α must be the largest of the α_i, say $\alpha = \alpha_k$, and similarly, β is the least of the β_j, say $\beta = \beta_\ell$. Since we are assuming $\alpha = \beta$, it follows that $\alpha_k = \beta_\ell$. Consider the entry $a_{k\ell}$. Then

$$\begin{aligned}
a_{k\ell} &\geq \min_j a_{kj} \\
&= \alpha_k \\
&= \beta_\ell \\
&= \max_i a_{i\ell} \\
&\geq a_{k\ell},
\end{aligned}$$

and hence $a_{k\ell} = \alpha_k = \beta_\ell = \alpha = \beta$.

Example 4.2 Show that the game defined by the matrix

$$A = \begin{bmatrix} 4 & 0 & -1 & 3 \\ 2 & 2 & 0 & -1 \\ 4 & 5 & 1 & 0 \\ 2 & 6 & 1 & 8 \\ -1 & -5 & 1 & 3 \end{bmatrix}$$

is strictly determined. Find the entry which is smallest in its row and largest in its column and compute the upper and lower values of the game. We compute that

$$\begin{aligned}
\alpha_1 &= \min_j a_{1j} \\
&= \min\{4, 0, -1, 3\} \\
&= -1.
\end{aligned}$$

Similarly,

$$\begin{aligned}
\alpha_2 &= -1, \\
\alpha_3 &= 0, \\
\alpha_4 &= 1, \\
\alpha_5 &= -5.
\end{aligned}$$

Hence

$$\begin{aligned} \alpha &= \max_i \alpha_i \\ &= \max\{-1, 0, 1, -5\} \\ &= 1 \\ &= \alpha_4. \end{aligned}$$

Also,

$$\begin{aligned} \beta_1 &= \max_i a_{i1} \\ &= \max\{4, 2, 4, 2, -1\} \\ &= 4. \end{aligned}$$

Similarly,

$$\begin{aligned} \beta_2 &= 6, \\ \beta_3 &= 1, \\ \beta_4 &= 8, \end{aligned}$$

and hence

$$\begin{aligned} \beta &= \min_j \beta_j \\ &= \min\{4, 6, 1, 8\} \\ &= 1 \\ &= \beta_3. \end{aligned}$$

Since $\alpha = \alpha_4 = \beta_3 = \beta = 1$, we see that the (4, 3) entry of A is equal to the common upper and lower values of the game and, moreover, 1 is the smallest entry in the fourth row and the largest entry in the third column.

Definition 4.1 ***Saddle point*** An entry in an $m \times n$ matrix A which is least in its row and greatest in its column is called a *saddle point* for the matrix A or for the game defined by A.

Of course, not every game has a saddle point, i.e., not every game is strictly determined.

Example 4.3 Consider the following game. Players R and C simultaneously show two fingers or one finger. If the total number of fingers exhibited is even, C wins the sum in dollars (i.e., a dollar for each finger) from R; if the total number of fingers is odd, R wins the sum in dollars from C. The payoff matrix for this game is

$$\begin{array}{cc} & \quad C \\ & \begin{array}{cc} \text{1 finger} & \text{2 fingers} \end{array} \\ R \; \begin{array}{c} \text{1 finger} \\ \text{2 fingers} \end{array} & \begin{bmatrix} -2 & 3 \\ 3 & -4 \end{bmatrix} \end{array}.$$

Observe that there is no entry in the matrix which is least in its row and greatest in its column. In fact,

$$\alpha_1 = -2,$$
$$\alpha_2 = -4,$$

and hence

$$\begin{aligned}\alpha &= \max\{-2, -4\}\\ &= -2\\ &= \alpha_1.\end{aligned}$$

Also,

$$\begin{aligned}\beta &= \min\{3, 3\}\\ &= 3\\ &= \beta_1 \text{ or } \beta_2.\end{aligned}$$

Thus the lower value of the game is strictly less than the upper value. If R decides to show the same number of fingers at every trial of this game, he will, of course, always show one finger. For, if he were to show two fingers all the time, C would get the point after a while and realize that he could consistently win 4 dollars by always showing two fingers also. Thus the best "pure strategy" for R is to minimize his losses by always showing one finger. Similarly the best pure strategy for C is to exhibit two fingers. Such pure strategies make this game rather uninteresting, and thus rational players would tend to mix their choices of rows and columns in expectation of winning more than they could with a pure strategy.

In order to discuss the notion of a *mixed strategy*, we consider a somewhat more general situation. Suppose that a game is played by two players, R and C, in which there are k outcomes denoted $u_1, \ldots, u_k$. Suppose that when outcome u_t occurs, C must pay R a sum of d_t dollars (d_t may be negative, indicating a payment from R to C). Imagine that this game is played a large number of times, say N times, and outcome u_t occurs a total number of n_t times, $t = 1, \ldots, k$. Then C pays R a sum of $n_1 d_1$ dollars for all outcomes u_1, and in general, C pays R a sum of $n_t d_t$ dollars for all outcomes u_t, $t = 1, \ldots, k$. Then the total amount that C pays R over the N plays of the game is the sum

$$n_1 d_1 + n_2 d_2 + \cdots + n_k d_k.$$

Since there are a total of N trials, the average or *expected* earning for R on any given trial is

$$\frac{n_1 d_1 + n_2 d_2 + \cdots + n_k d_k}{N}. \tag{5}$$

Now, if we set $p_t = \frac{n_t}{N}$, then (5) can be rewritten

(6) $$\sum_{t=1}^{k} p_t d_t.$$

The number p_t is the *frequency of occurrence* of u_t, $t = 1, \ldots, k$, among the N trials. Hence in order to compute the expected earnings on any given trial of the game, knowing the frequencies p_t, $t = 1, \ldots, k$, we simply compute the sum (6). If $U = \{u_1, \ldots, u_k\}$ and R^+ is the set of nonnegative real numbers, then the function $p\colon U \to R^+$ is clearly a probability measure if p is defined by $p(\{u_t\}) = p_t$, $t = 1, \ldots, k$. If we define a function f on U by $f(u_t) = d_t$, $t = 1, \ldots, k$, then the expected or average earnings of R on any given play of the game is

(7) $$\sum_{t=1}^{k} p(\{u_t\})f(u_t).$$

Denote the sum in (7) by $E(f)$. The reader will recall that in Exercise 22, Section 1.7, such a function f is called a *random variable* and $E(f)$ is called the *expectation* of f. If g is any other random variable, then in the same exercise it is noted that

$$E(f + g) = E(f) + E(g).$$

Example 4.4 Two fair coins are tossed and the following betting game is played by players R and C. If exactly two heads come up, C pays R one dollar; otherwise R pays C a half dollar. What can each player expect to win or lose in five trials of this game? Define a two-element sample space U as follows:

$$U = \{S, F\},$$

where S means that two heads come up, and F means that two heads do not come up. The probability of obtaining precisely two heads on any given toss of the two coins is $\frac{1}{4}$. Therefore define a probability measure p on U by

$$p(\{S\}) = \tfrac{1}{4},$$
$$p(\{F\}) = \tfrac{3}{4}.$$

Let f be a random variable defined on U by

$$f(S) = 1$$
$$f(F) = -\tfrac{1}{2},$$

that is, the value of f corresponds to the amount won or lost by player R. Then

$$\begin{aligned} E(f) &= p(\{S\})f(S) + p(\{F\})f(F) \\ &= \tfrac{1}{4}\cdot 1 + \tfrac{3}{4}\cdot(-\tfrac{1}{2}) \\ &= -\tfrac{1}{8}. \end{aligned}$$

In other words, player R will lose an average of $\frac{1}{8}$ of a dollar on any given trial, and hence in five trials he can expect to lose $\frac{5}{8}$ of a dollar.

Example 4.5 Recall the game matrix in Example 4.3.

$$A = \begin{bmatrix} -2 & 3 \\ 3 & -4 \end{bmatrix}$$

This matrix does not possess a saddle point, i.e., there is no element which is smallest in its row and largest in its column. We saw that the pure strategies for the two players R and C (i.e., R exhibiting one finger at each play, C exhibiting two fingers at each play) will guarantee certain minimum winnings for each of them, but the question is whether it is possible for R (or C) to do better by mixing his plays. Imagine that the game is played a large number of times. Let x_1 and x_2 denote the relative frequencies with which R chooses rows 1 and 2, respectively, and let y_1 and y_2 denote the relative frequencies with which C chooses columns 1 and 2, respectively. In a large number of trials, it is reasonable to assign the probability x_iy_j to the choice of row i and column j. Another way of saying this is that if row i is chosen with probability x_i and column j is chosen with probability y_j, then it is appropriate to assign the probability x_iy_j to the event that both row i and column j are chosen. This kind of assignment of probabilities to a compound event should be very familiar from Chapter 1. It is precisely analogous to the following situation. Suppose a fair die is rolled and a biased penny is tossed simultaneously. Suppose the penny comes up heads (here denoted by 1) twice as often as tails (here denoted by 2). The outcomes of this experiment are $(1, 1), \ldots, (1, 6), (2, 1), \ldots, (2, 6)$. The probability that heads comes up is $x_1 = 2/3$ and the probability that tails comes up is $x_2 = 1/3$; the probability that the number j comes up on a roll of the die is $y_j = 1/6$, $j = 1, \ldots, 6$. Thus it is sensible to assign the probability x_iy_j to the compound event that (i, j) occurs. Returning to the example, the payoff to R when row i and column j are chosen is a_{ij}, and thus the expected winnings for R are

$$\begin{aligned} (8) \quad & a_{11}x_1y_1 + a_{12}x_1y_2 + a_{21}x_2y_1 + a_{22}x_2y_2 \\ &= -2x_1y_1 + 3x_1y_2 + 3x_2y_1 - 4x_2y_2. \end{aligned}$$

Now, $x_1 + x_2 = y_1 + y_2 = 1$, $x_1 \geq 0$, $y_1 \geq 0$. Thus we can rewrite equation (8) as

(9) $$-2x_1y_1 + 3x_1(1 - y_1) + 3(1 - x_1)y_1 - 4(1 - x_1)(1 - y_1) = -12x_1y_1 + 7(x_1 + y_1) - 4.$$

Suppose that R chooses row 1 with probability 7/12. Then his expected winnings for any strategy y_1 that C uses can be computed from (9) by replacing x_1 by 7/12:

$$\begin{aligned} -12x_1y_1 + 7(x_1 + y_1) - 4 &= -12 \cdot \tfrac{7}{12}y_1 + 7(\tfrac{7}{12} + y_1) - 4 \\ &= -7y_1 + \tfrac{49}{12} + 7y_1 - 4 \\ &= \tfrac{1}{12}. \end{aligned}$$

This result is independent of y_1, i.e., R can guarantee an expected gain of $\frac{1}{12}$ whatever C does. Recall that with a pure strategy for R, i.e., a constant choice of one row, the most that R could win was -2, i.e., a loss of 2. Similarly, if $y_1 = \frac{7}{12}$, it is clear that the value of (9) is also $\frac{1}{12}$, and hence for any choice of strategy that R may play, C pays him an average amount of $\frac{1}{12}$, which is certainly better than the 3 he can expect to pay with the pure strategy.

Summarizing, R can rest assured that his expectation will be at least $\frac{1}{12}$ and, moreover, he cannot guarantee that this expectation will be any better than $\frac{1}{12}$, because C can take $y_1 = \frac{7}{12}$. Thus R might as well reconcile himself to showing one finger $\frac{7}{12}$ of the time. By a similar argument, the best that C can ensure for himself is a payment of $\frac{1}{12}$ to R. This game is favorable to R. But, as we have just seen, both players can do better with mixed strategies,

$$(x_1, x_2) = (\tfrac{7}{12}, \tfrac{5}{12})$$

and

$$(y_1, y_2) = (\tfrac{7}{12}, \tfrac{5}{12}),$$

than with a pure strategy.

The analysis in Example 4.5 suggests a general method for analyzing any game defined by an $m \times n$ matrix A.

Definition 4.2 ***Strategy*** A *strategy* for a player R in a game defined by an $m \times n$ matrix A is a vector $x \in R^m$ such that $\sum_{i=1}^{m} x_i = 1$ and $x \geq 0$. Similarly, a strategy for a player C is a vector $y \in R^n$ satisfying $\sum_{j=1}^{n} y_j = 1$ and $y \geq 0$. For a given pair of vectors x and y, the *expectation* (of R) is

(10) $$E(x, y) = \sum_{i=1}^{m} \sum_{j=1}^{n} x_i a_{ij} y_j.$$

The definition of a strategy in the general situation corresponds precisely to what we did in Example 4.5: we think of x_i as the probability with which R plays row i, $i = 1, \ldots, m$, and y_j as the probability with which C plays column j, $j = 1, \ldots, n$. As we saw, it is reasonable to assign the probability x_iy_j to the compound event consisting of the choice of row i by R and column j by C. Thus the number $E(x, y)$ is precisely the expected earnings for R for the choice of strategies x and y. Of course, the question arises as to precisely how R can play the game using a strategy x. A strategy can be determined in a number of ways. For example, if $m = 6$, R can decide on a choice of a row on any given play of the game by rolling a die and playing row i if i appears on the die, $i = 1, \ldots, 6$.

We must analyze what constitutes an *optimum* strategy for the two players. What R wants to do, of course, is to find a strategy x^* so that his expected winnings are as large as possible, regardless of what C does. In other words, he wants to find a strategy and the largest number w_R (representing winnings for R) such that for any strategy y for C,

$$E(x^*, y) \geq w_R. \tag{11}$$

We can state (11) in a somewhat different way: namely, R seeks a strategy x^* such that

$$\min_y E(x^*, y) \geq w_R,$$

and w_R is the largest such number. Thus w_R is the maximum for all strategies x of the minimum for all strategies y of the expected winnings of R, that is,

$$w_R = \max_x \min_y E(x, y).$$

We can similarly analyze the best thing for C to do. C wants to find a strategy y^* and a number w_C so that no matter what R does, C's expected payments to R, i.e., R's winnings, are as small as possible. That is, C wants a strategy y^* and a smallest number w_C such that

$$\max_x E(x, y^*) \leq w_C. \tag{12}$$

Thus w_C is the minimum for all strategies y of the maximum for all strategies x of $E(x, y)$, that is,

$$w_C = \min_y \max_x E(x, y).$$

Example 4.6 Find optimum strategies x^* and y^* for R and C and compute the numbers w_R and w_C for the game defined by the matrix

$$A = \begin{bmatrix} 2 & 3 & 5 \\ 7 & 5 & 2 \end{bmatrix}.$$

We can solve this problem graphically as follows. Let $x = (x_1, x_2)$ be a strategy for R, $x_1 + x_2 = 1$, $x \geq 0$. Suppose that C chooses column 1. Then the expected payoff to R is

$$\begin{aligned} 2x_1 + 7x_2 &= 2x_1 + 7(1 - x_1) \\ &= -5x_1 + 7. \end{aligned}$$

If C chooses column 2, then R's expected winnings are

$$\begin{aligned} 3x_1 + 5x_2 &= 3x_1 + 5(1 - x_1) \\ &= -2x_1 + 5, \end{aligned}$$

and if C chooses column 3, then R can expect to win

$$\begin{aligned} 5x_1 + 2x_2 &= 5x_1 + 2(1 - x_1) \\ &= 3x_1 + 2. \end{aligned}$$

Now, x_1 varies between 0 and 1, and if we graph the three lines

$$\begin{aligned} f_1(x_1) &= -5x_1 + 7, \\ f_2(x_1) &= -2x_1 + 5, \\ f_3(x_1) &= 3x_1 + 2, \end{aligned}$$

we obtain the following.

(13)

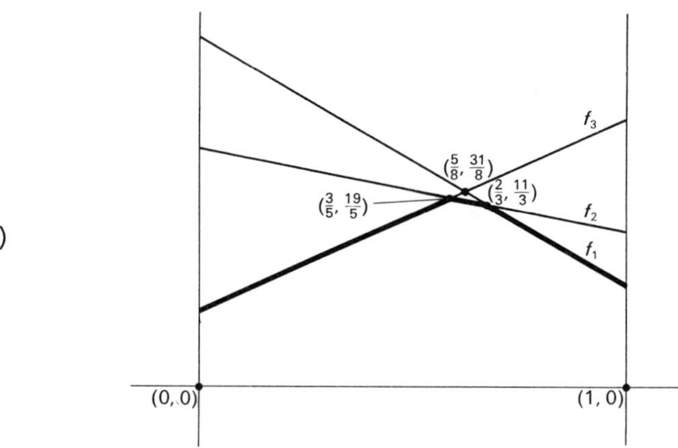

For any given number x_1, the most that R can hope to win regardless of what C does is the minimum of the three numbers $f_1(x_1), f_2(x_1)$, and $f_3(x_1)$. Thus, of course, an optimum choice of x_1 for R is that value which makes the minimum of the three numbers largest. Suppose we graph the function

$$f(x_1) = \min\{f_1(x_1), f_2(x_1), f_3(x_1)\}.$$

It is obvious from (13) that the graph of f is the indicated heavy line. The highest point on the graph of f is clearly $(\frac{3}{5}, \frac{19}{5})$. Therefore by using the strategy

$$x^* = (\tfrac{3}{5}, \tfrac{2}{5}),$$

R can maximize his expected winnings independently of what C does. Moreover, the value w_R of R's maximum expected winnings for this strategy is $\frac{19}{5}$.

Suppose we analyze the situation for C, assuming that R has chosen the strategy $x^* = (\frac{3}{5}, \frac{2}{5})$. That is, C is a completely rational player and knows that R is going to choose the strategy x^*. Now, for any $x = (x_1, x_2)$,

$$\begin{aligned} E(x, y) &= (2x_1 + 7x_2)y_1 + (3x_1 + 5x_2)y_2 + (5x_1 + 2x_2)y_3 \\ &= (-5x_1 + 7)y_1 + (-2x_1 + 5)y_2 + (3x_1 + 2)y_3 \\ &= f_1(x_1)y_1 + f_2(x_1)y_2 + f_3(x_1)y_3. \end{aligned} \tag{14}$$

Thus from (14) we have

$$\begin{aligned} E(x^*, y) &= f_1(\tfrac{3}{5})y_1 + f_2(\tfrac{3}{5})y_2 + f_3(\tfrac{3}{5})y_3 \\ &= 4y_1 + \tfrac{19}{5}(y_2 + y_3) \\ &= 4y_1 + \tfrac{19}{5}(1 - y_1) \\ &= \tfrac{1}{5}y_1 + \tfrac{19}{5}. \end{aligned} \tag{15}$$

Of course, C wants to minimize (15), which he can obviously do by choosing $y_1 = 0$. Thus in any optimum strategy, $y_1 = 0$. We now obtain an optimum strategy for C, knowing that $y_1 = 0$. Let $(0, y_2, y_3)$ be a strategy for C, and suppose R chooses row 1. Then the expected winnings for R are

$$\begin{aligned} 3y_2 + 5y_3 &= 3y_2 + 5(1 - y_2) \\ &= -2y_2 + 5. \end{aligned}$$

Similarly, if R chooses row 2, then R can expect to win

$$\begin{aligned} 5y_2 + 2y_3 &= 5y_2 + 2(1 - y_2) \\ &= 3y_2 + 2. \end{aligned}$$

Let

$$g_1(y_2) = -2y_2 + 5$$

and

$$g_2(y_2) = 3y_2 + 2.$$

For any given number y_2, the most that C can expect to lose is the largest of the two numbers $g_1(y_2)$ and $g_2(y_2)$. Since C wants to minimize this maximum loss, he will choose y_2 so that

$$g(y_2) = \max\{g_1(y_2), g_2(y_2)\}$$

is least. If we graph the functions g_1 and g_2, we see that the point of intersection, as well as $\min_{y_2} g(y_2)$, occurs for $y_2 = \frac{3}{5}$.

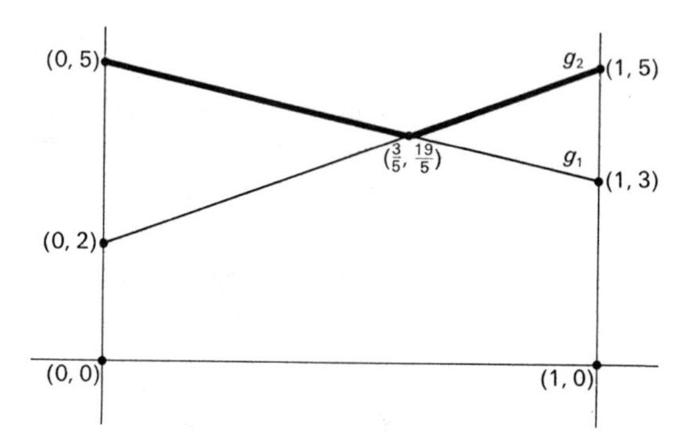

Now, $y_2 + y_3 = 1$ and hence $y_3 = \frac{2}{5}$. Thus the optimum strategy for C is $y^* = (0, \frac{3}{5}, \frac{2}{5})$, and $w_C = \frac{19}{5}$.

The result of the preceding example is perfectly general. There is a fundamental theorem due to von Neumann which can be stated as follows.

Theorem 4.1 *Let A be an $m \times n$ game matrix. There exists an optimum strategy x^* for player R, an optimum strategy y^* for player C, and a number w such that*

$$E(x^*, y^*) = w$$

and

$$\max_x \min_y E(x, y) = E(x^*, y^*) = \min_y \max_x E(x, y).$$

The proof of this result, although not particularly difficult, is rather long and depends on a result about polyhedra that takes us somewhat far afield. However, we can see quite easily that

$$\max_x \min_y E(x, y) \leq \min_y \max_x E(x, y).$$

For, obviously $\min_y E(x, y) \leq E(x, y) \leq \max_x E(x, y)$, and thus

$$\min_y E(x, y) \leq \max_x E(x, y). \tag{16}$$

The right side of the inequality in (16) does not depend on x, and so the maximum of the left side over all x can never exceed the right side, that is,

$$\max_x \min_y E(x, y) \leq \max_x E(x, y). \tag{17}$$

The left side of (17) is now a number depending on neither x nor y and hence is always less than or equal to the right side as y varies, that is,

$$\max_x \min_y E(x, y) \leq \min_y \max_x E(x, y). \blacksquare$$

We see from Theorem 4.1 that $w_R = w_C = w$, and thus any game defined by a rectangular matrix A can be *solved* in the following sense. It is possible to find optimum strategies x^* and y^* and a number w, called the *value of the game*, such that

$$E(x^*, y^*) = w, \tag{18}$$

and such that

$$E(x^*, y) \geq w \tag{19}$$

for all strategies y and

$$E(x, y^*) \leq w \tag{20}$$

for all strategies x.

Theorem 4.2 *For any strategies x and y,*

$$E(x, y) = (x, Ay), \tag{21}$$

where A is an $m \times n$ matrix defining the game and where the right side of (21) *is the inner product in R^m.*

We leave the proof of this theorem as Exercise 16.

Suppose w is the value of a game defined by an $m \times n$ matrix A. Of course, if R uses the optimum strategy x^*, then it follows that no matter what C does,

$$\begin{aligned} E(x^*, y) &= (x^*, Ay) \\ &\geq w. \end{aligned} \tag{22}$$

The inequality in (22) holds for all strategies y if and only if each of the following n inequalities hold:

$$(x^*, Ae_j) \geq w, \qquad j = 1, \ldots, n, \tag{23}$$

where e_j is the pure strategy corresponding to the choice of column j, i.e., the n-vector with 1 in position j and 0's elsewhere. For, if (22) holds for all strategies y, it of course holds for the strategies $e_1, \ldots, e_n$. Conversely, if each of the inequalities in (23) holds, then for any strategy $y = (y_1, \ldots, y_n) = \sum_{j=1}^{n} y_j e_j$, we have (using the linearity of the inner product)

$$\begin{aligned}(x^*, Ay) &= (x^*, A \sum_{j=1}^{n} y_j e_j) \\ &= (x^*, \sum_{j=1}^{n} y_j A e_j) \\ &= \sum_{j=1}^{n} y_j (x^*, Ae_j) \\ &\geq \sum_{j=1}^{n} y_j w \\ &= w \sum_{j=1}^{n} y_j \\ &= w,\end{aligned}$$

since $y_1 + \cdots + y_n = 1$. It is very easy to see that Ae_j is the jth column of A (see Exercise 21, Section 2.1),

$$\begin{aligned}Ae_j &= A^{(j)} \\ &= (a_{1j}, \ldots, a_{mj}).\end{aligned}$$

Thus, to solve a game we want to find the largest number w such that the n inequalities in (23) hold. Assume that this largest value w is positive. (We can do this by adding a fixed number k to every entry of the matrix A. The only effect of this will be to alter the value of the game by the value k. The sets of optimum strategies for R and C remain the same. See Exercise 15.) Divide both sides of the inequalities in (23) by w and set $x = \frac{1}{w} x^*$ to obtain

$$(x, Ae_j) \geq 1, \qquad j = 1, \ldots, n. \tag{24}$$

Denote $x = (x_1, \ldots, x_m)$ and $x^* = (x_1^*, \ldots, x_m^*)$ and observe that

$$\begin{aligned}\sum_{i=1}^{m} x_i &= \frac{1}{w} \cdot \sum_{i=1}^{m} x_i^* \\ &= \frac{1}{w},\end{aligned} \tag{25}$$

because x^* is a strategy. We want the largest possible value of w. According to (24) and (25), this means that we want to choose an

m-vector x, $x \geq 0$, such that

$$f(x) = \sum_{i=1}^{m} x_i = \frac{1}{w}$$

is minimal, where x is constrained to satisfy the inequalities in (24). But this is precisely the kind of problem we dealt with in the preceding sections on linear programming, and we have all the necessary methods for minimizing the function $f(x)$ on the polyhedron defined by the inequalities in (24).

Example 4.7 Solve the game defined by the matrix A in Example 4.6 according to the method indicated in the preceding discussion. The column vectors are (2, 7), (3, 5), (5, 2), and hence the inequalities in (24) become

$$\begin{aligned} 2x_1 + 7x_2 &\geq 1, \\ 3x_1 + 5x_2 &\geq 1, \\ 5x_1 + 2x_2 &\geq 1. \end{aligned} \tag{26}$$

We want to minimize the function $f(x) = x_1 + x_2$ on the set S of solutions to the system (26). If we graph the region S, we see that it is the intersection of the closed positive half-spaces defined by the lines

$$\begin{aligned} 2x_1 + 7x_2 &= 1, \\ 3x_1 + 5x_2 &= 1, \\ 5x_1 + 2x_2 &= 1. \end{aligned}$$

(27)

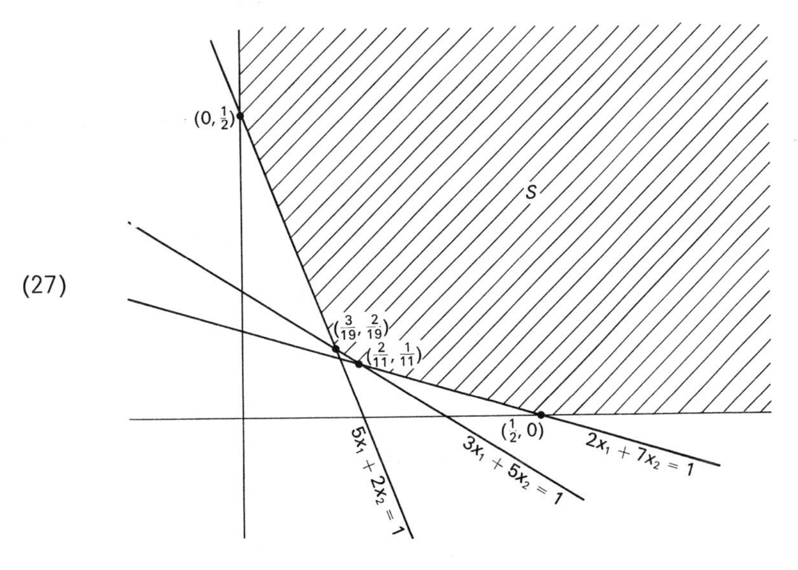

If we test the function f at the vertices of S, which are indicated in the graph (27), we find that the minimum of the function is taken on at the point $(\frac{3}{19}, \frac{2}{19})$, and this value is $\frac{5}{19}$. According to (25), the minimum of f is the reciprocal of w, that is,

$$\frac{1}{w} = \frac{5}{19},$$

so that

$$w = \tfrac{19}{5}.$$

This value is, of course, the same as the value of the game obtained in Example 4.6. Recall that $x = \dfrac{1}{w}x^*$, where x^* is an optimum strategy for R. Thus

$$\begin{aligned} x^* &= wx \\ &= \tfrac{19}{5}(\tfrac{3}{19}, \tfrac{2}{19}) \\ &= (\tfrac{3}{5}, \tfrac{2}{5}), \end{aligned}$$

coinciding with the optimum strategy that we obtained for R in Example 4.6.

We can obtain the optimum strategy for C by proceeding precisely as we did in Example 4.6. However, it is instructive to analyze the situation in the same way as we did in determining the optimum strategy for R. We know that the value of the game is $w = \frac{19}{5}$. If C uses his optimum strategy y^*, then the expected winnings must satisfy the inequality

$$\begin{aligned} E(x, y^*) &= (x, Ay^*) \\ &\leq w \end{aligned} \tag{28}$$

for any strategy x. By an argument analogous to the discussion immediately following (23) we can conclude that (28) is equivalent to the m conditions

$$(e_i, Ay^*) \leq w, \qquad i = 1, \ldots, m, \tag{29}$$

where e_i is the m-vector with 1 in position i and 0's elsewhere. Divide both sides of (29) by w, set $y = \dfrac{1}{w}y^*$ to obtain

$$(e_i, Ay) \leq 1, \qquad i = 1, \ldots, m, \tag{30}$$

and observe that

$$\sum_{j=1}^{m} y_j = \frac{1}{w}. \tag{31}$$

Thus we seek a nonnegative vector y which satisfies the m inequalities in (30) and the equation (31). For the matrix A, these inequalities become

$$
\begin{aligned}
2y_1 + 3y_2 + 5y_3 &\le 1, \\
7y_1 + 5y_2 + 2y_3 &\le 1, \\
y_1 + y_2 + y_3 &= \tfrac{5}{19}.
\end{aligned} \tag{32}
$$

The system (32) will have a nonnegative solution if we can obtain a nonnegative solution to the system of equations

$$
\begin{aligned}
2y_1 + 3y_2 + 5y_3 &= 1, \\
7y_1 + 5y_2 + 2y_3 &= 1 \\
y_1 + y_2 + y_3 &= \tfrac{5}{19}.
\end{aligned}
$$

If we reduce this system to Hermite normal form, we find that a nonnegative solution does indeed exist and is, in fact,

$$y = (0, \tfrac{3}{19}, \tfrac{2}{19}).$$

Thus the optimum strategy y^* for C is

$$y^* = (0, \tfrac{3}{5}, \tfrac{2}{5}),$$

which concides with the result we obtained in Example 4.6.

Quiz

Answer true or false.

1 The lower value of the game matrix

$$A = \begin{bmatrix} 1 & 0 & -2 \\ 0 & 3 & 7 \end{bmatrix}$$

is $\alpha = 0$.

2 The upper value of the matrix A in Question 1 is $\beta = 0$.

3 The matrix A in Question 1 has a saddle point.

4 In any matrix game, the upper value is always at least as large as the lower value.

5 If player R uses strategy x and player C uses strategy y, then it is always the case that $E(x, y) = 1$.

6 In Question 5, $E(x, y)$ represents the expected earnings for R for the choice of strategies x and y.

7 The game defined by the matrix

$$A = \begin{bmatrix} 1 & 0 \\ 2 & 3 \end{bmatrix}$$

is strictly determined.

8 A game is not strictly determined if there is no entry in the game matrix which is both a row maximum and a column minimum.

9 If x^* and x' are optimum strategies for R, then so is $\frac{1}{2}(x^* + x')$.

10 If $x = (x_1, x_2)$ and $y = (y_1, y_2, y_3)$ are strategies, then the vector

$$(x_1y_1, x_1y_2, x_1y_3, x_2y_1, x_2y_2, x_2y_3)$$

is a nonnegative 6-vector the sum of whose components is 1.

Exercises

1 Compute the lower and upper values of the games defined by the following matrices:

(a) $\begin{bmatrix} 2 & 1 \\ 3 & 5 \end{bmatrix}$;

(b) $\begin{bmatrix} 0 & 0 \\ 0 & 0 \end{bmatrix}$;

(c) $\begin{bmatrix} 1 & 0 \\ 0 & 1 \end{bmatrix}$;

(d) $\begin{bmatrix} 1 & -1 \\ -1 & 1 \end{bmatrix}$;

(e) $\begin{bmatrix} 0 & 1 \\ -3 & 10 \end{bmatrix}$;

(f) $\begin{bmatrix} -7 & -4 \\ 5 & 2 \end{bmatrix}$;

(g) $\begin{bmatrix} 3 & 5 \\ 4 & -2 \end{bmatrix}$;

(h) $\begin{bmatrix} 6 & -1 & 3 \\ 2 & -5 & 0 \end{bmatrix}$;

(i) $\begin{bmatrix} 7 & 0 \\ 0 & 1 \\ 5 & -2 \end{bmatrix}$;

(j) $\begin{bmatrix} 7 & 2 & 9 \\ 9 & 0 & 10 \\ 2 & 8 & -3 \end{bmatrix}$;

(k) $\begin{bmatrix} 2 & 7 & 3 & 8 & 5 \\ 0 & 1 & 6 & -2 & 0 \\ -1 & 6 & 1 & 3 & -1 \\ 2 & 2 & 0 & -1 & 7 \\ -3 & 1 & 2 & 0 & -1 \end{bmatrix}$;

(l) $\begin{bmatrix} 3 & 1 & 1 & 1 \\ 1 & 3 & 1 & 1 \\ 1 & 1 & 3 & 1 \\ 1 & 1 & 1 & 3 \end{bmatrix}$;

(m) $[1 \quad 0 \quad -2 \quad 3 \quad 5]$;

(n) $\begin{bmatrix} 25 & 17 & 123 & -57 \\ 34 & 279 & 1 & 78 \\ -19 & 93 & -112 & 0 \end{bmatrix}$;

(o) $\begin{bmatrix} 1 & 1 \\ 1 & 1 \end{bmatrix}$.

2 Using the graphical method discussed in Example 4.6, find optimum strategies for R and C and the value of the game in each of the following:

(a) $\begin{bmatrix} 4 & 0 \\ 0 & 2 \end{bmatrix}$;

(b) $\begin{bmatrix} 3 & 1 \\ 2 & 3 \end{bmatrix}$;

(c) $\begin{bmatrix} 10 & 3 & 4 & 12 \\ 3 & 8 & 4 & 2 \end{bmatrix}$;

(d) $\begin{bmatrix} 1 & 7 & 3 \\ 5 & 4 & 6 \end{bmatrix}$;

(e) $\begin{bmatrix} -2 & -5 \\ 1 & -1 \\ 0 & 2 \end{bmatrix}$.

3 Let A be an $m \times n$ matrix in which every column is the same.

(a) Show that $E(x, y)$ is independent of y.

(b) Compute the value of the game defined by A.

(Hint: $\max\limits_x \min\limits_y E(x, y) = \max\limits_x (u, x)$, where u is the m-tuple $A^{(1)} = \cdots = A^{(n)}$. Then $\max\limits_x (u, x)$ is the largest component of u.)

(c) Verify the equality

$$\max_x \min_y E(x, y) = \min_y \max_x E(x, y).$$

4 Let A be an $m \times n$ game matrix with a saddle point a_{st}, i.e., a_{st} is the least element in row s and the largest element in column t. Show that

$$\max_x \min_y E(x, y) = a_{st}$$

and that the pure strategies e_s and e_t are optimal for R and C, respectively. [Recall that

$$e_s = (0, \ldots, 0, \overset{\overset{s}{\downarrow}}{1}, 0, \ldots, 0) \in R^m, \; e_t = (0, \ldots, 0, \overset{\overset{t}{\downarrow}}{1}, 0, \ldots, 0) \in R^n.]$$

(Hint: Let y^* be an optimum strategy for C, and let w be the value of the game, i.e., $w = \min_y \max_x E(x, y) = \max_x \min_y E(x, y)$. Now for any strategy x, $(x, Ay^*) \leq w$ because $w = \max_x(x, Ay^*)$. It follows that each component of Ay^* can be at most w, $\sum_{j=1}^{n} a_{ij}y_i^* \leq w$, $i = 1, \ldots, m$. But if a convex combination of numbers is at most w, then surely the least of these numbers is at most w. Hence for each i, $\min_j a_{ij} \leq w$; in particular, if $i = s$ then $a_{st} = \min_j a_{sj} \leq w$. On the other hand, $(x^*, Ay) \geq w$ for any strategy y, because $w = \min_y(x^*, Ay)$. Now,

$$A^T x^* = \left(\sum_{i=1}^{m} a_{i1}x_i^*, \ldots, \sum_{i=1}^{m} a_{in}x_i^*\right),$$

so that

$$\begin{aligned}
(A^T x^*, y) &= \left(\sum_{i=1}^{m} a_{i1}x_i^*\right) y_1 + \cdots + \left(\sum_{i=1}^{m} a_{in}x_i^*\right) y_n \\
&= \sum_{j=1}^{n} \sum_{i=1}^{m} a_{ij}x_i^* y_j \\
&= (x^*, Ay).
\end{aligned}$$

Hence $(A^T x^*, y) \geq w$. It follows that each component of $A^T x^*$ must be at least w, i.e., $\sum_{i=1}^{n} a_{ij}x_i^* \geq w, j = 1, \ldots, n$. But if a convex combination of numbers is at least w, then the largest of the numbers must be at least w. Thus for each j, $\max_i a_{ij} \geq w$; in particular, if $j = t$ then $a_{st} = \max_i a_{it} \geq w$. Now we have

$$a_{st} \leq w \leq a_{st},$$

so a_{st} must, in fact, be equal to w.)

5 Let A be an $m \times n$ matrix. The ith row of A, $A_{(i)}$, is said to be a *recessive* row if there exists a row $A_{(k)}$ such that each entry of $A_{(i)}$ is less than or equal to the corresponding entry of $A_{(k)}$. This is written $A_{(i)} \leq A_{(k)}$. Show that in a game defined by the matrix A, R will never play $A_{(i)}$. (Hint: He can do better by playing $A_{(k)}$.)

6 Suppose the matrix in the preceding exercise contains a column $A^{(j)}$ such that there exists another column $A^{(k)}$ for which $A^{(k)} \leq A^{(j)}$, i.e., each component of $A^{(k)}$ is at most equal to the corresponding component of $A^{(j)}$. Then $A^{(j)}$ is said to be a *recessive* column. Show that C will never play $A^{(j)}$. (Hint: He can always do better with $A^{(k)}$, i.e., he pays less to R by choosing $A^{(k)}$.)

7 Show that in any game defined by a matrix A, all the recessive rows and columns of A can be deleted without affecting the value of the game.

8 Using the result of the preceding exercise, delete the recessive rows and columns in each of the following matrices and solve the resulting games:

(a) $[1 \quad 2 \quad 3]$;

(b) $\begin{bmatrix} 1 & -1 & 3 & 4 \\ 0 & -1 & 2 & 4 \\ 1 & 1 & 1 & 0 \end{bmatrix}$;

(c) $\begin{bmatrix} -1 & 1 & 5 \\ 6 & 3 & 6 \\ 2 & 7 & 8 \end{bmatrix}$;

(d) $\begin{bmatrix} 2 & -2 & 4 \\ -3 & 4 & 5 \\ 3 & -1 & 0 \end{bmatrix}$;

(e) $\begin{bmatrix} 1 & 2 & 3 & 4 \\ 4 & 1 & 2 & 3 \\ 1 & 1 & 1 & 1 \end{bmatrix}$.

9 Let A be an $m \times n$ matrix. Denote the inner products in R^m and R^n by the same notation: $(\, , \,)$. Show that for any vectors $x \in R^m$ and $y \in R^n$, $(x, Ay) = (A^T x, y)$. (Hint: Show that both expressions are equal to $\sum_{i=1}^{m} \sum_{j=1}^{n} x_i a_{ij} y_j$; see hint in Exercise 4.)

10 Show that if there exists a strategy y^o such that $Ay^o = 0 \in R^m$, then the value w of the game defined by the matrix A is at most 0, i.e., $w \leq 0$. (Hint: $\min_y(x, Ay) \leq (x, Ay^o) = (x, 0) = 0$. Hence $w = \max_x \min_y(x, Ay) \leq 0$.)

11 Show that if there exists a strategy x^o such that $A^T x^o = 0 \in R^n$, then the value w of the game defined by the matrix A is at least zero, i.e., $w \geq 0$. (Hint: $\max_x(x, Ay) \geq (x^o, Ay) = (A^T x^o, y) = (0, y) = 0$. Hence $w = \min_y \max_x(x, Ay) \geq 0$.)

12 Show that there exists a 3-vector x^o such that $A^T x^o = 0$, where

$$A = \begin{bmatrix} 0 & a & -b \\ -a & 0 & c \\ b & -c & 0 \end{bmatrix},$$

in which a, b, and c are positive numbers. Show that the same 3-vector satisfies $Ax^o = 0$. (Hint: Try $x^o = \left(\dfrac{1}{a + b + c}\right)(c, b, a)$.)

***13** Show that for the game defined by the matrix A in the preceding exercise, the optimum strategies for both R and C are the same and the value of the game is 0.

***14** The child's game of rock, scissors, and paper is played as follows. The two players, R and C, simultaneously exhibit a fist (rock), two fingers (scissors), or the palm of the hand (paper). Rock wins over scissors, scissors wins over paper, and paper wins over rock. In case both players exhibit the same thing, it is a tie. If the loser pays the winner one cent, show that the game matrix is

$$A = \begin{bmatrix} 0 & 1 & -1 \\ -1 & 0 & 1 \\ 1 & -1 & 0 \end{bmatrix}.$$

Find optimum strategies for both players and compute the value of the game. (Hint: See the preceding exercise.)

†15 Show that if a number k is added to each entry of the $m \times n$ matrix A, then the value of the game defined by A is increased by k, but the optimum strategies for the new game are the same as the optimum strategies for the old game. (Thus, in defining a game, we can always assume the entries of the matrix are positive numbers.)

†16 Prove Theorem 4.2.

17 Show that the set of optimum strategies for R in a game defined by an $m \times n$ matrix A is a convex subset of R^m. Similarly, show that the set of optimum strategies for C is a convex subset of R^n. (Hint: Let x^* and x' be two optimum strategies for R, and let θ satisfy $0 \leq \theta \leq 1$. Consider the strategy $\overline{x} = (1 - \theta)x^* + \theta x'$. (Why is $\overline{x}$ a strategy?) Let w be the value of the game. Then $w \leq (x^*, Ay)$ for any strategy y, and similarly, $w \leq (x', Ay)$. This is because Theorem 4.1 states that $\min_y (x^*, Ay) = w$.

But then

$$\begin{aligned} w &= (1 - \theta)w + \theta w \\ &\leq (1 - \theta)(x^*, Ay) + \theta(x', Ay) \\ &= ((1 - \theta)x^*, Ay) + (\theta x', Ay) \\ &= ((1 - \theta)x^* + \theta x', Ay) \\ &= (\overline{x}, Ay). \end{aligned}$$

Thus for any strategy y,

$$w \leq (\overline{x}, Ay)$$

and hence $\overline{x}$ is an optimum strategy for R. In other words, any convex combination of two optimum strategies for R is again an optimum strategy for R.)

18 Two opponents in a tennis game, R and C, observe the following facts about one another's game. If R stands inside the baseline when C serves a flat serve, R will win the point 40% of the time; if R stands outside the baseline on a flat serve, R will win the point 30% of the time. Similarly, if R receives a twist serve from inside the baseline, R wins the point 20% of the time, and standing outside the baseline, R wins the point 50% of the time. Find optimum strategies for R and C.

***19** Let k be a positive number. Show that if w is the value of the game defined by the $m \times n$ matrix A, then kw is the value of the game defined by kA. Show that the optimum strategies for both games are the same.

20 Let a game be defined by the 2×2 matrix

$$A = \begin{bmatrix} a & b \\ c & d \end{bmatrix},$$

in which every entry is positive. Show that A has a saddle point if the two lines defined by $ax_1 + cx_2 - 1 = 0$ and $bx_1 + dx_2 - 1 = 0$ do not intersect for positive values of x_1 and x_2. (Hint: In order to compute the value of the game defined by A, we minimize the function $f(x) = x_1 + x_2$ on the set S of solutions to the system of inequalities

$$\begin{aligned} ax_1 + cx_2 &\geq 1, \\ bx_1 + dx_2 &\geq 1. \end{aligned}$$

In graphing the two lines, there are two possibilities: either the lines intersect for positive x_1 and x_2, or they do not. Typical diagrams are pictured below.

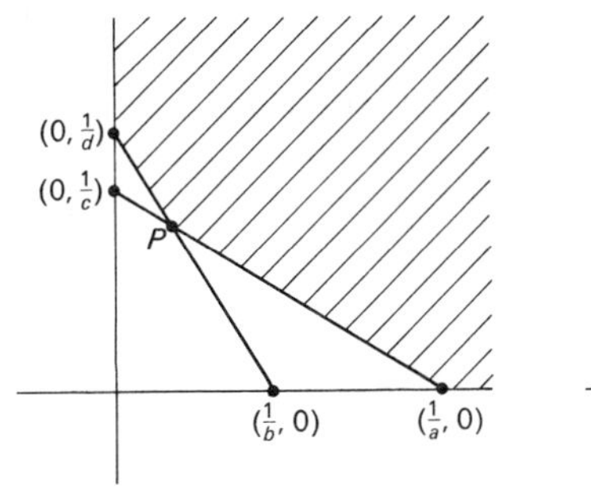

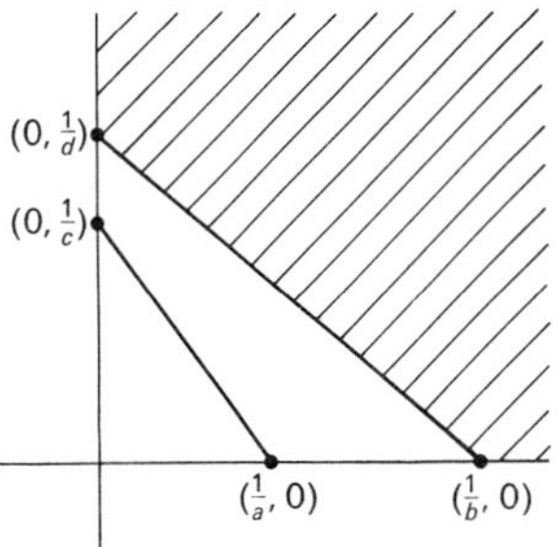

If the lines do not intersect, as in the second diagram, then $f(x)$ assumes its minimum value at $\left(0, \frac{1}{d}\right)$ or $\left(\frac{1}{b}, 0\right)$. This minimum value is $\frac{1}{d}$ or $\frac{1}{b}$. Hence the value of the game in these circumstances is $w = b$ or $w = d$. By checking the other case in which the lines do not intersect, we can similarly see that the other possibilities for the value of the game are $w = c$ or $w = a$. Thus if the lines do not intersect at a point with positive coordinates, it follows that the value of the game is one of the four values of the matrix.)

***21** Find the value w of the game defined by the 2×2 matrix A in Exercise 20 if the lines defined by $ax_1 + cx_2 = 1$ and $bx_1 + dx_2 = 1$ intersect. (Hint: In this case, a typical situation is given in the first diagram in the hint for Exercise 20. Find the coordinates of the point P and minimize the function $f(x)$ over the shaded region in the diagram.)

3.5

Markov Chains

In this section, we reconsider certain special classes of stochastic processes, a topic originally introduced in Section 1.7. Recall that a stochastic process is just a sequence of experiments in which each trial has a finite number of outcomes.

Example 5.1 Suppose an animal is being conditioned to respond to a given stimulus, and suppose that there are n distinct responses that are possible, $u_1, \ldots, u_n$. On each trial of an experiment, the animal is subjected to the same stimulus. Its response (one of the u_i) is recorded, and then the animal is given a prescribed reward determined by his response. For example, take $n = 2$. If baby says "da-da" (response u_1), he gets a cookie. If he does not say "da-da" (response u_2), he does not get a cookie. Suppose p_{11} is the probability that if he says "da-da" at a given stage of the experiment, he will say it at the next stage; and p_{21} is the probability that if he says "da-da" at a given stage, he will not say it at the next stage. On the other hand, let p_{12} be the probability that if he does not say "da-da" at one stage he will say it at the next; and p_{22} the probability that if he does not say "da-da" at a given stage, he will not say it at the next. A tree depicting the transition from the kth to the $(k + 1)$st stage of this experiment is shown in the following diagram.

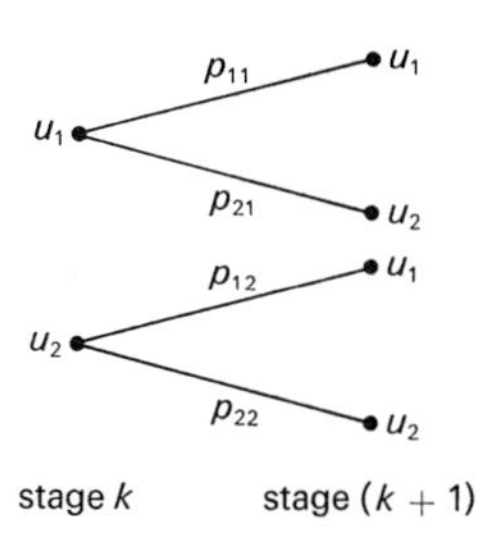

Suppose, moreover, that at the kth stage, the probability that response u_1 occurs is $x_1^{(k)}$ and the probability that response u_2 occurs is $x_2^{(k)}$. Then clearly the probability that response u_1 occurs at the $(k + 1)$st stage is

$$x_1^{(k+1)} = p_{11}x_1^{(k)} + p_{12}x_2^{(k)}. \tag{1}$$

Similarly, for u_2,

$$x_2^{(k+1)} = p_{21}x_1^{(k)} + p_{22}x_2^{(k)}. \tag{2}$$

If for any k we set $x^{(k)} = (x_1^{(k)}, x_2^{(k)})$, then the equations (1) and

(2) can be written

$$x^{(k+1)} = Px^{(k)},$$

where

$$P = \begin{bmatrix} p_{11} & p_{12} \\ p_{21} & p_{22} \end{bmatrix}.$$

In general, let p_{ij} denote the probability that if response u_j occurs at stage k, then response u_i occurs at stage $k + 1$, $i, j = 1, \ldots, n$.

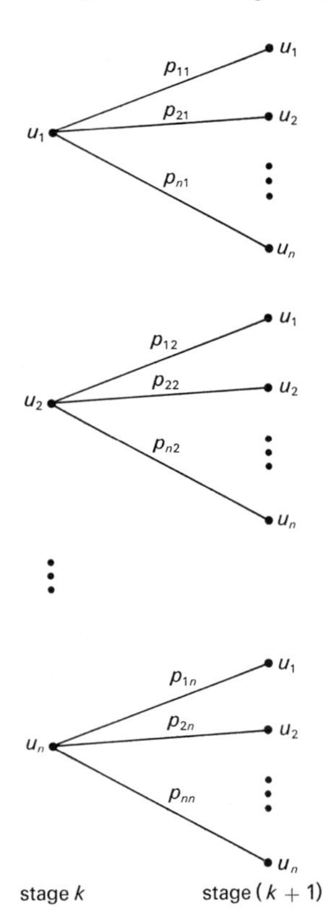

Let $x_i^{(k)}$ denote the probability that the response u_i occurs at stage k. We know from what we learned in the section on stochastic processes (Section 1.7) that

$$x_i^{(k+1)} = \sum_{j=1}^{n} p_{ij} x_j^{(k)}, \qquad i = 1, \ldots, n. \tag{3}$$

The equation in (3) states that we can compute the probability that u_i occurs at stage $k + 1$ as follows. Any one of the events $u_1, \ldots, u_n$ at stage k can lead to the event u_i with a probability $p_{ij}, j = 1, \ldots, n$, respectively. The probability that we start with u_j at stage k is $x_j^{(k)}$, and the probability that this results in u_i at stage $k + 1$ is therefore

$p_{ij}x_j^{(k)}$. Moreover, the event u_i can occur at stage $k + 1$ in any of n distinct ways: u_i can follow u_1, u_i can follow $u_2, \ldots, u_i$ can follow u_n. Hence the probability that event u_i occurs at stage $k + 1$ is just the sum of the probabilities $p_{ij}x_j^{(k)}$, $j = 1, \ldots, n$, i.e., the right side of (3).

The system of equations (3) can be summarized in a single matrix-vector equation,

$$x^{(k+1)} = Px^{(k)},$$

where P is the $n \times n$ matrix whose (i, j) entry is p_{ij} and $x^{(k)} = (x_1^{(k)}, x_2^{(k)}, \ldots, x_n^{(k)})$.

Observe that in the preceding example, the probability that u_i is the response at the $(k + 1)$st stage, given that u_j is the response at the kth stage, does not depend on k, i.e., the probability p_{ij} is independent of k. In this section, we are interested in the analysis of experiments in which the number of outcomes at each trial is the same, and in which the probability that the ith outcome occurs at a given trial (knowing that the jth outcome occurred at the previous trial) does not depend on the number of the trial.

Definition 5.1 ***Markov chain, state space, transition matrix, probability distribution*** Suppose that a sequence of experiments is performed in which there are n possible outcomes, $u_1, \ldots, u_n$, at each trial. If the probability p_{ij} that u_i occurs at the $(k + 1)$st trial, given that u_j occurred on the kth trial, does not depend on k, then the sequence of experiments is called a *Markov chain*. The set $\{u_1, \ldots, u_n\}$ is called the *state space*, and each u_i is called a *state*. Furthermore, the $n \times n$ matrix P with (i, j) entry p_{ij} is called the *transition matrix*, and the n-vector $x^{(k)}$, whose ith component is the probability $x_i^{(k)}$ that outcome u_i occurs at trial k, is called the kth *probability distribution* or *probability vector*.

We define $x^{(0)} = (x_1^{(0)}, \ldots, x_n^{(0)})$ to be the *initial* probability distribution in a Markov chain. The initial probability distribution simply tells us the probability of beginning in state u_i, $i = 1, \ldots, n$. For example, suppose that before rewarding the baby in Example 5.1 with any cookies, we know the chances are 1 in 100 that he will say "da-da." Then $(x_1^{(0)}, x_2^{(0)}) = (\frac{1}{100}, \frac{99}{100})$.

Observe that the sum of the entries in each column of the transition matrix P is 1, and each entry of P is nonnegative. Thus P is a column stochastic matrix as defined in Definition 5.1, Section 2.5. The $(k + 1)$st probability distribution $x^{(k+1)}$ is obtained from the

kth probability distribution by the fundamental equation we obtained above, namely,

$$x^{(k+1)} = Px^{(k)}. \tag{4}$$

Thus, setting $k = 0, 1, 2, \ldots$, we obtain

$$\begin{aligned} x^{(1)} &= Px^{(0)}, \\ x^{(2)} &= Px^{(1)}, \\ x^{(3)} &= Px^{(2)}, \\ &\vdots \\ x^{(k)} &= Px^{(k-1)}, \\ &\vdots \end{aligned} \tag{5}$$

If we replace $x^{(1)}$ in the second equation in (5) by its value given in the first equation, we see that

$$\begin{aligned} x^{(2)} &= P(Px^{(0)}) \\ &= P^2x^{(0)}. \end{aligned}$$

Similarly,

$$\begin{aligned} x^{(3)} &= P(P^2x^{(0)}) \\ &= P^3x^{(0)}, \end{aligned}$$

and, in general,

$$x^{(k)} = P^kx^{(0)}. \tag{6}$$

Thus the probability that a given event u_i occurs at a given stage depends only on the initial probability distribution $x^{(0)}$ and powers of the transition matrix P.

There are two general questions we will consider in studying Markov chains.

(a) To what extent does the initial situation, i.e., $x^{(0)}$, affect the outcome of the kth trial, for large k?
(b) What precisely is the outcome after k trials when k is large?

Example 5.2 Consider the 2×2 transition matrix

$$P = \begin{bmatrix} 0 & 1 \\ 1 & 0 \end{bmatrix}.$$

The student will convince himself that if k is even, $P^k = I_2$, and if k is odd, $P^k = P$. Thus from (6) we see that the kth probability distribution becomes

$$\begin{aligned} x^{(k)} &= P^kx^{(0)} \\ &= \begin{cases} x^{(0)} & \text{if k even,} \\ x^{(1)} & \text{if } k \text{ odd,} \end{cases} \end{aligned}$$

where $x^{(1)} = Px^{(0)} = (x_2^{(0)}, x_1^{(0)})$. Hence, in this case, the answer to question (a) above is that $x^{(0)}$ *does* affect what happens at the kth trial no matter how large k is.

It is clear in analyzing equation (6) that we need to know what happens to P^k for large k. As we saw in the preceding example, it is not necessarily the case that as k gets large P^k "gets close to" or "approaches" any particular matrix.

If every entry of the kth power of an $n \times n$ transition matrix P approaches the corresponding entry of a fixed $n \times n$ matrix A as k gets large, we say P^k *approaches* A, and we write

$$\lim_{k\to\infty} P^k = A. \tag{7}$$

The notation (7) is read, "*The limit of* P^k, *as* k *gets large, is the matrix* A."

To make this notion somewhat more precise, let $p_{ij}^{(k)}$ denote the (i, j) entry of P^k. As an example, suppose $p_{ij}^{(k)}$ is given by

$$p_{ij}^{(k)} = \frac{k}{k+1}. \tag{8}$$

What happens in this case to $p_{ij}^{(k)}$ as k gets large? The value of the fraction on the right side of (8) is unaltered if we divide numerator and denominator by k, $k > 0$, and thus

$$p_{ij}^{(k)} = \frac{1}{1+\dfrac{1}{k}}. \tag{9}$$

As k gets large, $\frac{1}{k}$ gets small, i.e., $\frac{1}{k}$ approaches 0, and hence we see from (9) that $p_{ij}^{(k)}$ approaches $\frac{1}{1+0} = 1$. As an example of a sequence of powers of a matrix that does not approach a limit, consider the (1, 2) entry in the transition matrix P in Example 5.2. Since $p_{12}^{(k)}$ is 0 or 1 according as k is even or odd, we see that

$$p_{12}^{(k)} = \frac{1+(-1)^{k+1}}{2}.$$

Observe that for any positive integer k, $p_{ij}^{(k)}$ is either 0 or 1 and hence does not approach any particular number.

If we look back at question (b), we see that in order to obtain an answer we must be able to compute

$$\lim_{k\to\infty} P^k.$$

Theorem 5.1 *Let P be an $n \times n$ column stochastic matrix, i.e., P is a transition matrix. If P^k approaches A, then:*

(a) *A is column stochastic;*
(b) *each column of A satisfies*

$$PA^{(j)} = A^{(j)}, \qquad j = 1, \ldots, n.$$

Proof

(a) Since P^k approaches A, we know that each entry of P^k approaches the corresponding entry of A. Moreover, since each entry of P^k is a nonnegative number, it follows that each entry of A is a nonnegative number. Each column sum of P^k is 1. (See Exercise 13, Section 2.5, and apply finite induction, taking $P = B = A$, $m = q = n$.) Since the entries of P^k approach the entries of A as k gets large, it follows that the column sums of A must be 1.

(b) For k very large, the entries of P^k and P^{k+1} are both close to the corresponding entries of A. But

$$P^{k+1} = P \cdot P^k. \tag{10}$$

Now, P^{k+1} approaches A and so does P^k, and hence, letting k grow arbitrarily large, we see from (10) that

$$A = PA. \tag{11}$$

But the jth column of PA is $PA^{(j)}, j = 1, \ldots, n$, and hence, matching columns on the two sides of (11), we have

$$PA^{(j)} = A^{(j)}. \blacksquare$$

Theorem 5.1 (b) is, in fact, a very efficient way of finding

$$\lim_{k \to \infty} P^k = A$$

when such a matrix exists.

Example 5.3 Assume that $\lim_{k \to \infty} P^k = A$ exists when

$$P = \begin{bmatrix} 1 & b \\ 0 & 1 - b \end{bmatrix},$$

$0 < b \leq 1$. (That this is indeed the case will be proved in Theorem 5.3.) Determine A explicitly from Theorem 5.1 (b), i.e., from

$$PA^{(1)} = A^{(1)} \tag{12}$$

and

$$PA^{(2)} = A^{(2)}. \tag{13}$$

To simplify the notation, let x be either $A^{(1)}$ or $A^{(2)}$. Then (12) and (13) have the form

$$Px = x,$$

or

$$(I_2 - P)x = 0.$$

If we write out this equation in terms of components, we obtain the system

$$\begin{aligned} -bx_2 &= 0, \\ bx_2 &= 0. \end{aligned}$$

Since b is not 0 we see that $x_2 = 0$, and since A is column stochastic, it follows that $x = (1, 0)$. Thus if the matrix A exists, both of its columns must be $(1, 0)$, and

$$A = \begin{bmatrix} 1 & 1 \\ 0 & 0 \end{bmatrix}.$$

Example 5.4 A gardner observes the probabilities with which seeds from white and red roses produce white and red roses. White roses result from seeds of white roses 70% of the time, and red roses result from seeds of white roses 30% of the time. On the other hand, white roses result from seeds of red roses 20% of the time, and red roses result from seeds of red roses 80% of the time. Given an initial population consisting of 40% white roses and 60% red roses, what will be the percentage distribution of white and red roses after a large number of generations? Suppose the distribution of white and red at the kth generation is

$$x^{(k)} = (x_1^{(k)}, x_2^{(k)}).$$

Initially we have

$$x^{(0)} = (.4, .6).$$

The transition from the kth generation to the $(k + 1)$st generation can be diagrammed in the usual manner. (See Section 1.7.)

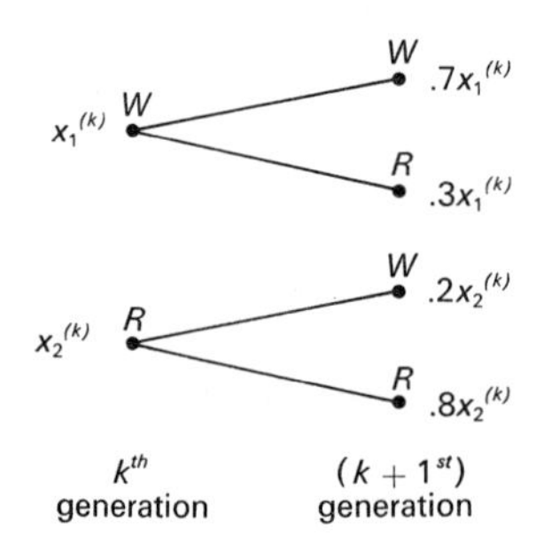

The expected color distribution of the $(k+1)$st generation is then

$$x_1^{(k+1)} = .7x_1^{(k)} + .2x_2^{(k)}$$

and

$$x_2^{(k+1)} = .3x_1^{(k)} + .8x_2^{(k)},$$

or

$$x^{(k+1)} = Px^{(k)},$$

where

$$P = \begin{bmatrix} .7 & .2 \\ .3 & .8 \end{bmatrix}.$$

Thus, as in the general discussion, we see that $x^{(k)}$ can be expressed in terms of P^k and $x^{(0)}$, i.e.,

$$x^{(k)} = P^k x^{(0)}. \tag{14}$$

Assume here that

$$\lim_{k\to\infty} P^k = A$$

exists. We shall shortly prove a general theorem showing that for matrices like P (i.e., column stochastic with positive entries) the limiting matrix A does indeed exist. According to Theorem 5.1, A must be column stochastic, and moreover, if x denotes either of the columns of A, then

$$Px = x. \tag{15}$$

If we let $x = (x_1, x_2)$ and write out (15) in terms of components, we obtain

$$\begin{aligned} .3x_1 - .2x_2 &= 0, \\ -.3x_1 + .2x_2 &= 0. \end{aligned}$$

Thus any solution to (15) must be of the form $(x_1, \frac{3}{2}x_1) = x_1(1, 1.5)$. But x is a column of the column stochastic matrix A, and thus the sum of the components must be 1. It follows that $x_1 = \frac{2}{5}$ and therefore that

$$x = (.4, .6).$$

Thus

$$A = \begin{bmatrix} .4 & .4 \\ .6 & .6 \end{bmatrix}.$$

If we refer to (14), we see that in the long run (i.e., for large k), the probability distribution $x^{(k)}$ approaches

$$\begin{aligned} Ax^{(0)} &= \begin{bmatrix} .4 & .4 \\ .6 & .6 \end{bmatrix} x^{(0)} \\ &= (.4, .6)x_1^{(0)} + (.4, .6)x_2^{(0)} \\ &= (.4, .6)(x_1^{(0)} + x_2^{(0)}) \\ &= (.4, .6). \end{aligned}$$

We have deliberately omitted the explicit values for $x^{(0)}$ in this last calculation to emphasize that the limiting distribution does not depend on the initial distribution $x^{(0)}$. In other words, no matter what the initial distribution of roses between red and white may be, the rose population will tend to stabilize at a ratio of 40% white and 60% red, given the above transition probabilities between generations.

Definition 5.2 ***Primitive matrix, regular markov chain*** Let P be an $n \times n$ column stochastic matrix. Then P is said to be *primitive* if some power of P has only positive entries. The Markov chain whose transition matrix is P is said to be *regular* when P is primitive.

Theorem 5.2 *If P is an $n \times n$ column stochastic primitive matrix and every entry of P^m is positive for some positive integer m, then every entry of P^k is positive for every integer $k \geq m$.*

Proof Let $Q = P^m$. It suffices to prove that $QP = P^{m+1}$ has positive entries. (Why? See Exercise 10.) Now the jth column of QP is

$$QP^{(j)} = \sum_{k=1}^{n} Q^{(k)} p_{kj}. \tag{16}$$

Observe that the right side of (16) is a convex combination of the vectors $Q^{(1)}, \ldots, Q^{(n)}$, and each of these vectors has only positive components. It follows that the right side of (16) is a vector with only positive components. (Why?) This completes the proof. ▮

What we need at this point is a theorem that tells us when the powers P^k of a column stochastic matrix approach a limit matrix A. The answer is provided by the following result.

Theorem 5.3 *Let P be an $n \times n$ primitive column stochastic matrix. Then the matrix*

$$\lim_{k \to \infty} P^k = A$$

exists. Moreover, every column of A is the same positive n-vector with no zero components.

Proof Before embarking on the formal aspects of this argument, consider the particular matrix

$$P = \begin{bmatrix} 0 & 1 \\ 1 & 0 \end{bmatrix}.$$

As we saw in Example 5.2, the powers P^k do not approach a limit. Moreover, no power of P has all of its entries positive, and so P does not fall under the purview of this theorem (i.e., P is not primitive). On the other hand, consider another matrix which is also not primitive:

$$P = \begin{bmatrix} 1 & \frac{1}{2} \\ 0 & \frac{1}{2} \end{bmatrix}.$$

The matrix P is a special case of the matrix in Example 5.3 with $b = \frac{1}{2}$. Let T be the matrix

$$T = \begin{bmatrix} 1 & -1 \\ 0 & 1 \end{bmatrix}.$$

Then we compute directly that

$$\begin{aligned} T^{-1}PT &= \begin{bmatrix} 1 & 1 \\ 0 & 1 \end{bmatrix}\begin{bmatrix} 1 & \frac{1}{2} \\ 0 & \frac{1}{2} \end{bmatrix}\begin{bmatrix} 1 & -1 \\ 0 & 1 \end{bmatrix} \\ &= \begin{bmatrix} 1 & 0 \\ 0 & \frac{1}{2} \end{bmatrix}. \end{aligned}$$

Call this latter matrix D, that is,

$$T^{-1}PT = D.$$

Hence for any positive integer k,

$$P^k = TD^kT^{-1}.$$

(Why? See Exercise 12.) Now,

$$D^k = \begin{bmatrix} 1 & 0 \\ 0 & \dfrac{1}{2^k} \end{bmatrix}$$

and hence D^k approaches the matrix

$$\Delta = \begin{bmatrix} 1 & 0 \\ 0 & 0 \end{bmatrix}.$$

It follows that P^k approaches

$$\begin{aligned} T\Delta T^{-1} &= \begin{bmatrix} 1 & -1 \\ 0 & 1 \end{bmatrix}\begin{bmatrix} 1 & 0 \\ 0 & 0 \end{bmatrix}\begin{bmatrix} 1 & 1 \\ 0 & 1 \end{bmatrix} \\ &= \begin{bmatrix} 1 & 1 \\ 0 & 0 \end{bmatrix}. \end{aligned}$$

We see from this second example that $\lim_{k\to\infty} P^k$ can exist even though P is not primitive. Observe that for any k, the (2, 1) entry of P^k is always 0 for the matrix

$$P = \begin{bmatrix} 1 & \frac{1}{2} \\ 0 & \frac{1}{2} \end{bmatrix}.$$

To proceed to the proof of the theorem, let m be a positive integer for which

$$S = P^m$$

has only positive entries (recall that P is primitive). Let σ be the least entry in S, let $R_i(k)$ be the largest entry in the ith row of S^k and $r_i(k)$ the smallest entry in the ith row of S^k, $i = 1, \ldots, n$. According to Exercise 12 in Section 3.3, we know that

$$R_i(k+1) - r_i(k+1) \leq (1 - 2\sigma)(R_i(k) - r_i(k)), \quad k = 1, 2, \ldots . \tag{17}$$

If we apply (17) repeatedly, starting with $k = 1$, we obtain

$$R_i(k+1) - r_i(k+1) \leq (1 - 2\sigma)^k(R_i(1) - r_i(1)). \tag{18}$$

According to Exercise 13 in Section 3.3,

$$\sigma \leq \frac{1}{n},$$

and hence

$$1 - 2\sigma \geq 1 - \frac{2}{n}.$$

Thus (since we are assuming $n > 1$)

$$1 - 2\sigma \geq 0.$$

But σ is strictly positive, and thus

$$0 \leq 1 - 2\sigma < 1.$$

Hence $(1 - 2\sigma)^k$ must approach zero, because we are taking higher and higher powers of a nonnegative number less than 1. Thus from (18) we see that

$$R_i(k+1) - r_i(k+1)$$

must approach 0 as k gets large. Hence the difference between the largest entry in the ith row of S^{k+1} and the least entry in the ith row of S^{k+1} approaches 0. (This alone does not prove that $R_i(k)$ and $r_i(k)$ tend to a common limit. For, they could both grow arbitrarily large as k increases and still stay close together.) However, if $s_{ij}^{(k+1)}$ is the (i, j) entry of S^{k+1}, then

$$s_{ij}^{(k+1)} = \sum_{t=1}^{n} s_{it}^{(k)} s_{tj}. \tag{19}$$

The matrix S is column stochastic (see Exercise 13, Section 2.5). The right-hand side of (19) is a convex combination of the numbers $s_{it}^{(k)}$, $t = 1, \ldots, n$. It therefore must exceed the least of these numbers $s_{it}^{(k)}$ and must also be no greater than the largest of these numbers. In symbols

$$r_i(k) \leq s_{ij}^{(k+1)} \leq R_i(k), \qquad j = 1, \ldots, n. \tag{20}$$

It follows from (20) that the largest and smallest of the numbers $s_{ij}^{(k+1)}$, $j = 1, \ldots, n$, must also satisfy the inequalities in (20), i.e.,

$$r_i(k) \leq r_i(k+1) \leq R_i(k)$$

and

$$r_i(k) \leq R_i(k+1) \leq R_i(k).$$

Thus we can conclude that

$$r_i(k) \leq r_i(k+1),$$
$$R_i(k+1) \leq R_i(k),$$

$k = 1, 2, 3, \ldots$. Hence

$$r_i(1) \leq r_i(2) \leq r_i(3) \leq \cdots$$

and

$$R_i(1) \geq R_i(2) \geq R_i(3) \geq \cdots.$$

Thus the numbers $r_i(k)$ steadily increase as k increases, the numbers $R_i(k)$ steadily decrease as k increases, and, as we remarked before, the difference

$$R_i(k) - r_i(k)$$

decreases in size as k grows large. Graphically the situation looks as follows.

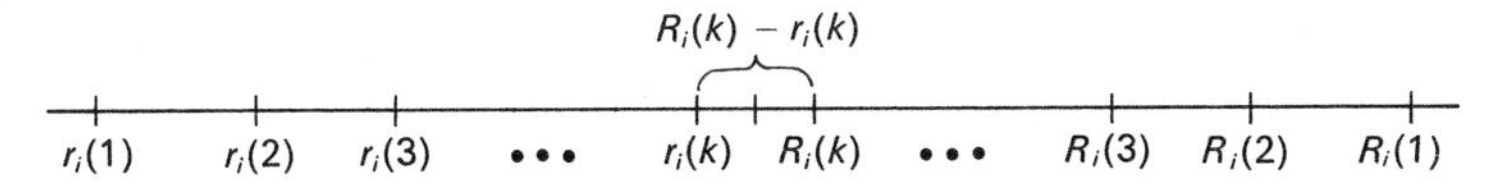

We see, then, that the difference between the largest and smallest entries in the ith row of S^k approaches 0 and, moreover, that the largest and smallest entries in each row must approach a common number. We have proved that as k gets large, the kth power of the matrix $S = P^m$, i.e., the matrix $(P^m)^k = P^{mk}$, approaches a limiting matrix A. Further, all the elements in the ith row of A are identical, $i = 1, \ldots, n$, i.e., every column is the same. If we look at the largest and smallest entries in the ith row of the kth power of P, we see by Exercise 12 in Section 3.3 that this difference does not increase as k increases. Moreover, the largest element in the ith row of P^k does not increase and the smallest element in the ith row of P^k does not decrease as k increases. However, the mth, the $(2m)$th, the $(3m)$th, ..., the (km)th of these differences strictly decrease to zero because these are precisely the numbers

$$R_i(k+1) - r_i(k+1) \leq (1-2\sigma)^k(R_i(1) - r_i(1)),$$

where σ is the least entry in P^m. It follows by precisely the argument that we used for S that P^k must approach the limiting matrix A.

The entries in $S = P^m$ are positive. On the other hand, we know that the least entry in P^k does not decrease with k, and hence the least entry in the limiting matrix A must be positive. This verifies the assertion that each column of A is the same *positive* n-vector with no zero components. ∎

Theorems 5.1 and 5.3 can be effectively combined to compute the limiting matrix of a primitive transition matrix, i.e., it is possible to analyze the behavior of a Markov chain after a large number of trials. For, if P is a primitive column stochastic matrix, we know by Theorem 5.3 that $\lim_{k\to\infty} P^k$ exists and is equal to a matrix A, each of whose columns is the same positive n-tuple x. Moreover, according to Theorem 5.1, this n-tuple x must satisfy

$$Px = x. \tag{21}$$

A non-zero vector x which satisfies (21) is called a *fixed* or a *characteristic vector* for the matrix P. The only ambiguity that remains is whether there is more than one n-vector which satisfies (21) when P is primitive. We resolve this minor dilemma in the next and final theorem.

Theorem 5.4 *Let P be an $n \times n$ column stochastic primitive matrix. Suppose that x and y are two positive n-vectors with no zero components which satisfy*

$$Px = x,$$
$$Py = y.$$

Assume, moreover, that the sums of the components in both x and y are 1, i.e., x and y are both probability distributions. Then $x = y$.

Proof Suppose that $x \neq y$. Then not all the ratios $\frac{x_i}{y_i}$ and $\frac{y_i}{x_i}$, $i = 1, \ldots, n$, can be 1. The smallest of these $2n$ ratios must be less than 1; call it θ. It is clear that we lose no generality in assuming

$$\theta = \frac{x_k}{y_k},$$

that is, that $\dfrac{x_k}{y_k}$ is the least of the $2n$ ratios. Consider the vector

$$z = x - \theta y.$$

Then for any $i = 1, \ldots, n$,

$$\begin{aligned} z_i &= x_i - \theta y_i \\ &= x_i - \left(\frac{x_k}{y_k}\right) y_i \\ &\geq x_i - \left(\frac{x_i}{y_i}\right) y_i \\ &= 0. \end{aligned} \tag{22}$$

The inequality in (22) follows from the fact that $\theta \leq \dfrac{x_i}{y_i}$, $y_i > 0$, $i = 1, \ldots, n$. Thus z is a nonnegative n-vector with $z_k = 0$. Suppose z is the zero vector. Then

$$\begin{aligned} 0 &= \sum_{i=1}^{n} z_i \\ &= \sum_{i=1}^{n} (x_i - \theta y_i) \\ &= \sum_{i=1}^{n} x_i - \theta \sum_{i=1}^{n} y_i \\ &= 1 - \theta, \end{aligned}$$

i.e., $\theta = 1$. But we are assuming $x \neq y$, and hence that $\theta < 1$. Thus z has at least one positive component. Now, P is a primitive matrix, and thus, for some positive integer m, $P^m = S$ has only positive entries. It follows immediately (see Exercise 15) from $Px = x$ that $P^m x = x$, and similarly, that $P^m y = y$. Hence

$$Sx = x$$

and

$$Sy = y.$$

We compute that

$$\begin{aligned} Sz &= S(x - \theta y) \\ &= Sx - \theta Sy \\ &= x - \theta y \\ &= z, \end{aligned} \tag{23}$$

that is, z is a fixed vector of $S = P^m$. From (23) it follows that

$$z = \sum_{j=1}^{n} S^{(j)} z_j. \tag{24}$$

Thus the right side of (24) is a sum of nonnegative multiples of the positive n-tuples $S^{(j)}$ and not all the coefficients are zero, so that z is a positive n-vector with no zero components. Yet we know that $z_k = x_k - y_k = x_k - \frac{x_k}{y_k} y_k = 0$. This is a contradiction and thus $x = y$. ▮

We are now in a position to give a complete answer to questions (a) and (b) for regular Markov chains. In answer to (a), the initial distribution $x^{(0)}$ has no effect on the long term outcome. The answer to question (b) is that the long term outcome is precisely the vector x which is the common column of the limiting matrix A. For, from (6) we see that $x^{(k)}$ approaches the vector $Ax^{(0)}$ for large k and

$$\begin{aligned} Ax^{(0)} &= \sum_{j=1}^{n} A^{(j)} x_j^{(0)} \\ &= \sum_{j=1}^{n} x x_j^{(0)} \\ &= x \sum_{j=1}^{n} x_j^{(0)} \\ &= x. \end{aligned}$$

Thus the long term outcome is completely determined by the transition matrix P. Moreover, we can compute the vector x without computing the powers P^k of the transition matrix, for we know from our results that the common column x of $\lim_{k \to \infty} P^k = A$ is the unique probability vector x which satisfies

$$Px = x.$$

Example 5.5 Assume that the population of women is divided into three classes: those who are overweight at 40, those who are underweight at 40, and those who are normal at 40. Call these three conditions u_1, u_2, and u_3, respectively. Suppose, moreover, that it has been observed that these weight characteristics are transferred from mother to daughter as follows.

(25)

		Mother u_1	Mother u_2	Mother u_3
Daughter	u_1	70%	30%	15%
Daughter	u_2	20%	50%	60%
Daughter	u_3	10%	20%	25%

Thus, for example, 60% of the daughters of women who are of normal weight at age 40 are observed to be underweight at age 40, while 25% of them are observed to be of normal weight. Suppose we are given an initial distribution of women into these three classes, say

$$x^{(0)} = (x_1^{(0)}, x_2^{(0)}, x_3^{(0)}).$$

After a large number of generations, find the distribution into the three classes. We might as well assume $x^{(0)}$ is a probability vector, i.e., that the components represent the percentage of women in each of the three classes. We can reformulate the table in (25) as a transition matrix:

$$P = \begin{bmatrix} .70 & .30 & .15 \\ .20 & .50 & .60 \\ .10 & .20 & .25 \end{bmatrix}.$$

Since every entry of P is positive, it follows that P is primitive and thus $\lim_{k\to\infty} P^k = A$ exists. Moreover, the columns of A are identical and, if x denotes their common value,

$$Px = x. \tag{26}$$

The equation in (26) becomes

$$\begin{aligned} .30x_1 - .30x_2 - .15x_3 &= 0, \\ -.20x_1 + .50x_2 - .60x_3 &= 0, \\ -.10x_1 - .20x_2 + .75x_3 &= 0. \end{aligned}$$

Add the first equation to the third, and then add the second equation to the third to obtain the equivalent system

$$\begin{aligned} .30x_1 - .30x_2 - .15x_3 &= 0, \\ -.20x_1 + .50x_2 - .60x_3 &= 0, \\ 0 &= 0. \end{aligned}$$

Multiplying the first equation by $\frac{100}{15}$ and the second equation by 10, we obtain

$$\begin{aligned} 2x_1 - 2x_2 - x_3 &= 0, \\ -2x_1 + 5x_2 - 6x_3 &= 0. \end{aligned}$$

(We have dropped the third equation, since it adds no new information.) Add the first equation to the second:

$$\begin{aligned} 2x_1 - 2x_2 - x_3 &= 0, \\ 3x_2 - 7x_3 &= 0. \end{aligned}$$

Thus $x_2 = \frac{7}{3}x_3$ and

$$\begin{aligned} x_1 &= x_2 + \tfrac{1}{2}x_3 \\ &= \tfrac{7}{3}x_3 + \tfrac{1}{2}x_3 \\ &= \tfrac{17}{6}x_3. \end{aligned}$$

Therefore

$$x = x_3(\tfrac{17}{6}, \tfrac{7}{3}, 1).$$

We want to choose x_3 so that the sum of the components is 1. Hence the resulting vector x is

$$x = (\tfrac{17}{37}, \tfrac{14}{37}, \tfrac{6}{37}) = (.46, .38, .16) \qquad \text{(approximately)}.$$

Thus we see that regardless of the initial distribution into overweight, underweight, and normal classes, after a large number of generations the population of women of age 40 will tend to stabilize at 46% overweight, 38% underweight, and 16% normal.

Example 5.6 Suppose that a gambler enters a game with a capital of N dollars, $N > 1$. Suppose, moreover, that he bets \$1 at each trial of the game, and that he decides in advance to quit either when he reaches a stage at which he has \$1 left or when he accumulates $M > N$ dollars. Assume that it is known that the probability of his winning on any given trial of the game is θ, $0 \leq \theta \leq 1$. Two questions can be asked about this situation.

(a) Given an integer p, what is the probability that he has \$1 left and therefore must quit after at most p trials?
(b) What is the probability that he has M dollars after at most p trials?

To analyze this problem, we must define an appropriate state space. This will simply be the set of numbers $\{1, 2, \ldots, M\}$, denoting his capital in dollars at any given stage of the game. Suppose, for example, that he has \$3 at the beginning of the kth trial of the game, i.e., he is in "state 3." Since he bets \$1 at each trial with a probability θ of winning, at the $(k+1)$st trial he will have \$4 with the probability θ and \$2 with the probability $1 - \theta$. On the other hand, if he has \$$M$ at the beginning of the kth trial, he will have \$$M$ at the $(k+1)$st trial, because he has decided in advance to quit when his capital reaches \$$M$. We can fill out the $M \times M$ transition matrix as follows.

$$(27) \qquad P = \begin{array}{c} \\ 1 \\ 2 \\ 3 \\ \cdot \\ \cdot \\ \cdot \\ M-1 \\ M \end{array} \begin{array}{c} \begin{array}{cccccc} 1 & 2 & 3 & \cdots & M-1 & M \end{array} \\ \begin{bmatrix} 1 & 1-\theta & 0 & \cdots & 0 & 0 \\ 0 & 0 & 1-\theta & & \cdot & \cdot \\ 0 & \theta & 0 & \cdots & \cdot & \cdot \\ \cdot & \cdot & & \cdots & \cdot & \cdot \\ \cdot & \cdot & & \cdots & 1-\theta & 0 \\ \cdot & \cdot & & & 0 & 0 \\ 0 & 0 & 0 & \cdots & \theta & 1 \end{bmatrix} \end{array}$$

In the first column, the only non-zero entry is a 1 in the (1, 1) position. In column k, $k = 2, \ldots, M - 1$, the only non-zero entries are $1 - \theta$ in row $k - 1$ and θ in row $k + 1$. The last column is all zeros except for the 1 in row M. The initial probability distribution is the vector

$$x^{(0)} = (0, \ldots, 0, \overset{\overset{N}{\downarrow}}{1}, 0, \ldots, 0),$$

that is, the gambler starts with $\$N$, in "state N." Consider the tree diagram representing the transition from the kth to the $(k + 1)$st trial.

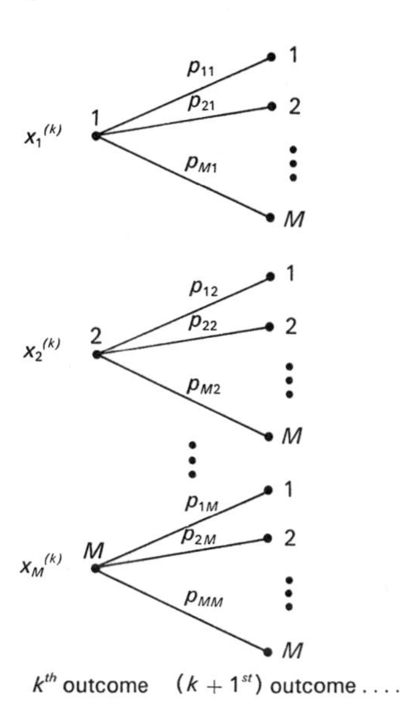

Suppose the probability that the gambler has \$1 at the end of the kth trial is $x_1^{(k)}, \ldots,$ the probability that he has $\$M$ at the end of the kth trial $x_M^{(k)}$, i.e., $x^{(k)} = (x_1^{(k)}, \ldots, x_M^{(k)})$ is the probability distribution for the kth trial. The probability that he has \$1 at the end of the $(k + 1)$st trial is then

$$p_{11}x_1^{(k)} + p_{12}x_2^{(k)} + \cdots + p_{1M}x_M^{(k)};$$

the probability that he has \$2 at the end of the $(k + 1)$st trial is

$$p_{21}x_1^{(k)} + p_{22}x_2^{(k)} + \cdots + p_{2M}x_M^{(k)};$$

and continuing, the probability that he has $\$M$ at the end of the $(k + 1)$st trial is

$$p_{M1}x_1^{(k)} + p_{M2}x_2^{(k)} + \cdots + p_{MM}x_M^{(k)},$$

where p_{ij} is the (i, j) entry in the matrix P in (27). Thus

$$x_i^{(k+1)} = \sum_{j=1}^{M} p_{ij} x_j^{(k)}, \qquad i = 1, \ldots, M.$$

The fact that $x^{(0)}$ has a 1 as its Nth component is just the statement that the gambler starts with \$$N$, i.e., it is assumed that he has \$$N$ initially. Now, in general

$$x^{(k)} = P^k x^{(0)}.$$

Thus we see that the probability that the gambler has \$1 left after p trials is just the first component of $x^{(p)}$, and this can be explicitly computed in any specific example. Similarly, if we are interested in the probability that he has stopped gambling because he has achieved his goal of having \$$M$ after p trials, then we simply look at the Mth component of $x^{(p)}$. We remark that the matrix P is *not* primitive (see Exercise 21).

Quiz

Answer true or false.

1 The matrix

$$\begin{bmatrix} 1 & \frac{1}{2} \\ 0 & \frac{1}{2} \end{bmatrix}$$

is primitive.

2 Any $n \times n$ matrix with positive entries is primitive.

3 The matrix

$$\begin{bmatrix} 1 & 1 \\ 1 & 1 \end{bmatrix}$$

is a transition matrix.

4 The vector $x = (0, \frac{1}{2}, -\frac{1}{2}, 1, 0)$ is a probability vector.

5 If the sum of the entries in each column of an $n \times n$ matrix is 1, then the matrix is a transition matrix.

6 The matrix

$$P = \begin{bmatrix} \frac{1}{2} & 1 \\ \frac{1}{2} & 0 \end{bmatrix}$$

is primitive.

7 The matrix P in the preceding question satisfies $\lim_{k \to \infty} P^k = A$, where

$$A = \begin{bmatrix} \frac{2}{3} & \frac{2}{3} \\ \frac{1}{3} & \frac{1}{3} \end{bmatrix}.$$

8 If x and y are probability vectors in R^n and θ is a number, $0 \leq \theta \leq 1$, then $(1 - \theta)x + \theta y$ is a probability vector.

9 There exists a nontrivial linear combination of the vectors (1, 2, 1) and (4, 1, 3) which has nonnegative components and at least one zero component.

10 The Markov chain with transition matrix

$$P = \begin{bmatrix} 0 & \frac{2}{3} \\ 1 & \frac{1}{3} \end{bmatrix}$$

is regular.

Exercises

1 Identify the probability vectors among the following:

(a) (0, 0, 1); (b) (−1, 0, 2);
(c) $(\frac{1}{2}, \frac{1}{4}, \frac{1}{8})$; (d) $(\frac{1}{2}, \frac{1}{4}, \frac{1}{4})$;
(e) $(\frac{1}{2}, \frac{1}{2}, 1)$; (f) $(-\frac{2}{3}, -\frac{1}{3}, 0)$;
(g) $(-\frac{2}{3}, \frac{2}{3}, 1)$; (h) (1, 1, 1);
(i) $(\frac{1}{3}, \frac{1}{3}, \frac{1}{3})$; (j) (2, 3, −4);
(k) (1, 0, 0, 0); (l) $(\frac{1}{2}, 0, \frac{1}{2}, 0, -\frac{1}{2}, -\frac{1}{2}, 1)$.

2 Identify the transition matrices among the following:

(a) $\begin{bmatrix} \frac{1}{2} & -\frac{1}{2} \\ \frac{1}{2} & \frac{3}{2} \end{bmatrix}$; (b) $\begin{bmatrix} 1 & \frac{1}{2} & \frac{1}{3} \\ 0 & \frac{1}{2} & \frac{2}{3} \end{bmatrix}$;

(c) $\begin{bmatrix} 1 & 0 & 0 \\ 0 & 1 & 0 \\ 0 & 0 & 1 \end{bmatrix}$; (d) $\begin{bmatrix} 1 & 0 & 0 \\ 0 & \frac{1}{2} & \frac{1}{2} \\ \frac{1}{3} & 0 & \frac{2}{3} \end{bmatrix}$;

(e) $\begin{bmatrix} \frac{1}{4} & \frac{1}{2} & 0 \\ \frac{1}{4} & 0 & 0 \\ \frac{1}{2} & \frac{1}{2} & 0 \\ 0 & 0 & 1 \end{bmatrix}$; (f) $\begin{bmatrix} 1 & \frac{1}{2} & \frac{1}{3} \\ -1 & \frac{1}{4} & \frac{1}{3} \\ 1 & \frac{1}{4} & \frac{1}{3} \end{bmatrix}$;

(g) $\begin{bmatrix} -1 & 1 & 1 \\ 1 & -1 & 1 \\ 1 & 1 & -1 \end{bmatrix}$; (h) $\begin{bmatrix} 0 & \frac{1}{2} & \frac{1}{2} \\ \frac{1}{2} & 0 & \frac{1}{2} \\ \frac{1}{2} & \frac{1}{2} & 0 \end{bmatrix}$;

(i) $\begin{bmatrix} \frac{1}{4} & 0 & \frac{1}{9} \\ \frac{1}{3} & \frac{3}{8} & 0 \\ \frac{1}{6} & \frac{1}{3} & \frac{2}{9} \\ \frac{1}{4} & \frac{7}{24} & \frac{2}{3} \end{bmatrix}$; (j) $\begin{bmatrix} 0 & \frac{1}{3} & \frac{1}{5} & \frac{3}{4} & \frac{1}{6} \\ \frac{1}{2} & 0 & \frac{2}{5} & 0 & \frac{1}{6} \\ \frac{1}{4} & \frac{1}{3} & 0 & 0 & \frac{1}{6} \\ \frac{1}{4} & \frac{1}{3} & \frac{2}{5} & \frac{1}{4} & \frac{1}{2} \end{bmatrix}$;

(k) $\begin{bmatrix} 1 & 1 \\ 1 & 1 \end{bmatrix}$; (l) $\begin{bmatrix} 27 & -31 \\ -26 & 32 \end{bmatrix}$;

(m) $\begin{bmatrix} 1 \\ 0 \\ 0 \\ 0 \\ 0 \end{bmatrix}$; (n) $[1 \quad 1 \quad 1 \quad 1 \quad 1]$;

(o) $\begin{bmatrix} 0 & 1 & 0 \\ 1 & 0 & 0 \\ 0 & 0 & 1 \end{bmatrix}$.

3 For each of the following transition matrices P, decide whether $\lim_{k\to\infty} P^k$ exists. Find the limiting matrix when it exists.

(a) $\begin{bmatrix} 0 & 1 \\ 1 & 0 \end{bmatrix}$; (b) $\begin{bmatrix} 1 & 0 & 0 \\ 0 & 0 & 1 \\ 0 & 1 & 0 \end{bmatrix}$;

(c) $\begin{bmatrix} 0 & 1 & 0 \\ 0 & 0 & 1 \\ 1 & 0 & 0 \end{bmatrix}$; (d) $\begin{bmatrix} 1 & 0 & 0 \\ 0 & 1 & 0 \\ 0 & 0 & 1 \end{bmatrix}$;

(e) $\begin{bmatrix} 0 & \frac{1}{4} & \frac{1}{2} \\ 1 & \frac{3}{4} & \frac{1}{3} \\ 0 & 0 & \frac{1}{6} \end{bmatrix}$; (f) $\begin{bmatrix} 0 & \frac{1}{2} & \frac{1}{2} \\ \frac{1}{2} & 0 & \frac{1}{2} \\ \frac{1}{2} & \frac{1}{2} & 0 \end{bmatrix}$;

(g) $\begin{bmatrix} 0 & \frac{2}{3} & 0 \\ \frac{1}{2} & \frac{1}{3} & 1 \\ \frac{1}{2} & 0 & 0 \end{bmatrix}$; (h) $\begin{bmatrix} 1 & 0 & 0 & 0 \\ 0 & \frac{1}{3} & \frac{1}{3} & \frac{1}{3} \\ 0 & \frac{1}{3} & \frac{1}{3} & \frac{1}{3} \\ 0 & \frac{1}{3} & \frac{1}{3} & \frac{1}{3} \end{bmatrix}$;

(i) $\begin{bmatrix} \frac{2}{3} & \frac{2}{3} \\ \frac{1}{3} & \frac{1}{3} \end{bmatrix}$; (j) $\begin{bmatrix} \frac{7}{9} & 1 \\ \frac{2}{9} & 0 \end{bmatrix}$.

4 Let P be the transition matrix

$$P = \begin{bmatrix} 1-a & b \\ a & 1-b \end{bmatrix}.$$

where $0 < a < 1, 0 < b < 1$. Show that the limiting matrix $A = \lim_{k\to\infty} P^k$ is given by

$$A = \begin{bmatrix} \dfrac{b}{a+b} & \dfrac{b}{a+b} \\ \dfrac{a}{a+b} & \dfrac{a}{a+b} \end{bmatrix}.$$

5 Larry dates one of two girls, Jane or Sue. If he dates Jane one day, then he will date Sue the next day with probability .2. The probability that he will date Sue for two consecutive days is .7. In the long run, how often does Larry date each girl? (Hint: Let $U = \{u_1, u_2\}$ be the state space in which u_1 is the event of dating Jane, and u_2 is the event of dating Sue. The transition matrix is

$$P = \begin{bmatrix} .8 & .3 \\ .2 & .7 \end{bmatrix}.)$$

6 It is observed in a population of laboratory mice that among three commercial brands of feed, A, B, and C:

(i) the mice will never eat the same feed twice in a row;
(ii) if the mice eat A at one feeding, they will only eat B at the next feeding;
(iii) if the mice eat B at one feeding, they will eat A at the next feeding with a probability of .6, and C at the next feeding with a probability of .4;

(iv) if the mice eat C at one feeding, then at the next feeding they will eat A with probability $\frac{1}{2}$ and B with probability $\frac{1}{2}$.

In what ratios should A, B, and C be purchased for a large number of feedings? (Hint: The transition matrix from one feeding to the next is

$$P = \begin{bmatrix} 0 & .6 & .5 \\ 1 & 0 & .5 \\ 0 & .4 & 0 \end{bmatrix}.)$$

7 Two jars, I and II, contain two vanilla and three chocolate cookies, respectively. Junior (remember him?) takes a cookie from each jar, but because of fear of being discovered he puts them back, interchanging them in his haste. What is the probability that there are two chocolate cookies in the first jar after three such interchanges? After a large number of interchanges, what is the probability that there are two chocolate cookies in the first jar? (Hint: There are 3 states, u_1, u_2, and u_3, for the distribution of the cookies in the two jars that can arise by such interchanges.

	jar I	jar II
u_1	$2v$	$3c$
u_2	$1v$, $1c$	$2c$, $1v$
u_3	$2c$	c, $2v$

The probability of going from u_1 to u_2 with one interchange is 1, the probability of going from u_1 to u_3 with one interchange is 0, the probability of going from u_2 to u_1 is $\frac{1}{6}$ (Why?), etc. Thus the transition matrix is

$$P = \begin{matrix} & \begin{matrix} u_1 & u_2 & u_3 \end{matrix} \\ \begin{matrix} u_1 \\ u_2 \\ u_3 \end{matrix} & \begin{bmatrix} 0 & \frac{1}{6} & 0 \\ 1 & \frac{1}{2} & \frac{2}{3} \\ 0 & \frac{1}{3} & \frac{1}{3} \end{bmatrix} \end{matrix}.$$

The initial probability distribution is $x^{(0)} = (1, 0, 0)$.)

8 Assume that in a pinball machine, there are four replay buttons, I, II, III, and IV. If the pinball is hit by I, it will go to II with a probability of .3, to III with a probability of .4, to IV with a probability of .1, and it will return to I with a probability of .2. The probabilities of the pinball hitting other combinations of the replays are displayed in the table below.

	I	II	III	IV
I	.2	.3	.2	.1
II	.3	.3	.2	.4
III	.4	.2	.3	.3
IV	.1	.2	.3	.2

In the long run, what are the probabilities with which the pinball hits each of the replays?

9 Assume that the probability that a pretty mother has a pretty daughter is .6, and that a homely mother has a homely daughter is .2. What is the probability of a pretty woman being the great grandmother of a homely girl? If we assume the population of women initially is equally divided into pretty ones and homely ones, what will the probability distribution into pretty ones and homely ones be after 4 generations? After a large number of generations, what will be the distribution? Does it make any difference whether Eve was pretty or homely?

†10 Answer the question posed in the second sentence of the proof of Theorem 5.2.

11 Let x be a nonnegative n-vector, $x \neq 0$. Show that if Q is an $n \times n$ matrix with positive entries, then Qx is an n-vector with positive entries.

†12 Let T be a nonsingular $n \times n$ matrix and let P be an arbitrary $n \times n$ matrix. Show that for any positive integer k,

$$(T^{-1}PT)^k = T^{-1}P^kT.$$

13 Show that if x is a fixed vector for the $n \times n$ matrix P, then $I_n - P$ does not have an inverse. (Hint: If $I_n - P$ has an inverse, then since $(I_n - P)x = 0$, it follows that $x = 0$.)

***14** Let P be an $n \times n$ doubly stochastic primitive matrix. Prove that $\lim_{n\to\infty} P^n = J_n$, where J_n is the matrix each of whose entries is $\frac{1}{n}$ ·

†15 Let x be a fixed vector for the $n \times n$ matrix P. Show that for any positive integer m, x is a fixed vector for P^m.

16 For each of the following pairs of vectors x and y, find a linear combination which has nonnegative components and at least one zero component:

(a) $x = (1, 1, 2, 3)$, $y = (2, 1, 1, 1)$;
(b) $x = (9, 16, 7, 1, 2)$, $y = (1, 3, 5, 7, 9)$;
(c) $x = (3, 1, 2)$, $y = (3, 2, 1)$;
(d) $x = (2, 1, 7, 6, 4)$, $y = (1, 2, 9, 3, 7)$;
(e) $x = (1, 2)$, $y = (2, 1)$.

***17** Suppose that a gambler starts a game with a capital of \$2. Assume that he bets \$1 at each trial of the game, and that he decides in advance to quit either when he has only \$1 left or when he has accumulated \$5. Assume, moreover, that the probability of winning at any particular trial is $\frac{1}{2}$. After 3 trials, what is the probability that he will quit? What is the probability that his capital is at least \$4 after 5 trials of the game?

***18** The weather bureau in Santa Barbara has observed that if it is sunny on any given day, then the probability that it will be sunny the next day is .8, the probability that it will be foggy is .1, and the probability that it will be rainy is .1. It is also observed that if it is rainy on any given day, it is equally probable that it will be sunny, foggy, or rainy the next day, whereas if it is foggy on a given day, the probability is .5 that it will be sunny and .5 that it will be foggy the next day.

(a) If it is sunny on Monday, what is the probability that it will be foggy on the following Friday?

(b) If it is foggy on Wednesday, what is the probability that it will be rainy or foggy on either the following Saturday or the following Sunday?

(c) On January 1 in the year 2500, what is the probability that it will be rainy (approximately)?

***19** Let P be an $n \times n$ primitive column stochastic matrix and let S be any $n \times n$ column stochastic matrix. Show that if $0 < \theta \leq 1$, then the matrix

$$Q = \theta P + (1 - \theta)S$$

is primitive column stochastic. (Hint: Q^m is a sum of nonnegative matrices, one of which is $\theta^m P^m$.)

†20 Show that a lower (upper) triangular matrix can never be primitive.

21 Show that the matrix P in (27) can never be primitive. (Hint: Show that the powers of any matrix with 0's in the first column below the (1, 1) entry must also have 0's in these positions.)

answers to quizzes and selected exercises

chapter 1

1.1 Quiz

1 True. See Table (7).

2 False. See Table (6).

3 True.

p	$\sim p$	$p \vee \sim p$
T	F	T
F	T	T

4 True. Let p be "$1 = 0$," which is F; let q be "$2 = 1$," which is F. Then $p \rightarrow q$ is T.

5 True. Let p be "$1 = 0$," which is F; let q be "$2 = 2$," which is T. Then $p \rightarrow q$ is T.

6 False. That is the definition of a conjunction.

7 True.

8 True.

9 False. For example, $p \wedge \sim p$.

10 True. A truth table for a compound statement f formed from $p_1, p_2, \ldots, p_n$ has 2^n rows.

1.1 Exercises

1 (d) p_1: Scientists are deferred.
p_2: The present draft law is in the national interest.

$$f\colon\ ((p_2 \rightarrow p_1) \wedge \sim p_1) \rightarrow \sim p_2.$$

(h) This statement is valid.

2 (b)

p	q	r	$q \wedge r$	(5) $p \vee (q \wedge r)$	(6) $p \vee q$	(7) $p \vee r$	(8) (6) $\wedge$ (7)	(5) $\leftrightarrow$ (8)
T	T	T	T	T	T	T	T	T
T	T	F	F	T	T	T	T	T
T	F	T	F	T	T	T	T	T
T	F	F	F	T	T	T	T	T
F	T	T	T	T	T	T	T	T
F	T	F	F	F	T	F	F	T
F	F	T	F	F	F	T	F	T
F	F	F	F	F	F	F	F	T

4

p	q	r	f
T	T	T	F
T	T	F	T
T	F	T	F
T	F	F	F
F	T	T	F
F	T	F	F
F	F	T	F
F	F	F	F

f: $p \wedge q \wedge \sim r$ by Theorem 1.4.

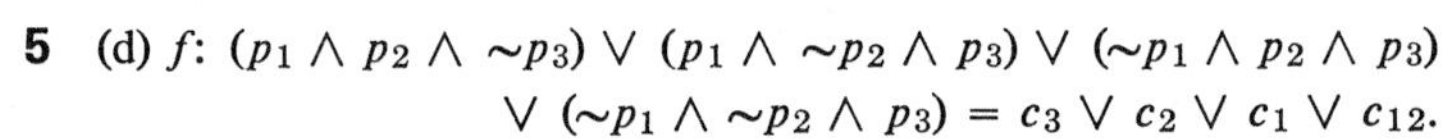

5 (d) f: $(p_1 \wedge p_2 \wedge \sim p_3) \vee (p_1 \wedge \sim p_2 \wedge p_3) \vee (\sim p_1 \wedge p_2 \wedge p_3) \vee (\sim p_1 \wedge \sim p_2 \wedge p_3) = c_3 \vee c_2 \vee c_1 \vee c_{12}$.

6 (a)

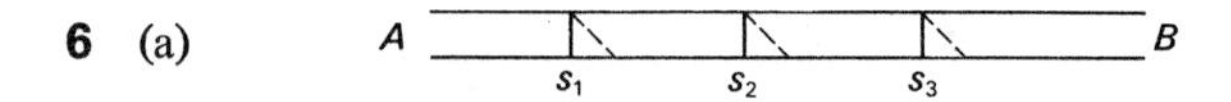

(b)

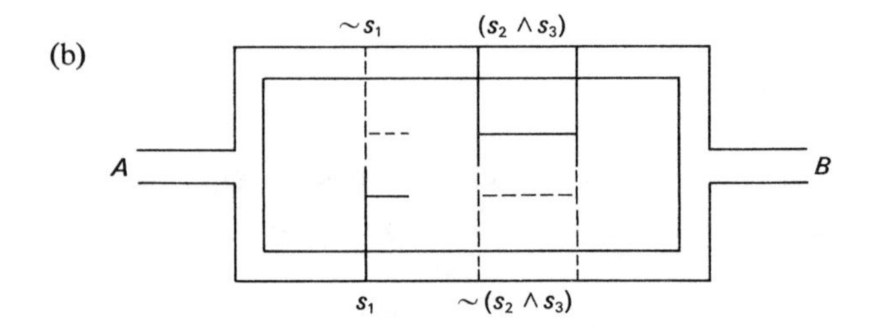

11

p	q	$p \to q$	$\sim p \vee q$	$(p \to q) \leftrightarrow (\sim p \vee q)$
T	T	T	T	T
T	F	F	F	T
F	T	T	T	T
F	F	T	T	T

12 The equivalent statement is

$$\sim[\sim(\sim p \vee q) \vee \sim(\sim q \vee p)].$$

1.2 *Quiz*

1 True. Any real number between 0 and 1 can be expanded to (at least) five places and have its fourth place examined.

2 True. Formula (7) in Section 1.2 implies

$$\sim \bigwedge_{x \in S} (p_1(x) \leftrightarrow p_2(x)) \leftrightarrow \exists_{x \in S} \sim (p_1(x) \leftrightarrow p_2(x)).$$

Hence if we can prove that

$$\sim(p_1(x) \leftrightarrow p_2(x)) \leftrightarrow (\sim p_1(x) \wedge p_2(x)) \vee (p_1(x) \wedge \sim p_2(x)),$$

we can conclude that

$$\sim \bigwedge_{x \in S} (p_1(x) \leftrightarrow p_2(x)) \leftrightarrow \exists_{x \in S} [(\sim p_1(x) \wedge p_2(x)) \vee (p_1(x) \wedge \sim p_2(x))].$$

But the first equivalence follows easily from an examination of the truth table.

3 True. This is Theorem 2.1 for two sets.

4 False. Take $U = \{1, 2, 3, 4, 5\}$, $S_1 = \{1, 2, 3\}$, $S_2 = \{2, 3, 4\}$. Then

$$\begin{aligned} S_1 \cap S_2 &= \{2, 3\}, \\ (S_1 \cap S_2)' &= \{1, 4, 5\}, \\ S_1' &= \{4, 5\}, \\ S_2' &= \{1, 5\}, \\ S_1' \cap S_2' &= \{5\} \neq (S_1 \cap S_2)'. \end{aligned}$$

5 True. Let p be the statement $x \in \emptyset$, and let q be the statement $x \in S$. Then for any set S and any element x in S, p has truth value F and q has truth value T. Hence the implication $p \to q$ has truth value T for any S, any $x \in S$. But $p \to q$ is just the statement, "If x is in $\emptyset$ then x is in S," that is, $\emptyset \subset S$.

6 False.

7 False. The set has two distinct elements: a and $\{a\}$.

8 True.

9 True. $\{a, \{a, b\}\} \cup \{a, b\} = \{a, b, \{a, b\}\}$. Hence the number of elements in the above set is 3.

10 False. Again let $U = \{1, 2, 3, 4, 5\}$, $S_1 = \{1, 2, 3\}$, $S_2 = \{2, 3, 4\}$. Then

$$\begin{aligned} \nu(S_1 \cap S_2') &= \nu(\{1\}) = 1, \\ \nu(S_1) &= 3, \\ \nu(S_2) &= 3, \\ \nu(S_1) - \nu(S_2) &= 0 \\ &\neq \nu(S_1 \cap S_2'). \end{aligned}$$

1.2 *Exercises*

1 (c)

$$\begin{aligned} S_1 \cap S_2 &= \{6\}, \\ (S_1 \cap S_2)' &= \{1, 2, 3, 4, 5, 7, 8, 9, 10\}, \\ S_3' &= \{1, 2, 3, 9, 10\}, \end{aligned}$$

and so

$$(S_1 \cap S_2)' \cap S_3' = \{1, 2, 3, 9, 10\}.$$

5 (a) There exists a woman who has blue eyes and is pretty.
(e) There exists a woman who is blonde, blue-eyed, and pretty but not overweight.

10 (e)
$$\begin{aligned} x \in X \cap (Y \cup Z) &\leftrightarrow x \in X \wedge (x \in Y \cup Z) \\ &\leftrightarrow x \in X \wedge (x \in Y \vee x \in Z) \\ &\leftrightarrow (x \in X \wedge x \in Y) \vee (x \in X \wedge x \in Z) \\ &\leftrightarrow (x \in X \cap Y) \vee (x \in X \cap Z) \\ &\leftrightarrow x \in (X \cap Y) \cup (X \cap Z). \end{aligned}$$

11 (e) $A \subset B \rightarrow B \cup A' = U$:
$$\begin{aligned} A \subset B &\rightarrow B' \cap A = \emptyset \\ &\rightarrow (B' \cap A)' = \emptyset' \\ &\rightarrow (B')' \cup A' = U \\ &\rightarrow B \cup A' = U; \end{aligned}$$

$B \cup A' = U \rightarrow A \subset B$:
$$\begin{aligned} B \cup A' = U &\rightarrow (B \cup A')' = U' \\ &\rightarrow B' \cap (A')' = \emptyset \\ &\rightarrow B' \cap A = \emptyset \\ &\rightarrow A \subset B. \end{aligned}$$

12 (g)

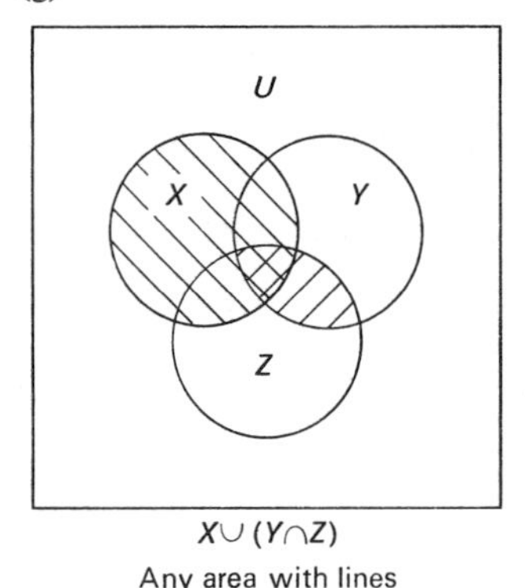

$X \cup (Y \cap Z)$
Any area with lines

$(X \cup Y) \cap (X \cup Z)$
Cross hatched area

13 No.

$$\begin{aligned} n \in Z &\rightarrow n = 48t \qquad (t \in U) \\ &\rightarrow n = 6(8t) \wedge n = 8(6t) \\ &\rightarrow n \in Y \wedge n \in X \\ &\rightarrow n \in X \cap Y. \end{aligned}$$

Therefore $Z \subset X \cap Y$. But $24 \in X$ and $24 \in Y$, so $24 \in X \cap Y$. However, $24 \notin Z$, hence $X \cap Y$ is not a subset of Z.

14 Let S_1 be the set of students who take English, S_2 the set of students who take math, and S_3 the set of students who take physics. Then

$$\begin{aligned} \nu(S_1 \cup S_2 \cup S_3) &= 100, \\ \nu(S_1 \cap S_2 \cap S_3) &= 12, \\ \nu(S_1 \cap S_2 \cap S_3') &= 22, \\ \nu(S_1 \cap S_2' \cap S_3) &= 3, \\ \nu(S_1' \cap S_2 \cap S_3) &= 7. \end{aligned}$$

Then

$$\begin{aligned} &\nu(S_1 \cap S_2' \cap S_3') + \nu(S_1' \cap S_2 \cap S_3') + \nu(S_1' \cap S_2' \cap S_3) \\ &\quad = \nu(S_1 \cup S_2 \cup S_3) - [\nu(S_1 \cap S_2 \cap S_3) + \nu(S_1 \cap S_2 \cap S_3') \\ &\qquad\qquad + \nu(S_1 \cap S_2' \cap S_3) + \nu(S_1' \cap S_2 \cap S_3)] \\ &\quad = 100 - (12 + 22 + 3 + 7) \\ &\quad = 100 - 44 \\ &\quad = 56. \end{aligned}$$

Hence 56 students take just one subject.

17 $\bigcup_{n=1}^{\infty} S_n$ is the set of all positive rational numbers in the interval [0, 1]. Clearly $\bigcup_{n=1}^{\infty} S_n$ is contained in this set. Conversely, any positive rational number in [0, 1] is of the form p/q, p and q positive integers, $p \leq q$, where p and q have no common factors. Hence $p/q \in S_q \subset \bigcup_{n=1}^{\infty} S_n$.

The sets S_n are pairwise disjoint. In fact, if there exist two distinct positive integers m and n such that $S_m \cap S_n \neq \emptyset$, then there is an element a/m in S_m which is also in S_n and so can be expressed in the form b/n, where a and b are positive integers. Hence $\frac{a}{m} = \frac{b}{n}$, a and m have no common factors, and b and n have no common factors. From $\frac{a}{m} = \frac{b}{n}$ we get $na = mb$. Since a divides the left side of this latter equality, it must divide the right side. But a has no factors in common with m, hence a must divide b. Similarly, b divides a. Hence $a = b$. It follows that $m = n$ and thus $S_m = S_n$.

1.3 *Quiz*

1 True. By definition, $\emptyset$ contains no elements.

2 True. $\{\emptyset\}$ contains one element: the set $\emptyset$.

3 True. Even though $x \neq y$, the set $\{(x, y)\}$ contains just one element: the ordered pair (x, y).

4 False. Because $(1)^2 = 1$ and $(-1)^2 = 1$, $\operatorname{rng} R = \{1\}$.

5 False. R is not an equivalence relation because it is not reflexive.

6 True. The difference of two odd numbers or two even numbers will be even, so if i and j are both odd or both even, the $(i, j)^{\text{th}}$ entry of $A(R)$ will be 1; otherwise it will be 0.

7 True. For all $x \in X$,
$$\begin{aligned} gf(x) &= g(f(x)) \\ &= g((x, x)) \\ &= 1. \end{aligned}$$

8 False. The two pairs (0, 1) and (0, 0) in g have the same first member and different second members, hence g cannot be a function.

9 True. Let $f: N_2 \to Y = \{y_1, y_2\}$. Then the only possible 2-permutations are $f = \{(1, y_1), (2, y_2)\}$ or $f = \{(1, y_2), (2, y_1)\}$.

10 True. Let $f: N_3 \to Y = \{y_1, y_2\}$. Then all 3-samples are:

$$\begin{aligned} &\{(1, y_1), (2, y_1), (3, y_1)\}, \ \{(1, y_1), (2, y_1), (3, y_2)\}, \\ &\{(1, y_1), (2, y_2), (3, y_1)\}, \ \{(1, y_1), (2, y_2), (3, y_2)\}, \\ &\{(1, y_2), (2, y_1), (3, y_1)\}, \ \{(1, y_2), (2, y_1), (3, y_2)\}, \\ &\{(1, y_2), (2, y_2), (3, y_1)\}, \ \{(1, y_2), (2, y_2), (3, y_2)\}. \end{aligned}$$

1.3 *Exercises*

1 (c) Function.

(f) Not a function, since (0, 2) and (0, 3) are in the set.

2 (f) $f = \{(1, 1), (2, 4), (-1, -1)\}$.

3 (c) Not an equivalence relation. R is not reflexive, symmetric, or transitive.
(f) R is an equivalence relation.

7 (a) $R = \{(1, 1), (1, 2), (1, 3), (1, 4), (3, 1), (3, 3)\}$.

9 $Y = X \times X = \{(0, 0), (0, 1), (1, 0), (1, 1)\}$. Then

$$\begin{aligned} Y \times Y &= \{(0, 0), (0, 1), (1, 0), (1, 1)\} \times \{(0, 0), (0, 1), (1, 0), (1, 1)\} \\ &= \{((0, 0), (0, 0)), ((0, 0), (0, 1)), ((0, 0), (1, 0)), ((0, 0), (1, 1)), \\ &\quad ((0, 1), (0, 0)), ((0, 1), (0, 1)), ((0, 1), (1, 0)), ((0, 1), (1, 1)), \\ &\quad ((1, 0), (0, 0)), ((1, 0), (0, 1)), ((1, 0), (1, 0)), ((1, 0), (1, 1)), \\ &\quad ((1, 1), (0, 0)), ((1, 1), (0, 1)), ((1, 1), (1, 0)), ((1, 1), (1, 1))\}. \end{aligned}$$

Hence $\nu(Y \times Y) = 16$.

11 In this case $A(R)$ is the matrix with a 1 in the (i, j) position if an impulse travels from neuron j to neuron i. Assume that if an impulse travels from N_i to N_j and from N_j to N_k, then it travels from N_i to N_k, i.e., R is transitive. Hence

$$A(R) = \begin{bmatrix} 1 & 1 & 0 & 0 & 1 & 1 & 0 \\ 1 & 1 & 0 & 0 & 1 & 1 & 0 \\ 0 & 0 & 1 & 1 & 0 & 0 & 1 \\ 0 & 0 & 1 & 1 & 0 & 0 & 1 \\ 1 & 1 & 0 & 0 & 1 & 1 & 0 \\ 1 & 1 & 0 & 0 & 1 & 1 & 0 \\ 0 & 0 & 1 & 1 & 0 & 0 & 1 \end{bmatrix}.$$

13 (a) R is reflexive since for all $x \in X$, $x - x = 0$, which is always divisible by 3. R is also symmetric, since if $x - y$ is divisible by 3, $y - x = -(x - y)$ is divisible by 3. Finally, R is transitive, since if $x - y$ is divisible by 3, say $x - y = 3m$, and $y - z$ is divisible by 3, say $y - z = 3n$, then $x - z = x - y + y - z = (x - y) + (y - z) = 3m + 3n = 3(m + n)$, which is obviously divisible by 3.
(b) Suppose $[x_1] = [x_2]$. Then $y \in [x_2]$ implies $y \in [x_1]$, in particular $x_2 \in [x_1]$, so $(x_1, x_2) \in R$. Conversely, suppose $(x_1, x_2) \in R$. If $y \in [x_2]$ then $(x_2, y) \in R$, so $(x_1, y) \in R$, which implies $y \in [x_1]$. Hence $[x_2] \subset [x_1]$. On the other hand, if $y \in [x_1]$ then $(x_1, y) \in R$, i.e., $(y, x_1) \in R$, so that $(y, x_2) \in R$, i.e., $(x_2, y) \in R$, which implies $y \in [x_2]$. Hence $[x_2] \subset [x_1]$.
(c) Upon dividing an integer by 3, there is either a remainder of 0, 1, or 2. Hence an element of X is either in [0], [1], or [2], so $X \subset [0] \cup [1] \cup [2]$. But since $[x] \subset X$ for any $x \in X$, it is clear that $[0] \cup [1] \cup [2] \subset X$. Hence $X = [0] \cup [1] \cup [2]$. Now suppose $[0] \cap [1] \neq \emptyset$; let $x \in [0] \cap [1]$. Then $(0, x) \in R$ and $(x, 1) \in R$, which implies $(0, 1) \in R$, which is clearly ridiculous since $0 - 1 = -1$ is not divisible by 3. Thus $[0] \cap [1] = \emptyset$, and similarly, $[0] \cap [2] = [1] \cap [2] = \emptyset$.

(d) By part (b), $[x_1] = [x_2]$ if and only if $(x_1, x_2) \in R$, i.e., $x_1 - x_2$ is divisible by 3. Now suppose $x_1 = 3m_1 + r_1$ and $x_2 = 3m_2 + r_2$, where r_1 and r_2 are 0, 1, or 2. Then $x_1 - x_2 = 3(m_1 - m_2) + (r_1 - r_2)$, which is divisible by 3. Hence $r_1 - r_2$ must be divisible by 3. But since r_1 and r_2 are 0, 1, or 2, $|r_1 - r_2|$ is less than 3, and so must be 0. Therefore $r_1 = r_2$ and x_1 and x_2 have the same remainder upon division by 3.

14 (b)

$$A(R) = \begin{bmatrix} 1 & 0 & 0 & 0 & 0 & 0 & 0 \\ 0 & 0 & 1 & 0 & 0 & 0 & 0 \\ 0 & 0 & 0 & 0 & 1 & 0 & 0 \\ 0 & 0 & 0 & 0 & 0 & 0 & 1 \end{bmatrix}.$$

16 (a) Suppose $A(f)$ is the incidence matrix for $f: X \to X$. Then every $x \in X$ is the first member of exactly one ordered pair in f. Hence each column of $A(f)$ has exactly one 1 and all other entries 0.

Conversely, if each column of $A(f)$ has exactly one 1 and $n - 1$ zeros, then each element in X is the first member of exactly one ordered pair in f. Hence $A(f)$ is the incidence matrix of a function on X to X.

(b) Suppose $A(f)$ is the incidence matrix for a 1-1 function $f: X \to X$. Then $A(f)$ cannot have more than one 1 in each row. For suppose the ith row, $1 \leq i \leq n$, contained two 1's. Then there would be two elements of X, say x_j and x_k, such that $(x_j, x_i) \in f$ and $(x_k, x_i) \in f$. But then f would not be $1 - 1$. Thus, since each row of $A(f)$ has at most one 1 and using (a) and the fact that $A(f)$ is $n \times n$, we see that $A(f)$ has exactly one 1 and $n - 1$ zeros in each row and in each column.

18 (a) If $A(R)$ is the matrix with (i, j) entry 1 if p_i exhibits the symptoms of d_j, then the fact that any k of the patients exhibit at least k symptoms implies that any set of k columns will together contain 1's in at least k rows.

(b) If there are three 1's, no two in the same row or column, this means there are three different patients, each exhibiting a symptom of a different disease.

19 Using the theory in Example 3.9, we see that the matrix we want is in fact the composition of the two given matrices:

$$\begin{bmatrix} 0 & 1 & 0 \\ 1 & 1 & 1 \\ 1 & 1 & 0 \\ 0 & 0 & 1 \end{bmatrix} \begin{bmatrix} 1 & 0 & 1 & 0 & 1 \\ 1 & 0 & 0 & 0 & 1 \\ 1 & 0 & 1 & 1 & 0 \end{bmatrix} = \begin{matrix} \\ m_1 \\ m_2 \\ m_3 \\ m_4 \end{matrix} \begin{matrix} \begin{matrix} q_1 & q_2 & q_3 & q_4 & q_5 \end{matrix} \\ \begin{bmatrix} 1 & 0 & 0 & 0 & 1 \\ 3 & 0 & 2 & 1 & 2 \\ 2 & 0 & 1 & 0 & 2 \\ 1 & 0 & 1 & 1 & 0 \end{bmatrix} \end{matrix}.$$

For example, the (2, 1) entry is 3, i.e., q_1 has been exposed to m_2 through three different people. This is true because both q_1 and m_2 have been exposed to p_1, p_2, and p_3.

21 We must prove that $(f^{-1}f)(x) = I_X(x)$ for every $x \in X$. But by (14), $f^{-1}(f(x)) = x$ for every $x \in X$, so $(f^{-1}f)(x) = f^{-1}(f(x)) = x = I_X(x)$ for every $x \in X$. Hence, $f^{-1}f = I_X$. Similarly, $ff^{-1} = I_Y$.

22 (d)

$$gf(x) = g\left(\frac{x}{1 + x^2}\right) = \frac{\left(\frac{x}{1 + x^2}\right)}{1 + \left(\frac{x}{1 + x^2}\right)^2} = \frac{x(1 + x^2)}{(1 + x^2)^2 + x^2}.$$

1.4 *Quiz*

1 True. By definition, $4! = 4 \cdot 3 \cdot 2 \cdot 1 = 24$.

2 False. The binomial coefficient $\binom{n}{k}$, $1 \leq k \leq n$, has the value

$$\binom{n}{k} = \frac{n!}{(n - k)!k!}.$$

Evaluating the two coefficients in this question, we get:

$$\binom{n}{0} = \frac{n!}{0!(n - 0)!} = \frac{n!}{1 \cdot n!} = 1,$$

$$\binom{n}{1} = \frac{n!}{1!(n - 1)!} = \frac{n!}{1 \cdot (n - 1)!} = n.$$

We have used the facts that $0! = 1$ and $1! = 1$. It then becomes clear that the original question is false since it does not hold for any value of $n \in N$ except $n = 1$.

3 True. $$\binom{0}{0} = \frac{0!}{0!(0 - 0)!} = \frac{1}{1 \cdot 1} = 1.$$

4 False. The correct number of 3-samples of a 4-element set is 4^3. This result follows from a straight-forward application of Theorem 4.2(a).

5 False. We know from Theorem 4.2(e) that the number of subsets of an n-set is 2^n. Since in this case $X = \{a, b\}$ is a 2-set, the number of subsets of X is $2^2 = 4$. They are in fact $\emptyset$, $\{a\}$, $\{b\}$, $\{a, b\}$.

6 True. See Theorem 4.2(c).

7 False.

8 True. As in Question 2,

$$\binom{n + 1}{0} = \frac{(n + 1)!}{0!(n + 1)!} = 1.$$

Also,

$$\binom{n}{n} = \frac{n!}{n!(n - n)!} = \frac{n!}{n!0!} = 1.$$

Hence

$$\binom{n}{n} = \binom{n + 1}{0}.$$

9 True. We again appeal to Theorem 4.2, parts (c) and (d). From part (c) we know that the number of r-combinations of an n-set is $\binom{n}{r}$. Part (d) tells us that the number of r-selections of an n-set is $\binom{n+r-1}{r}$. In order to verify that the statement is correct we need only show that

$$\binom{n+r-1}{r} \geq \binom{n}{r}.$$

Now, $$\binom{n+r-1}{r} = \frac{(n+r-1)(n+r-2)\cdots(n+1)(n)}{r!}$$

and $$\binom{n}{r} = \frac{n(n-1)\cdots(n-r+1)}{r!}.$$

Noticing that $(n+r-1)(n+r-2)\ldots(n+1)(n) \geq n(n-1)\ldots(n-r+1)$ when $1 \leq r \leq n$, the result follows.

A more direct but possibly less clear argument can be made from Theorem 4.1, parts (b) and (c). In part (b) we notice that none of the k_i's may be equal, whereas in part (c) some may. From this it follows that there are more sequences that satisfy the conditions of (c) than satisfy the conditions of (b), and hence the number of r-selections of an n-set exceeds the number of r-combinations of an n-set, where $1 \leq r \leq n$.

10 True. This fact follows from Theorem 4.2(c), which tells us that the number of r-combinations (or r-subsets) of an n-set is $\binom{n}{r}$. In this case $r = n$, so we need to know the value of $\binom{n}{n}$. But by the discussion in Question 8 above, we know that $\binom{n}{n} = 1$.

1.4 *Exercises*

1 (f) First evaluate $\binom{4}{2}$:

$$\binom{4}{2} = \frac{4!}{2!(4-2)!} = \frac{4 \cdot 3 \cdot 2 \cdot 1}{2 \cdot 1 \cdot 2 \cdot 1} = \frac{4 \cdot 3}{2 \cdot 1} = 6.$$

Hence

$$\binom{7}{\binom{4}{2}} = \binom{7}{6} = \frac{7!}{6!(7-6)!} = \frac{7 \cdot 6 \cdots 2 \cdot 1}{6 \cdots 2 \cdot 1 \cdot 1} = 7.$$

(g) $$\frac{1}{0!} = \frac{1}{1} = 1.$$

2 (c) We use part (b) of Theorem 4.2, with $n = 6$ and $r = 3$:

$$\frac{6!}{(6-3)!} = \frac{6 \cdot 5 \cdot 4 \cdot 3 \cdot 2 \cdot 1}{3 \cdot 2 \cdot 1} = 120.$$

(e) We use part (d) of Theorem 4.2, with $n = 2$ and $r = 7$:

$$\binom{2+7-1}{7} = \binom{8}{7} = \frac{8 \cdot 7 \cdot 6 \cdot 5 \cdot 4 \cdot 3 \cdot 2 \cdot 1}{7 \cdot 6 \cdot 5 \cdot 4 \cdot 3 \cdot 2 \cdot 1 \cdot 1} = 8.$$

3 (c) The sequence is (3, 2, 1). In this case f is a 3-permutation.

(i) The sequence is (3, 2, 1, 1). Here f is a 4-sample.

4 (d) We will count the number of "nondecreasing numbers" not exceeding 299 by counting those between 100 and 199, and then between 200 and 299. If a number is to be "nondecreasing" and between 100 and 199, there are 5 choices for the second digit (the 10's place), and also for the third digit (the 1's place). Hence we want to count the number of 2-selections of a 5-set. This number we know to be

$$\binom{5+2-1}{2} = \frac{6!}{2!4!} = \frac{6 \cdot 5}{2 \cdot 1} = 15.$$

Similarly, if a number is to be "nondecreasing" and between 200 and 299, there are 4 choices for the second and third digits. This time we must count the number of 2-selections of a 4-set:

$$\binom{4+2-1}{2} = \frac{5!}{2!3!} = \frac{5 \cdot 4}{2 \cdot 1} = 10.$$

Thus the desired answer is

$$15 + 10 = 25.$$

5 (e) For $k = 1$, $p(n)$ becomes

$$1(1!) = (1 + 1)! - 1,$$

which is obviously true.

Now assume k is a number for which $p(k)$ is true, that is,

$$1 \cdot (1!) + \cdots + k(k!) = (k + 1)! - 1.$$

We will show $p(k) \to p(k + 1)$ has truth value T. Now,

$$\begin{aligned} 1 \cdot (1)! + \cdots &+ k(k!) + (k+1)[(k+1)!] \\ &= (k+1)! - 1 + (k+1)[(k+1)!] \\ &= (k+1)![1 + (k+1)] - 1 \\ &= (k+1)!(k+2) - 1 \\ &= (k+2)! - 1 \\ &= [(k+1) + 1]! - 1. \end{aligned}$$

In other words, using the truth of $p(k)$, we have deduced the truth of $p(k + 1)$, thus proving the truth of the implication.

6 (b)

$$\binom{n}{r} + \binom{n}{r+1} = \frac{n!}{r!(n-r)!} + \frac{n!}{(r+1)!(n-r-1)!}$$
$$= \frac{n!}{(n-r)!}\left[\frac{1}{r!} + \frac{n-r}{(r+1)!}\right]$$
$$= \frac{n!}{(n-r)!}\left[\frac{r+1}{(r+1)!} + \frac{n-r}{(r+1)!}\right]$$
$$= \frac{n!}{(n-r)!}\left[\frac{n+1}{(r+1)!}\right]$$
$$= \frac{(n+1)!}{(n-r)!(r+1)!}$$
$$= \binom{n+1}{r+1}.$$

8 For $k = 1$,

$$3^{2k} - 1 = 3^2 - 1 = 9 - 1 = 8,$$

which is indeed divisible by 8. Hence $p(1)$ is true. Now suppose k is a number such that $3^{2k} - 1$ is divisible by 8. We prove that the difference

$$f(k+1) - f(k)$$

is divisible by 8, where $f(k) = 3^{2k} - 1$. Then the sum

$$[f(k+1) - f(k)] + f(k) = f(k+1)$$

will be divisible by 8. Now,

$$\begin{aligned} f(k+1) - f(k) &= 3^{2(k+1)} - 3^{2k} \\ &= 3^{2k+2} - 3^{2k} \\ &= 3^{2k}3^2 - 3^{2k} \\ &= 3^{2k}(3^2 - 1) \\ &= 3^{2k} \cdot 8, \end{aligned}$$

and the truth of the implication $p(k) \rightarrow p(k+1)$ follows.

10 Let us denote the three girls by G_1, G_2, G_3, and the three boys by B_1, B_2, B_3. We set up the incidence matrix for this situation where the (i, j) entry is 0 if G_j and B_i have not been introduced and 1 if they have.

$$\begin{array}{c} \\ B_1 \\ B_2 \\ B_3 \end{array} \begin{array}{c} \begin{array}{ccc} G_1 & G_2 & G_3 \end{array} \\ \begin{bmatrix} 1 & 1 & 0 \\ 0 & 1 & 1 \\ 1 & 0 & 1 \end{bmatrix} \end{array}$$

We can easily see that only if the matrix is in this form (i.e., precisely two 1's in each row and each column) will each girl have been introduced to two boys and each boy to two girls. Then we know the only possible pairings are either

$$(G_1, B_1), (G_2, B_2), (G_3, B_3) \qquad \text{or} \qquad (G_1, B_3), (G_2, B_1), (G_3, B_2).$$

13 This problem becomes more transparent if we think of lining up 4 boys and 4 girls so that they alternate. If we start with a boy, we have four choices for the first position, then four choices of a girl for the second position, 3 choices of a boy for the third position and 3 choices of a girl for the fourth position, etc. That is, there are

$$4 \cdot 4 \cdot 3 \cdot 3 \cdot 2 \cdot 2 \cdot 1 \cdot 1 = 4!4! = 576$$

possible arrangements.*

16 (a) The composition of the committee requires that no more than one person be chosen from among those who dislike one another. If we choose no one from this subset, there is exactly one way to choose the committee of five from the remaining five. If we choose exactly one of the three from the subset, then we must choose four members from the remaining five. Hence the number of ways to choose this committee is

$$1 + \binom{3}{1}\binom{5}{4} = 16.$$

(b) In this case we must have all three of the special members on the committee. If we have all three, then we must choose two members from the remaining five. Hence the number of ways this committee can be formed is

$$\binom{5}{2} = 10.$$

20 For $k = 1$ we get

$$1 \cdot 2 = 2 > \frac{1}{3} = \frac{1^3}{3};$$

hence $p(1)$ is true. Suppose k is a number for which $p(k)$ is true, that is,

$$1 \cdot 2 + 2 \cdot 3 + \cdots + k(k+1) > \frac{k^3}{3}.$$

Then

$$\begin{aligned} 1 \cdot 2 + \cdots + k(k+1) + (k+1)(k+2) &> \frac{k^3}{3} + (k+1)(k+2) \\ &= \frac{k^3 + 3(k+1)(k+2)}{3} \\ &= \frac{k^3 + 3k^2 + 9k + 6}{3} \\ &> \frac{k^3 + 3k^2 + 3k + 1}{3} \\ &= \frac{(k+1)^3}{3}, \end{aligned}$$

that is, $p(k+1)$ is true.

*If we start with a girl we also obtain 576 arrangements. Since the table is round, shifting ahead through any one of the 8 positions does not alter the relative arrangement of the boys and the girls. Thus the number of arrangements is $\frac{2 \cdot 576}{8} = 144$.

1.5 *Quiz*

1 True. If $p + q = n$, then $\binom{n}{p\ q} = \frac{n!}{p!q!} = \frac{n!}{p!(n-p)!} = \binom{n}{p}$.

2 True.

$$\binom{6}{3\ 2\ 1} = \frac{6!}{3!2!1!} = \frac{6!}{1!2!3!} = \binom{6}{1\ 2\ 3}.$$

3 True.

$$\prod_{i=1}^{3} i^i = 1^1 \cdot 2^2 \cdot 3^3 = 108.$$

4 False.

$$\sum_{i=1}^{3} i^i = 1^1 + 2^2 + 3^3 = 32.$$

5 True. By Theorem 5.1 we know that the number of ways of doing this is

$$\binom{5}{3\ 2} = \binom{5}{2}.$$

6 False.

$$\binom{3}{1\ 1\ 1} = \frac{3!}{1!1!1!} = 3! = 6.$$

7 False. See Question 6.

8 True.

$$\sum_{i=1}^{17} 2 = \underbrace{2 + 2 + \cdots + 2}_{17 \text{ times}} = 34.$$

9 True.

$$\prod_{i=1}^{10} 2^3 = \underbrace{2^3 \cdot 2^3 \cdots 2^3}_{10 \text{ times}} = 2^{30}.$$

10 True.

$$\sum_{i=1}^{3} \frac{i^2}{i} = \sum_{i=1}^{3} i = 1 + 2 + 3 = 6.$$

1.5 *Exercises*

1 (d)

$$\binom{4}{2\ 2\ 0} = \frac{4!}{2!2!0!} = \frac{4 \cdot 3 \cdot 2}{2 \cdot 2 \cdot 1} = 6.$$

(i)

$$\binom{5}{1\ 1\ 1\ 1\ 1} = 5! = 120.$$

2 (i) Expanding by the multinomial formula we get

$$\left(\sum_{i=1}^{n} \frac{1}{x_i}\right)^r = \sum_{k_1+\cdots+k_n=r} \binom{r}{k_1\ \cdots\ k_n} \prod_{i=1}^{n} \left(\frac{1}{x_i}\right)^{k_i}.$$

The coefficient of $\prod_{i=1}^{n}\left(\frac{1}{x_i}\right)^{k_i}$ is $\binom{r}{k_1 \cdots k_n}$.

Our problem requires that $k_1 = k_2 = \cdots = k_r = 1$ and $k_{r+1} = k_{r+2} = \cdots = k_n = 0$ when $r \leq n$. Applying this fact, we get

$$\binom{r}{1 \cdots 1 \cdot 0 \cdots 0} = r!,$$

where there are r ones and $n - r$ zeros. Clearly it must be that $r \leq n$, otherwise the notation $\prod_{i=1}^{r} \frac{1}{x_i}$ is meaningless.

(k) Writing out the multinomial expansion, we see that

$$\begin{aligned}(x_1 - x_2 + x_3 + 1)^{10} &= \sum_{k_1+k_2+k_3+k_4=10} \binom{10}{k_1\ k_2\ k_3\ k_4} x_1^{k_1}(-x_2)^{k_2} x_3^{k_3} 1^{k_4} \\ &= \sum_{k_1+k_2+k_3+k_4=10} (-1)^{k_2} \binom{10}{k_1\ k_2\ k_3\ k_4} x_1^{k_1} x_2^{k_2} x_3^{k_3}.\end{aligned}$$

We want to know the coefficient when $k_1 = 9$, $k_2 = k_3 = 0$, and $k_4 = 1$, that is,

$$(-1)^0 \binom{10}{9\ 0\ 0\ 1} = 10.$$

6 Recall that a deck of cards contains 12 face cards, 40 no-face cards, 4 aces, 4 kings.

(a) This is the number of 2-subsets of a 12-set, or

$$\binom{12}{2} = 66.$$

(b) This is the number of 2-subsets of a 40-set, or

$$\binom{40}{2} = 780.$$

(c) If the sum of the two cards is to be less than 5, it can be either 4, 3, or 2. We simply count the number of ways this can happen by looking at the following table.

A, A - - - - sum is 2
$A, 2$ - - - - sum is 3
$\left.\begin{matrix} A, 3 \\ 2, 2 \end{matrix}\right\}$ - - - - sum is 4

There are 4 aces in the deck and hence $\binom{4}{2} = 6$ ways to form a sum equal to 2. Similarly, there are $4 \cdot 4 = 16$ ways to obtain a sum of 3 and $\binom{4}{2} = 6$ plus $4 \cdot 4 = 16$ ways to obtain a sum of 4. The total number of 2-card hands whose sum is less than 5 is therefore 44.

(d) In a 2-card hand, to have at least one ace means to have either one ace or two aces. The number of 1-ace 2-card hands is $4 \cdot 48 = 192$ and the number of 2-ace 2-card hands is $\binom{4}{2} = 6$, yielding a total of 198 ways that a player can have at least one ace in a 2-card hand.

(e) The number of no-king 2-card hands is $\binom{48}{2} = 1128$.

9 Clearly we want 5, 5, and 6 students in the three discussion groups. Hence the number of ways this can be done is

$$\binom{16}{5\ 5\ 6}.$$

10 (c)

$$\sum_{j=1}^{3}\left(\prod_{i=1}^{j} i\right)^{j} = \sum_{j=1}^{3}(j!)^{j} = (1!)^{1} + (2!)^{2} + (3!)^{3} = 221.$$

(d)

$$\prod_{j=1}^{3}\left(\sum_{i=1}^{j} i\right)^{j} = (1)(1+2)^{2}(1+2+3)^{3} = 1944.$$

(e)

$$\sum_{i=1}^{3} i\left(\prod_{j=1}^{i} j^{2}\right) = 1\cdot 1^{2} + 2(1^{2}\cdot 2^{2}) + 3(1^{2}\cdot 2^{2}\cdot 3^{2}) = 117.$$

(f)

$$\prod_{i=1}^{2}\left(\prod_{j=1}^{i}\left(\sum_{k=1}^{j} k\right)\right) = \left[\prod_{j=1}^{1}\left(\sum_{k=1}^{j} k\right)\right]\left[\prod_{j=1}^{2}\left(\sum_{k=1}^{j} k\right)\right] = \left[\sum_{k=1}^{1} k\right]\left[\sum_{k=1}^{1} k\right]\left[\sum_{k=1}^{2} k\right] = 1\cdot 1\cdot 3 = 3.$$

(r)

$$\prod_{i=1}^{4} x_i^{j} = x_1^{j}\cdot x_2^{j}\cdot x_3^{j}\cdot x_4^{j} = (x_1x_2x_3x_4)^{j}.$$

(s)

$$\prod_{i=1}^{4} x^{ij} = x^{1\cdot j}\cdot x^{2\cdot j}\cdot x^{3\cdot j}\cdot x^{4\cdot j} = x^{10j}.$$

11 (f)
$$\prod_{k_1+k_2=1} \binom{1}{k_1\ k_2} = \binom{1}{1\ 0}\binom{1}{0\ 1} = 1.$$

(g)
$$\sum_{k_1+k_2+k_3=3} \binom{3}{k_1\ k_2\ k_3} \prod_{i=1}^{3} x_i^{k_i}$$
$$= \sum_{k_1+k_2+k_3=3} \binom{3}{k_1\ k_2\ k_3} x_1^{k_1}x_2^{k_2}x_3^{k_3}.$$

There are ten terms in the above sum which correspond to the columns of the following table.

	1	2	3	4	5	6	7	8	9	10
k_1	0	0	3	0	1	1	0	2	2	1
k_2	0	3	0	1	0	2	2	0	1	1
k_3	3	0	0	2	2	0	1	1	0	1

Each column sum is equal to 3. Hence the total sum becomes

$$x_1^3 + x_2^3 + x_3^3 + 3(x_2x_3^2 + x_1x_3^2 + x_1x_2^2 + x_2^2x_3 + x_1^2x_3 + x_1^2x_2) + 6x_1x_2x_3.$$

12 Recall that Theorem 5.2 states that if r is a positive integer, then

$$\left(\sum_{i=1}^{r} x_i\right)^n = \sum_{k_1+\cdots+k_r=n} \binom{n}{k_1 \cdots k_r} \prod_{i=1}^{r} x_i^{k_i}.$$

Taking $r = 2$, we get

$$(x_1 + x_2)^n = \sum_{k_1+k_2=n} \binom{n}{k_1\ k_2} x_1^{k_1}x_2^{k_2}.$$

Now, since $k_1 + k_2 = n$ we have $k_2 = n - k_1$. Therefore

$$(x_1 + x_2)^n = \sum_{k_1=0}^{n} \binom{n}{k_1} x_1^{k_1}x_2^{n-k_1}.$$

29 Using the method of Example 5.11, we look at the expansion of

$$(x_1 + 2x_2 + 3x_3 + 4x_4 + 5x_5 + 6x_6 + 5x_7 + 4x_8 + 3x_9 + 2x_{10} + x_{11})^2.$$

We seek the coefficient of x_3x_4. In this case, $k_3 = k_4 = 1$ and the rest of the k_i are 0. Hence we have

$$\binom{2}{1\ 1}(3x_3)(4x_4) = 24x_3x_4.$$

Therefore there are 24 ways for a 4 and a 5 to come up in some order.

1.6 *Quiz*

1 False. See equation (1) at the beginning of this section.

2 True. Recall that the number of subsets of a set with n elements is 2^n, and that we are considering only those sigma fields containing all subsets of U.

3 False. By definition, an elementary event is a set consisting of exactly one element.

4 True. See the discussion immediately preceding Example 6.2.

5 True. This is Theorem 6.1 (f).

6 False. The probability $p(X|Y)$ is defined only if $p(Y) \neq 0$, and $Y \neq \emptyset$ does not imply that $p(Y) \neq 0$.

7 True. This is Theorem 6.2, the Multiplication Theorem.

8 True. If $p(X)p(Y) \neq 0$, then $p(X) \neq 0 \neq p(Y)$, and hence both conditional probabilities are defined.

9 True. Recall that $0 \leq p(X) \leq 1$ for all $X \subset U$. Therefore $p(X)^2 \leq p(X)$.

10 False. Since $q(U) = 1 - p(U) = 0$, the function q does not satisfy part (i) of Definition 6.2.

1.6 *Exercises*

2 (a) Let

$$U = \{(\text{Monday}), (\text{Tuesday}), (\text{Wednesday}), (\text{Thursday}), (\text{Friday}), (\text{Saturday}), (\text{Sunday})\}.$$

We want each of the elementary events to be equally probable, i.e., we want U to be an equi-probable space. Hence define $p(Y) = \frac{1}{7}$, where Y is any elementary event in U. Let $X \subset U$ be the subset of U consisting of those days of the week that have more than six letters in their spellings:

$$\begin{aligned} X &= \{(\text{Tuesday}), (\text{Wednesday}), (\text{Thursday}), (\text{Saturday})\} \\ &= \{(\text{Tuesday})\} \cup \{(\text{Wednesday})\} \cup \{(\text{Thursday})\} \cup \{(\text{Saturday})\}. \end{aligned}$$

Then $p(X) = \frac{1}{7} + \frac{1}{7} + \frac{1}{7} + \frac{1}{7} = \frac{4}{7}$.

(d) Denote the red marbles by r_1 and r_2, the white marbles by w_1 and w_2, and the black marbles by b_1, b_2, b_3, and b_4. Then let

$$U = \{r_1, r_2, w_1, w_2, b_1, b_2, b_3, b_4\},$$

where the probability space is assumed to be equi-probable. If we take

$$X = \{w_1, w_2, b_1, b_2, b_3, b_4\}$$

the desired probability is

$$p(X) = 6 \cdot \tfrac{1}{8} = \tfrac{3}{4}.$$

9 (a) There are $5! = 120$ possible seatings of the students, and exactly one of these seatings is correct, i.e., all students in assigned seats. Hence the probability that all five students take their assigned seats is $1/120$.

(b) Let p_1 denote the probability that exactly one student takes his assigned seat, ..., p_5 the probability that all five take their assigned seats. Then the probability that no student takes his assigned seat is

$$1 - (p_1 + \cdots + p_5).$$

Hence we must find the numbers $p_1, \ldots, p_5$. As seen above, $p_5 = 1/120$. Since if four students take the right seats, five must, $p_4 = 0$. If three students are correctly seated, there is only one way the remaining two can take the wrong seats. Since there are $\binom{5}{3}$ ways the first three can be chosen, three students can be correctly seated in $\binom{5}{3} \cdot 1 \cdot 1 \cdot 1 \cdot 1 = 10$ ways. Hence

$$p_3 = \frac{10}{120} = \frac{1}{12}.$$

If two students are correctly seated, it can be checked that there are only two ways the remaining three can all sit in unassigned seats. Since there are $\binom{5}{2}$ ways to choose the first two, there are $\binom{5}{2} \cdot 1 \cdot 1 \cdot 2 = 20$ ways to seat exactly two students correctly. Hence

$$p_2 = \frac{20}{120} = \frac{1}{6}.$$

The number of ways to seat one student correctly is the same as the number of ways to seat four incorrectly, and this is the original problem again, except for a smaller number of students. As above, we find there are $4! = 24$ possible ways of seating all four, one way of seating all four correctly, no ways of seating exactly three correctly, $\binom{4}{2} = 6$ ways of seating exactly two correctly, and $\binom{4}{1} \cdot 2 = 8$ ways of seating exactly one correctly. Hence there are $\binom{5}{1} \cdot (24 - 15) = 45$ ways that four students can take unassigned seats, so that

$$p_1 = \frac{45}{120} = \frac{3}{8}.$$

Thus the desired probability is

$$
\begin{aligned}
1 - (p_1 + p_2 + p_3 + p_4 + p_5) & \\
&= 1 - \left(\frac{45}{120} + \frac{20}{120} + \frac{10}{120} + 0 + \frac{1}{120}\right) \\
&= 1 - \left(\frac{76}{120}\right) \\
&= \frac{44}{120}.
\end{aligned}
$$

(c) This probability is simply the sum of the probabilities

$$p_1 + p_2 + p_3 + p_4 + p_5 = \frac{76}{120}.$$

(d) As calculated above, $p_2 = \frac{1}{6}$.

10 From Example 2.12, Section 2, we know we can express $X \cup Y$ as a disjoint union:

$$X \cup Y = (X \cap Y') \cup (X \cap Y) \cup (X' \cap Y).$$

Applying parts (d) and (f) of Theorem 6.1, we have

$$
\begin{aligned}
p(X \cup Y) &= p(X \cap Y') + p(X \cap Y) + p(X' \cap Y) \\
&= p(X) - p(X \cap Y) + p(X \cap Y) + p(Y) - p(X \cap Y) \\
&= p(X) + p(Y) - p(X \cap Y).
\end{aligned}
$$

11 Let $U = \{(i, j) \mid i, j = 1, \ldots, 6\}$, and assume U is equi-probable. Let X be the event consisting of all pairs whose sum is even, and let Y be the event consisting of all pairs whose sum is greater than 5. Then

$$
\begin{aligned}
p(X \mid Y) &= \frac{p(X \cap Y)}{p(X)} \\
&= \frac{\frac{14}{36}}{\frac{18}{36}} \\
&= \tfrac{7}{9}.
\end{aligned}
$$

15 Let U be the ethnic group, let X_1 be the event that a person is a woman, and X_2 the event that a person is a man. Then $X_1 \cap X_2 = \emptyset$, $X_1 \cup X_2 = U$. Let Y be the event that a person is overweight. Then we know that

$$
\begin{aligned}
p(X_1) &= .55, \\
p(X_2) &= .45, \\
p(Y \mid X_1) &= .90, \\
p(Y \mid X_2) &= .70.
\end{aligned}
$$

We are interested in finding the probability of X_1 given Y. Since the conditions satisfy Theorem 6.3, we can use Formula (11) for $n = 2$ to get

$$
\begin{aligned}
p(X_1 \mid Y) &= \frac{p(X_1)p(Y \mid X_1)}{p(X_1)p(Y \mid X_1) + p(X_2)p(Y \mid X_2)} \\
&= \frac{(.55)(.9)}{(.55)(.9) + (.45)(.7)} \\
&= \frac{.495}{.81} \\
&= \frac{99}{162}.
\end{aligned}
$$

18 Let $n = 2$ and $U = X_1 \cup X_1'$ in Theorem 6.3, Formula (11), and observe that

$$p(X_1 \mid X) = \frac{p(X_1)p(X \mid X_1)}{p(X_1)p(X \mid X_1) + p(X_1')p(X \mid X_1')}.$$

19 Let X_1 be the event that a random person has cancer and X the event that in a random population, a person will test positively for cancer. Then applying the result of Exercise 10, we get

$$
\begin{aligned}
p(X_1 \mid X) &= \frac{p(X_1)p(X \mid X_1)}{p(X_1)p(X \mid X_1) + p(X_1')p(X \mid X_1')} \\
&= \frac{(.02)(.75)}{(.02)(.75) + (.98)(.15)} \\
&= .09.
\end{aligned}
$$

1.7 *Quiz*

1 False. There are 8 branches.

2 False. See diagram (4) and the discussion preceding it.

3 True. See the discussion immediately following Definition 7.1.

4 False. See the discussion immediately following Definition 7.1.

5 True. We show that $p((X_1 \cup X_2) \cap X_3) = p(X_1 \cup X_2)p(X_3)$:

$$
\begin{aligned}
p((X_1 \cup X_2) \cap X_3) &= p((X_1 \cap X_3) \cup (X_2 \cap X_3)) \\
&= p(X_1 \cap X_3) + p(X_2 \cap X_3) - p(X_1 \cap X_2 \cap X_3) \\
&= p(X_1)p(X_3) + p(X_2)p(X_3) - p(X_1)p(X_2)p(X_3) \\
&= [p(X_1) + p(X_2) - p(X_1)p(X_2)]p(X_3) \\
&= p(X_1 \cup X_2)p(X_3).
\end{aligned}
$$

Apply the result of Question 3.

6 True. We must show that $p(X_1 \cap X_2) = p(X_1)p(X_2)$ implies that $p(X_1' \cap X_2') = p(X_1')p(X_2')$:

$$\begin{aligned} p(X_1' \cap X_2') &= p((X_1 \cup X_2)') \\ &= 1 - p(X_1 \cup X_2) \\ &= 1 - [p(X_1) + p(X_2) - p(X_1)p(X_2)] \\ &= [1 - p(X_1)] - p(X_2)[1 - p(X_1)] \\ &= [1 - p(X_1)][1 - p(X_2)] \\ &= p(X_1')p(X_2'). \end{aligned}$$

7 False. Choose independent events X_1 and X_2 such that $p(X_1) \neq 0 \neq p(X_2)$. Then

$$p(X_1 \cap X_2) = p(X_1)p(X_2) \neq 0.$$

But if $X_1 \cap X_2 = \emptyset$ we would have

$$p(X_1 \cap X_2) = p(\emptyset) = 0.$$

8 False. See Question 7.

9 True. This is a direct application of Theorem 7.2 with $n = 3$, $k_1 = 2$, $k_2 = 1$, $x_1 = \frac{1}{2}$, $x_2 = \frac{1}{2}$.

10 False. The probability of obtaining no heads in 3 tosses is the probability of obtaining 3 tails in three tosses. Again applying Theorem 7.2 with $n = 3$, $k_1 = 0$, $k_2 = 3$, $x_1 = \frac{1}{2}$, and $x_2 = \frac{1}{2}$, that number is

$$\binom{3}{0\ 3}\left(\frac{1}{2}\right)^0\left(\frac{1}{2}\right)^3 = \frac{3!}{3!}\cdot\frac{1}{2^3} = \frac{1}{8}.$$

1.7 *Exercises*

1 (c) Define the sample space U to be

$$U = \{(x_1, x_2, x_3) \mid x_1, x_2, x_3 \text{ are } B \text{ or } W\}.$$

Now define a probability measure p by the probability at the end of each branch:

$$\begin{aligned} p(\{(W, W, W)\}) &= \tfrac{1}{35}, \\ p(\{(W, W, B)\}) &= \tfrac{4}{35}, \\ p(\{(W, B, W)\}) &= \tfrac{4}{35}, \\ p(\{(W, B, B)\}) &= \tfrac{6}{35}, \\ p(\{(B, W, W)\}) &= \tfrac{4}{35}, \\ p(\{(B, W, B)\}) &= \tfrac{6}{35}, \\ p(\{(B, B, W)\}) &= \tfrac{6}{35}, \\ p(\{(B, B, B)\}) &= \tfrac{4}{35}. \end{aligned}$$

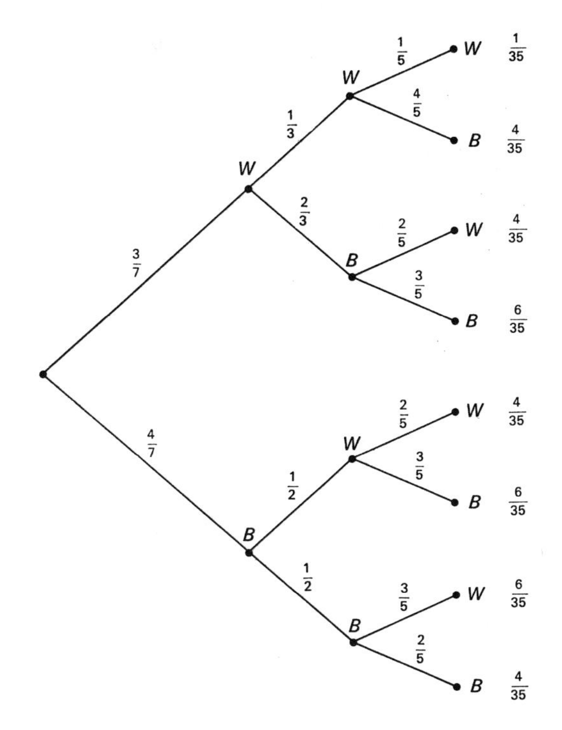

(d) Let U be defined as in part (c) and again use as probability measure the probabilities listed at the end of each branch in the tree below.

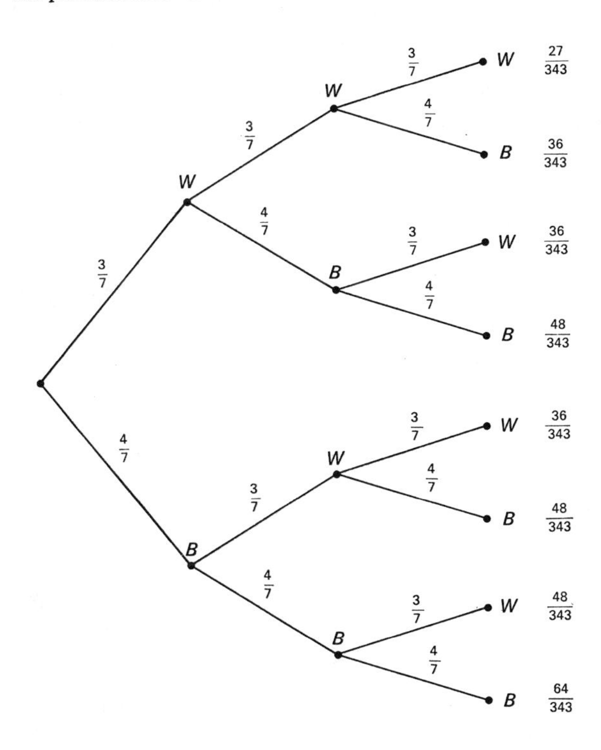

4

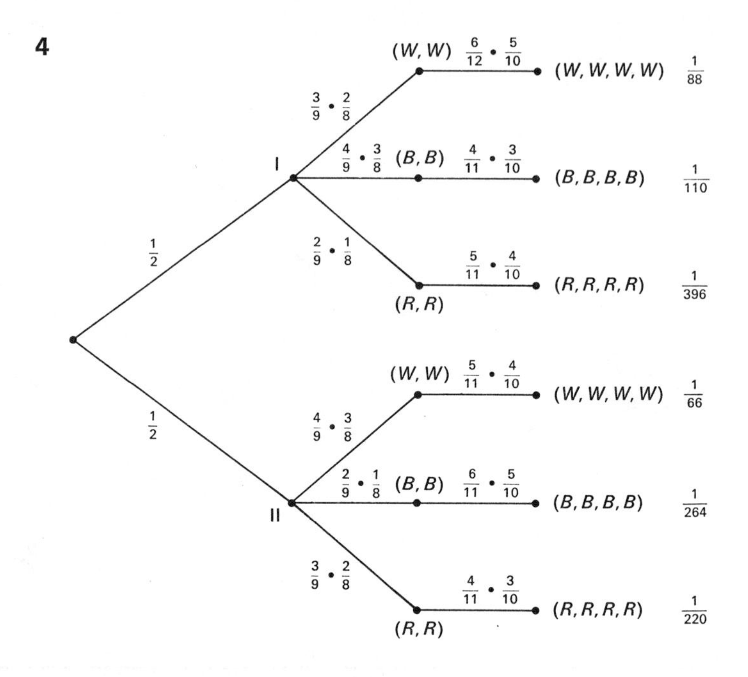

In this tree diagram we have shown only those branches which matter. Hence the probability that all four marbles are of the same color is the sum of the above probabilities.

8 By Theorem 7.2 we have $p(\{k \text{ successes in } n \text{ trials}\}) = \binom{n}{k} x_1^k x_2^{n-k}$.

11 By Theorem 6.1,

$$p(X_1 \cap X_2') = p(X_1) - p(X_1 \cap X_2).$$

Therefore

$$\begin{aligned} p(X_1 \cap X_2') &= p(X_1) - p(X_1)p(X_2) \\ &= p(X_1)[1 - p(X_2)] \\ &= p(X_1)p(X_2'). \end{aligned}$$

Similarly,

$$p(X_1' \cap X_2) = p(X_1')p(X_2).$$

Finally,

$$\begin{aligned} p(X_1' \cap X_2') &= p((X_1 \cup X_2)') \\ &= 1 - p(X_1 \cup X_2) \\ &= 1 - p((X_1 \cap X_2) \cup (X_1 \cap X_2') \cup (X_1' \cap X_2)) \\ &= 1 - [p(X_1 \cap X_2) + p(X_1 \cap X_2') + p(X_1' \cap X_2)] \\ &= 1 - [p(X_1)p(X_2) + p(X_1)p(X_2') + p(X_1')p(X_2)] \\ &= 1 - [p(X_1)(p(X_2) + p(X_2')) + p(X_1')p(X_2)] \end{aligned}$$

$$
\begin{aligned}
&= 1 - [p(X_1)p(X_2 \cup X_2') + p(X_1')p(X_2)] \\
&= 1 - [p(X_1)p(U) + p(X_1')p(X_2)] \\
&= [1 - p(X_1)] - p(X_1')p(X_2) \\
&= p(X_1') - p(X_1')p(X_2) \\
&= p(X_1')[1 - p(X_2)] \\
&= p(X_1')p(X_2').
\end{aligned}
$$

12 Here

$$X_1 = \{(H, H, H), (H, H, T), (H, T, H), (H, T, T)\}$$

and

$$X_2 = \{(H, H, H), (T, T, T)\}.$$

Then $X_1 \cap X_2 = \{(H, H, H)\}$, so that

$$
\begin{aligned}
p(X_1 \cap X_2) &= p(\{(H, H, H)\}) \\
&= \tfrac{1}{2} \cdot \tfrac{1}{2} \cdot \tfrac{1}{2} \\
&= \tfrac{1}{8}.
\end{aligned}
$$

Also,

$$
\begin{aligned}
p(X_1) &= p(\{(H, H, H)\}) + p(\{(H, H, T)\}) + p(\{(H, T, H)\}) \\
&\quad + p(\{(H, T, T)\}) \\
&= \tfrac{1}{2} \cdot \tfrac{1}{2} \cdot \tfrac{1}{2} + \tfrac{1}{2} \cdot \tfrac{1}{2} \cdot \tfrac{1}{2} + \tfrac{1}{2} \cdot \tfrac{1}{2} \cdot \tfrac{1}{2} + \tfrac{1}{2} \cdot \tfrac{1}{2} \cdot \tfrac{1}{2} \\
&= 4(\tfrac{1}{8}) \\
&= \tfrac{1}{2},
\end{aligned}
$$

and

$$
\begin{aligned}
p(X_2) &= p(\{(H, H, H)\}) + p(\{(T, T, T)\}) \\
&= \tfrac{1}{2} \cdot \tfrac{1}{2} \cdot \tfrac{1}{2} + \tfrac{1}{2} \cdot \tfrac{1}{2} \cdot \tfrac{1}{2} \\
&= 2(\tfrac{1}{8}) \\
&= \tfrac{1}{4}.
\end{aligned}
$$

Hence

$$
\begin{aligned}
p(X_1)p(X_2) &= \tfrac{1}{2} \cdot \tfrac{1}{4} \\
&= \tfrac{1}{8} \\
&= p(X_1 \cap X_2).
\end{aligned}
$$

Hence X_1 and X_2 are independent events.

22 (a)

$$
\begin{aligned}
E(1 - f) &= \sum_{i=1}^{n} p(\{u_i\})(1 - f)(u_i) \\
&= \sum_{i=1}^{n} p(\{u_i\})(1 - f(u_i)) \\
&= \sum_{i=1}^{n} p(\{u_i\}) - \sum_{i=1}^{n} p(\{u_i\})f(u_i) \\
&= 1 - E(f).
\end{aligned}
$$

(b)
$$\begin{aligned} E(f+g) &= \sum_{i=1}^{n} p(\{u_i\})(f+g)(u_i) \\ &= \sum_{i=1}^{n} p(\{u_i\})(f(u_i)+g(u_i)) \\ &= \sum_{i=1}^{n} p(\{u_i\})f(u_i) + \sum_{i=1}^{n} p(\{u_i\})g(u_i) \\ &= E(f)+E(g). \end{aligned}$$

(d) Define a sample space $U = \{S, F\}$, where S means that a sum of 3 or 4 shows on the two dice, and F means this does not happen. Define $p(\{S\}) = \frac{5}{36}$, $p(\{F\}) = \frac{31}{36}$. Define a random variable f by $f(S) = 3$, $f(F) = -2$. Then

$$\begin{aligned} E(f) &= p(\{S\})f(S) + p(\{F\})f(F) \\ &= \tfrac{5}{36}\cdot 3 - \tfrac{31}{36}\cdot 2 \\ &= -1\tfrac{11}{36}. \end{aligned}$$

Hence on one toss, player I can expect to lose $1\frac{11}{36}$ dollars, so that in ten tosses of the dice, player I can expect to lose $13\frac{1}{18}$ dollars. (Player II can expect to win that much in ten tosses.)

---

chapter 2

2.1 Quiz

1 True. $0u = 0(u_1, \ldots, u_m) = (0, \ldots, 0) = 0_m$.

2 True. The vectors u and v are functions, $u\colon N_n \to R$, $v\colon N_n \to R$. Two functions are equal if and only if they have the same value at each element in their common domain. Thus $u = v$ if and only if $u_i = v_i$, $i = 1, \ldots, n$.

3 True. $(cd)v = ((cd)v_1, \ldots, (cd)v_n) = (c(dv_1), \ldots, c(dv_n)) = c(dv)$.

4 False. The vector $-v$ has the following geometric representation.

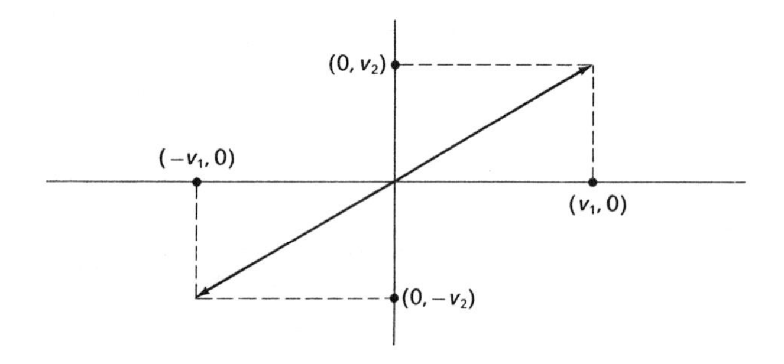

5 False. Consider the matrix $A = \begin{bmatrix} 1 & 1 \\ -1 & -1 \end{bmatrix}$.

6 False. The product of a 2×3 and a 3×2 matrix is a 2×2 matrix.

7 False. If A is $n \times m$ and B is $m \times n$, $m \neq n$, then AB is $n \times n$.

8 True. The (i, j) entry of $A + (-A)$ is $a_{ij} + (-a_{ij}) = 0$ for all i and j.

9 True. $B - A = -(A - B)$ if and only if the corresponding entries of the two matrices are the same. The (i, j) entry of $B - A$ is $b_{ij} - a_{ij}$, while the (i, j) entry of $-(A - B)$ is

$$-(a_{ij} - b_{ij}) = -a_{ij} + b_{ij} = b_{ij} - a_{ij}.$$

10 True. Since $I_m A = A$ is true for any $m \times m$ matrix A, it is in fact true for $A = I_m$.

2.1 *Exercises*

1 (d) $\begin{bmatrix} 25 & 15 & -5 \\ 37 & 17 & 3 \\ -12 & -7 & 18 \end{bmatrix}$ (g) $\begin{bmatrix} 0 & 0 & 1 \\ 0 & 1 & 0 \\ 1 & 0 & 0 \end{bmatrix}$

(i) $\begin{bmatrix} 3 & 2 & 1 \\ 2 & 5 & 7 \\ 4 & 3 & 2 \end{bmatrix}$ (o) 0_{44}

4 According to the hint, $AB^{(j)} = A^{(1)}b_{1j} + A^{(2)}b_{2j} + \cdots + A^{(n)}b_{nj} = (a_{11}b_{1j} + a_{12}b_{2j} + \cdots + a_{1n}b_{nj}, a_{21}b_{1j} + a_{22}b_{2j} + \cdots + a_{2n}b_{nj}, \ldots, a_{n1}b_{1j} + \cdots + a_{nn}b_{nj})$. Hence the ith component of $AB^{(j)}$, which is in fact c_{ij}, is

$$a_{i1}b_{1j} + a_{i2}b_{2j} + \cdots + a_{in}b_{nj} = \sum_{k=1}^{n} a_{ik}b_{kj}.$$

5 Let $v_j = (v_{1j}, v_{2j}, \ldots, v_{nj})$, $j = 1, 2, \ldots, p$. Let $u = (u_1, u_2, \ldots, u_n)$ be the sum $\sum_{j=1}^{p} v_j$. Then $u_i = \sum_{j=1}^{p} v_{ij}$, $i = 1, 2, \ldots, n$.

Now,

$$\begin{aligned} Au &= \sum_{i=1}^{n} A^{(i)}u_i \\ &= \sum_{i=1}^{n} \left(A^{(i)} \sum_{j=1}^{p} v_{ij} \right) \\ &= \sum_{i=1}^{n} \sum_{j=1}^{p} A^{(i)}v_{ij} \\ &= \sum_{j=1}^{p} \sum_{i=1}^{n} A^{(i)}v_{ij} \\ &= \sum_{j=1}^{p} Av_j. \end{aligned}$$

6 The ith components of the following vectors are given by

$$(c(Av))_{(i)} = \left(c\left(\sum_{k=1}^{n} A^{(k)}v_k\right)\right)_{(i)} = \left(\sum_{k=1}^{n} c(A^{(k)}v_k)\right)_{(i)}$$

$$= \sum_{k=1}^{n} c(a_{ik}v_k);$$

$$(A(cv))_{(i)} = \left(\sum_{k=1}^{n} A^{(k)}(cv_k)\right)_{(i)} = \sum_{k=1}^{n} a_{ik}(cv_k);$$

$$((Av)c)_{(i)} = \left(\left(\sum_{k=1}^{n} A^{(k)}v_k\right)c\right)_{(i)} = \sum_{k=1}^{n} (a_{ik}v_k)c;$$

$$(A(vc))_{(i)} = \left(\sum_{k=1}^{n} A^{(k)}(v_kc)\right)_{(i)} = \sum_{k=1}^{n} a_{ik}(v_kc);$$

$$((Ac)v)_{(i)} = \left(\sum_{k=1}^{n} (A^{(k)}c)v_k\right)_{(i)} = \sum_{k=1}^{n} (a_{ik}c)v_k.$$

Since a_{ik}, v_k, and c are scalars, the ith components of $c(Av)$, $A(cv)$, $(Av)c$, $A(vc)$, and $(Ac)v$ are equal, i.e., $c(Av) = A(cv) = (Av)c = A(vc) = (Ac)v$.

12 (a) $A_{(1)} = [a_{11}\ a_{12}\ a_{13}\ a_{14}]$,
$A_{(2)} = [a_{21}\ a_{22}\ a_{23}\ a_{24}]$,
$A_{(3)} = [a_{31}\ a_{32}\ a_{33}\ a_{34}]$.

(e) $(B^T)_{(1)} = [b_{11}\ b_{21}\ b_{31}\ b_{41}]$.

(m) $((A^T)_{(1)})^T = \begin{bmatrix} a_{11} \\ a_{21} \\ a_{31} \end{bmatrix}$.

(o) $(((B^T)_{(1)})^T)_{(2)} = [b_{21}]$.

20 One of the required linear combinations is

$$v_1(1, 0, 0, 0) + v_2(0, 1, 0, 0) + v_3(0, 0, 1, 0) + v_4(0, 0, 0, 1) = (v_1, v_2, v_3, v_4).$$

In general, if

$$\lambda_1(1, 0, 0, 0) + \lambda_2(0, 1, 0, 0) + \lambda_3(0, 0, 1, 0) + \lambda_4(0, 0, 0, 1) = (v_1, v_2, v_3, v_4)$$

is any linear combination, then

$$(\lambda_1, \lambda_2, \lambda_3, \lambda_4) = (v_1, v_2, v_3, v_4),$$

that is, $\lambda_i = v_i$, $i = 1, 2, 3, 4$. Therefore there is exactly one solution.

21 Let $\delta_{ij} = \begin{cases} 0 \text{ if } i \neq j, \\ 1 \text{ if } i = j. \end{cases}$ Then $E^j = \begin{bmatrix} \delta_{1j} \\ \vdots \\ \delta_{nj} \end{bmatrix}$. Hence

$$AE^j = \sum_{k=1}^{n} A^{(k)}\,\delta_{kj} = A^{(j)}.$$

22 Let δ_{ij} be defined as in Exercise 21. Then $E_i = [\delta_{i1}, \ldots, \delta_{im}]$. Now,

$$(E_iA)^{(k)} = E_iA^{(k)} = \sum_{j=1}^{m} \delta_{ij}a_{jk} = a_{ik},$$

$k = 1, \ldots, n$. Therefore

$$E_iA = [a_{i1}\ a_{i2}\ \ldots\ a_{in}] = A_{(i)}.$$

2.2 *Quiz*

1 True. $R = \{(1, 1)\}$, i.e., none of (1, 2), (2, 1), or (2, 2) are in R.

2 True. The function $f: X \to Y$, where $X = \{x_1, x_2, x_3\}$, $Y = \{y_1, y_2\}$, is defined by $f(x_i) = y_1$, $i = 1, 2, 3$.

3 False. In this case three pairs in the relation have the same first element.

4 False. The matrix

$$\begin{bmatrix} 1 & 0 & 0 \\ 0 & 0 & 1 \end{bmatrix}$$

cannot be the incidence matrix of a function because the second column does not contain a 1.

5 True. Because f is a function, its incidence matrix has precisely one 1 in each column. Because f is a 1-1 function, each row has at most one 1 in it. But because the matrix has the same number of rows as columns, each row also has precisely one 1 in it.

6 True. If $X = \{x_1, x_2, x_3\}$, the matrix

$$\begin{bmatrix} 1 & 1 \\ 0 & 0 \\ 0 & 0 \end{bmatrix}$$

corresponds to the function $f(1) = x_1$, $f(2) = x_1$, which is indeed a 2-selection of X.

7 True. The matrix $A(f)$ corresponds to the function $f(x_1) = x_2$, $f(x_2) = x_3$, $f(x_3) = x_1$, which is clearly 1-1 onto. A quick calculation shows that f^{-1} is defined by $f^{-1}(x_1) = x_3, f^{-1}(x_2) = x_1, f^{-1}(x_3) = x_2$. Hence

$$A(f^{-1}) = \begin{bmatrix} 0 & 1 & 0 \\ 0 & 0 & 1 \\ 1 & 0 & 0 \end{bmatrix}.$$

8 False. Consider

$$A_1 = \begin{bmatrix} 1 & 1 \\ 1 & -1 \end{bmatrix}, \qquad A_2 = \begin{bmatrix} 1 & 1 \\ 1 & 1 \end{bmatrix}.$$

Then

$$C = A_1 A_2 = \begin{bmatrix} 2 & 2 \\ 0 & 0 \end{bmatrix}.$$

9 True. Assume A is $n \times n$. By assumption, $a_{ij} > 0$ for all i, j. We will prove by induction that any positive power of A has only positive entries. Suppose we know A^{p-1} has all entries positive, $p > 1$. Then the (i, j) entry of A^p is

$$\sum_{k=1}^{n} (A^{p-1})_{ik} a_{kj}.$$

We know a_{kj} is positive, and by assumption, the (i, k) entry of A^{p-1} is positive. Hence $(A^{p-1})_{ik} a_{kj}$ is positive, $k = 1, \ldots, n$, and therefore the (i, j) entry of A^p is positive.

10 True. More generally, by applying Theorem 2.2 with $X_1 = \cdots = X_{p+1} = X$, $R_1 = \cdots = R_p = R$, and $A_1 = \cdots = A_p = A$, we know the (i, j) entry of A^p is the number of p-step connections from the jth element of X to the ith.

2.2 *Exercises*

1

$$A = \begin{bmatrix} 3 & 2 & 0 \\ 2 & 1 & 1 \\ 0 & 1 & 2 \end{bmatrix}, \qquad B = \begin{bmatrix} 2 & 1 & 0 \\ 2 & 1 & 0 \end{bmatrix};$$

$$BA = \begin{bmatrix} 8 & 5 & 1 \\ 8 & 5 & 1 \end{bmatrix}.$$

Examining BA, we find that the number of roads connecting

x_1 and z_1 is 8,
x_2 and z_1 is 5,
x_3 and z_1 is 1,
x_1 and z_2 is 8,
x_2 and z_2 is 5,
x_3 and z_2 is 1.

2 Using the same notation as in Theorem 2.1, if the jth column of A_p consists entirely of zeros, then for each $k = 1, 2, \ldots, n_p$, $a_{kj}^{(p)} = 0$. The product in equation (6) is then zero for all i, and hence so is their sum. Therefore the jth column of C consists entirely of 0's.

If $A_1, A_2, \ldots, A_p$ are incidence matrices for relations $R_1, R_2, \ldots, R_p$, since the jth column of A_p consists entirely of zeros, no element in X_2 is related to $x_j \in X_1$. The above result implies that when this is the case, then there is no p-step connection from $x_j \in X_1$ to any element in X_{p+1}.

3 See answer to Quiz Question 10.

6 As an aid to the reader's intuition, we first compute (c).
(c) There are six diagonals of S:

$$(1, 5, 9), \quad (2, 6, 7), \quad (3, 4, 8),$$
$$(3, 5, 7), \quad (2, 4, 9), \quad (1, 6, 8).$$

If we place ones in the positions corresponding to each diagonal entry and zeros elsewhere, we have six matrices as follows:

$$\begin{bmatrix} 1 & 0 & 0 \\ 0 & 1 & 0 \\ 0 & 0 & 1 \end{bmatrix}, \quad \begin{bmatrix} 0 & 1 & 0 \\ 0 & 0 & 1 \\ 1 & 0 & 0 \end{bmatrix}, \quad \begin{bmatrix} 0 & 0 & 1 \\ 1 & 0 & 0 \\ 0 & 1 & 0 \end{bmatrix},$$
$$\begin{bmatrix} 0 & 0 & 1 \\ 0 & 1 & 0 \\ 1 & 0 & 0 \end{bmatrix}, \quad \begin{bmatrix} 0 & 1 & 0 \\ 1 & 0 & 0 \\ 0 & 0 & 1 \end{bmatrix}, \quad \begin{bmatrix} 1 & 0 & 0 \\ 0 & 0 & 1 \\ 0 & 1 & 0 \end{bmatrix}.$$

These six matrices are incidence matrices of 1-1 functions from a 3-set onto itself. Therefore there is a 1-1 correspondence between the set of permutations of a 3-set and the set of all diagonals of S.

Part (a) now becomes obvious upon replacing 3 by n. Since there are $n!$ permutations on an n-set, (b) is a direct result of (a).

10 Let f be defined by $f(1) = 3$, $f(2) = 1$, $f(3) = 2$, $f(4) = 5$, $f(5) = 4$. Then

$$A(f) = \begin{bmatrix} 0 & 1 & 0 & 0 & 0 \\ 0 & 0 & 1 & 0 & 0 \\ 1 & 0 & 0 & 0 & 0 \\ 0 & 0 & 0 & 0 & 1 \\ 0 & 0 & 0 & 1 & 0 \end{bmatrix}, \qquad A^5 = \begin{bmatrix} 0 & 0 & 1 & 0 & 0 \\ 1 & 0 & 0 & 0 & 0 \\ 0 & 1 & 0 & 0 & 0 \\ 0 & 0 & 0 & 0 & 1 \\ 0 & 0 & 0 & 1 & 0 \end{bmatrix} \neq I_5.$$

14 Let a_{ij} denote the (i, j) entry of the incidence matrix for this problem. Then we know:

$$\text{(i)}\ a_{ij} + a_{ji} = 1, \qquad i, j = 1, 2, 3;$$
$$\text{(ii)}\ \qquad a_{ii} = 0, \qquad i = 1, 2, 3.$$

There are only 8 matrices of this kind:

$$\begin{bmatrix} 0 & 1 & 1 \\ 0 & 0 & 1 \\ 0 & 0 & 0 \end{bmatrix}, \quad \begin{bmatrix} 0 & 1 & 1 \\ 0 & 0 & 0 \\ 0 & 1 & 0 \end{bmatrix}, \quad \begin{bmatrix} 0 & 1 & 0 \\ 0 & 0 & 1 \\ 1 & 0 & 0 \end{bmatrix}, \quad \begin{bmatrix} 0 & 1 & 0 \\ 0 & 0 & 0 \\ 1 & 1 & 0 \end{bmatrix},$$
$$\begin{bmatrix} 0 & 0 & 1 \\ 1 & 0 & 1 \\ 0 & 0 & 0 \end{bmatrix}, \quad \begin{bmatrix} 0 & 0 & 1 \\ 1 & 0 & 0 \\ 0 & 1 & 0 \end{bmatrix}, \quad \begin{bmatrix} 0 & 0 & 0 \\ 1 & 0 & 1 \\ 1 & 0 & 0 \end{bmatrix}, \quad \begin{bmatrix} 0 & 0 & 0 \\ 1 & 0 & 0 \\ 1 & 1 & 0 \end{bmatrix}.$$

In all but two of these matrices, there is a column which contains two 1's, and the 1's are off-diagonal entries. If the (i, j) entry is interpreted as signifying whether or not c_j pecks c_i, then if the jth column contains two 1's, the jth chicken is dominant. The two matrices which do not satisfy this nice property are

$$A_1 = \begin{bmatrix} 0 & 1 & 0 \\ 0 & 0 & 1 \\ 1 & 0 & 0 \end{bmatrix} \quad \text{and} \quad A_2 = \begin{bmatrix} 0 & 0 & 1 \\ 1 & 0 & 0 \\ 0 & 1 & 0 \end{bmatrix}.$$

In this case

$$A_1 + A_1^2 = \begin{bmatrix} 0 & 1 & 1 \\ 1 & 0 & 1 \\ 1 & 1 & 0 \end{bmatrix} \quad \text{and} \quad A_2 + A_2^2 = \begin{bmatrix} 0 & 1 & 1 \\ 1 & 0 & 1 \\ 1 & 1 & 0 \end{bmatrix},$$

and in both cases the matrix plus its square has ones in its off-diagonal entries. But this simply means all of the three chickens are "dominant."

2.3 *Quiz*

1 True.

$$A + A^T = \begin{bmatrix} 0 & 0 & 0 \\ 1 & 0 & 0 \\ 1 & 1 & 0 \end{bmatrix} + \begin{bmatrix} 0 & 1 & 1 \\ 0 & 0 & 1 \\ 0 & 0 & 0 \end{bmatrix} = \begin{bmatrix} 0 & 1 & 1 \\ 1 & 0 & 1 \\ 1 & 1 & 0 \end{bmatrix} = J_3 - I_3.$$

2 False. Since A satisfies Theorem 3.1, then according to equation (3) in the proof, every main diagonal entry of $A + A^2$ must be zero.

3 False. If this were the case, the column sums would have to be 3.

4 False. The matrix

$$\begin{bmatrix} 0 & 0 & 0 \\ 1 & 0 & 0 \\ 0 & 1 & 0 \end{bmatrix}$$

corresponds to the relation $R = \{(x_1, x_2), (x_2, x_3)\}$, where $(x_2, x_1) \notin R$, $(x_3, x_2) \notin R$, and $(x_i, x_i) \notin R$, $i = 1, 2, 3$. Since neither of (x_1, x_3) and (x_3, x_1) is in R, this is not an asymmetric relation.

5 False. If $(i, j) \in R$ then $i - j$ is even, so that $-(i - j) = j - i$ is also even, i.e., $(j, i) \in R$. Also $(i, i) \in R$ for all i, since $i - i = 0$ is considered even.

6 True. If $(i, j) \in R$ then $i < j$; hence $j > i$, i.e., $(j, i) \notin R$. Similarly, since $i = i$, $(i, i) \notin R$.

7 False. If x_1 is the mother of x_2, then x_2 cannot be the mother of x_1, and no one can be his own mother. Hence if $(x_1, x_2) \in R$, then $(x_2, x_1) \notin R$, and $(x_i, x_i) \notin R$ for all i. However, if x_1 and x_2 are both male, then neither (x_1, x_2) nor (x_2, x_1) belongs to R.

8 False. The matrix $\frac{1}{2}J_n - \frac{1}{2}I_n$ satisfies $A + A^T = J_n - I_n$, but it is not an incidence matrix for any relation.

9 True.

$$J_3^2 = 3J_3,$$
$$J_3^3 = 3J_3 \cdot J_3 = 9J_3,$$
$$J_3^4 = 9J_3 \cdot J_3 = 27J_3.$$

10 True.

$$\begin{aligned}(J_3 - I_3)^3 &= J_3^3 - 3J_3^2 I_3 + 3J_3 I_3^2 - I_3^3\\ &= 9J_3 - 9J_3 + 3J_3 - I_3\\ &= 3J_3 - I_3.\end{aligned}$$

2.3 *Exercises*

1 (e) In general, it can be proved by mathematical induction on m that

$$J_n^m = n^{(m-1)} J_n.$$

Hence

$$J_2^{15} = 2^{14} J_2.$$

(See Exercise 10 (b).)

4 As suggested in the hint, $A + A^T = J_n - I_n$, where $A = A(R)$. Hence by Theorem 3.1, there is a row in $A + A^2$ which has $n - 1$ positive entries. By investigating the proof of the theorem, we find that the zero entry is on the main diagonal. Hence any off-diagonal entry in this row (say corresponding to $x \in X$) is greater than or equal to 1. Therefore in any off-diagonal position, either A or A^2 (or both) must have an entry greater than or equal to one. If, for $y \neq x$, the "(y, x) entry" of A is positive (in fact 1), then $(x, y) \in R$. If the (y, x) entry of A^2 is positive, then there is a 2-step connection between x and y, i.e., there is a $z \in X$ such that $(x, z) \in R$ and $(z, y) \in R$.

9 Let R be the relation on the community defined by $(x, y) \in R$ if y loves x. Then R is an asymmetric relation and so $A = A(R)$ satisfies the hypothesis of Theorem 3.1. Thus there is a row of $A + A^2$ containing $n - 1$ positive entries (greater than or equal to 1), and the zero entry is on the main diagonal. If this row corresponds to person p, then either the "(p, x) entry" of A or the "(p, x) entry" of A^2 (or both), $p \neq x$, is a positive number. If there is a positive number in the (p, x) position of A, it must be a 1, and x loves p. If there is a positive number in the (p, x) position of A^2, then there exists at least one 2-step connection between x and p, i.e., there is a person y such that x loves y and y loves p.

10 (a) Applying Exercise 4, Section 2.1,

$$c_j = \sum_{k=1}^{m} (J_m)_{ik} a_{kj} = \sum_{k=1}^{m} a_{kj},$$

which is simply the sum of the entries in column j of A.

(b) The proof will be by induction on p. When $p = 1$, the assertion is obvious. When $p = 2$,

$$\begin{aligned}(J_m^2)_{ij} &= \sum_{k=1}^{m} (J_m)_{ik}(J_m)_{kj}\\ &= \sum_{k=1}^{m} 1\\ &= m.\end{aligned}$$

Hence $J_m^2 = mJ_m$. Suppose that

$$J_m^{k-1} = m^{k-2}J_m$$

holds for some $k > 1$. Multiplying both sides of the equation by J_m, we see that

$$\begin{aligned} J_m^k &= m^{k-2}J_m^2 \\ &= m^{k-2}(mJ_m) \\ &= m^{k-1}J_m, \end{aligned}$$

and the induction step is complete.

(c) Applying Theorem 2.1 for $p = 3$, the (i, j) entry of J_mAJ_n is

$$\sum_{\alpha=1}^{m}\sum_{\beta=1}^{n}(J_m)_{i\alpha}a_{\alpha\beta}(J_n)_{\beta j} = \sum_{\alpha=1}^{m}\sum_{\beta=1}^{n}a_{\alpha\beta} = s.$$

(d) Let e_{nj} and e_{mj} denote the jth components of e_n and e_m, respectively. Then $e_{nj} = e_{mj} = 1$. If $\sum_{j=1}^{n} a_{ij} = 1$ for $i = 1, \ldots, m$, then the kth component of Ae_n is

$$\sum_{j=1}^{n} a_{kj}e_{nj} = \sum_{j=1}^{n} a_{kj} = 1, \qquad k = 1, \ldots, m.$$

Conversely, if $Ae_n = e_m$, then

$$1 = (e_m)_k = (Ae_n)_k = \sum_{j=1}^{n} a_{kj}e_{nj} = \sum_{j=1}^{n} a_{kj}.$$

(e) By part (d), we have

$$A^Te_m = e_n$$

if and only if

$$\sum_{j=1}^{m}(A^T)_{ij} = 1$$

if and only if

$$\sum_{j=1}^{m} a_{ji} = 1,$$

that is, if and only if every column sum of A is 1.

2.4 *Quiz*

1 True. Let $\{x_1, x_2\}$ be a 2-element set. The only possible ways of choosing two different non-empty subsets are as follows:

$$\{x_1\},\ \{x_2\};$$
$$\{x_1\},\ \{x_1, x_2\};$$
$$\{x_1, x_2\},\ \{x_2\}.$$

In each case, the sets possess an SDR.

2 True. See Example 4.3 (a).

3 True. Suppose $n \leq m$. Then a diagonal of A is a choice of n entries of A lying in all n rows and n distinct columns. Hence the number of diagonals of A is equal to the number of n-permutations of the set N_m. Since A^T is $m \times n$, a diagonal of A^T is a choice of n entries of A^T lying in all n columns and n distinct rows. Thus, the number of diagonals of A^T is also equal to the number of n-permutations of N_m. If $m \leq n$, we follow the same procedure.

4 True. Suppose for each n-permutation σ of N_n the diagonals $(a_{1\sigma(1)}, a_{2\sigma(2)}, \ldots, a_{n\sigma(n)})$ and $(b_{1\sigma(1)}, b_{2\sigma(2)}, \ldots, b_{n\sigma(n)})$ of two $n \times n$ matrices A and B are equal. Then $a_{i\sigma(i)} = b_{i\sigma(i)}$ for $i = 1, \ldots, n$. Since every entry of an $n \times n$ matrix is contained in some diagonal, we have $A = B$.

5 False. This matrix has submatrices [1 2], [1], and [2].

6 False. See Example 4.7.

7 True. Let $(a_{1\sigma(1)}, a_{2\sigma(2)}, a_{3\sigma(3)})$ be a diagonal of this matrix. In order that this diagonal contain no zeros, we must have $a_{1\sigma(1)} = a_{12}$ and $a_{3\sigma(3)} = a_{32}$. But then $\sigma(1) = \sigma(3) = 2$, which is impossible since σ is a 3-permutation. Hence every diagonal contains a 0.

8 False. The matrix

$$\begin{bmatrix} 1 & 1 \\ 1 & 1 \end{bmatrix}$$

is not situated at the intersection of any set of rows and columns of the matrix in Question 7.

9 True. The 2×2 submatrices are the following:

$$\begin{bmatrix} 0 & 1 \\ 1 & 1 \end{bmatrix}, \begin{bmatrix} 0 & 1 \\ 0 & 1 \end{bmatrix}, \begin{bmatrix} 1 & 1 \\ 0 & 1 \end{bmatrix}, \begin{bmatrix} 0 & 0 \\ 1 & 1 \end{bmatrix}, \begin{bmatrix} 0 & 0 \\ 0 & 0 \end{bmatrix},$$
$$\begin{bmatrix} 1 & 1 \\ 0 & 0 \end{bmatrix}, \begin{bmatrix} 1 & 0 \\ 1 & 1 \end{bmatrix}, \begin{bmatrix} 1 & 0 \\ 1 & 0 \end{bmatrix}, \begin{bmatrix} 1 & 1 \\ 1 & 0 \end{bmatrix}.$$

10 True. See Exercise 10.

2.4 *Exercises*

1 (c) [1], [2], [3], [4], [5], [6], [1 2], [1 3], [2 3], [4 5], [4 6], [5 6], $\begin{bmatrix} 1 \\ 4 \end{bmatrix}, \begin{bmatrix} 2 \\ 5 \end{bmatrix}, \begin{bmatrix} 3 \\ 6 \end{bmatrix}$, [1 2 3], [4 5 6], $\begin{bmatrix} 1 & 2 \\ 4 & 5 \end{bmatrix}, \begin{bmatrix} 2 & 3 \\ 5 & 6 \end{bmatrix}, \begin{bmatrix} 1 & 3 \\ 4 & 6 \end{bmatrix}, \begin{bmatrix} 1 & 2 & 3 \\ 4 & 5 & 6 \end{bmatrix}$.

3 The following are all the SDR's for S_1, S_2, and S_3:

$$(1, 2, 3);\ (4, 2, 1);\ (4, 2, 3);\ (4, 3, 1).$$

4 (b) $(a_{11}, a_{22}) = (1, 4)$; $(a_{11}, a_{32}) = (1, 1)$; $(a_{21}, a_{12}) = (3, 2)$;

$(a_{21}, a_{32}) = (3, 1)$; $(a_{31}, a_{12}) = (-1, 2)$; $(a_{31}, a_{22}) = (-1, 4)$.

6 Let $X = \{p_1, \ldots, p_5\}$ be the set of personality characteristics and $Y = \{c_1, \ldots, c_6\}$ the set of children. Define a relation R on X to Y by

$$R = \{(p_j, c_i) \mid p_j \in X \wedge c_i \in Y \wedge \text{characteristic } p_j \text{ is exhibited by child } c_i\}.$$

Then

$$A(R) = \begin{bmatrix} 0 & 1 & 0 & 1 & 1 \\ 0 & 0 & 0 & 1 & 0 \\ 1 & 0 & 1 & 0 & 0 \\ 1 & 0 & 0 & 0 & 1 \\ 1 & 0 & 1 & 0 & 0 \\ 0 & 0 & 0 & 0 & 1 \end{bmatrix}.$$

Choosing a group of 5 children who exhibit all 5 of the characteristics among them is equivalent to finding a 5×5 submatrix of $A(R)$ which has a 1 in every column. Notice that the first row must always be in such a submatrix, since it is the only row with a 1 in column 2. There are five 5×5 submatrices of $A(R)$ which include the first row of $A(R)$:

$$\begin{bmatrix} 0 & 1 & 0 & 1 & 1 \\ 1 & 0 & 1 & 0 & 0 \\ 1 & 0 & 0 & 0 & 1 \\ 1 & 0 & 1 & 0 & 0 \\ 0 & 0 & 0 & 0 & 1 \end{bmatrix}, \begin{bmatrix} 0 & 1 & 0 & 1 & 1 \\ 0 & 0 & 0 & 1 & 0 \\ 1 & 0 & 0 & 0 & 1 \\ 1 & 0 & 1 & 0 & 0 \\ 0 & 0 & 0 & 0 & 1 \end{bmatrix}, \begin{bmatrix} 0 & 1 & 0 & 1 & 1 \\ 0 & 0 & 0 & 1 & 0 \\ 1 & 0 & 1 & 0 & 0 \\ 1 & 0 & 1 & 0 & 0 \\ 0 & 0 & 0 & 0 & 1 \end{bmatrix},$$

$$\begin{bmatrix} 0 & 1 & 0 & 1 & 1 \\ 0 & 0 & 0 & 1 & 0 \\ 1 & 0 & 1 & 0 & 0 \\ 1 & 0 & 0 & 0 & 1 \\ 0 & 0 & 0 & 0 & 1 \end{bmatrix}, \begin{bmatrix} 0 & 1 & 0 & 1 & 1 \\ 0 & 0 & 0 & 1 & 0 \\ 1 & 0 & 1 & 0 & 0 \\ 1 & 0 & 0 & 0 & 1 \\ 1 & 0 & 1 & 0 & 0 \end{bmatrix}.$$

Each of these has a 1 in each column, hence there are five groups of five children with the desired property.

10 Let $d = (r_1, \ldots, r_m)$ be an SDR for sets $S_1, \ldots, S_m$. Fix k, $1 \leq k \leq m$, and choose k integers $i_1, \ldots, i_k$ such that

$$1 \leq i_1 < i_2 < \cdots < i_k \leq m.$$

Now, $S_{i_1} \cup S_{i_2} \cup \cdots \cup S_{i_k} \supset \{r_{i_1}, r_{i_2}, \ldots, r_{i_k}\}$, so

$$\nu(S_{i_1} \cup S_{i_2} \cup \cdots \cup S_{i_k}) \geq \nu(\{r_{i_1}, r_{i_2}, \ldots, r_{i_k}\}).$$

But no two of the r_{i_j}, $j = 1, \ldots, k$, can be equal, since d is an m-permutation. Hence $\nu(\{r_{i_1}, r_{i_2}, \ldots, r_{i_k}\}) = k$ and

$$\nu(S_{i_1} \cup S_{i_2} \cup \cdots \cup S_{i_k}) \geq k.$$

13 No. Construct an incidence matrix A for this exercise as in Example 4.6. We see in this case that each column of A has exactly k ones, $k < n$, and each row of A has at least one 1. Again we want to find a diagonal of A consisting entirely of 1's, but this is not necessarily possible. For example, suppose $n = 4$ and $k = 2$. The following matrix has precisely two 1's in each column and at least one 1 in each row:

$$A = \begin{bmatrix} 0 & 1 & 1 & 1 \\ 0 & 1 & 1 & 1 \\ 1 & 0 & 0 & 0 \\ 1 & 0 & 0 & 0 \end{bmatrix}.$$

However, A has no diagonal consisting entirely of 1's, as can be easily checked.

14 Define a 5×5 matrix A such that the (i, j) entry of A is 1 if there is a one-way road from city c_j to city c_i and 0 if there is not. Then each column and each row of A contains precisely three 1's. We are again looking for a diagonal of 1's. We know by Example 4.6, however, that any $n \times n$ matrix with precisely k ones in each row and each column, $k \leq n$, has a diagonal of 1's. Hence A has such a diagonal, say corresponding to the 5-permutation σ: $(a_{1\sigma(1)}, \ldots, a_{5\sigma(5)})$. In fact, A has yet another diagonal of 1's! For suppose we consider the matrix $B = A - P$, where P is the incidence matrix of σ. Then B has two 1's in each row and each column. By Example 4.6, B also has a diagonal consisting entirely of 1's. But any diagonal of 1's of B is a diagonal of 1's of A, distinct from $(a_{1\sigma(1)}, \ldots, a_{5\sigma(5)})$. Hence it is possible to tour all five cities, visiting each city precisely once, and it can be done in more than one way.

2.5 *Quiz*

1 True. They are (1, 2, 3), (1, 3, 2), and (2, 3, 1).

2 False. The sum of the entries in the third column is 0.

3 True. In Theorem 5.1, let $n = 5$, $m = 3$, and $t = 4$.

4 True. In Theorem 5.1, let $n = 10$, $m = 4$ and $t = 7$.

5 True. Each row and each column sum is 1.

6 False. Let $n = 2$, $A = \begin{bmatrix} 1 & 0 \\ 1 & 0 \end{bmatrix}$, and $B = \begin{bmatrix} 1 & 1 \\ 0 & 0 \end{bmatrix}$. Then A is row stochastic and B is column stochastic, but $\frac{1}{2}A + \frac{1}{2}B = \begin{bmatrix} 1 & \frac{1}{2} \\ \frac{1}{2} & 0 \end{bmatrix}$ is not doubly stochastic.

7 True. In the previous section (Example 4.6), we showed that such a matrix has a diagonal of 1's. Hence its term rank is 4.

8 True. By definition, an $m \times n$ matrix has m rows and n columns.

9 False. Let $A = \begin{bmatrix} 1 & 0 & 0 \\ 0 & 1 & 0 \\ 0 & 0 & 1 \end{bmatrix}$ and $B = \begin{bmatrix} 1 & 1 & 1 \\ 1 & 1 & 1 \\ 0 & 0 & 0 \end{bmatrix}$. Then the term rank of A is 3, while that of B is 2.

10 True. Let $A = \begin{bmatrix} a_{11} & a_{12} \\ a_{21} & a_{22} \end{bmatrix}$. Then

$$AA^T = \begin{bmatrix} a_{11}^2 + a_{12}^2 & a_{11}a_{21} + a_{12}a_{22} \\ a_{21}a_{11} + a_{22}a_{12} & a_{21}^2 + a_{22}^2 \end{bmatrix}.$$

If the term rank of A is 0, then A is the zero matrix and hence so is AA^T. If the term rank of A is 1, then A contains at least one 1. But then the (1, 1) entry or the (2, 2) entry of AA^T is positive, so the term rank of AA^T is at least 1. Finally, if the term rank of A is 2, then either a_{11} and a_{22} are both 1's, or a_{12} and a_{21} are both 1's. In either case, the (1, 1) and (2, 2) entries of AA^T are positive.

2.5 *Exercises*

1 (c) Here there are 4 sets, each containing at least 6 elements. Then by Theorem 5.1, there are at least $\dfrac{6!}{2!} = 360$ SDR's.

5 From Exercise 3, we know that $a_{11}a_{22} + a_{12}a_{21} = s^2 + (1 - s)^2$, where $0 \leq s \leq 1$. We will first determine for which numbers s equality holds; i.e., we will solve the equation $s^2 + (1 - s)^2 = \frac{1}{2}$ for s. Using the quadratic formula, we see that $s = \frac{1}{2}$ is the only solution. Hence equality holds if and only if

$$A = \begin{bmatrix} \frac{1}{2} & \frac{1}{2} \\ \frac{1}{2} & \frac{1}{2} \end{bmatrix}.$$

Now, $s^2 + (1 - s)^2 \geq \frac{1}{2}$ if and only if $2s^2 - 2s + \frac{1}{2} \geq 0$. If we graph the equation $y = 2s^2 - 2s + \frac{1}{2}$, $0 \leq s \leq 1$, we see that the graph is a parabola which opens upward and which intersects the s-axis only once, at $s = \frac{1}{2}$. Hence $y \geq 0$ for all s.

7 According to Theorem 5.2, all the 1's in A are contained in one line of A. The sum of the entries of A is thus equal to the sum of the entries in that one line of A, which can be at most n.

10 In Example 4.6, we used Theorem 4.2 to show that for any positive integer k, if a (0, 1) matrix has exactly k ones in each row and each column, then the matrix has at least one diagonal of 1's. Hence A has at least one diagonal of 1's. If we think of A as an incidence matrix for sets S_1, S_2, S_3, S_4, and S_5, we see from Theorem 5.1 that A has at least $3! = 6$ diagonals of 1's. Since $\prod_{i=1}^{5} a_{i\sigma(i)} = 1$ for each 5-permutation σ corresponding to a diagonal of 1's, the result follows.

12 For any $m \times n$ matrix A, if $1 \leq i \leq m$ and $1 \leq j \leq n$, the (i, j) entry of AJ_n is

$$\sum_{k=1}^{n} a_{ik}(J_n)_{kj} = \sum_{k=1}^{n} a_{ik},$$

which is the ith row sum of A. Also, the (i, j) entry of J_mA is

$$\sum_{k=1}^{m} (J_m)_{ik}a_{kj} = \sum_{k=1}^{m} a_{kj},$$

which is the jth column sum of A. Hence, if A has nonnegative entries, every entry of AJ_n is 1 if and only if A is row stochastic. Similarly, every entry of J_mA is 1 if and only if A is column stochastic.

13 Let $1 \leq i \leq m$. For $j = 1, \ldots, n$, the (i, j) entry of AB is

$$\sum_{k=1}^{n} a_{ik}b_{kj}.$$

The ith row sum of AB is therefore

$$\sum_{j=1}^{q}\left(\sum_{k=1}^{n} a_{ik}b_{kj}\right).$$

But

$$\begin{aligned}\sum_{j=1}^{q}\left(\sum_{k=1}^{n} a_{ik}b_{kj}\right) &= \sum_{k=1}^{n} a_{ik}\left(\sum_{j=1}^{q} b_{kj}\right)\\ &= \sum_{k=1}^{n} a_{ik}\cdot 1\\ &= 1,\end{aligned}$$

where the second and third equalities hold because B and A are row stochastic. It is obvious that AB is row stochastic. Suppose A is an $m \times n$ column stochastic matrix and B is an $n \times q$ column stochastic matrix. Then

$$\sum_{i=1}^{m}\left(\sum_{k=1}^{n} a_{ik}b_{kj}\right)$$

is the jth column sum of AB, and

$$\begin{aligned}\sum_{i=1}^{m}\left(\sum_{k=1}^{n} a_{ik}b_{kj}\right) &= \sum_{k=1}^{n}\left(\sum_{i=1}^{m} a_{ik}\right)b_{kj}\\ &= \sum_{k=1}^{n} 1\cdot b_{kj}\\ &= 1.\end{aligned}$$

Hence AB is a column stochastic matrix. If A and B are both doubly stochastic matrices, the fact that AB is doubly stochastic follows from the two preceding results.

16 From Exercise 16, Section 1.3, we know that $A(f)$ is the incidence matrix for a 1-1 function from an n-element set to itself if and only if each row and each column of $A(f)$ contains precisely one 1 and $n - 1$ zeros. Also, we know that an incidence matrix with one 1 in each column uniquely determines a function. Hence the results follow.

22 By Theorem 5.2, the term rank of A is equal to the minimal number of lines in A that contain all the 1's in A. But every line of A is a line of A^T, and vice versa. Hence the term rank of A^T is equal to the term rank of A.

23 The matrix A must have exactly two 1's in each row and in each column. We know that such a matrix has a diagonal of 1's, and thus it must have term rank n.

26 Define t subsets $T_1, \ldots, T_t$ of the set of integers $\{s + 1, \ldots, m\}$ as follows: an integer $i \in T_j$ if and only if the (i, j) entry of A is 1. No T_j can be empty, for if this were the case, all the 1's in A could be covered by the first s rows and the $t - 1$ columns numbered $1, 2, \ldots, j - 1, j + 1, \ldots, t$. Then the 1's in A could be covered by $s + t - 1 = \ell - 1$ lines, contradicting the minimality of ℓ. Now suppose no SDR exists for the sets $T_1, \ldots, T_t$. Then by Theorem 5.1, there exist k of the sets, say $T_{j_1}, \ldots, T_{j_k}$, such that the union of these sets contains fewer than k elements. Hence in columns $j_1, \ldots, j_k$, the 1's appear in fewer than k rows, say rows $i_1, \ldots, i_q$, $1 \leq q < k$. Consider the following set of lines in A: rows $1, \ldots, s$, rows $i_1, \ldots, i_q$, and the $t - k$ columns obtained by deleting columns $j_1, \ldots, j_k$ from the first t columns. This set of lines contains all the 1's in A. But there are

$$\begin{aligned} s + q + t - k &= s + t - (k - q) \\ &= \ell - (k - q) \\ &< \ell \end{aligned}$$

lines in this set, again contradicting the minimality of ℓ. Hence the sets $T_1, \ldots, T_t$ possess an SDR, so A_3 contains a diagonal of t ones.

27 We want to find the minimal number of lines which contain all the 1's in the matrix. By Theorem 5.2, we want to find the term rank k of the matrix. First note that at the intersections of columns 2, 3, 4, and 7 with rows 1, 4, 5, and 6 is a 4×4 submatrix of 0's, where $4 + 4 = 8 = 7 + 1$. Hence by Theorem 4.2, every diagonal of the matrix contains a 0. Therefore $k < 7$. Now observe that the 6×6 submatrix at the intersections of rows 1, 2, 3, 5, 6, 7 with columns 1, 2, 3, 5, 6, 7 has a diagonal of 1's. Hence $k = 6$ and we can conclude that the minimum number of people who should receive psychiatric counselling is 6.

2.6 *Quiz*

1 False. The system

$$\begin{aligned} x_1 + x_2 + x_3 &= 0 \\ x_1 + x_2 + x_3 &= 1 \end{aligned}$$

has no solution.

2 True. If $x = (x_1, x_2)$ is any vector in R^2, then $x = cu + dv$ where $c = \frac{1}{5}(x_1 + x_2)$ and $d = \frac{1}{5}(2x_2 - 3x_1)$.

3 True. *Any* vector $x = (x_1, x_2)$ in R^2 can be written as a linear combination of the vectors $(1, -1)$ and $(0, 1)$. In fact, writing

$$\begin{aligned}(x_1, x_2) &= c(1, -1) + d(0, 1)\\ &= (c, -c) + (0, d)\\ &= (c, -c + d)\end{aligned}$$

we get $c = x_1$ and $d = x_1 + x_2$. Since this is true for any vector in R^2, it is true for the solution vectors of the given equation.

4 False. The solution vectors to the given systems are respectively $(\frac{3}{2}, -\frac{1}{2})$ and $(1, -1)$ and hence the systems are not equivalent.

5 False. Let $A = \begin{bmatrix} 1 & 1 & 1 \\ 1 & 0 & 0 \end{bmatrix}$. Then $(0, 1, -1)$ is a solution to the system $Ax = 0_2$.

6 True. See Definition 6.3.

7 False. The first non-zero element in the second row is not 1.

8 True. The matrix in Question 7 is obtained by multiplying the second row of the matrix in Question 6 by -1.

9 False. If the zero matrix is the Hermite normal form of a non-zero matrix A, then they are row equivalent. But this is not possible, since elementary row operations cannot change a zero matrix to a non-zero matrix.

10 True. Using the result of Exercise 12 in this section, if A and B are two 2×2 matrices and A is equivalent to B, then the row rank of A is equal to the row rank of B. Now, the row rank of A is either 0 (in which case $A = B = 0_{22}$), 2 (in which case $A = B = I_2$), or 1. If the row rank of A and B is 1, they are of the form

$$A = \begin{bmatrix} 1 & a \\ 0 & 0 \end{bmatrix}, \qquad B = \begin{bmatrix} 1 & b \\ 0 & 0 \end{bmatrix}.$$

Since A and B are row equivalent, it is clear that they can be equivalent by a type (iii) operation only, i.e., multiplication of row 1 by a constant. But in order to preserve the 1 in the (1, 1) positions of A and B, that constant must be a 1. Hence $a = b$ and $A = B$.

2.6 *Exercises*

2 (e)
$$A = \begin{bmatrix} 2 & 3 & -4 & 1 & 0 \\ 1 & 0 & -2 & 1 & -1 \\ 2 & -7 & 0 & 0 & 1 \end{bmatrix},$$
$$x = (x_1, x_2, x_3, x_4, x_5),$$

and

$$b = (0, 0, 0).$$

3 (i) Add -2, -3, and -4 times row 1 to rows 2, 3, and 4, respectively, to produce the matrix

$$A = \begin{bmatrix} 1 & 4 & 5 & 3 \\ 0 & -5 & -5 & -5 \\ 0 & -10 & -10 & -10 \\ 0 & -15 & -15 & -15 \end{bmatrix}.$$

Add -2 times row 2 to row 3 and -3 times row 2 to row 4, and then multiply row 2 by $-\frac{1}{5}$ to produce

$$A_1 = \begin{bmatrix} 1 & 4 & 5 & 3 \\ 0 & 1 & 1 & 1 \\ 0 & 0 & 0 & 0 \\ 0 & 0 & 0 & 0 \end{bmatrix}.$$

Adding -4 times row 2 to row 1 reduces the matrix to Hermite normal form:

$$H = \begin{bmatrix} 1 & 0 & 1 & -1 \\ 0 & 1 & 1 & 1 \\ 0 & 0 & 0 & 0 \\ 0 & 0 & 0 & 0 \end{bmatrix}.$$

5 (a) Let $\alpha x_1 + \beta x_2 + \gamma x_3 = c$ be the equation of the required plane. Then each of the three given vectors satisfies this equation and we get the following system:

$$\begin{aligned} \gamma &= c, \\ -\alpha + 3\beta + 5\gamma &= c, \\ \alpha - \beta \qquad &= c. \end{aligned}$$

Adding the second and third equations, we have

$$2\beta + 5\gamma = 2c,$$

or

$$\beta = \tfrac{1}{2}(2c - 5\gamma) = \tfrac{1}{2}(2c - 5c) = -\tfrac{3}{2}c.$$

Then the third equation gives

$$\alpha = c + \beta = c - \tfrac{3}{2}c = -\tfrac{1}{2}c.$$

Thus the equation of the plane containing the given vectors is

$$x_1 + 3x_2 - 2x_3 = -2.$$

(b) The vector $0_3 = (0, 0, 0)$ does not satisfy the equation of the plane in (a). Hence the three vectors in (a) together with the zero vector do not lie in a plane.

6 Using the notation of Example 6.7, we have

$$a_{11} = 5, \quad a_{12} = 7, \quad a_{21} = 6, \quad a_{22} = 4, \quad s_1 = 10, \quad s_2 = 15,$$
$$b_1 = 12, \quad b_2 = 13.$$

If x_{ij} denotes the amount shipped from h_i to o_j, $i = 1, 2$, $j = 1, 2$, then we have

(1) $$x_{11} + x_{12} = 10,$$
$$x_{21} + x_{22} = 15$$

and

(2) $$x_{11} + x_{21} = 12,$$
$$x_{12} + x_{22} = 13.$$

The total cost of shipping is

(3) $$c = 5x_{11} + 7x_{12} + 6x_{21} + 4x_{22}.$$

Therefore the problem amounts to finding all nonnegative solutions to the equations (1) and (2) and then finding which of these solutions cause the function c in (3) to have the least value.

7 (a) If $c_1(1, 0) + c_2(-1, -1) = (0, 0)$ then $(c_1 - c_2, -c_2) = (0, 0)$. But this implies that $c_2 = 0$ and $c_1 = 0$. Thus the two given vectors are linearly independent.

(b) Let c be any non-zero number. Then

$$0(1, 0) + 0(-1, -1) + c(0, 0) = (0, 0).$$

Hence the given vectors are linearly dependent.

(c) These three vectors are linearly independent because

$$c_1(1, 0, 0) + c_2(0, 1, 0) + c_3(0, 0, 1) = (0, 0, 0)$$

implies

$$(c_1, c_2, c_3) = (0, 0, 0).$$

But this is possible if and only if $c_1 = c_2 = c_3 = 0$.

(d) The given vectors are linearly independent. For if

$$c_1(1, 0, 0, 0) + c_2(0, -1, 0, 0) + c_3(0, 0, 1, 0) + c_4(0, 0, 0, -1) = 0$$

then

$$(c_1, -c_2, c_3, -c_4) = (0, 0, 0, 0),$$

which implies

$$c_1 = c_2 = c_3 = c_4 = 0.$$

(e) Since

$$(1, 1) + (2, 2) - (3, 3) = (0, 0),$$

the given vectors are linearly dependent.

8 (i) Since for any pair of vectors u, v in R^n we have $u + v = v + u$, a linear combination $\sum_{i=1}^{m} c_i u_i$ results in the same vector if we interchange any two of the summands. Hence the linear dependence or independence of U is not affected by an interchange of two vectors.

(ii) Without loss of generality, assume u_1 is multiplied by a non-zero constant, and let

$$U_1 = \{cu_1, u_2, \ldots, u_m\}, \qquad c \neq 0.$$

We know that U is linearly dependent if and only if there exist constants $c_1, \ldots, c_m$, not all zero, such that

$$\sum_{i=1}^{m} c_i u_i = 0.$$

We can also find a constant $c_1' = \dfrac{c_1}{c}$ (this is possible since $c \neq 0$) such that

$$\begin{aligned} c_1' c u_1 + \sum_{i=2}^{m} c_i u_i &= c_1 u_1 + \sum_{i=2}^{m} c_i u_i \\ &= \sum_{i=1}^{m} c_i u_i \\ &= 0, \end{aligned}$$

making U_1 linearly dependent also. Now suppose U is linearly independent. Taking U_1 as above, suppose there exist constants $c_1, \ldots, c_m$ such that

$$c_1(cu_1) + c_2 u_2 + \cdots + c_m u_m = 0.$$

But then the linear independence of U implies $c_1 c = c_2 = \cdots = c_m = 0$, and since $c \neq 0$; we have $c_1 = 0$. Hence U_1 is linearly independent. Since $u_1 = \dfrac{1}{c}(cu_1)$, the linear dependence or independence of U_1 likewise implies that of U by interchanging the roles of U and U_1.

(iii) Let c be a non-zero constant and let

$$U_1 = \{u_1, \ldots, u_j, \ldots, u_{k-1}, cu_j + u_k, u_{k+1}, \ldots, u_m\}.$$

Then U_1 is linearly dependent if and only if

$$c_1 u_1 + \cdots + c_j u_j + \cdots + c_{k-1} u_{k-1} + c_k(cu_j + u_k) + c_{k+1} u_{k+1} + \cdots + c_m u_m = 0, \tag{4}$$

where not all of $c_1, \ldots, c_m$ are zero. Now (4) can be written

$$c_1 u_1 + \cdots + (c_j + cc_k) u_j + c_{j+1} u_{j+1} + \cdots + c_k u_k + \cdots + c_m u_m = 0. \tag{5}$$

If all the coefficients of the u_i in (5) are 0, then $c_k = 0$ and $c_j + cc_k = 0$, which implies $c_j = 0$. Hence all of $c_1, \ldots, c_m$ are zero. But this contradicts our assumption in (4). Therefore (5) shows that U is linearly dependent.

On the other hand, let us say U is linearly independent and $c_1, \ldots, c_m$ are constants such that equation (4) holds. Since (4) can be rewritten in the form (5), the linear independence of U implies

$$c_1 = \cdots = c_j + cc_k = c_{j+1} = \cdots = c_k = \cdots = c_m = 0.$$

Then, as shown above, $c_1 = \cdots = c_m = 0$, i.e., U_1 is also linearly independent. Since $c_j = (c_j + cc_k) - cc_k$, the linear independence or dependence of U_1 implies that of U by interchanging the roles of U and U_1.

9 (a) The equation $c_1(1, 1) + c_2(1, -1) = 0$ implies $c_1 + c_2 = 0$ and $c_1 - c_2 = 0$. Adding these two equations we get $c_1 = 0$ and $c_2 = 0$. Thus the two vectors are linearly independent and so $\rho(U) = 2$.

(b) This set contains the zero vector and so is linearly dependent (see Exercise 7 (b)). Now the vector $(2, 2) = 2(1, 1)$ and therefore the vectors $(2, 2)$ and $(1, -1)$ are linearly independent (see Exercises 8(ii) and 9(a)). Thus $\rho(U) = 2$.

(c) Observe that

$$(1, 0, 0) + (0, 1, 0) - (1, 1, 0) = 0.$$

Therefore these three vectors are linearly dependent, while the first two vectors are linearly independent (see Exercise 7(c)). Therefore $\rho(U) = 2$.

(d) The first three vectors are linearly independent (see Exercise 7(c)), while any four vectors make a linearly dependent set because

$$(1, 0, 0) + 2(0, 0, 1) - (1, 0, 2) = 0$$

and

$$(0, 1, 0) - (0, 0, 1) + (0, -1, 1) = 0.$$

Thus $\rho(U) = 3$.

(e) Notice that

$$(1, 0) + (0, 1) = (1, 1) = \tfrac{1}{2}(2, 2) = \tfrac{1}{3}(3, 3) = -(-1, -1).$$

Therefore since $(1, 0)$ and $(0, 1)$ are linearly independent, we have $\rho(U) = 2$.

10 Suppose $u_1 = 0$. Then for any non-zero constant c,

$$cu_1 + 0u_2 + 0u_3 + \cdots + 0u_m = 0.$$

Hence U is linearly dependent, i.e., the number of linearly independent vectors in U cannot exceed $m - 1$.

11 In each of the following parts, A will always denote the given matrix.

(a) $\rho(A) = 2$ because $(1, 0)$ and $(0, 1)$ are linearly independent.

(b) $\rho(A) = 2$ (see Exercise 7 (c)).

(c) $A_{(3)} = 0$, therefore $\rho(A) < 3$ (see Exercise 10). However, $A_{(1)}$ and $A_{(2)}$ are linearly independent because $c_1A_{(1)} + c_2A_{(2)} = 0$ implies $(0, c_1, c_2, -c_2) = 0$, which is possible only when $c_1 = c_2 = 0$. Hence $\rho(A) = 2$.

(d) The number of non-zero rows is 3, therefore $\rho(A) \leq 3$. If $c_1A_{(1)} + c_2A_{(2)} + c_3A_{(3)} = 0$ then

$$(0, c_1, 3c_1, c_1, c_2, c_3, 4c_1 + 3c_2 + 2c_3) = 0,$$

which implies $c_1 = c_2 = c_3 = 0$. Thus $\rho(A) = 3$.

(e) The number of non-zero rows is 4, therefore $\rho(A) \leq 4$. If

$$c_1A_{(1)} + c_2A_{(2)} + c_3A_{(3)} + c_4A_{(4)} = 0, \tag{6}$$

then, in particular, each of the 3rd, 5th, 8th, and 12th components in the left side of (6) is 0. That is, $c_1 = c_2 = c_3 = c_4 = 0$, and thus $\rho(A) = 4$.

(f) Since $A_{(1)}$, $A_{(2)}$, and $A_{(3)}$ are the only non-zero rows and since they are linearly independent, $\rho(A) = 3$.

(g) Since $A_{(1)} - A_{(2)} + A_{(3)} = 0$, $\rho(A) \leq 2$. If $c_1A_{(1)} + c_3A_{(3)} = 0$, then $(c_1, c_1, c_1 + c_3) = 0$ which implies $c_1 = c_3 = 0$. Hence $\rho(A) = 2$.

12 A type I elementary operation interchanges two rows and a type III elementary operation multiplies a row in the set

$$U = \{A_{(1)}, A_{(2)}, \ldots, A_{(m)}\}$$

by a non-zero constant.

By Exercise 8, parts (i) and (ii), neither of these operations changes the linear dependence or independence of the vectors in U, and hence the rank of the matrix is not affected.

13 Let r be the number of non-zero rows in H, and suppose the vectors $e_1, \ldots, e_r$ appear in columns $n_1, \ldots, n_r$. We know that $\rho(H) \leq r$. If $c_1A_{(1)} + \cdots + c_rA_{(r)}$ is 0, then the components $n_1, \ldots, n_r$ must be 0, i.e., each of the numbers $c_1, \ldots, c_r$ is 0. Therefore $A_{(1)}, \ldots, A_{(r)}$ are linearly independent and hence $\rho(H) = r$.

14 The matrix H is obtained from A by performing repeated elementary row operations on A. Therefore if $A_1, A_2, \ldots, H$ are the matrices obtained at successive steps of the process (see proof of Theorem 6.1), then by Exercise 12 we get $\rho(A) = \rho(A_1) = \cdots = \rho(H)$.

15 (a) Add -1 times row 1 to row 2 and -2 times row 1 to row 3; add row 2 to row 1 and -1 times row 2 to rows 3 and 4; multiply row 2 by -1 and add row 3 to it; then multiplying row 3 by -1

reduces the matrix to Hermite normal form:

$$\begin{bmatrix} 1 & 0 & 2 & 0 \\ 0 & 1 & -1 & 0 \\ 0 & 0 & 0 & 1 \\ 0 & 0 & 0 & 0 \end{bmatrix}.$$

Clearly $\rho(A) = 3$.

16 Let H be the hermite normal form of A. Since $\rho(A) = n$, we know (see Exercise 14) that $\rho(H) = n$. Hence there are n columns, say $n_1, \ldots, n_n$, such that $H^{(n_i)} = e_i$. But H has only n columns, therefore $n_1 = 1$, $n_2 = 2, \ldots, n_n = n$. Thus column i of H is the n-vector e_i, $i = 1, \ldots, n$, i.e., $H = I_n$. Hence A is row equivalent to I_n.

17 Since $\rho(A) = n$, we know by the previous exercise that the Hermite normal form of A is I_n. Hence the system $Ax = 0$ is equivalent to the system $I_n x = 0$; but the only solution to this latter system is the zero vector.

18 Denote each $u_i \in R^m$ by $u_i = (u_{i1}, \ldots, u_{im})$, $i = 1, \ldots, p$. Then the matrix A defined in the hint has the form

$$A = \begin{bmatrix} u_{11} & \ldots & u_{1m} \\ \vdots & & \vdots \\ u_{p1} & \ldots & u_{pm} \end{bmatrix},$$

i.e., A is $p \times m$. If U is a linearly independent set, then $\rho(A) = p$ by definition. Therefore the Hermite normal form of A must include $e_1, \ldots, e_p$ among its columns. But this implies $p \leq m$, contradicting the hypothesis that $p > m$. Hence U must be linearly dependent.

2.7 *Quiz*

1 False. The homogeneous system associated with the given equations is $Ax = 0$ where

$$A = \begin{bmatrix} 2 & 0 & 1 \\ 1 & 1 & 0 \end{bmatrix}.$$

2 True. The zero vector is always a solution of a homogeneous system of linear equations.

3 False. The Hermite normal form of the augmented matrix is

$$\begin{bmatrix} 1 & 1 & 1 & 1 \\ 0 & 0 & 0 & 0 \\ 0 & 0 & 0 & 0 \end{bmatrix},$$

and therefore any vector of the form $(x_1, x_2, 1 - (x_1 + x_2))$ is a solution, where x_1 and x_2 are any numbers.

4 False. The Hermite normal form of the given matrix is

$$\begin{bmatrix} 1 & 1 & 0 \\ 0 & 0 & 1 \\ 0 & 0 & 0 \end{bmatrix},$$

which may be obtained by performing the following sequences of elementary row operations: (a) add -1 times row 1 to row 2; (b) multiply row 2 by -1; (c) add -1 times row 2 to rows 1 and 3.

5 False. The Hermite normal form of J_n is the matrix with first row $(1, 1, \ldots, 1)$ and every other entry 0.

6 True. Add -1 times row 1 to row 2 of the first matrix and then multiply row 2 by $\frac{1}{2}$. Then adding row 2 to row 1 gives the other matrix.

7 True. Since I_n is an elementary matrix, we can write $A = I_nA$.

8 True. If A is row equivalent to B and B is row equivalent to C then there exist elementary matrices $E_1, \ldots, E_k$ and $F_1, \ldots, F_\ell$ such that $A = E_k \cdots E_1B$ and $B = F_\ell \cdots F_1C$. But this implies

$$A = E_k \cdots E_1 \cdot F_\ell \cdots F_1C.$$

9 False. Let

$$A = \begin{bmatrix} 0 & -1 \\ 1 & 0 \end{bmatrix}$$

and

$$B = I_2.$$

Then

$$A^{-1} = \begin{bmatrix} 0 & 1 \\ -1 & 0 \end{bmatrix},$$

and A and B are nonsingular. However,

$$A^2 + B^2 = \begin{bmatrix} -1 & 0 \\ 0 & -1 \end{bmatrix} + \begin{bmatrix} 1 & 0 \\ 0 & 1 \end{bmatrix} = \begin{bmatrix} 0 & 0 \\ 0 & 0 \end{bmatrix},$$

which is indeed singular.

10 True. It is easily verified that the product of the two matrices is I_2.

2.7 *Exercises*

1 (f) Perform the following sequence of elementary row operations on the augmented matrix

$$\begin{bmatrix} 1 & 2 & -2 & 1 & -1 & -1 \\ 2 & 3 & 1 & -4 & 2 & -1 \\ 3 & 5 & -1 & 0 & 1 & -1 \\ 1 & 1 & 3 & -2 & 3 & -1 \end{bmatrix}.$$

(i) Add -2 times row 1 to row 2, -3 times row 1 to row 3,

and -1 times row 1 to row 4.
(ii) Multiply row 2 by -1.
(iii) Add row 2 to rows 3 and 4, and add -2 times row 2 to row 1.
(iv) Multiply row 3 by $\frac{1}{3}$.
(v) Add 11 times row 3 to row 1, -6 times row 3 to row 2, and -3 times row 3 to row 4.

The resulting matrix is

$$\begin{bmatrix} 1 & 0 & 8 & 0 & 7 & \frac{14}{3} \\ 0 & 1 & -5 & 0 & -4 & -3 \\ 0 & 0 & 0 & 1 & 0 & \frac{1}{3} \\ 0 & 0 & 0 & 0 & 0 & -2 \end{bmatrix},$$

which shows that the row rank of the coefficient matrix is 3. Since $c_4 \neq 0$, the given system of equations has no solution.

3 The matrix $I_n + P$ is reduced to the matrix I_n by performing $n - 1$ elementary row operations, namely, adding -1 times row n to row $n - 1$, then adding -1 times row $n - 1$ to row $n - 2$, etc. The same operations when performed on I_n in the same order yield the matrix whose (i, j) entry is 0 if $i > j$ and $(-1)^{i+j}$ if $i \leq j$, and this matrix is precisely $(I_n + P)^{-1}$.

4 (b) The following sequence of elementary row operations reduces the given matrix to I_4.
(i) Interchange the first and second rows.
(ii) Add 2 times row 1 to row 2, 4 times row 1 to row 3, and add row 1 to row 4.
(iii) Add -1 times row 2 to row 3.
(iv) Add -2 times row 3 to row 1, and -4 times row 3 to row 2.
(v) Add -3 times row 4 to row 1, -7 times row 4 to row 2, and add 2 times row 4 to row 3.

This sequence of operations when performed on I_4 produces the matrix

$$\begin{bmatrix} 2 & -6 & -2 & -3 \\ 5 & -13 & -4 & -7 \\ -1 & 4 & 1 & 2 \\ 0 & 1 & 0 & 1 \end{bmatrix},$$

which is the inverse of the given matrix.

10 In Theorem 7.6 it was proved that if A is nonsingular then $A^{-1} = E_k E_{k-1} \cdots E_1$ where $E_1, \ldots, E_k$ are elementary matrices. From this it follows that

$$\begin{aligned} A &= (A^{-1})^{-1} \\ &= (E_k E_{k-1} \cdots E_1)^{-1} \\ &= (E_{k-1} E_{k-2} \cdots E_1)^{-1} E_k^{-1} \\ &\vdots \\ &= E_1^{-1} E_2^{-1} \cdots E_k^{-1}, \end{aligned}$$

where $E_1^{-1}, \ldots, E_k^{-1}$ are also elementary matrices.

14 Suppose A and B are nonsingular. Then, by Theorem 7.3(a), AB is nonsingular. Conversely, suppose A is singular. Then the last row of H, the Hermite normal form of A, consists entirely of zeros. Now, $A = E_k \cdots E_1 H$ for some elementary matrices $E_1, \ldots, E_k$. Therefore $AB = E_k \cdots E_1 HB$, i.e., $E_1^{-1} E_2^{-1} \cdots E_k^{-1}(AB) = HB$. Now, if AB were nonsingular then, again by Theorem 7.3(a), $E_1^{-1} \cdots E_k^{-1} AB$ would be nonsingular, i.e., HB would be nonsingular. But this is impossible because the last row of HB is zero. Thus if A is singular then so is AB. Also, if B is singular then B^T is singular (Theorem 7.3 (b)), and so the last row in the Hermite normal form of B^T is zero. Using the same argument as above, it can be proved that $(AB)^T = B^T A^T$ is singular, hence AB is singular. Thus if AB is nonsingular then both A and B are nonsingular, since we have proved the contrapositive to be true.

15 If $\prod_{i=1}^{n} a_{ii} \neq 0$ then $a_{ii} \neq 0$, $i = 1, 2, \ldots, n$. Multiply the ith row of A by $\dfrac{1}{a_{ii}}$, $i = 1, \ldots, n$. Add suitable multiples of row 1 to the other rows to make the first entry in each of the last $n - 1$ rows zero. Then add suitable multiples of row 2 to the last $(n - 2)$ rows so that the second entry in each of the rows $3, \ldots, n$ is zero. The continuation of this process reduces A to I_n. Therefore, by Theorem 7.6, A is nonsingular. To obtain A^{-1} we perform the same sequence of elementary operations on I_n, and so the (i, i) entry of A^{-1} is $\dfrac{1}{a_{ii}}$, $i = 1, 2, \ldots, n$.

17 Suppose A is a nonsingular lower triangular matrix. The elementary row operations we perform on A to reduce it to I_n consist of either multiplying a row by a non-zero number or adding a certain multiple of a row to some row which is below it, and both of these operations preserve the lower triangular property. Since I_n is lower triangular, the corresponding elementary matrices are lower triangular, and hence so are their inverses. The same argument applies for an upper triangular matrix. The answer then follows from the answer to Exercise 10.

19 Using Theorem 3.2 in Section 2.3, we have $A(f)A(f^{-1}) = A(ff^{-1}) = A(e) = I_n$, where e denotes the identity permutation on X. This implies that $A(f)$ is nonsingular and $(A(f))^{-1} = A(f^{-1})$.

22 (b) Since

$$AE = ((AE)^T)^T = (E^T A^T)^T$$

and E^T is again an elementary matrix, $E^T A^T$ results from A^T by performing elementary row operations on A^T. Hence $(E^T A^T)^T$ results from $(A^T)^T = A$ by performing elementary column operations on $(A^T)^T = A$. This also can be verified by direct computation of AE, where E is an elementary matrix.

23 Let B be the Hermite normal form of A. Then columns $n_1, \ldots, n_r$ of B will be $e_1, \ldots, e_r$, respectively, and rows $r + 1, \ldots, n$ will consist entirely of 0's. Perform a sequence of elementary column operations of type II on B to make every entry in the first row except the $(1, n_1)$ entry equal to 0. Next, by type II elementary column operations reduce every entry in the second row except the $(2, n_2)$ entry to 0, and so forth. Finally, interchange columns 1 and n_1, 2 and n_2, ..., r and n_r in order. The resulting matrix is D.

24 Let P be the elementary matrix such that $PA = B$ as in Exercise 23. Let Q be the product of elementary matrices corresponding to the column operations described above, in the same order. Then $(PA)Q = D$.

25 (b) The given matrix is nonsingular (see Exercise 4 (d)). Therefore its rank is 5 and hence its required form is I_5.

---

chapter **3**

3.1 *Quiz*

1 False. By Definition 1.1 we know that the equation of the line through a in the direction of u is $x = a + tu$, t a real number.

2 False. Consider the two non-zero vectors $(1, 0)$ and $(0, 1)$, whose inner product is zero.

3 True. Their inner product is zero:

$$((2, 3), (-3, 2)) = (2)(-3) + (3)(2) = 0.$$

This is the requirement for orthogonality given by Definition 1.3.

4 True. See the discussion preceding and including equation (13).

5 True. Given a hyperplane Π in R^n defined by $(x, u) = c$, the two open half-spaces correspond to the inequalities $(x, u) > c$ and $(x, u) < c$. Since no element x of R^n can satisfy both inequalities simultaneously, the intersection of the two half-spaces must be the null set.

6 True. If Π is the hyperplane in Question 5, the two closed half-spaces are $(x, u) \geq c$ and $(x, u) \leq c$. An element $x \in R^n$ can satisfy both inequalities if and only if $(x, u) = c$; hence the intersection of $(x, u) \geq c$ and $(x, u) \leq c$ is the hyperplane Π.

7 True. Again, if Π is the hyperplane in Question 5, the positive open half-space of Π is $(x, u) > c$ and the positive closed half-space of Π is $(x, u) \geq c$. An element $x \in R^n$ can satisfy both inequalities if and only if $(x, u) > c$; hence the intersection of the two half-spaces is the positive open half-space.

8 True. If Π is the hyperplane in Question 5, the positive open half-space of Π is $(x, u) > c$ and the negative closed half-space of Π is $(x, u) \leq c$. Since no element x of R^n can satisfy both inequalities, the intersection of $(x, u) > c$ and $(x, u) \leq c$ is the null set.

9 False. Consider the two line segments s_1 and s_2 joining the points $(0, 0)$, $(1, 1)$ and $(0, 1)$, $(0, 2)$, respectively. From the graph below, it can be seen that these segments do not intersect and are not parallel.

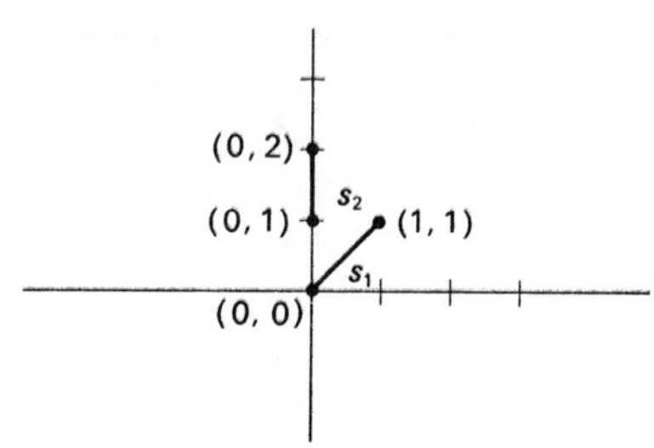

10 False. We know that any point x lying between two points a and b can be expressed in the form

$$x = (1 - t)a + tb, \qquad 0 \leq t \leq 1.$$

Letting $(2, 2)$ and $(3, 3)$ be the points a and b, respectively, we have

$$x = (1 - t)(2, 2) + t(3, 3).$$

Now, if the point $(1, 1)$ did lie between a and b then the value of t would have to be between 0 and 1. But

$$\begin{aligned}(1, 1) &= (1 - t)(2, 2) + t(3, 3) \\ &= (2(1 - t) + 3t, 2(1 - t) + 3t) \\ &= (2 + t, 2 + t);\end{aligned}$$

thus $2 + t = 1$ and $t = -1$. However this violates the condition that $0 \leq t \leq 1$. Therefore the point $(1, 1)$ does not lie between points $(2, 2)$ and $(3, 3)$.

3.1 *Exercises*

1 (e) $x = (1, 1, 1) + t(1, 1, 1)$.

(j) The fact that the line ℓ is parallel to the two planes means that it is orthogonal to their normal vectors u and v. We can rewrite the equations of the two given planes:

$$\begin{aligned}(u, x) &= 1, \quad \text{where} \quad u = (1, 1, 1); \\ (v, x) &= 0, \quad \text{where} \quad v = (1, 1, -1).\end{aligned}$$

Then since ℓ is orthogonal to both u and v, the components of a point $y = (y_1, y_2, y_3)$ on any vector in the direction of ℓ must satisfy the equations

$$y_1 + y_2 + y_3 = 0$$

and

$$y_1 + y_2 - y_3 = 0.$$

Adding the equations we find that $y_1 = -y_2$ and $y_3 = 0$. Hence the line ℓ is in the direction of the vector $w = (1, -1, 0)$. The parametric equation of the line is therefore

$$x = (1, 2, 3) + t(1, -1, 0).$$

(k) We already know from (j) that ℓ is in the direction of the vector $w = (1, -1, 0)$. Since ℓ is the line of intersection of the two planes, it must go through a point a which lies in both planes. Since $a = (a_1, a_2, a_3)$ must satisfy the two equations $x_1 + x_2 + x_3 = 1$ and $x_1 + x_2 - x_3 = 0$, we get

$$a_1 + a_2 + a_3 = 1,$$
$$a_1 + a_2 - a_3 = 0.$$

Adding the above equations, we find that $a_2 = \frac{1}{2} - a_1$ and $a_3 = \frac{1}{2}$. Thus we can take a to be the point $a = (1, -\frac{1}{2}, \frac{1}{2})$. Hence the parametric equation for ℓ is

$$x = (1, -\tfrac{1}{2}, \tfrac{1}{2}) + t(1, -1, 0).$$

(l) The line $x_1 + x_2 = -1$ has as its normal $u = (1, 1)$. Thus our new line ℓ will be orthogonal to u and pass through the point $(10, 10)$. Now, ℓ will be in the direction of any vector that is orthogonal to u, for example, $w = (1, -1)$. Therefore we can write the parametric equation of ℓ:

$$x = (10, 10) + t(1, -1),$$

where t is real.

2 (c)

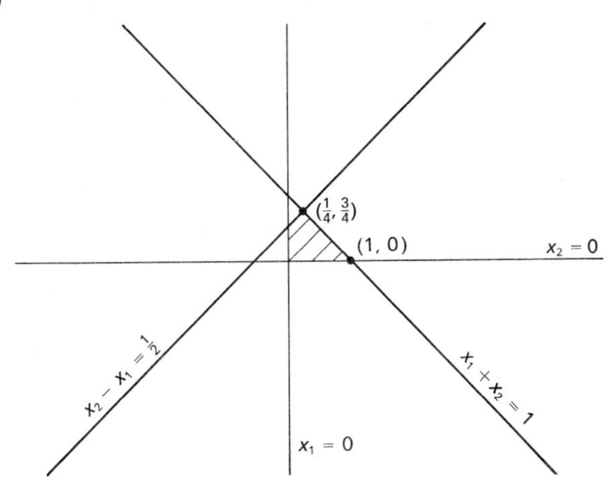

4 We know from Definition 1.2 that if u and v are in R^2,

$$(u, v) = u_1v_1 + u_2v_2.$$

(a) When $u = v$,

$$\begin{aligned}(u, u) &= u_1u_1 + u_2u_2\\ &= u_1^2 + u_2^2.\end{aligned}$$

Since both of u_1^2 and u_2^2 are nonnegative, we know that their sum will also be nonnegative, that is,

$$u_1^2 + u_2^2 \geq 0.$$

Hence $(u, u) = 0$ if and only if $u_1 = u_2 = 0$.

(b)

$$\begin{aligned}(u, v) &= u_1v_1 + u_2v_2\\ &= v_1u_1 + v_2u_2\\ &= (v, u).\end{aligned}$$

(c)

$$\begin{aligned}(u, v + w) &= u_1(v_1 + w_1) + u_2(v_2 + w_2)\\ &= u_1v_1 + u_1w_1 + u_2v_2 + u_2w_2\\ &= (u_1v_1 + u_2v_2) + (u_1w_1 + u_2w_2)\\ &= (u, v) + (u, w).\end{aligned}$$

5 We know from Example 1.7 that the distance from the point $a = (a_1, a_2)$ to the line ℓ whose equation is $(x, u) = c$ is

$$d = \left[\frac{(c - (a, u))^2}{(u, u)}\right]^{1/2}.$$

(c) Following the procedure in Example 1.5, we can find the point of intersection of the two given lines by solving

$$(2, 3) + t_1(1, -1) = (1, 4) + t_2(2, 2)$$

for t_1 and t_2 real. Equating components,

$$\begin{aligned}2 + t_1 &= 1 + 2t_2,\\ 3 - t_1 &= 4 + 2t_2,\end{aligned}$$

or simplifying,

$$\begin{aligned}2t_2 - t_1 &= 1,\\ 2t_2 + t_1 &= -1.\end{aligned}$$

Solving the two equations simultaneously, we get

$$t_2 = 0 \quad \text{and} \quad t_1 = -1.$$

Therefore the point of intersection of the two lines is

$$(1, 4) + 0(2, 2) = (1, 4).$$

Then

$$d = \left(\frac{[3 - ((1, 4), (2, \frac{1}{2}))]^2}{((2, \frac{1}{2}), (2, \frac{1}{2}))}\right)^{1/2}$$

$$= \left(\frac{[3 - (1 \cdot 2 + 4 \cdot \frac{1}{2})]^2}{(2 \cdot 2 + \frac{1}{2} \cdot \frac{1}{2})}\right)^{1/2}$$
$$= \left(\frac{[3 - 4]^2}{4 + \frac{1}{4}}\right)^{1/2}$$
$$= \left(\frac{(-1)^2}{\frac{17}{4}}\right)^{1/2}$$
$$= \frac{2}{\sqrt{17}}.$$

6 (a) We know that the line ℓ given by $x = a + tv$ goes through the point a in the direction of the vector v. We want first to rewrite the equation of the line in the form $(x, u) = c$. We know that the inner product of v with the normal u to ℓ must be zero,

$$(u, v) = u_1v_1 + u_2v_2 = 0;$$

hence choose $u = (v_2, -v_1)$. The equation of ℓ will therefore be of the form

$$(x, u) = v_2x_1 - v_1x_2 = c.$$

We already know that the point $a = (a_1, a_2)$ lies on ℓ, therefore we can find the value of c:

$$c = v_2a_1 - v_1a_2.$$

We now must find the point z on ℓ that is closest to the origin, i.e., that point which lies at the intersection of ℓ and ℓ', the perpendicular to ℓ through (0, 0). Since ℓ' is in the direction of u and goes through (0, 0), its equation is

$$x = (0, 0) + tu$$
$$= tu.$$

The point of intersection of ℓ and ℓ' is determined by that value of t for which $x = tu$ satisfies the equation of the line ℓ:

$$(tu, u) = c.$$

We then know from Example 1.7 how to find the point of intersection:

$$z = (0, 0) + \left(\frac{c - ((0, 0), u)}{(u, u)}\right) u$$
$$= \left(\frac{(v_2a_1 - v_1a_2)}{v_2^2 + v_1^2}\right)(v_2, -v_1)$$
$$= \left(\frac{v_2a_1 - v_1a_2}{v_1^2 + v_2^2}\right)(v_2, -v_1)$$
$$= \left(\frac{v_2^2a_1 - v_1v_2a_2}{v_1^2 + v_2^2}, \frac{v_1^2a_2 - v_1v_2a_1}{v_1^2 + v_2^2}\right).$$

(b) Following the procedure in part (a), we have

$$x = (2, 3) + t(1, 1),$$

where $a = (2, 3)$ and $v = (1, 1)$, so let $u = (1, -1)$ and let

$$\begin{aligned} c &= v_2 a_1 - v_1 a_2 \\ &= 2 - 3 \\ &= -1. \end{aligned}$$

Thus we can rewrite the equation of the line to be $(x, u) = -1$, where $u = (1, -1)$. Then the point of intersection of the line with the perpendicular through the point (0, 0) is

$$\begin{aligned} z &= \left(\frac{-1}{(-1)^2 + 1^2}\right)(1, -1) \\ &= \frac{-1}{2}(1, -1) \\ &= (-\tfrac{1}{2}, \tfrac{1}{2}). \end{aligned}$$

(c) Again referring back to Example 1.7, we are given the line $(u, x) = 3$, where $u = (1, \frac{1}{2})$, but in this case we are given the point $(3, -2)$ rather than the origin. Hence the point of intersection is

$$\begin{aligned} z &= (3, -2) + \left(\frac{3 - ((3, -2), (1, \frac{1}{2}))}{((1, \frac{1}{2}), (1, \frac{1}{2}))}\right)(1, \tfrac{1}{2}) \\ &= (3, -2) + \left(\frac{3 - (3 - 1)}{1 + \frac{1}{4}}\right)(1, \tfrac{1}{2}) \\ &= (3, -2) + \left(\frac{1}{\frac{5}{4}}\right)(1, \tfrac{1}{2}) \\ &= (3, -2) + (\tfrac{4}{5}, \tfrac{2}{5}) \\ &= \left(\frac{19}{5}, \frac{-8}{5}\right). \end{aligned}$$

12 From the information given in the problem, we can form the inequalities

$$\begin{aligned} x_1 - x_2 &\geq 10, \\ x_1 - [100 - (x_1 + x_2)] &\leq 40, \\ [100 - (x_1 + x_2)] - x_2 &\leq 0, \end{aligned}$$

where x_1, x_2, and $[100 - (x_1 + x_2)]$ are the quantities of each of the ingredients x, y, z, respectively. We can rewrite the inequalities in the following form:

$$\begin{aligned} x_1 - x_2 &\geq 10, \\ 2x_1 + x_2 &\leq 140, \\ x_1 + 2x_2 &\geq 100. \end{aligned}$$

The function we wish to minimize (and maximize) is

$$\begin{aligned} f((x_1, x_2)) &= 6x_1 + 3x_2 + 4[100 - (x_1 + x_2)] \\ &= 2x_1 - x_2 + 400. \end{aligned}$$

Graphing the inequalities, we find that the region determined by them has vertices (40, 30), (50, 40), and (60, 20).

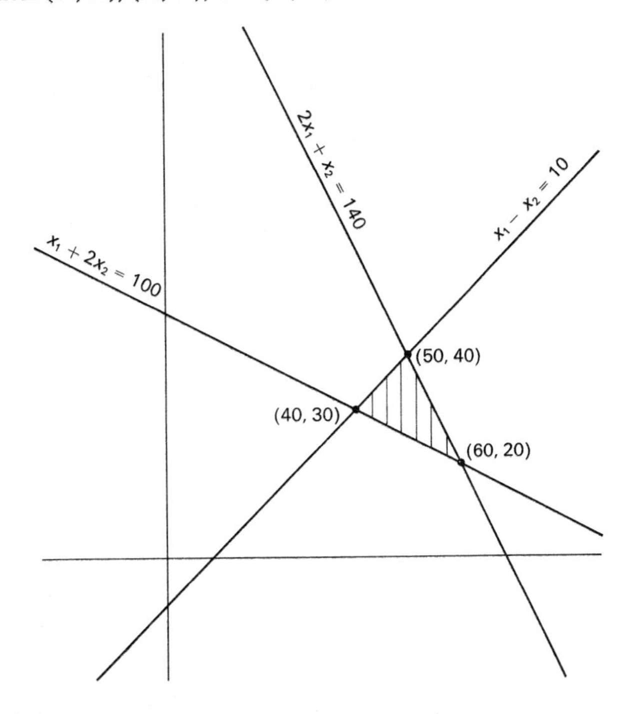

Then, testing f on each of the vertices of the region:

$$\begin{aligned} f((50, 40)) &= 100 - 40 + 400 \\ &= 460, \\ f((60, 20)) &= 120 - 20 + 400 \\ &= 500, \\ f((40, 30)) &= 80 - 30 + 400 \\ &= 450, \end{aligned}$$

we see that the least cost of the mixture is \$450 and that the greatest cost of the mixture is \$500.

14 (a) Let Π denote the hyperplane $(u, x) = c$. If both a and b lie in the closed positive half-space of Π, then $(u, a) = \alpha \geq c$ and $(u, b) = \beta \geq c$. If a point x lies on the line segment joining a and b, x can be written $x = (1 - t)a + tb$, where $0 \leq t \leq 1$.

Hence we have

$$\begin{aligned}(u, x) &= (u, (1-t)a + tb)\\ &= (1-t)(u, a) + t(u, b)\\ &= (1-t)\alpha + t\beta\\ &\geq (1-t)c + tc\\ &= c,\end{aligned}$$

that is, x is the closed positive half-space of Π. Since x was arbitrary, we know the entire line segment joining a and b lies in the closed positive half-space of Π.

(b) If a and b are in the open positive half-space of Π, then $(u, a) = \alpha > c$ and $(u, b) = \beta > c$. Then if x is on the line segment joining a and b, there is a t satisfying $0 \leq t \leq 1$ such that $x = (1-t)a + tb$. Since

$$\begin{aligned}(u, x) &= (u, (1-t)a + tb)\\ &= (1-t)(u, a) + t(u, b)\\ &= (1-t)\alpha + t\beta\\ &> (1-t)c + tc\\ &= c,\end{aligned}$$

x is in the open positive half-space of Π.

(c) The open negative half-space of Π is given by $(u, x) < c$; suppose $(u, a) = \alpha < c$ and $(u, b) = \beta < c$. Then if $x = (1-t)a + tb$, $0 \leq t \leq 1$,

$$\begin{aligned}(u, x) &= (u, (1-t)a + tb)\\ &= (1-t)\alpha + t\beta\\ &< (1-t)c + tc\\ &= c.\end{aligned}$$

(d) The closed negative half-space of Π is given by $(u, x) \leq c$; suppose $(u, a) = \alpha \leq c$ and $(u, b) = \beta \leq c$. Then if $x = (1-t)a + tb$, $0 \leq t \leq 1$,

$$\begin{aligned}(u, x) &= (1-t)\alpha + t\beta\\ &\leq (1-t)c + tc\\ &= c.\end{aligned}$$

20 (a) Let $a = (a_1, a_2)$ and $b = (b_1, b_2)$. We know that any point x on the line segment between a and b satisfies the equation

$$x = (1-\theta)a + \theta b,$$

where $0 \leq \theta \leq 1$. Then the square of the distance from x to a is

$$\begin{aligned}&(x-a, x-a)\\ &\quad= ((1-\theta)a_1 + \theta b_1 - a_1)^2 + ((1-\theta)a_2 + \theta b_2 - a_2)^2\\ &\quad= (\theta b_1 - \theta a_1)^2 + (\theta b_2 - \theta a_2)^2\end{aligned}$$

$$= \theta^2[(b_1 - a_1)^2 + (b_2 - a_2)^2]$$
$$= \theta^2(b - a, b - a).$$

In other words, the distance from x to a is θ of the distance from b to a. Thus,

$$x = (1 - \theta)(a_1, a_2) + \theta(b_1, b_2)$$
$$= ((1 - \theta)a_1 + \theta b_1, (1 - \theta)a_2 + \theta b_2).$$

(b) The point $\frac{2}{3}$ of the distance from $a = (1, 1)$ to $b = (2, 3)$ is

$$x = ((1 - \tfrac{2}{3})1 + (\tfrac{2}{3})2, (1 - \tfrac{2}{3})1 + (\tfrac{2}{3})3)$$
$$= (\tfrac{1}{3} + \tfrac{4}{3}, \tfrac{1}{3} + \tfrac{6}{3})$$
$$= (\tfrac{5}{3}, \tfrac{7}{3}).$$

Following the procedure in Example 1.4, we see that the required line must go through $(\frac{5}{3}, \frac{7}{3})$ in the direction of a vector w which is perpendicular to $u = (2, 3) - (1, 1) = (1, 2)$. If we pick w to be $(2, -1)$, then the equation of the line is $x = (\frac{5}{3}, \frac{7}{3}) + t(2, -1)$, where t assumes all real values.

3.2 *Quiz*

1 True. See proof of Theorem 2.1 (c).

2 True. This follows immediately from formula (5).

3 True. If $S_1 \subset \{u \mid \|u - a\| \leq r_1\}$ and $S_2 \subset \{u \mid \|u - b\| \leq r_2\}$ then $S_1 \cap S_2 \subset \{u \mid \|u - a\| \leq r\}$, where $r = \|b - a\| + r_1 + r_2$.

4 False. If S_1, S_2, and r are as in Question 3, then

$$S_1 \cup S_2 \subset \{u \mid \|u - a\| \leq r\}.$$

5 False. Take $c = -2$.

6 True. If $c > 0$ then both $|c|$ and c would be positive and their sum would be positive. But this contradicts the fact that $|c| + c = 0$. Therefore $c \leq 0$.

7 True. See Example 2.2.

8 False. See Theorem 2.2.

9 True. See Theorem 2.3.

10 True. Suppose $a \in \bigcap_{A \in \mathfrak{A}} A$, $b \in \bigcap_{A \in \mathfrak{A}} A$. Then $a \in A$ and $b \in A$ for all $A \in \mathfrak{A}$, and since each set A is convex, $(1 - t)a + tb \in A$ for all $0 \leq t \leq 1$. Thus the line segment joining a and b is contained in each $A \in \mathfrak{A}$ and hence in the intersection. It follows that $\bigcap_{A \in \mathfrak{A}} A$ is a convex set.

3.2 *Exercises*

3 (d) It is clear that if $u = (u_1, u_2, u_3) \in R^3$ is such that $u_3 = \dfrac{1}{u_1 u_2}$, then u will always be in the given set. The distance of u from the origin is

$$\|u\| = \left(u_1^2 + u_2^2 + \frac{1}{u_1^2 u_2^2}\right)^{1/2}.$$

As $|u_1|$ and $|u_2|$ increase, $\dfrac{1}{u_1^2 u_2^2}$ approaches zero; hence by taking $|u_1|$ and $|u_2|$ large enough, $\|u\|$ can be made larger than any fixed positive real number. Thus the set $\{u \mid |u_1 u_2 u_3| \leq 1\}$ cannot be contained in any sphere centered at 0; since any sphere is contained in a sphere centered at 0, the set is not bounded.

4 We know that the length of the projection of the vector v in the direction of a unit vector u is $|(u, v)|$.

(d) If

$$u = \left(\frac{1}{\sqrt{3}}, 0, \frac{1}{\sqrt{3}}, 0, \frac{1}{\sqrt{3}}\right),$$

then

$$\begin{aligned}\|u\| &= \left[\left(\frac{1}{\sqrt{3}}\right)^2 + 0^2 + \left(\frac{1}{\sqrt{3}}\right)^2 + 0^2 + \left(\frac{1}{\sqrt{3}}\right)^2\right]^{1/2}\\ &= (\tfrac{1}{3} + 0 + \tfrac{1}{3} + 0 + \tfrac{1}{3})^{1/2}\\ &= 1.\end{aligned}$$

Hence u is a unit vector. Therefore the projection of $v = (3, -1, 0, 4, -7)$ in the direction of u is

$$\begin{aligned}|(u, v)| &= \left|\frac{1}{\sqrt{3}} \cdot 3 + 0 \cdot (-1) + \frac{1}{\sqrt{3}} \cdot 0 + 0 \cdot 4 + \frac{1}{\sqrt{3}} \cdot (-7)\right|\\ &= \left|\frac{3}{\sqrt{3}} - \frac{7}{\sqrt{3}}\right| = \left|-\frac{4}{\sqrt{3}}\right| = \frac{4}{\sqrt{3}}.\end{aligned}$$

6 We know from formula (19) that the distance from a to $(x, u) = c$ is given by

$$d = \frac{|c - (a, u)|}{\|u\|}.$$

(g)

$$\begin{aligned}\|u\| &= (2^2 + 4^2 + 6^2)^{1/2}\\ &= (4 + 16 + 36)^{1/2}\\ &= (56)^{1/2}\\ &= 2\sqrt{14}\,;\end{aligned}$$

$$(a, u) = (1 \cdot 2 + 0 \cdot 4 + 0 \cdot 6) = 2;$$

$$d = \frac{|10 - 2|}{2\sqrt{14}} = \frac{8}{2\sqrt{14}} = \frac{4\sqrt{14}}{14} = \frac{2\sqrt{14}}{7}.$$

9 (b) Take $n = 2$. Then $a = (-\frac{3}{2}, 0)$ and $b = (\frac{3}{2}, 0)$ are in the given set; however, the point $(0, 0) = \frac{1}{2}a + \frac{1}{2}b$ is obviously on the segment between a and b but not in the set. Thus it is not convex.

(f) Let $a = (a_1, \ldots, a_n)$ and suppose $u = (u_1, \ldots, u_n)$ and $v = (v_1, \ldots, v_n)$ are two points in R^n satisfying $|(u, a)| \leq 1$ and $|(v, a)| \leq 1$. Then if $0 \leq t \leq 1$,

$$\begin{aligned} |(tu + (1 - t)v, a)| &= |t(u, a) + (1 - t)(v, a)| \\ &\leq |t(u, a)| + |(1 - t)(v, a)| \\ &= t|(u, a)| + (1 - t)|(v, a)| \\ &\leq t + (1 - t) \\ &= 1. \end{aligned}$$

Therefore the set $\{u \mid u \in R^n \wedge |(u, a)| \leq 1\}$ is convex.

11 (d) We can write

$$\begin{aligned} x_1 + x_2 + x_4 &= 4, \\ -x_2 - x_3 + x_5 &= -4, \\ x_1 + x_3 + x_6 &= 5, \end{aligned}$$

where $x_4 = 4 - (x_1 + x_2) \geq 0$, $x_5 = -4 + x_2 + x_3 \geq 0$, and $x_6 = 5 - (x_1 + x_3) \geq 0$. Then

$$A = \begin{bmatrix} 1 & 1 & 0 & 1 & 0 & 0 \\ 0 & -1 & -1 & 0 & 1 & 0 \\ 1 & 0 & 1 & 0 & 0 & 1 \end{bmatrix}.$$

12 (c)

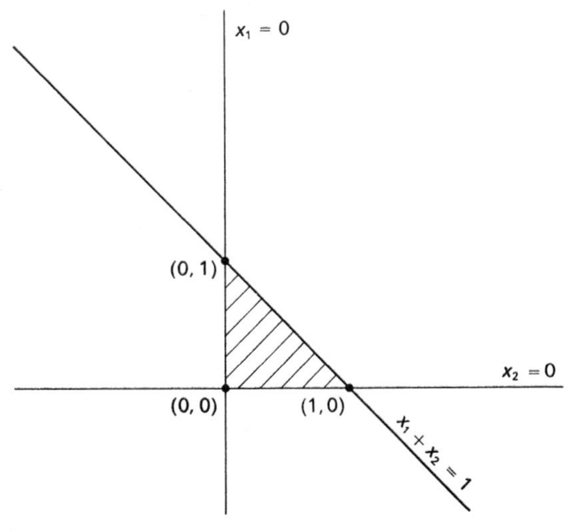

We know from Definition 2.2 that a vertex is an extreme point, thus the extreme points for the set are (0, 1), (1, 0), and (0, 0).

13 (a) We first rewrite the system:

$$\begin{aligned} -2x_1 - x_2 &\le -3, \\ x_1 - 2x_2 &\le -1, \\ x_2 &\le 3, \end{aligned}$$

and then introduce slack variables x_3, x_4, x_5 to write the system in equality form:

$$\begin{aligned} -2x_1 - x_2 + x_3 &= -3, \\ x_1 - 2x_2 + x_4 &= -1, \\ x_2 + x_5 &= 3, \\ x_3 &\ge 0, \\ x_4 &\ge 0, \\ x_5 &\ge 0. \end{aligned}$$

Following the procedure of Example 2.9, we now write the system in matrix-vector notation:

$$\begin{bmatrix} -2 & -1 & 1 & 0 & 0 \\ 1 & -2 & 0 & 1 & 0 \\ 0 & 1 & 0 & 0 & 1 \end{bmatrix} x = \begin{bmatrix} -3 \\ -1 \\ 3 \end{bmatrix},$$

where $x = (x_1, x_2, x_3, x_4, x_5)$. We can reduce this system to its Hermite normal form:

$$\begin{bmatrix} 1 & 0 & 0 & 1 & 2 \\ 0 & 1 & 0 & 0 & 1 \\ 0 & 0 & 1 & 2 & 5 \end{bmatrix} x = \begin{bmatrix} 5 \\ 3 \\ 10 \end{bmatrix}.$$

We will determine all extreme points by considering all subsystems of three equations in three unknowns obtained by setting any two of the x_i equal to zero.

(i) $x_1 = x_2 = 0$:

$$\begin{aligned} x_4 + 2x_5 &= 5, \\ x_5 &= 3, \\ x_3 + 2x_4 + 5x_5 &= 10. \end{aligned}$$

The solution to this system is

$$x_3 = -3, \qquad x_4 = -1, \qquad x_5 = 3.$$

Since x_3 and x_4 are negative, we have no extreme point.

(ii) $x_1 = x_3 = 0$:

$$\begin{aligned} x_4 + 2x_5 &= 5, \\ x_2 + x_5 &= 3, \\ 2x_4 + 5x_5 &= 10. \end{aligned}$$

The solution to this system is

$$x_2 = 3, \qquad x_4 = 5, \qquad x_5 = 0.$$

The columns (0 1 0), (1 0 2), and (2 1 5) are linearly independent; hence we obtain (0, 3, 0, 5, 0) as an extreme point.

(iii) $x_1 = x_4 = 0$:

$$\begin{aligned} 2x_5 &= 5, \\ x_2 + x_5 &= 3, \\ x_3 + 5x_5 &= 10. \end{aligned}$$

The solution is

$$x_2 = \tfrac{1}{2}, \qquad x_3 = -\tfrac{5}{2}, \qquad x_5 = \tfrac{5}{2}.$$

Since x_3 is negative, we do not have an extreme point.

(iv) $x_1 = x_5 = 0$:

$$\begin{aligned} x_4 &= 5, \\ x_2 &= 3, \\ x_3 + 2x_4 &= 10, \end{aligned}$$

which is satisfied by

$$x_2 = 3, \qquad x_3 = 0, \qquad x_4 = 5.$$

The columns (0 1 0), (0 0 1), and (1 0 2) are linearly independent, so (0, 3, 0, 5, 0) is an extreme point.

(v) $x_2 = x_3 = 0$:

$$\begin{aligned} x_1 + x_4 + 2x_5 &= 5, \\ x_5 &= 3, \\ 2x_4 + 5x_5 &= 10 \end{aligned}$$

yields the solution

$$x_1 = \tfrac{3}{2}, \qquad x_4 = -\tfrac{5}{2}, \qquad x_5 = 3,$$

but since x_4 is negative we have no extreme point.

(vi) $x_2 = x_4 = 0$:

$$\begin{aligned} x_1 + 2x_5 &= 5, \\ x_5 &= 3, \\ x_3 + 5x_5 &= 10 \end{aligned}$$

implies

$$x_1 = -1, \qquad x_3 = -5, \qquad x_5 = 3,$$

and since x_1 and x_3 are negative, we have no extreme point.

(vii) $x_2 = x_5 = 0$:

The corresponding system has no solutions.

(viii) $x_3 = x_4 = 0$:

$$\begin{aligned} x_1 + 2x_5 &= 5, \\ x_2 + x_5 &= 3, \\ 5x_5 &= 10, \end{aligned}$$

and we get $x_1 = 1$, $x_2 = 1$, $x_5 = 2$ as a solution. Since the columns $(1 \ 0 \ 0)$, $(0 \ 1 \ 0)$, and $(2 \ 1 \ 5)$ are linearly independent, $(1, 1, 0, 0, 2)$ is an extreme point.

(ix) $x_3 = x_5 = 0$:

$$\begin{aligned} x_1 + x_4 &= 5, \\ x_2 &= 3, \\ 2x_4 &= 10 \end{aligned}$$

has as a solution

$$x_1 = 0, \qquad x_2 = 3, \qquad x_4 = 5.$$

The columns $(1 \ 0 \ 0)$, $(0 \ 1 \ 0)$, and $(1 \ 0 \ 2)$ are linearly independent, thus we have as an extreme point $(0, 3, 0, 5, 0)$.

(x) $x_4 = x_5 = 0$:

$$\begin{aligned} x_1 &= 5, \\ x_2 &= 3, \\ x_3 &= 10. \end{aligned}$$

The columns $(1 \ 0 \ 0)$, $(0 \ 1 \ 0)$, and $(0 \ 0 \ 1)$ are linearly independent, thus we have the extreme point $(5, 3, 10, 0, 0)$.

The set of extreme points for our system of inequalities is $\{(0, 3), (1, 1), (5, 3)\}$.

(d) We can rewrite the system as follows:

$$\begin{aligned} -2x_1 + x_2 &\leq 0, \\ \tfrac{1}{2}x_1 - x_2 &\leq 0, \\ x_1 + 2x_2 &\leq 3, \\ -x_1 - 2x_2 &\leq -1. \end{aligned}$$

Introducing slack variables x_3, x_4, x_5, and x_6, we write the system in equality form:

$$\begin{aligned} -2x_1 + x_2 + x_3 &= 0, \\ \tfrac{1}{2}x_1 - x_2 + x_4 &= 0, \\ x_1 + 2x_2 + x_5 &= 3, \\ -x_1 - 2x_2 + x_6 &= -1, \end{aligned}$$

where $x_3 \geq 0$, $x_4 \geq 0$, $x_5 \geq 0$, and $x_6 \geq 0$. Reducing the system to Hermite normal form we obtain

$$\begin{aligned} x_1 + x_4 - \tfrac{1}{2}x_6 &= \tfrac{1}{2}, \\ x_2 - \tfrac{1}{2}x_4 - \tfrac{1}{4}x_6 &= \tfrac{1}{4}, \\ x_3 + \tfrac{5}{2}x_4 - \tfrac{3}{4}x_6 &= \tfrac{3}{4}, \\ x_5 + x_6 &= 2. \end{aligned}$$

(i) $x_1 = x_2 = 0$:

$$\begin{aligned} x_4 - \tfrac{1}{2}x_6 &= \tfrac{1}{2}, \\ -\tfrac{1}{2}x_4 - \tfrac{1}{4}x_6 &= \tfrac{1}{4}, \\ x_3 + \tfrac{5}{2}x_4 - \tfrac{3}{4}x_6 &= \tfrac{3}{4}, \\ x_5 + x_6 &= 2, \end{aligned}$$

and we get $x_3 = 0$, $x_4 = 0$, $x_5 = 3$, $x_6 = -1$. But x_6 is negative, thus we have no extreme point.

(ii) $x_1 = x_3 = 0$:

$$\begin{aligned} x_4 - \tfrac{1}{2}x_6 &= \tfrac{1}{2}, \\ x_2 - \tfrac{1}{2}x_4 - \tfrac{1}{4}x_6 &= \tfrac{1}{4}, \\ \tfrac{5}{2}x_4 - \tfrac{3}{4}x_6 &= \tfrac{3}{4}, \\ x_5 + x_6 &= 2, \end{aligned}$$

and we get $x_2 = 0$, $x_4 = 0$, $x_5 = 3$, $x_6 = -1$. But again x_6 is negative and we have no extreme point.

(iii) $x_1 = x_4 = 0$:

$$\begin{aligned} -\tfrac{1}{2}x_6 &= \tfrac{1}{2}, \\ x_2 - \tfrac{1}{4}x_6 &= \tfrac{1}{4}, \\ x_3 - \tfrac{3}{4}x_6 &= \tfrac{3}{4}, \\ x_5 + x_6 &= 2, \end{aligned}$$

and we get $x_2 = 0$, $x_3 = 0$, $x_5 = 3$, $x_6 = -1$. Once again x_6 is negative and we get no extreme point.

(iv) $x_1 = x_5 = 0$:

$$\begin{aligned} x_4 - \tfrac{1}{2}x_6 &= \tfrac{1}{2}, \\ x_2 - \tfrac{1}{2}x_4 - \tfrac{1}{4}x_6 &= \tfrac{1}{4}, \\ x_3 + \tfrac{5}{2}x_4 - \tfrac{3}{4}x_6 &= \tfrac{3}{4}, \\ x_6 &= 2, \end{aligned}$$

and we get $x_2 = \tfrac{3}{2}$, $x_3 = -\tfrac{3}{2}$, $x_4 = \tfrac{3}{2}$, $x_6 = 2$. This time x_3 is negative, so we still do not get an extreme point.

(v) $x_1 = x_6 = 0$:

$$\begin{aligned} x_4 &= \tfrac{1}{2}, \\ x_2 - \tfrac{1}{2}x_4 &= \tfrac{1}{4}, \\ x_3 + \tfrac{5}{2}x_4 &= \tfrac{3}{4}, \\ x_5 &= 2, \end{aligned}$$

and we get $x_2 = \tfrac{1}{2}$, $x_3 = -\tfrac{1}{2}$, $x_4 = \tfrac{1}{2}$, $x_5 = 2$. Here x_3 is negative, so we have no extreme point.

(vi) $x_2 = x_3 = 0$:

$$\begin{aligned} x_1 + x_4 - \tfrac{1}{2}x_6 &= \tfrac{1}{2}, \\ -\tfrac{1}{2}x_4 - \tfrac{1}{4}x_6 &= \tfrac{1}{4}, \\ \tfrac{5}{2}x_4 - \tfrac{3}{4}x_6 &= \tfrac{3}{4}, \\ x_5 + x_6 &= 2, \end{aligned}$$

and we get $x_1 = 0$, $x_4 = 0$, $x_5 = 3$, $x_6 = -1$, but since x_6 is negative we have no extreme point.

(vii) $x_2 = x_4 = 0$:

$$\begin{aligned} x_1 - \tfrac{1}{2}x_6 &= \tfrac{1}{2}, \\ -\tfrac{1}{4}x_6 &= \tfrac{1}{4}, \\ x_3 - \tfrac{3}{4}x_6 &= \tfrac{3}{4}, \\ x_5 + x_6 &= 2, \end{aligned}$$

and we get $x_1 = 0$, $x_3 = 0$, $x_5 = 3$, $x_6 = -1$, but since x_6 is negative we have no extreme point.

(viii) $x_2 = x_5 = 0$:

$$\begin{aligned} x_1 + x_4 - \tfrac{1}{2}x_6 &= \tfrac{1}{2}, \\ -\tfrac{1}{2}x_4 - \tfrac{1}{4}x_6 &= \tfrac{1}{4}, \\ x_3 + \tfrac{5}{2}x_4 - \tfrac{3}{4}x_6 &= \tfrac{3}{4}, \\ x_6 &= 2, \end{aligned}$$

and we get $x_1 = 3$, $x_3 = 6$, $x_4 = -\tfrac{3}{2}$, $x_6 = 2$, but since x_4 is negative we still do not have an extreme point.

(ix) $x_2 = x_6 = 0$:

$$\begin{aligned} x_1 + x_4 &= \tfrac{1}{2}, \\ -\tfrac{1}{2}x_4 &= \tfrac{1}{4}, \\ x_3 + \tfrac{5}{2}x_4 &= \tfrac{3}{4}, \\ x_5 &= 2, \end{aligned}$$

and we get $x_1 = 1$, $x_3 = 2$, $x_4 = -\tfrac{1}{2}$, $x_5 = 2$, but since x_4 is negative we have no extreme point.

(x) $x_3 = x_4 = 0$:

$$\begin{aligned} x_1 - \tfrac{1}{2}x_6 &= \tfrac{1}{2}, \\ x_2 - \tfrac{1}{4}x_6 &= \tfrac{1}{4}, \\ -\tfrac{3}{4}x_6 &= \tfrac{3}{4}, \\ x_5 + x_6 &= 2, \end{aligned}$$

and we get $x_1 = 0$, $x_2 = 0$, $x_5 = 3$, $x_6 = -1$, but since x_6 is once more negative, there is no extreme point.

(xi) $x_3 = x_5 = 0$:

$$\begin{aligned} x_1 + x_4 - \tfrac{1}{2}x_6 &= \tfrac{1}{2}, \\ x_2 - \tfrac{1}{2}x_4 - \tfrac{1}{4}x_6 &= \tfrac{1}{4}, \\ \tfrac{5}{2}x_4 - \tfrac{3}{4}x_6 &= \tfrac{3}{4}, \\ x_6 &= 2, \end{aligned}$$

and we get $x_1 = \frac{3}{5}$, $x_2 = \frac{6}{5}$, $x_4 = \frac{9}{10}$, $x_6 = 2$, and we finally obtain an extreme point

$$(\tfrac{3}{5}, \tfrac{6}{5}, 0, \tfrac{9}{10}, 0, 2).$$

(xii) $x_3 = x_6 = 0$:

$$\begin{aligned} x_1 + \phantom{\tfrac{1}{2}}x_4 &= \tfrac{1}{2}, \\ x_2 - \tfrac{1}{2}x_4 &= \tfrac{1}{4}, \\ \tfrac{5}{2}x_4 &= \tfrac{3}{4}, \\ x_5 &= 2, \end{aligned}$$

and we get $x_1 = \frac{1}{5}$, $x_2 = \frac{2}{5}$, $x_4 = \frac{3}{10}$, $x_5 = 2$, so we have another extreme point

$$(\tfrac{1}{5}, \tfrac{2}{5}, 0, \tfrac{3}{10}, 2, 0).$$

(xiii) $x_4 = x_5 = 0$:

$$\begin{aligned} x_1 - \tfrac{1}{2}x_6 &= \tfrac{1}{2}, \\ x_2 - \tfrac{1}{4}x_6 &= \tfrac{1}{4}, \\ x_3 - \tfrac{3}{4}x_6 &= \tfrac{3}{4}, \\ x_6 &= 2, \end{aligned}$$

and we get $x_1 = \frac{3}{2}$, $x_2 = \frac{3}{4}$, $x_3 = \frac{9}{4}$, $x_6 = 2$, and we obtain the extreme point

$$(\tfrac{3}{2}, \tfrac{3}{4}, \tfrac{9}{4}, 0, 0, 2).$$

(xiv) $x_4 = x_6 = 0$:

$$\begin{aligned} x_1 &= \tfrac{1}{2}, \\ x_2 &= \tfrac{1}{4}, \\ x_3 &= \tfrac{3}{4}, \\ x_5 &= 2, \end{aligned}$$

and we immediately get the extreme point

$$(\tfrac{1}{2}, \tfrac{1}{4}, \tfrac{3}{4}, 0, 2, 0).$$

(xv) $x_5 = x_6 = 0$:

The corresponding system has no solutions.

The set of extreme points of the region satisfying the system of inequalities is thus

$$\{(\tfrac{3}{5}, \tfrac{6}{5}), (\tfrac{1}{5}, \tfrac{2}{5}), (\tfrac{3}{2}, \tfrac{3}{4}), (\tfrac{1}{2}, \tfrac{1}{4})\}.$$

3.3 Quiz

1 True. $(0, 0) = \frac{1}{2}(-1, -1) + \frac{1}{2}(1, 1)$.

2 True. $S = \{u \in R^n \mid u = ta^1 + (1 - t)a^2\} = H(a^1, a^2)$.

3 True.

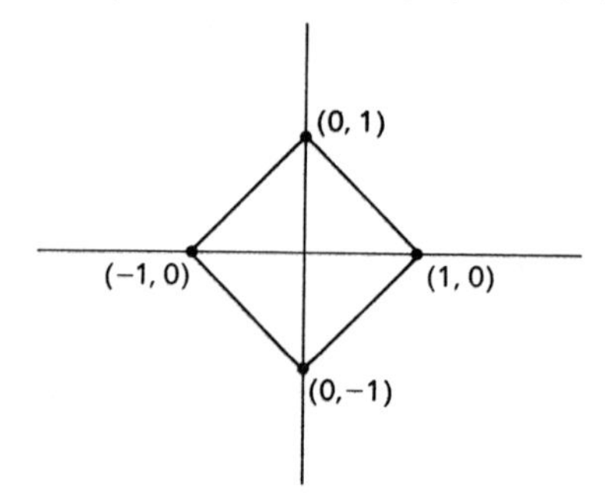

4 False. Let $x = (0, 0)$ and $\alpha = 2$. Then $f(\alpha x) = 1$ while $\alpha f(x) = 2$.

5 True. $h(x) = (f + g)x = f(x) + g(x) = (u, x) + (v, x) = (u + v, x)$.

6 False. Not unless S is bounded.

7 True. Let $a = (a_1, \ldots, a_n) \in S^n$, $b = (b_1, \ldots, b_n) \in S^n$. Then if $0 \leq \theta \leq 1$,

$$\begin{aligned}\sum_{i=1}^{n} [\theta a_i + (1 - \theta)b_i] &= \theta \sum_{i=1}^{n} a_i + (1 - \theta) \sum_{i=1}^{n} b_i \\ &= \theta + (1 - \theta) \\ &= 1.\end{aligned}$$

Therefore $\theta a + (1 - \theta)b \in S^n$, and S^n is a convex polyhedron spanned by $e_i = (0, \ldots, 0, \overset{\overset{i}{\downarrow}}{1}, 0, \ldots, 0)$, $i = 1, \ldots, n$.

8 False. $H(a^1, a^2)$ is the line segment between (2, 1) and (1, 2), while $H(a^3, a^4)$ is the segment between (1, 1) and (2, 2).

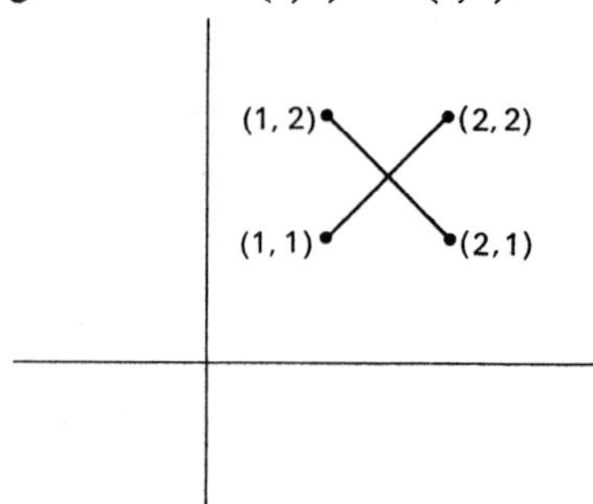

9 True. Clearly $H(a^1, a^2) \subset H(a^1, a^2, a^3)$. But since $a^3 \in H(a^1, a^2)$, then $H(a^1, a^2, a^3) \subset H(a^1, a^2)$.

10 True. Since $H(b^1, \ldots, b^r) = H(a^1, \ldots, a^k)$, each $b^i \in H(a^1, \ldots, a^k)$.

3.3 Exercises

2 (c) $v = \frac{1}{4}a^1 + \frac{1}{12}a^2 + \frac{2}{3}a^3$.

(e) Let α, β, γ be nonnegative real numbers such that $\alpha + \beta + \gamma = 1$ and $v = \alpha a^1 + \beta a^2 + \gamma a^3$. We obtain a system of simultaneous linear equations as follows:

$$\begin{aligned} \alpha + \beta + \gamma &= 1, \\ \alpha + 3\beta + 5\gamma &= \tfrac{5}{2}, \\ \alpha - 2\beta + \gamma &= \tfrac{1}{4}. \end{aligned}$$

Solving the system we obtain $\alpha = \frac{1}{2}$, $\beta = \frac{1}{4}$, and $\gamma = \frac{1}{4}$. Therefore the solution is

$$v = \tfrac{1}{2}a^1 + \tfrac{1}{4}a^2 + \tfrac{1}{4}a^3.$$

4 Suppose that v can be expressed as follows:

$$v = \alpha_1 a^1 + \alpha_2 a^2 + \alpha_3 a^3,$$

where $\alpha_1 + \alpha_2 + \alpha_3 = 1$ and $\alpha_i \geq 0$, $i = 1, 2, 3$. Setting $\alpha_2 = 1 - (\alpha_1 + \alpha_3)$, we get

$$v = a^2 - \alpha_1(a^2 - a^1) + \alpha_3(a^3 - a^2).$$

If v can also be expressed $v = \sigma_1 a^1 + \sigma_2 a^2 + \sigma_3 a^3$, then as before, we can express v in terms of σ_1 and σ_3:

$$v = a^2 - \sigma_1(a^2 - a^1) + \sigma_3(a^3 - a^2).$$

Setting $a^2 - \sigma_1(a^2 - a^1) + \sigma_3(a^3 - a^2) = a^2 - \alpha_1(a^2 - a^1) + \alpha_3(a^3 - a^2)$ we obtain

$$(\sigma_1 - \alpha_1)(a^2 - a^1) + (\alpha_3 - \sigma_3)(a^3 - a^2) = 0.$$

By the assumption that $a^2 - a^1$ and $a^3 - a^2$ are linearly independent, we have

$$\sigma_1 - \alpha_1 = 0 \quad \text{and} \quad \alpha_3 - \sigma_3 = 0,$$

i.e., $\sigma_1 = \alpha_1$ and $\sigma_3 = \alpha_3$. Therefore $\sigma_2 = \alpha_2$, and the expression is unique.

5 (a) The diagram of $H(a^1, a^2, a^3)$ is the triangle with vertices a^1, a^2, a^3.

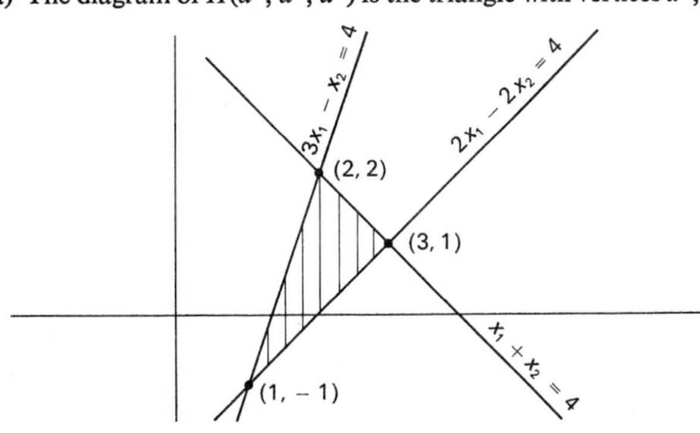

Then $H(a^1, a^2, a^3)$ is the intersection of the half-spaces $(x, u_1) \geq 4$, $u_1 = (3, -1)$; $(x, u_2) \leq 4$, $u_2 = (1, 1)$; and $(x, u_3) \leq 4$, $u_3 = (2, -2)$.

6 Let x and y be two points in the polyhedron $H(a^1, \ldots, a^k)$. Then there exist $\alpha_i \geq 0$, $\beta_i \geq 0$, $i = 1, \ldots, k$, such that $\sum_{i=1}^{k} \alpha_i = \sum_{i=1}^{k} \beta_i = 1$, and $x = \sum_{i=1}^{k} \alpha_i a^i$, $y = \sum_{i=1}^{k} \beta_i a^i$. Let z be any point of the line segment between x and y. Then there exist $\alpha \geq 0$, $\beta \geq 0$ such that $\alpha + \beta = 1$ and $z = \alpha x + \beta y$. Hence we get $z = \sum_{i=1}^{k} \sigma_i a^i$, where $\sigma_i = \alpha\alpha_i + \beta\beta_i$, $i = 1, \ldots, k$, and

$$\begin{aligned}\sum_{i=1}^{k} \sigma_i &= \sum_{i=1}^{k} (\alpha\alpha_i + \beta\beta_i) \\ &= \alpha \sum_{i=1}^{k} \alpha_i + \beta \sum_{i=1}^{k} \beta_i \\ &= \alpha + \beta \\ &= 1.\end{aligned}$$

Therefore $z \in H(a^1, \ldots, a^k)$. Thus $H(a^1, \ldots, a^k)$ is convex.

7 (a) Let x and y be any two vectors in $H(X)$. Then there exist $a^1, a^2, \ldots, a^m$; $b^1, b^2, \ldots, b^n$ in X and $\alpha_1, \alpha_2, \ldots, \alpha_m$; $\beta_1, \beta_2, \ldots, \beta_n$ nonnegative numbers such that

$$\begin{aligned}\sum_{i=1}^{m} \alpha_i = 1 &= \sum_{j=1}^{n} \beta_j, \\ x &= \sum_{i=1}^{m} \alpha_i a^i, \\ y &= \sum_{j=1}^{n} \beta_j b^j.\end{aligned}$$

Let z be any point on the line segment between x and y. Then there exist $\alpha \geq 0$, $\beta \geq 0$ such that $\alpha + \beta = 1$ and $z = \alpha x + \beta y$.

Hence

$$z = \sum_{i=1}^{m} \alpha\alpha_i a^i + \sum_{j=1}^{n} \beta\beta_j b^j$$

and

$$\begin{aligned}\sum_{i=1}^{m} \alpha\alpha_i + \sum_{j=1}^{n} \beta\beta_j &= \alpha \sum_{i=1}^{m} \alpha_i + \beta \sum_{j=1}^{n} \beta_j \\ &= \alpha + \beta \\ &= 1.\end{aligned}$$

Consequently $z \in H(a^1, a^2, \ldots, a^m, b^1, b^2, \ldots, b^n)$, thus $z \in H(X)$. Therefore $H(X)$ is convex.

(b) The fact that $X \subset H(X)$ is trivial. Now let x be any point of

$H(X)$. Then x is a convex combination of finitely many elements of X. Since $X \subset S$, x is in fact a convex combination of finitely many elements of S. Since S is convex, x must belong to S. Hence $H(X) \subset S$.

(c) If $X = H(X)$, $H(X)$ is convex by part (a), and therefore X is convex. Conversely, if X is convex, replacing S by X in part (b), we have $X \subset H(X) \subset S = X$, i.e., $X = H(X)$.

(d) By part (a), $H(X)$ is convex. Therefore by part (c), $H(H(X)) = H(X)$.

8 Clearly the extreme points of the domain on which f is defined are the vectors $e_1, e_2, \ldots, e_n$, where $e_i = (0, 0, \ldots, 0, \overset{\overset{i}{\downarrow}}{1}, 0, \ldots, 0)$. By Theorem 3.3, the largest (smallest) of the numbers $f(e_i)$, $i = 1, 2, \ldots, n$, will be the maximum (minimum) value of f.

(a) $f(e_1) = 2, f(e_2) = 3$. Thus 2 is minimum and 3 is maximum.

(e) Let $g(x) = x_1 + 2x_2 + x_3$. Then the minimum of g is 1, the maximum is 2. Since g is a nonnegative function, $f = g^2$ attains its maximum (minimum) if and only if g does. Therefore the minimum of f is $1^2 = 1$ and the maximum is $2^2 = 4$.

(f) The minimum of the function $g(x) = x_1 + 2x_2 - x_3 - x_4$ is -1 and the maximum is 2. Thus the minimum of f is 4 and the maximum of f is 7.

9 Clearly the extreme points of $H(a^1, a^2, \ldots, a^k)$ form a subset of $\{a^1, \ldots, a^k\}$. Thus by Theorem 3.3 it is sufficient to find the maximum and minimum of the values $f(a^i)$ for $i = 1, \ldots, k$.

(d) $f(a^1) = f(a^2) = f(a^3) = f(a^4) = f(a^5) = 0$. Therefore the maximum and minimum values are both equal to zero. Consequently, f is a constant function on $H(a^1, \ldots, a^5)$.

15

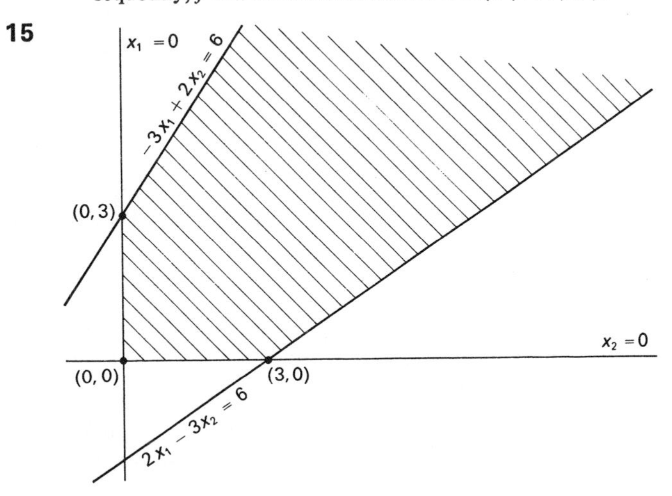

The function $f(x) = 3x_1 + 5x_2$ does not have a maximum on S, since the point $(t, t) \in S$ for all positive values t and the value of $f((t, t)) = 8t$ can be arbitrarily large.

If the additional inequalities $x_1 \leq 10$, $x_2 \leq 5$ are included, then the domain of f is bounded. Theorem 3.3 is now applicable. The extreme points are $(0, 0)$, $(0, 3)$, $(3, 0)$, $(\frac{4}{3}, 5)$, $(10, \frac{14}{3})$, $(10, 5)$; and $f((0, 0)) = 0$, $f((0, 3)) = 15$, $f((3, 0)) = 9$, $f((\frac{4}{3}, 5)) = 29$, $f((10, \frac{14}{3})) = \frac{160}{3}$, $f((10, 5)) = 55$. Thus the maximum of f is 55 and the minimum of f is 0.

16 Let x_1 and x_2 be the numbers of type I and type II rackets, respectively, manufactured each hour. Then x_1, $2x_1$, and $3x_1$ are the amounts of A, S, and M used in constructing the type I rackets, while $2x_2$, x_2, x_2 are the amounts of A, S, and M used in constructing the type II rackets, respectively. The inequalities

$$\begin{aligned} x_1 + 2x_2 &\leq 16, \\ 2x_1 + x_2 &\leq 11, \\ 3x_1 + x_2 &\leq 15, \\ x_1 &\geq 0, \\ x_2 &\geq 0 \end{aligned}$$

are the conditions imposed on the production; the amount earned per hour is

$$f(x) = 30x_1 + 50x_2,$$

$x = (x_1, x_2)$. The region defined by the above inequalities is a polyhedron whose vertices are $(0, 0)$, $(5, 0)$, $(0, 8)$, $(2, 7)$, $(4, 3)$. If we evaluate f on each of these vertices we see that the maximum is achieved at $(2, 7)$. Hence each hour's production should consist of 2 Faults and 7 Snowshoes in order that the amount earned is maximal.

17 The availability of the three ingredients determines the following set of inequalities:

$$\begin{aligned} 2x_1 + 3x_2 &\leq 150, \\ 3x_1 + x_2 &\leq 120, \\ x_1 + 2x_2 &\leq 90, \\ x_1 &\geq 0, \\ x_2 &\geq 0, \end{aligned}$$

where x_1 and x_2 are the amounts produced of Nostrils Forever and Enchanted Sinus, respectively. The cost function we wish to maximize is

$$f(x) = f((x_1, x_2)) = 3.5x_1 + 5x_2.$$

Now, the above inequalities determine a convex region with vertices $(0, 0)$, $(30, 30)$, $(40, 0)$, and $(0, 45)$; and

$$\begin{aligned} f((0, 0)) &= 0, \\ f((30, 30)) &= 105 + 150 = 255, \\ f((40, 0)) &= 140, \\ f((0, 45)) &= 225. \end{aligned}$$

Hence the Olfactory Factory will maximize the cost function by producing an equal quantity of each scent.

18 Let x_1, x_2, and x_3 be the amounts shipped from plant I to d_1, d_2, and d_3, respectively, and let x_4, x_5, and x_6 be the amounts shipped from plant II to d_1, d_2, and d_3. Now, the conditions of the problem state that

$$\begin{aligned} x_1 + x_4 &= 250, \\ x_2 + x_5 &= 600, \\ x_3 + x_6 &= 350. \end{aligned}$$

Moreover,

$$\begin{aligned} x_1 + x_2 + x_3 &\leq 700, \\ x_4 + x_5 + x_6 &\leq 500. \end{aligned}$$

Neither of the latter two inequalities can be strict since this would imply that the sum of all the x_i would be strictly less than 1200. However, if we add the first three equations, we see that the sum of all the x_i is precisely 1200. Now, the cost function to be minimized is

$$f(x) = 3x_1 + 2x_2 + 5x_3 + 5x_4 + 3x_5 + 7x_6,$$

where $x = (x_1, \ldots, x_6)$ and x satisfies the preceding system with the last two inequalities replaced by equalities. If we solve the resulting system by reduction to Hermite normal form, we see that the general solution vector is of the form

$$(x_5 + x_6 - 250, 600 - x_5, 350 - x_6, 500 - x_5 - x_6, x_5, x_6).$$

If we evaluate f on a vector of this type, we see that the cost reduces to $4700 - x_5$. Hence the cost will be minimum when x_5 is maximum. We know that $x_5 \leq 500$, the capacity of plant II. Moreover, if $x_5 = 500$, then since $x_4 + x_5 + x_6 = 500$, it follows that $x_4 = x_6 = 0$. Then from the first three equations we get $x_1 = 250$, $x_2 = 100$, and $x_3 = 350$.

19 (a) Since f is linear, we have

$$f((1 - t)x + ty) = (1 - t)f(x) + tf(y),$$

which meets the requirement of a convex function.

(b) Let $x, y \in R^1$ and $\alpha, \beta \geq 0$, with $\alpha + \beta = 1$. Then

$$\begin{aligned} [\alpha f(x) + \beta f(y)] - f(\alpha x + \beta y) & \\ &= \alpha x^2 + \beta y^2 - (\alpha x + \beta y)^2 \\ &= \alpha(1 - \alpha)x^2 + \beta(1 - \beta)y^2 - 2\alpha\beta xy \\ &= \alpha\beta x^2 + \beta\alpha y^2 - 2\alpha\beta xy \\ &= \alpha\beta(x - y)^2 \\ &\geq 0. \end{aligned}$$

Therefore $f(x) = x^2$ is a convex function.

(c) Let $\alpha, \beta \geq 0$ and $\alpha + \beta = 1$. Let x, y be vectors in R^n. Then

$$\begin{aligned} f(\alpha x + \beta y) & \\ &= \varphi_1(\alpha x_1 + \beta y_1) + \cdots + \varphi_n(\alpha x_n + \beta y_n) \\ &\leq [\alpha\varphi_1(x_1) + \beta\varphi_1(y_1)] + \cdots + [\alpha\varphi_n(x_n) + \beta\varphi_n(y_n)] \\ &= \alpha[\varphi_1(x_1) + \cdots + \varphi_n(x_n)] + \beta[\varphi_1(y_1) + \cdots + \varphi_n(y_n)] \\ &= \alpha f(x) + \beta f(y), \end{aligned}$$

which shows that f is a convex function on R^n.

(d) Let $x, y \in S$ and $\alpha, \beta \geq 0, \alpha + \beta = 1$. Then

$$\begin{aligned}(f + g)(\alpha x + \beta y) &= f(\alpha x + \beta y) + g(\alpha x + \beta y)\\ &\leq [\alpha f(x) + \beta f(y)] + [\alpha g(x) + \beta g(y)]\\ &= \alpha[f(x) + g(x)] + \beta[f(y) + g(y)]\\ &= \alpha(f + g)(x) + \beta(f + g)(y),\end{aligned}$$

which shows that $f + g$ is a convex function.

(e) Let $S = R^1$ and let $f(x) = x$; $g(x) = -x$. Then clearly f and g are both convex functions. However, $h(x) = f(x)g(x)$ is not convex, since $0 = \frac{1}{2}(1) + \frac{1}{2}(-1)$ is a convex combination of 1 and -1, but $h(0) = 0 > \frac{1}{2}(1)(-1) + \frac{1}{2}(-1)(1) = \frac{1}{2}h(1) + \frac{1}{2}h(-1)$.

(f) Let $x, y \in S$ and $\alpha, \beta \geq 0, \alpha + \beta = 1$. Since f is a convex function,

$$f(\alpha x + \beta y) \leq \alpha f(x) + \beta f(y) \leq \alpha h(x) + \beta h(y).$$

Similarly, $g(\alpha x + \beta y) \leq \alpha h(x) + \beta h(y)$. Thus

$$\begin{aligned}h(\alpha x + \beta y) &= \max[f(\alpha x + \beta y), g(\alpha x + \beta y)]\\ &\leq \alpha h(x) + \beta h(y).\end{aligned}$$

(g) Let x be any point in $H(a^1, \ldots, a^k)$. Then there exist $\sigma_i \geq 0$, $i = 1, 2, \ldots, k$, such that $\sum_{i=1}^{k} \sigma_i = 1$ and $x = \sum_{i=1}^{k} \sigma_i a^i$. Suppose f is a convex function. Then

$$\begin{aligned}f(x) &= f\left(\sum_{i=1}^{k} \sigma_i a^i\right)\\ &\leq \sum_{i=1}^{k} [\sigma_i f(a^i)]\\ &\leq \sum_{i=1}^{k} [\sigma_i \max_{1 \leq j \leq k} f(a^j)]\\ &= \max_{1 \leq j \leq k} f(a^j) \sum_{i=1}^{k} \sigma_i\\ &= \max_{1 \leq j \leq k} f(a^j).\end{aligned}$$

20 Let $T(T')$ be the set of all extreme points of $S(S')$, and let $\hat{T}$ be the set of all vectors obtained by chopping off the last m components of the vectors in T'. Then Theorem 3.2 tells us that $T \subset \hat{T}$. Conversely, let

$$u = (u_1, u_2, \ldots, u_n) \in \hat{T}.$$

Then there exist nonnegative numbers $u_{n+1}, \ldots, u_{n+m}$ such that $u' = (u_1, u_2, \ldots, u_n, u_{n+1}, \ldots, u_{n+m}) \in T'$:

$$u_{n+k} = b_k - \sum_{j=1}^{n} a_{kj} u_j, \qquad k = 1, \ldots, m.$$

We want to show that u is an extreme point of S. Suppose that $u = (1-t)w + tv$, $0 < t < 1$, where $w = (w_1, \ldots, w_n) \in S$ and $v = (v_1, \ldots, v_n) \in S$. Therefore $Aw \leq b$ and $Av \leq b$, that is,

$$\sum_{j=1}^{n} a_{ij}w_j \leq b_i$$

and

$$\sum_{j=1}^{n} a_{ij}v_j \leq b_i,$$

$i = 1, \ldots, m$. For any $x = (x_1, \ldots, x_n) \in S$, define

$$x_{n+k} = b_k - \sum_{j=1}^{n} a_{kj}x_j \geq 0, \quad k = 1, \ldots, m.$$

Then, if $w' = (w_1, \ldots, w_n, w_{n+1}, \ldots, w_{n+m})$ and $v' = (v_1, \ldots, v_n, v_{n+1}, \ldots, v_{n+m})$,

$$[A\colon I_m]w' = b$$

and

$$[A\colon I_m]v' = b.$$

Hence w' and v' are points of S'. Furthermore, $u_i = (1-t)w_i + tv_i$ for $i = 1, \ldots, n$, and for $k = 1, \ldots, m$,

$$\begin{aligned}(1-t)w_{n+k} + tv_{n+k} &= b_k - \sum_{j=1}^{n} a_{kj}[(1-t)w_j + tv_j] \\ &= b_k - \sum_{j=1}^{n} a_{kj}u_j \\ &= u_{n+k}.\end{aligned}$$

Therefore $u' = (1-t)w' + tv'$, $0 < t < 1$, which contradicts the fact that u' is an extreme point of S'. Therefore u must be an extreme point of S and so $u \in T$. We have thus shown that $\hat{T} = T$.

3.4 *Quiz*

1 True. Here $\alpha_1 = -2$, $\alpha_2 = 0$, so $\alpha = \max\{-2, 0\} = 0$.

2 False. In this case $\beta_1 = 1$, $\beta_2 = 3$, $\beta_3 = 7$, and $\beta = \min\{1, 3, 7\} = 1 \neq 0$.

3 False. As shown above, $\alpha \neq \beta$.

4 True. See formula (1) and the discussion following it.

5 False. If, in Example 4.3, R and C use pure strategies, then $x = (1, 0)$, $y = (0, 1)$, and $E(x, y) = 3$.

6 True. See the discussion following Definition 4.2.

7 True. For this game, $\alpha = \beta = a_{21} = 2$.

8 False. A saddle point is a row minimum and a column maximum, which might still exist.

9 True. We know $w \leq E(x^*, y)$ and $w \leq E(x', y)$, for all strategies y. Therefore

$$\begin{aligned} E(\tfrac{1}{2}(x^* + x'), y) &= (\tfrac{1}{2}(x^* + x'), Ay) \\ &= \tfrac{1}{2}[(x^*, Ay) + (x', Ay)] \\ &\geq \tfrac{1}{2}(w + w) \\ &= w. \end{aligned}$$

10 True. $\displaystyle\sum_{i=1}^{2}\sum_{j=1}^{3} x_i y_j = \sum_{i=1}^{2} x_i\left(\sum_{j=1}^{3} y_j\right) = \sum_{i=1}^{2} x_i = 1.$

3.4 *Exercises*

1 (a) $\alpha = \beta = 3$. (b) $\alpha = \beta = 0$. (c) $\alpha = 0, \beta = 1$.
(d) $\alpha = -1, \beta = 1$. (e) $\alpha = \beta = 0$. (f) $\alpha = \beta = 2$.
(g) $\alpha = 3, \beta = 4$. (h) $\alpha = \beta = -1$. (i) $\alpha = 0, \beta = 1$.
(j) $\alpha = 2, \beta = 8$. (k) $\alpha = \beta = 2$. (l) $\alpha = 1, \beta = 3$.
(m) $\alpha = \beta = -2$. (n) $\alpha = 1, \beta = 34$. (o) $\alpha = \beta = 1$.

2 (b) If C chooses column 1, then R's expected winning is

$$3x_1 + 2x_2 = x_1 + 2.$$

If C chooses column 2, then R's expected winning is

$$x_1 + 3x_2 = 3 - 2x_1.$$

The highest point on the graph of the function $\min\{(x_1 + 2), (3 - 2x_1)\}, 0 \leq x_1 \leq 1$, is at $(\frac{1}{3}, \frac{7}{3})$.

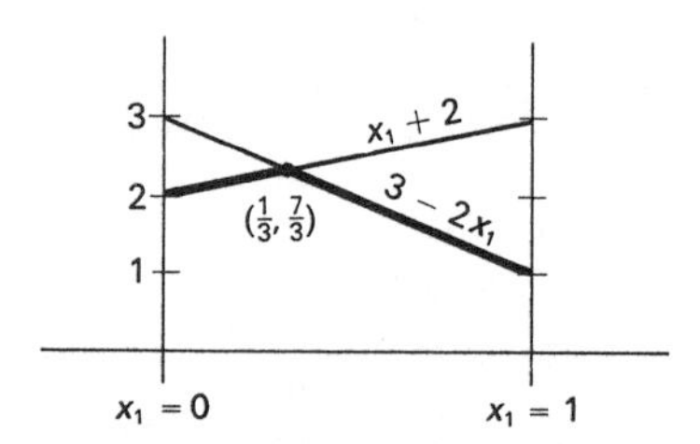

Therefore $x^* = (\frac{1}{3}, \frac{2}{3})$ is an optimum strategy for R. Observe that

$$E(x^*, y) = (\tfrac{1}{3}, \tfrac{2}{3})\begin{bmatrix} 3 & 1 \\ 2 & 3 \end{bmatrix}\begin{pmatrix} y_1 \\ y_2 \end{pmatrix} = \tfrac{7}{3}.$$

Therefore any strategy y for C is optimum.

(d) For column 1, R's expected winning is

$$x_1 + 5x_2 = 5 - 4x_1.$$

For column 2, it is

$$7x_1 + 4x_2 = 4 + 3x_1,$$

and for column 3, it is

$$3x_1 + 6x_2 = 6 - 3x_1.$$

The highest point on the graph of the function $\min\{(5 - 4x_1), (4 + 3x_1), (6 - 3x_1)\}$, $0 \le x_1 \le 1$, is at $(\frac{1}{7}, \frac{31}{7})$.

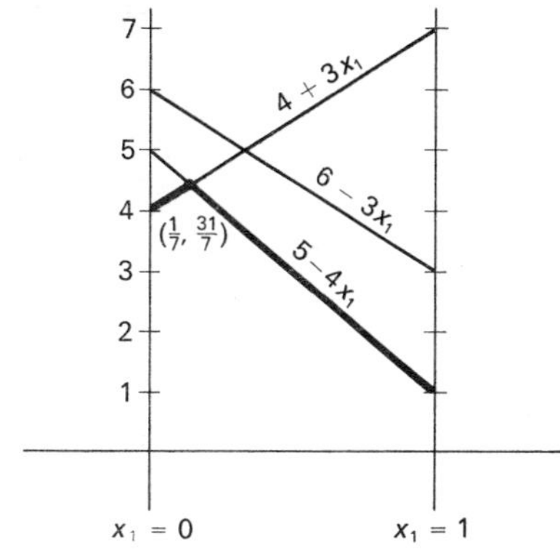

Therefore an optimum strategy for R is $x^* = (\frac{1}{7}, \frac{6}{7})$. Observe that

$$E(x^*, y) = (\tfrac{1}{7}, \tfrac{6}{7})\begin{bmatrix} 1 & 7 & 3 \\ 5 & 4 & 6 \end{bmatrix}\begin{pmatrix} y_1 \\ y_2 \\ y_3 \end{pmatrix}$$
$$= \tfrac{1}{7}(31y_1 + 31y_2 + 39y_3)$$
$$= \tfrac{1}{7}(31 + 8y_3).$$

Since $y_3 \ge 0$, the minimum will be attained at $y_3 = 0$. Thus in any optimum strategy for C, $y_3 = 0$. Let $(y_1, 1 - y_1, 0)$ be a strategy for C. If R chooses row 1, he can expect to win

$$y_1 + 7(1 - y_1) = 7 - 6y_1,$$

and if R chooses row 2 he can expect to win

$$5 + 4(1 - y_1) = 4 + y_1.$$

The lowest point on the graph of the function $\max\{7 - 6y_1, 4 + y_1\}$, $0 \le y_1 \le 1$, is at $(\frac{3}{7}, \frac{31}{7})$.

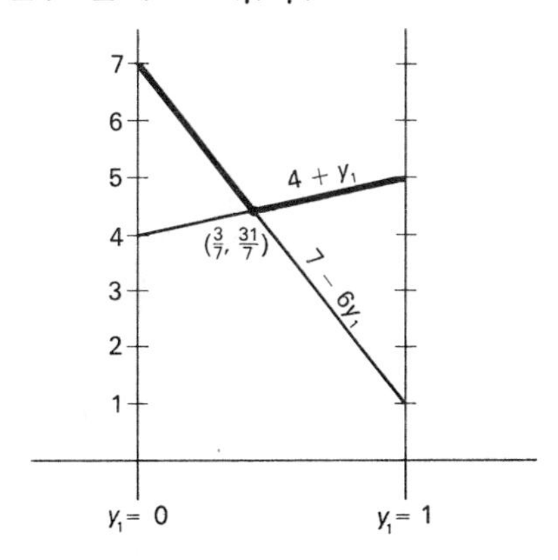

Hence an optimum strategy for C is $y^* = (\frac{3}{7}, \frac{4}{7}, 0)$, and the value of the game is

$$E(x^*, y^*) = \frac{31}{7}.$$

(e) In order to simplify the problem, we will find C's optimum strategy first. For row 1, C's expected winnings are

$$-2y_1 - 5y_2 = -5 + 3y_1,$$

for row 2 they are

$$y_1 - y_2 = -1 + 2y_1,$$

and for row 3 they are

$$2y_2 = 2 - 2y_1.$$

The lowest point on the graph of the function $\max\{-5 + 3y_1, -1 + 2y_1, 2 - 2y_1\}, 0 \leq y_1 \leq 1$, is at the point $(\frac{3}{4}, \frac{1}{2})$.

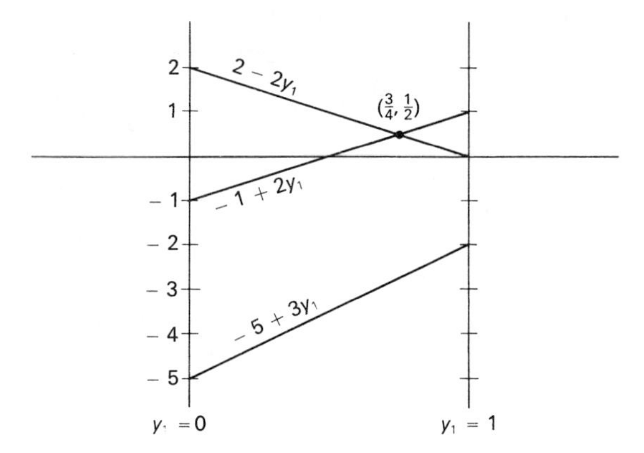

Hence an optimum strategy for C is $y^* = (\frac{3}{4}, \frac{1}{4})$. Then

$$\begin{aligned} E(x, y^*) &= (x_1, x_2, x_3)\begin{bmatrix} -2 & -5 \\ 1 & -1 \\ 0 & 2 \end{bmatrix}\begin{pmatrix} \frac{3}{4} \\ \frac{1}{4} \end{pmatrix} \\ &= \tfrac{3}{4}(-2x_1 + x_2) + \tfrac{1}{4}(-5x_1 - x_2 + 2x_3) \\ &= -\tfrac{11}{4}x_1 + \tfrac{1}{2}x_2 + \tfrac{1}{2}(1 - x_1 - x_2) \\ &= \tfrac{1}{2} - \tfrac{13}{4}x_1. \end{aligned}$$

Since $E(x, y^*)$ is maximum when $x_1 = 0$, any optimum strategy for R will be of the form $(0, x_2, 1 - x_2)$. Then when C chooses column 1, R can expect to win

$$x_2 + 0(1 - x_2) = x_2,$$

and when C chooses column 2, R can expect to win

$$-x_2 + 2(1 - x_2) = 2 - 3x_2.$$

The highest point on the graph of the function $\min\{x_2, 2 - 3x_2\}$ is at $(\frac{1}{2}, \frac{1}{2})$.

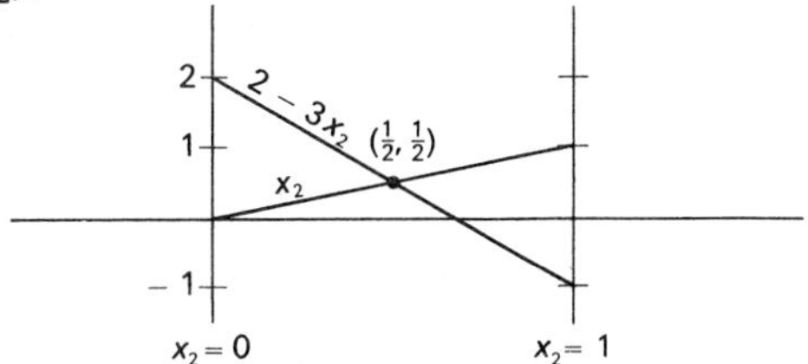

Thus an optimum strategy for R is $x^* = (0, \frac{1}{2}, \frac{1}{2})$, and the value of the game is

$$E(x^*, y^*) = \tfrac{1}{2}.$$

3 (a) Since every column is the same, we can assume that $a_{ij} = a_i$ for all j. The expectation is given by Theorem 4.2 as

$$\begin{aligned} E(x, y) &= (x, Ay) \\ &= \sum_{i=1}^{m} \sum_{j=1}^{n} x_i a_{ij} y_j \\ &= \sum_{i=1}^{m} \sum_{j=1}^{n} (x_i a_i y_j) \\ &= \left(\sum_{i=1}^{m} x_i a_i\right)\left(\sum_{j=1}^{n} y_j\right) \\ &= \sum_{i=1}^{m} x_i a_i. \end{aligned}$$

Therefore $E(x, y)$ is independent of y.

(b) See hint.

(c)

$$\begin{aligned} \max_x \min_y E(x, y) &= \max_x \left(\sum_{i=1}^{m} x_i a_i\right) \\ &= \max_x E(x, y) \\ &= \min_y \max_x E(x, y). \end{aligned}$$

13 By Exercise 12, there exists a vector x^o such that $A^T x^o = 0$. Therefore by Exercise 11, the value w of the game defined by the matrix A is nonnegative, i.e., $w \geq 0$. Observe that $Ay^o = 0$, where

$$y^o = \frac{1}{a + b + c}(c, b, a).$$

Thus by Exercise 10, $w \leq 0$. Consequently, $w = 0$. Hence the optimum strategies for R and C are x^o and y^o, respectively. But

$$x^o = y^o = \frac{1}{a + b + c}(c, b, a).$$

14 Let the first column denote the choice of a fist for C, the second the choice of two fingers, and the third the choice of an open hand. Let the first, second, and third rows denote corresponding choices for R. Then the game matrix is

$$A = \begin{bmatrix} 0 & 1 & -1 \\ -1 & 0 & 1 \\ 1 & -1 & 0 \end{bmatrix}.$$

The matrix A is clearly in the form of the matrix in Exercise 13, hence the value of the game is zero and optimum strategies for R and C are

$$x^o = \tfrac{1}{3}(1, 1, 1)$$

and

$$y^o = \tfrac{1}{3}(1, 1, 1),$$

respectively.

15 Let $A = (a_{ij})$ be $m \times n$ and $B = (b_{ij}) = (a_{ij} + k)$. Then if w_A and w_B are the values of the games defined by A and B, and $E_A(x, y)$ and $E_B(x, y)$ the corresponding expectations,

$$\begin{aligned}
w_B &= \max_x \min_y E_B(x, y) \\
&= \max_x \min_y \sum_{i=1}^{m} \sum_{j=1}^{n} x_i b_{ij} y_j \\
&= \max_x \min_y \sum_{i=1}^{m} \sum_{j=1}^{n} x_i (a_{ij} + k) y_j \\
&= \max_x \min_y \left(\sum_{i=1}^{m} \sum_{j=1}^{n} x_i a_{ij} y_j + \sum_{i=1}^{m} \sum_{j=1}^{n} k x_i y_j \right) \\
&= \max_x \min_y \left(\sum_{i=1}^{m} \sum_{j=1}^{n} x_i a_{ij} y_j + k \sum_{i=1}^{m} x_i \left(\sum_{j=1}^{n} y_j \right) \right) \\
&= \max_x \min_y \left(\sum_{i=1}^{m} \sum_{j=1}^{n} x_i a_{ij} y_j + k \sum_{i=1}^{m} x_i \right) \\
&= \left(\max_x \min_y \sum_{i=1}^{m} \sum_{j=1}^{n} x_i a_{ij} y_j \right) + k \\
&= w_A + k.
\end{aligned}$$

Also, if x^* is an optimum strategy for R for the game determined by B, then

$$\begin{aligned}
w_B &\le E_B(x^*, y) \\
&= \sum_{i=1}^{m} \sum_{j=1}^{n} x_i^* b_{ij} y_j \\
&= \sum_{i=1}^{m} \sum_{j=1}^{n} x_i^* a_{ij} y_j + \sum_{i=1}^{m} \sum_{j=1}^{n} k x_i^* y_j \\
&= E_A(x^*, y) + k.
\end{aligned}$$

Hence

$$w_A = w_B - k \le E_A(x^*, y),$$

i.e., x^* is also an optimum strategy for the game determined by A. Conversely, if x^* is an optimum strategy for the game determined by A, then

$$\begin{aligned} E_B(x^*, y) &= E_A(x^*, y) + k \\ &\geq w_A + k \\ &= w_B, \end{aligned}$$

so that x^* is also an optimum strategy for the game determined by B. Hence the optimum strategies for R with respect to the matrix B are the same as those for the matrix A. By analogous calculations it can be shown that the optimum strategies for C are the same for both games.

18 The matrix for this game is

$$A = \begin{bmatrix} 4 & 2 \\ 3 & 5 \end{bmatrix},$$

where row 1 corresponds to R standing in front of the baseline, row 2 to R standing behind the baseline, column 1 to C delivering a flat serve, and column 2 to C delivering a twist serve. The entries indicate the percentage of points R wins multiplied by 10. (See Exercise 19.) Let $x = (x_1, x_2)$ be a strategy for R. If C chooses column 1, the expectation for R is

$$4x_1 + 3x_2 = 3 + x_1,$$

and if C chooses column 2, it is

$$2x_1 + 5x_2 = 5 - 3x_1.$$

The highest point on the graph of the function $\min\{3 + x_1, 5 - 3x_1\}$ is at $(\frac{1}{2}, \frac{7}{2})$.

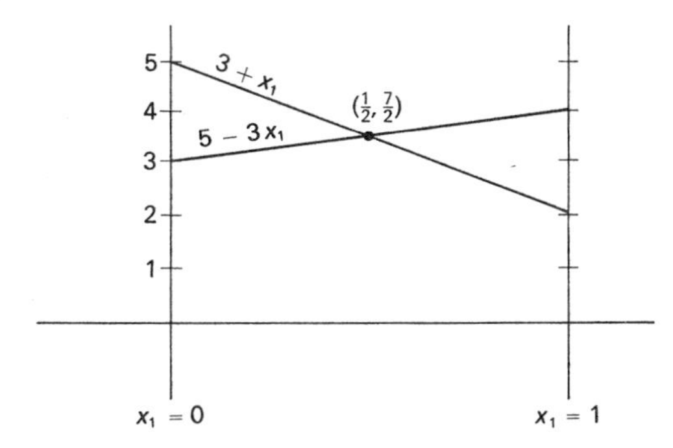

Hence an optimum strategy for R is $x^* = (\frac{1}{2}, \frac{1}{2})$. Now let $y = (y_1, y_2)$ be a strategy for C. If R chooses row 1, C's expectation is

$$4y_1 + 2y_2 = 2 + 2y_1,$$

and if R chooses row 2, it is

$$3y_1 + 5y_2 = 5 - 2y_1.$$

Graphing the function $\max\{2 + 2y_1, 5 - 2y_1\}$, the highest point is at $(\frac{3}{4}, \frac{7}{2})$.

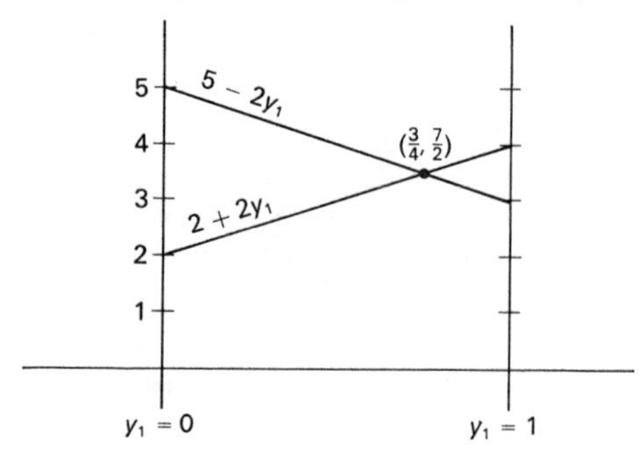

Hence an optimum strategy for C is $(\frac{3}{4}, \frac{1}{4})$.

19 Let $A = (a_{ij})$ be $m \times n$ and let $B = (b_{ij}) = (ka_{ij})$, $k > 0$. Let w_A and w_B denote the values of the games defined by A and B, and $E_A(x, y)$ and $E_B(x, y)$ the corresponding expectations. Then

$$\begin{aligned} w_B &= \max_x \min_y E_B(x, y) \\ &= \max_x \min_y \sum_{i=1}^{m} \sum_{j=1}^{n} x_i b_{ij} y_j \\ &= \max_x \min_y \sum_{i=1}^{m} \sum_{j=1}^{n} x_i (ka_{ij}) y_j \\ &= k\left(\max_x \min_y \sum_{i=1}^{m} \sum_{j=1}^{n} x_i a_{ij} y_j\right) \\ &= kw_A. \end{aligned}$$

If x^* is an optimum strategy for R for the game determined by B, then

$$\begin{aligned} w_B &\leq E_B(x^*, y) \\ &= \sum_{i=1}^{m} \sum_{j=1}^{n} x_i^* b_{ij} y_j \\ &= k \sum_{i=1}^{m} \sum_{j=1}^{n} x_i^* a_{ij} y_j \\ &= kE_A(x^*, y). \end{aligned}$$

Hence

$$w_A = \frac{w_B}{k} \leq E_A(x^*, y)$$

(since $k > 0$), i.e., x^* is an optimum strategy for R in the game determined by A. Similarly, if x^* is an optimum strategy for R with respect to A, then

$$\begin{aligned} w_B &= kw_A \\ &\leq kE_A(x^*, y) \end{aligned}$$

$$= k \sum_{i=1}^{m} \sum_{j=1}^{n} x_i^* a_{ij} y_j$$

$$= \sum_{i=1}^{m} \sum_{j=1}^{n} x_i^*(ka_{ij}) y_j$$

$$= E_B(x^*, y),$$

and x^* is an optimum strategy for R with respect to B. Thus the optimum strategies determined by A and B for R are equal and the same can be shown to be true for the optimum strategies for C.

3.5 *Quiz*

1 False. The (2, 1) entry of any power of the matrix will always be zero.

2 True. All positive powers of the matrix, including the first power (i.e., the matrix itself) will have positive entries.

3 False. The columns do not sum to 1.

4 False. A probability vector has nonnegative components only.

5 False. Each entry of the matrix must be nonnegative.

6 True. Every entry of

$$P^2 = \begin{bmatrix} \frac{3}{4} & \frac{1}{2} \\ \frac{1}{4} & \frac{1}{2} \end{bmatrix}$$

is positive.

7 True. We can check that $PA^{(1)} = A^{(1)}$. By Theorem 5.1 we know that if $\lim_{k \to \infty} P^k = A$ exists, then $PA^{(1)} = A^{(1)}$. Since P is primitive column stochastic, Theorem 5.3 tells us that A does exist, and by Theorem 5.4, A is unique. Hence it must be that $\lim_{k \to \infty} P^k = A$.

8 True. Since $x_i \geq 0$, $y_i \geq 0$, $i = 1, \ldots, n$, we know that $(1 - \theta)x_i + \theta y_i \geq 0$, $i = 1, \ldots, n$. Furthermore,

$$\sum_{i=1}^{n} [(1 - \theta)x_i + \theta y_i] = (1 - \theta) \sum_{i=1}^{n} x_i + \theta \sum_{i=1}^{n} y_i$$

$$= (1 - \theta) + \theta$$

$$= 1.$$

9 True. Using the method in the proof of Theorem 5.4, we must find θ, the smallest of the ratios $\frac{y_k}{x_k}$ or $\frac{x_k}{y_k}$, $k = 1, 2, 3$, where $x = (1, 2, 1)$ and $y = (4, 1, 3)$. Clearly $\theta = \frac{1}{4}$ is smallest. Then the vector $z = x - \theta y$ will have a zero component. In fact,

$$z = (1, 2, 1) - (1, \tfrac{1}{4}, \tfrac{3}{4})$$

$$= (0, \tfrac{7}{4}, \tfrac{1}{4}).$$

10 True. Since every entry of P^2 is positive, P is primitive.

3.5 Exercises

2 "Yes" will indicate the matrix is a transition matrix, "no" will indicate it is not.

(a) no. (b) no. (c) yes. (d) no.
(e) no. (f) no. (g) no. (h) yes.
(i) no. (j) no. (k) no. (l) no.
(m) no. (n) no. (o) yes.

3 "Yes" will indicate $\lim_{k\to\infty} P^k$ exists, "no" will indicate it does not.

(a) no. (b) no. (c) no.
(d) yes; $A = I_3$.

(e) yes; $$A = \begin{bmatrix} \frac{1}{5} & \frac{1}{5} & \frac{1}{5} \\ \frac{4}{5} & \frac{4}{5} & \frac{4}{5} \\ 0 & 0 & 0 \end{bmatrix}.$$

(f) yes; $$A = \begin{bmatrix} \frac{1}{3} & \frac{1}{3} & \frac{1}{3} \\ \frac{1}{3} & \frac{1}{3} & \frac{1}{3} \\ \frac{1}{3} & \frac{1}{3} & \frac{1}{3} \end{bmatrix}.$$

(g) yes; $$A = \begin{bmatrix} \frac{1}{3} & \frac{1}{3} & \frac{1}{3} \\ \frac{1}{2} & \frac{1}{2} & \frac{1}{2} \\ \frac{1}{6} & \frac{1}{6} & \frac{1}{6} \end{bmatrix}.$$

(h) yes; the limiting matrix A is P itself.

(i) yes; $$A = \begin{bmatrix} \frac{2}{3} & \frac{2}{3} \\ \frac{1}{3} & \frac{1}{3} \end{bmatrix}.$$

(j) yes; $$A = \begin{bmatrix} \frac{9}{11} & \frac{9}{11} \\ \frac{2}{11} & \frac{2}{11} \end{bmatrix}.$$

4 Since P is primitive and column stochastic, $\lim_{k\to\infty} P^k = A$ exists and satisfies $PA^{(i)} = A^{(i)}$, $i = 1, 2$, where, in fact, $A^{(1)} = A^{(2)}$. Denote $A^{(1)}$ by (x_1, x_2). Then $PA^{(1)} = A^{(1)}$ is equivalent to the two simultaneous equations

$$(1 - a)x_1 + bx_2 = x_1,$$
$$ax_1 + (1 - b)x_2 = x_2.$$

These equations will be satisfied by any vector of the form $\lambda(b, a)$, where λ is a positive scalar. But since A is column stochastic, we know $\lambda a + \lambda b = 1$, or $\lambda = \dfrac{1}{a + b}$. Hence A is the matrix

$$A = \begin{bmatrix} \dfrac{a}{a+b} & \dfrac{a}{a+b} \\ \dfrac{b}{a+b} & \dfrac{b}{a+b} \end{bmatrix}.$$

9 Let the transition matrix be

$$P = \begin{bmatrix} .6 & .8 \\ .4 & .2 \end{bmatrix},$$

where columns one and two correspond to pretty and homely mothers, respectively, and rows one and two to pretty and homely daughters. Then if $x_0 = (x_1, x_2)$ is the initial distribution of the first generation of the population into pretty girls (x_1) and homely girls (x_2), then $P_0^k x$ is the distribution of the $(k + 1)$st generation into pretty and homely girls. The probability of a pretty woman being the great grandmother of a homely girl will be the (2, 1) entry of P^3. But P^3 is just

$$\begin{aligned} P^3 &= \begin{bmatrix} .6 & .8 \\ .4 & .2 \end{bmatrix}^3 \\ &= \begin{bmatrix} .6 & .8 \\ .4 & .2 \end{bmatrix}\begin{bmatrix} .68 & .64 \\ .32 & .36 \end{bmatrix} \\ &= \begin{bmatrix} .66 & .67 \\ .34 & .33 \end{bmatrix}. \end{aligned}$$

Hence the probability that a homely girl has a pretty great-grandmother is .34. If we take $x_0 = (\frac{1}{2}, \frac{1}{2})$, the probability distribution after four generations will be

$$\begin{aligned} P^3 x_0 &= \begin{bmatrix} .66 & .67 \\ .34 & .33 \end{bmatrix}\begin{pmatrix} \frac{1}{2} \\ \frac{1}{2} \end{pmatrix} \\ &= (.665, .335), \end{aligned}$$

i.e., $66\frac{1}{2}\%$ of the women will be pretty and $33\frac{1}{2}\%$ will be homely. In order to find the distribution after a large number of generations, we must find $\lim_{k\to\infty} P^k = A$ (which exists, since P is primitive and column stochastic). Solving for A as in Exercise 4, we find

$$A = \begin{bmatrix} \frac{2}{3} & \frac{2}{3} \\ \frac{1}{3} & \frac{1}{3} \end{bmatrix}.$$

Hence after a large number of generations, $\frac{2}{3}$ of the girls will be pretty and $\frac{1}{3}$ will be homely. It actually does not matter *what* Eve looked like, since the limiting matrix is independent of the initial distribution.

14 We know that if P is column stochastic then $\lim_{k\to\infty} P^k$ is column stochastic (see proof of Theorem 5.1). There is an identical proof showing that if P is row stochastic then $\lim_{k\to\infty} P^k$ is row stochastic. Hence if P is doubly stochastic then $\lim_{k\to\infty} P^k = A$ is also doubly stochastic. But in addition we have the requirement that $A^{(1)} = A^{(2)} = \cdots = A^{(n)}$. Denote this common n-vector by $(a_1, a_2, \ldots, a_n)$. Then

$$a_1 + a_2 + \cdots + a_n = 1$$

and

$$na_i = 1, \qquad i = 1, \ldots, n,$$

that is,

$$a_i = \frac{1}{n}, \qquad i = 1, \ldots, n.$$

Hence $A = J_n$.

16 (a) $z = 3y - x$;
(b) $z = 9y - x$;
(c) $z = 2x - y$;
(d) $z = 2x - y$;
(e) $z = 2x - y$.

17 Following the format of Example 5.6, the transition matrix will be 5×5:

$$P = \begin{bmatrix} 1 & \frac{1}{2} & 0 & 0 & 0 \\ 0 & 0 & \frac{1}{2} & 0 & 0 \\ 0 & \frac{1}{2} & 0 & \frac{1}{2} & 0 \\ 0 & 0 & \frac{1}{2} & 0 & 0 \\ 0 & 0 & 0 & \frac{1}{2} & 1 \end{bmatrix}.$$

Since he starts with a capital of \$2, the initial distribution is

$$x^{(0)} = (0, 1, 0, 0, 0).$$

The probability that he will quit after three trials is

$$x_1^{(3)} + x_5^{(3)} = (P^3x^{(0)})_1 + (P^3x^{(0)})_5.$$

Now,

$$P^3 = \begin{bmatrix} 1 & \frac{5}{8} & \frac{1}{4} & \frac{1}{8} & 0 \\ 0 & 0 & \frac{1}{4} & 0 & 0 \\ 0 & \frac{1}{4} & 0 & \frac{1}{4} & 0 \\ 0 & 0 & \frac{1}{4} & 0 & 0 \\ 0 & \frac{1}{8} & \frac{1}{4} & \frac{5}{8} & 1 \end{bmatrix},$$

and hence

$$P^3x^{(0)} = (\tfrac{5}{8}, 0, \tfrac{1}{4}, 0, \tfrac{1}{8}).$$

Thus the probability that he will quit after three trials is

$$\begin{aligned}(P^3x^{(0)})_1 + (P^3x^{(0)})_5 &= \tfrac{5}{8} + \tfrac{1}{8} \\ &= \tfrac{3}{4}.\end{aligned}$$

The probability that he has won at least \$4 after five trials is

$$(P^5x^{(0)})_4 + (P^5x^{(0)})_5.$$

We can calculate that

$$P^5 = \begin{bmatrix} 1 & \frac{11}{16} & \frac{3}{8} & \frac{3}{16} & 0 \\ 0 & 0 & \frac{1}{8} & 0 & 0 \\ 0 & \frac{1}{8} & 0 & \frac{1}{8} & 0 \\ 0 & 0 & \frac{1}{8} & 0 & 0 \\ 0 & \frac{3}{16} & \frac{3}{8} & \frac{11}{16} & 1 \end{bmatrix},$$

and hence $P^5x^{(0)} = (\frac{11}{16}, 0, \frac{1}{8}, 0, \frac{3}{16})$. Therefore the probability that he will have at least \$4 after five trials is $\frac{3}{16}$.

Index

A CATALOG OF SELECTED

DOVER BOOKS

IN SCIENCE AND MATHEMATICS

HANDBOOK OF MATHEMATICAL FUNCTIONS WITH FORMULAS, GRAPHS, AND MATHEMATICAL TABLES, edited by Milton Abramowitz and Irene A. Stegun. Vast compendium: 29 sets of tables, some to as high as 20 places. 1,046pp. 8 × 10½. 61272-4 Pa. $22.95

MATHEMATICAL METHODS IN PHYSICS AND ENGINEERING, John W. Dettman. Algebraically based approach to vectors, mapping, diffraction, other topics in applied math. Also generalized functions, analytic function theory, more. Exercises. 448pp. 5⅜ × 8¼. 65649-7 Pa. $8.95

A SURVEY OF NUMERICAL MATHEMATICS, David M. Young and Robert Todd Gregory. Broad self-contained coverage of computer-oriented numerical algorithms for solving various types of mathematical problems in linear algebra, ordinary and partial, differential equations, much more. Exercises. Total of 1,248pp. 5⅜ × 8½. Two volumes. Vol. I 65691-8 Pa. $14.95
Vol. II 65692-6 Pa. $14.95

TENSOR ANALYSIS FOR PHYSICISTS, J.A. Schouten. Concise exposition of the mathematical basis of tensor analysis, integrated with well-chosen physical examples of the theory. Exercises. Index. Bibliography. 289pp. 5⅜ × 8½.
65582-2 Pa. $7.95

INTRODUCTION TO NUMERICAL ANALYSIS (2nd Edition), F.B. Hildebrand. Classic, fundamental treatment covers computation, approximation, interpolation, numerical differentiation and integration, other topics. 150 new problems. 669pp. 5⅜ × 8½. 65363-3 Pa. $14.95

INVESTIGATIONS ON THE THEORY OF THE BROWNIAN MOVEMENT, Albert Einstein. Five papers (1905–8) investigating dynamics of Brownian motion and evolving elementary theory. Notes by R. Fürth. 122pp. 5⅜ × 8½.
60304-0 Pa. $4.95

NUMERICAL METHODS FOR SCIENTISTS AND ENGINEERS, Richard Hamming. Classic text stresses frequency approach in coverage of algorithms, polynomial approximation, Fourier approximation, exponential approximation, other topics. Revised and enlarged 2nd edition. 721pp. 5⅜ × 8½. 65241-6 Pa. $14.95

AN INTRODUCTION TO STATISTICAL THERMODYNAMICS, Terrell L. Hill. Excellent basic text offers wide-ranging coverage of quantum statistical mechanics, systems of interacting molecules, quantum statistics, more. 523pp. 5⅜ × 8½. 65242-4 Pa. $11.95

ELEMENTARY DIFFERENTIAL EQUATIONS, William Ted Martin and Eric Reissner. Exceptionally clear, comprehensive introduction at undergraduate level. Nature and origin of differential equations, differential equations of first, second and higher orders. Picard's Theorem, much more. Problems with solutions. 331pp. 5⅜ × 8½. 65024-3 Pa. $8.95

STATISTICAL PHYSICS, Gregory H. Wannier. Classic text combines thermodynamics, statistical mechanics and kinetic theory in one unified presentation of thermal physics. Problems with solutions. Bibliography. 532pp. 5⅜ × 8½.
65401-X Pa. $11.95

CATALOG OF DOVER BOOKS